Hurrah For Georgia!

The History of the 38th Georgia Regiment

by D. Gary Nichols

Hurrah For Georgia!
The History of the 38th Georgia Regiment
G. Dale Nichols

Dedication

This book is dedicated to the all soldiers that served in the 38th Georgia Volunteer Infantry Regiment and their descendants, who will ensure they are never forgotten. It is also especially dedicated to my ancestors who served in Company F of the 38th Georgia Regiment and provided the inspiration for this book:

Private Henry Carter Partain - Enlisted April 6th, 1862, retired due to disability June 24th, 1862

Private Lindsey Bonaparte Partain - Enlisted April 6th, 1862, last on roll November 1864

Private Wiley Powell - Enlisted July 29th, 1862, died of disease at Lynchburg, Virginia, January 24th, 1864.

Corporal William Joseph Powell - Enlisted November 15th, 1861, killed in action at Gettysburg, July 1st, 1863.

Private James Lewis Powell, enlisted November 15th, 1861, wounded at Gaines Mill June 27th, 1862, sent home to Elbert County, Georgia, died there October 8th, 1863.

Acknowledgements

I would like to thank all the people who helped make this book possible. First, I want to thank my family, my beautiful wife Amy and my sons Joshua and Caleb, for tolerating my obsession in writing this book over the past eight years. A special thanks goes to Mr. John Davis for the letters of Simpson Hagood; Dr. Keith Bohannon, University of West Georgia; Mr. Gregory C. White, author of *A History of the 31st Georgia Regiment*; Colonel Pharris Deloach Johnson, author of *Under the Southern Cross*. Mr. Don Ernsberger, author of *Meade's Breakthrough at Fredericksburg*. Mr. Keith Jones, author of *The Boys of Diamond Hill*; Mr. Bruce H. Jackson for his Emory University research; Ms. Gail Gross for her research at Emory University; Ms. Linda Seymour Byers for the photo of Jonathan Green Seymour; Mr. Roy Hunter for the photo of Samuel H. Braswell; Mr. Peter J. Camouche Jr. for the image of the painting of George Washington Gaddy; Ms. Joan Bounds for letters of George Washington Gaddy; Mr. Jonathan Hopper and Ken Harris for the photo of the Hopper brothers; Mr. Chandler Eavenson for the photo of James C. Hall; Ms. Amy C. Parker for the photo of William Absolam Booth; Mr. Donnie Whitmire for information on the Chestatee Artillery Company and the hundreds of descendants of 38th Georgia Regiment soldiers who have provided information on their ancestors. If you helped in some way but don't see your name here, thank you and my apologies.

Table of Contents

Chapter 1 - The Beginning – 1861 1

Chapter 2 - To Savannah! 5

Chapter 3 - On To Virginia! 10

Chapter 4 - The Battle of Gaines Mill - June 27th, 1862 15

Chapter 5 - The 2nd Manassas Campaign, July - Aug 1862 27

Chapter 6 - The Bloodiest Day – Antietam, September 17th, 1862 43

Chapter 7 - The Battle of Shepherdstown, or Boteler's Ford September 19 & 20th, 1862 57

Chapter 8 - The Battle of Fredericksburg, Sep - Dec 1862 62

Chapter 9 - Chancellorsville Campaign, Jan - May 1863 79

Chapter 10 - The Battle of 2nd Winchester, June, 1863 99

Chapter 11 - The Road to Gettysburg, July 1st, 1863 108

Chapter 12 - The Battle of Wilderness, Day One – May 5th, 1864 121

Chapter 13 - The Wilderness, Day Two - May 6th, 1864 135

Chapter 14 - The Battles of Spotsylvania Court House - May 10th, 1864 141

Chapter 15 - Spotsylvania Court House – May 12th 1864 146

Chapter 16 - From Spotsylvania To Cold Harbor, May - June 1864 160

Chapter 17 - From Cold Harbor to Washington, D. C., June - July 1864 165

Chapter 18 - Return to the Shenandoah Valley, Aug - Sept 1864 176

Chapter 19 - The Battle of 3rd Winchester (or Opequon Creek) September 19th, 1864 181

Chapter 20 - The Battle of Fisher's Hill – September 21-22, 1864 192

Chapter 21 - The Valley in Flames, September - October, 1864 197

Chapter 22 - Battle of Cedar Creek – October 19th, 1864....................................200

Chapter 23 - In the Trenches at Petersburg, Dec 1864 - April 1865........................211

Chapter 24 - The Road to Appomattox Court House, April 2nd - 12th 1865..........224

Chapter 25 - The Surrender and Prisoners of War..229

Chapter 26 - Deaths, Deserters & "Galvanized Yankees"..238

Chapter 27 - Reunions And Life After The War..243

Appendix - Roster of the 38th Georgia Regiment...247

End Notes...343

Index..363

Prologue

The election of Abraham Lincoln as President prompted South Carolina to secede from the Union on Dec. 20th, 1860 and three other states soon followed: Mississippi, Florida, and Alabama. Georgia was the 5th southern state to secede and passed an ordinance of secession on Jan. 19th, 1861. Louisiana and Texas soon joined the five southern states. On Mar. 6th, 1861, Confederate Congress issued a call for 100,000 volunteers to serve for 12 months in enforcing secession.

On Apr. 12th, 1861, South Carolina batteries opened fire on Fort Sumter, a Union fort in Charleston Harbor, igniting the war. After Fort Sumter, President Abraham Lincoln issued a call for 75,000 volunteers to serve 90 days to put down the "insurrection." Four other states joined the Southern Confederacy after the battle of Fort Sumter; Virginia, Arkansas, Tennessee, and North Carolina, these eleven states formed the Confederate States of America. President Lincoln again called for 42,000 more soldiers in early May, to serve 3 years to suppress the "rebellion." The US Congress approved Lincoln's actions and authorized 500,000 additional volunteers in July.

The first significant battle of the war was the Battle of Bull Run, fought on July 21, 1861, near Manassas, Virginia. It was a Confederate victory and the state of Georgia contributed, losing among others, Colonel Francis S. Bartow, a prominent Georgia political leader. Bartow became a martyred symbol for Georgia and several military companies were named in his honor, including "The Bartow Avengers," which later became Company K of the 38th Georgia Regiment.

The prominent men of Georgia continued to prepare for war by recruiting and organizing new regiments. The first published notice of the forming of Wright's Legion occurred on September 14th, 1861, when a small article was printed in the Columbus Daily Enquirer announcing: "Wright's Legion, a new military organization, authorized by the Confederate government is being formed." Wright's Legion was re-designated the 38th Georgia Volunteer Infantry Regiment in early 1862.

This is the history of the 38th Georgia Regiment.

Foreword

I've used the words of the men of the 38th Georgia to tell their story when possible. These voices leave no doubt as to their meaning; however, gaps exist in many cases when the words of the men of the 38th Georgia were unrecorded, or lost to history. In those cases, I've used the writings of soldiers from other regiments of the Georgia Brigade to fill the gaps. These other regiments of the Georgia Brigade, the 13th, 26th, 31st, 60th, 61st Georgia and later in 1864, the 12th Georgia Battalion, often served side by side with the soldiers of the 38th Georgia and were rarely more than a few hundred yards away.

Chapter 1
The Beginning – 1861

Author's rendering of the flag of the "Murphey Guards" of DeKalb County, Ga., based on historical description.

The first troops from Georgia responded to the call to arms in the spring of 1861, but companies continued to form from all over Georgia. Men from northern and central Georgia responded to the second major call to arms, culminating in the forming of "Wright's Legion," later known as the 38th Georgia Volunteer Infantry Regiment. Wright's Legion was named in honor of Augustus R. Wright of Rome, who was subsequently elected Colonel. Many newly formed companies were ordered to report to Camp Kirkpatrick, a training camp near Atlanta. Camp Kirkpatrick was located two miles west of Decatur, Ga., and four miles east of Atlanta, on the Georgia railroad.

A "Legion" was organized as a single military command, combining all three military arms of the time: infantry, cavalry, and artillery. There was discussion of increasing Wright's Legion to 20 companies. Two companies of artillery were initially part of the Legion, but the cavalry companies were never assigned and the Legion never exceeded 11 companies at any one time. The Legion organization proved to be impractical and the two artillery companies were later detached and replaced by infantry companies, leaving Wright's Legion as purely an infantry regiment with ten companies.

Three companies of Wright's Legion formed from the men of DeKalb County, Georgia. Captain John Yancy Flowers formed the Murphey Guards. "They came from the upper part of the county, near Doraville. This company was named in memory of Honorable Charles Murphey, of DeKalb County, a prominent lawyer and member of Congress, but recently deceased. The people of the county, a large share being contributed by Mr. and Mrs. Milton A. Candler, and Mr. and Mrs. Ezekiel Mason, had uniformed the company. Mrs. Candler gave the banner, upon which was inscribed, The God of Jacob is With Us." This company became too large and was divided on April 1st, 1862. The new company was commanded by

Captain John Rankin, of Stone Mountain, and was designated, "The McCullough Rifles." Captain William Wright formed The Bartow Avengers and was from the lower part of DeKalb County about South River.[1] The Jo Thompson Artillery was raised by Lewis J. Parr and Cornelius Hanleiter. Hanleiter recalled the forming of the company, “The Company was formed through the personal efforts of Lewis J. Parr and myself, after the departure for the seat of war of several regiments formed in Upper Georgia, and at a time when the martial spirit was rapidly dying out; principally, I believe, on account of the difficulty in procuring arms and other necessary equipments. We were ordered to Camp Kirkpatrick, for instruction and drill.” [2] At the time these companies were assembling at Camp Kirkpatrick, still other companies were gathering at Augusta, Georgia. Citizens were attempting to gather ten companies in Augusta to form a regiment to go to Virginia. The effort failed and only five companies were raised. Lieutenant George H. Lester, of The Tom Cobb Infantry (TCI), from Oglethorpe County, remembered meeting in Augusta, “On the 4th day of Sept, 1861, the TCIs started for the city of Augusta, Ga. (being the third company of volunteers from Oglethorpe, County.) It being previously arranged that the regiment was to form there and proceed to Virginia. We arrived the following day (Sept 5th, 1861) and marched out with the other companies in a few days, The ‘Battey Guards,’ from Jefferson County, commanded by Captain William H. Battey, arrived and five companies were formed.”[3] Lieutenant Lester noted men’s disgust in failing to raise their regiment, “The dissatisfaction among the officers and men was outspoken; indeed, some of our officers thought the war would be over without their ever having an opportunity of covering themselves with glory and honor, or even striking a blow for their country. They spoke of leaving at once and joining some other Georgia regiment in Virginia, as privates. Colonel Augustus R. Wright, of Rome, Ga., was forming a legion at Camp Kirkpatrick, between Atlanta and Decatur, and proposed to accept our company in the legion, and in the early part of the month of Oct we left Augusta and arrived at Camp Kirkpatrick, together with the ‘Battey Guards’ and ‘Ben Hill Guards,’ and were mustered into the Confederate service, by Captain Bomar, the mustering officer.” [4] Wright’s Legion was originally comprised of ten companies, with two

BATTEY GUARDS.—This fine company, encamped near the city for some time past, had a parade in our streets yesterday afternoon. We were much pleased with their appearance as they filed past our office. Their precision in marching, the correctness of their evolutions, and their general soldierly bearing, won general admiration. Capt. BATTEY may well feel proud of his company.

Augusta Chronicle, Wednesday, Oct. 2, 1861, Page 3

COL. A. R WRIGHT, ROME GA.

companies designated artillery companies. These two companies were The Chestatee Artillery from Forsyth County, under Captain Thomas H. Bomar and the Jo Thompson Artillery, from Fulton County, under Captain Cornelius R. Hanleiter. Captain William L. McLeod raised and organized Company C, The McLeod Artillery. This company was intended to be organized as an artillery company, but the effort to acquire the necessary cannon failed. The company was renamed the “Ben Hill Guards" and served as infantry.

WRIGHT LEGION.—An election was held a few a few days ago at the camps of the Wright Legion, when A. R. Wright was chosen Colonel, and Geo. W. Lee Lieut. Colonel, unanimously. We are informed that the Legion will probably be composed of twenty companies—Colonel Wright having authority to increase it to that amount. The ten companies already mustered in, will report for the scene of their service in a few days.—*Atlanta Confederacy*, 13*th*.

Savannah Republican, Oct. 15, 1861 -- page 1

The following officers were elected to lead the legion at conception: Augustus R. Wright, of Rome, Ga., was elected Colonel of the Regiment. George. W. Lee, of Atlanta, was elected Lieutenant Col., Lewis J. Parr, of Atlanta, was elected Maj., John H. Sherrod, of Swainesboro, Ga., was elected as adjutant, Barney D. Lee, of Atlanta, was Sergeant Major, Dr. William J. Arrington, M. D., of Louisville, Ga., was regimental surgeon, John M. Quinn, of Rome, Ga., was commissary, William H. Jernigan, Lexington, Ga., was quartermaster, and John Harvey Mashburn was regimental chaplain. The following ten companies composed Wright’s Legion at conception, later known as the 38th Georgia Volunteer Infantry Regiment: [5]

1. Company A, "Ben Hill Guards," Emanuel County. Captain, William L. McLeod; Lieutenants, Miller A. Wright, John A. Williamson and Jacob P. Pughsley. Company mustered into the service Oct. 1st, 1861.
2. Company B, "Goshen Blues," Elbert County. Captain, Robert P. Eberhart; Lieutenants, William T. Andrews, James C. Hall, and John Oglesby. Company mustered into the service on Oct. 15th, 1861.
3. Company C, known as Murphey Guards, of DeKalb county. Captain, John Y. Flowers; Lieutenants Alfred J. H. Pool, John J. Marabel and William A. C. Miller. Company mustered into the service Sept. 26th, 1861. It was divided Apr. 1st, 1862 and part of it became Co. D.
4. Company D, "Milton Guards," Milton County. Captain, George W. McCleskey; Lieutenants, Andrew J. McMakin, Andrew J. Phillips and Joseph J. Maddox. Company was mustered into the service Oct. 6th, 1861. Note: Milton County, GA was absorbed by Fulton County, GA in 1932.
5. Company E, "Battey Guards," of Jefferson County. Captain, William. H. Battey; Lieutenants, John W. Brinson, Isaac C. Vaughn and Levin W. Farmer. Company mustered into the service on Oct. 1st, 1861.
6. Company F, "Jo Thompson Artillery," Fulton County. Captain, Cornelius R. Hanleiter; Lieutenants, Augustus A. Shaw, Elijah J. Craven and William P.

McDaniel. This company was mustered into the service on Sept. 26th, 1861. This company was detached from the 38th Georgia Regiment on June 10th, 1862 and assigned to the Army Of Tennessee, C. S. A. This company never returned to serve with the 38th Georgia Infantry Regiment.

7. Company G, "Bartow Avengers," DeKalb County. Captain, William Wright; Lieutenants, Julius J. Gober, Gustin E. Goodwin and George W. Stubbs. This company was mustered into the service Sept. 26th, 1861.

8. Company H, "Chestatee Artillery," Forsyth County. Captain, Thomas H. Bomar; Lieutenants, Truman H. Sandford, Samuel E. Taylor and William Hendrix. This company was mustered into the service Oct. 13th, 1861. This company as detached from the 38th Georgia Regiment on June 10th, 1862 and remained at Savannah. The company was ordered to re-join the 38th Georgia Infantry Regiment, on May 5th, 1864, and served as infantry until the end of the war.

9. Company M, "Thornton Line Volunteers," Hart County. Captain, John C. Thornton; Lieutenants, John H. H. Teasley, Benager T. Brown and Jackson O. Maxwell. Company mustered into the service Oct. 15th, 1861.

10. Company K, "Tom Cobb Infantry," Oglethorpe County. Captain, James D. Matthews; Lieutenants, George H. Lester, John J. Daniel and Charles A. Hawkins. Company Mustered into the service Sept. 29th, 1861.

Chapter 2
To Savannah!

Wright's Legion was drilled every day at Camp Kirkpatrick, learning how to march, and practicing the maneuvers required of an infantry regiment; in company and battalion formation, and performing dress parades. They were ordered to report to Savannah, Georgia in October of 1861. Captain Hanleiter wrote of the events leading up to the regiment's departure for Savannah, *"Whilst awaiting our equipments and arms, and receiving and entertaining visitors, at our luxurious camp near Decatur, and enjoying ourselves only as newly fledged soldiers can with the "blare of infantry and roar of cannon" a long way off," we very unexpectedly, late one fine October afternoon, received orders to cook up three days' rations, and be prepared to take a train that would be ready early next morning for Richmond!*

Our camp was crowded with the wives, sweethearts, and friends of the officers and men, and all had been as joyous as if at a Mayday picnic. But soon after the promulgation of the order, which was understood to be imperative, the men began to "hustle," the women and girls hastily departed in anguish of heart and tears. However, by nine o'clock that night everything necessary for the "onward movement" was cooked, packed, and ready. About an hour later, greatly to our joy, another order was received; changing our destination to our own loved Savannah. By nine o'clock the following morning the train, consisting of box and cattle cars, backed down from Atlanta, and we embarked. We were detained at Atlanta until about 2 p.m, and reached Savannah on Sunday morning soon alter day-break, after the coldest and most fatiguing ride we ever experienced."

"The Legion was marched out to a point about five and a half miles on the Skidaway Island Shell Road and camped...the camp being named Camp Bartow...doing duty as the emergency seemed to demand on Skidaway Island ...and various other localities around and in the city." [1] The regiment was still unarmed and the Tom Cobb Infantry was, "allowed the privilege of returning home for three days, to gather up all the arms we could. On our return....the order met us to lay by our old rifles and shot guns, as the 'Fingal' has arrived in Savannah, and her cargo consisted mainly of Enfield rifles and other arms and equipment."[2]

The Fingal was an 800-ton blockade-runner, sea steamer, purchased by the Confederate government and contracted to transport war supplies to the southern states. The Fingal slipped into Savannah harbor on Nov. 14th, 1861, with over 13,000 Enfield Rifles, accouterments and other much needed war supplies. On November 19th, 1861, the eight infantry companies were issued, "Enfield Rifles – a magnificent gun with sabre bayonet, and a liberal supply of cartridges from Europe." The men were also issued "a knapsack, haversack, cross and circular belts, cartridge box, cap pouch. These articles were also manufactured in London and brought over on the steamer "Fingal." One thousand rounds of ball cartridges and caps were issued to each Captain with instructions to issue them at the slightest alarm."[3] The two artillery companies remained unarmed, but agreed to be temporarily armed as infantry companies and were issued arms and equipment on Nov. 22nd.

Many soldiers fell ill at Camp Bartow during November of 1861, but death was so novel that when it occurred, it was noted in meticulous detail. On Nov 29th, Captain C. R. Hanleiter, of the Jo Thompson Artillery, wrote: *"Assisted this afternoon in preparing for burial, the body of Seaborn J. Cash, a Private in the Goshen Blues, who departed this life at Camp Bartow, this morning at 7 o'clock, after an illness of one week of typhoid fever. After placing the body in the coffin, it was borne to the head of the street occupied by the tents of the Blues, where three volleys were fired over it by a detachment from that company - then the coffin was placed in the wagon and taken to Savannah for transportation to the late residence of the deceased, in Elbert County. The scene at the Camp an impressive and sorrowful one, yet there was little or no order observed by the men who crowded around. Farwell poor Cash, thou art now far away from the sound of the guns of friend or foes, and I trust in a happier realm!"* [4]

Private Seaborn J. Cash, of Elbert Co., Georgia, was the first recorded death in the 38th Georgia, but hundreds more would soon follow. Twenty-four other members of the 38th Georgia died of disease while the regiment was at Savannah, between November 1861 and June 1862. Sixteen soldiers died at camps near Savannah, or in local hospitals. Eight others died at hospitals in Augusta, Ga.[5] The funeral rites afforded Private Cash would later seem rather elaborate, since they included a coffin and his body was returned to his family home in Elbert Co. Death

Confederate Blockade Runner *Fingal*

would soon be an everyday occurrence and such grand treatment of the dead would soon become virtually unknown. A few months later, many men killed on the battlefield were lucky to receive a simple blanket as a funeral shroud.

Lieutenant Lester, of the Tom Cobb Infantry, wrote of a memorable Christmas eve march and life at Skidaway Island. *"At night, on the 24th of Dec,*

1861, the order was given to march on Skidaway Island, as there were twenty-seven vessels which had that evening come into Warsaw Sound; and Skidaway with our work and army were in imminent danger. We ate our Christmas breakfast at Modena Junction, and some will remember the fare was scant. Modena Junction was neither a village nor a town- not a shanty or house – but an unbroken pine forest, where one road intersected another. We staid on the island during the day; in the evening marched back to our camp, weary with our march through the deep sand. The works, bomb-proofs and mounted guns were the first some of us had ever seen 'of the glorious pomp and circumstance.'

A few weeks after this, we were marched over the Long Bridge, on Skidaway, and remained there the balance of the winter – not for the purpose of charging on the enemy's gunboats and war Vessels- though a furious attack was made on the oyster beds, which were daily charged, and the oysters were slaughtered by the thousands. We were put to work on the fortifications. Our encampment was on the 'Waring Place,' which had been deserted by all but a few old Negros. It was fine Sea Island cotton plantation. When the troops evacuated the island, Dr. Waring found his houses and plantation in a bad condition for farming – everything was torn to pieces "by the ruthless hand of war. We were kept busy drilling, working on the sand bag fortifications and picketing the various posts on the island that were required to be picketed by the legion. We had a splendid lookout in a large magnolia tree, which was sixty feet high.

The first our regiment ever saw General Robert E. Lee, he came down from the city of Savannah with General A. R. Lawton, and went up on the lookout with his telescope, to view the enemy's vessels in Warsaw Sound. Captain (then Sergt.) J. I. Callaway (Jonathan I. Callaway), of the Tom Cobb Infantry, shot the first Federal that attempted to invade Georgia soil, from this lookout. A boat, with some naval officers and seamen, had left the fleet, for the purpose of reconnoitering Skidaway Island. Sergt. Callaway was the officer that day on the lookout, and when the boat had glided into the proper range, he pulled down on them. One man fell at the crack of his gun. For a moment his companions were all confusion and excitement; when the boat shot through the waters, and was out of the range of the rest of the shots that were fired.

After General R. E. Lee had viewed the situation, and his military eye had detected the error of any attempt to hold the island, the guns were dismounted at night, logs were placed in the positions occupied by them, and the guns were shipped on board of boats and carried up the river to Fort Jackson, Fort Boggs, and other places near the city of Savannah. All of our works were abandoned, and the labors of thousands of men, for six or eight months, were left. It was with regret that we saw our long and fruitless toil would avail nothing. We left the island and encamped on the Isle of Hope, near Parkersville. The legion was more immediately under the command of General H. W. Mercer, Captain George A. Mercer was his Adjutant and better officers never entered the Confederate service, or more perfect gentlemen." [6]

Colonel Augustus R. Wright resigned his commission on February 14th, 1862, as he was elected to serve in the Confederate congress. George W. Lee was elected Colonel to serve as the commander of the 38th Georgia Regiment.[7]

The majority of the soldiers who died in Savannah died of disease, but one soldier was accidently shot and killed by a fellow soldier. Private Andrew J. Ables, of the Jo Thompson Artillery, encountered Private Raymond B. Payne who was on his way back to his tent from dress parade, on March 5th, 1862. Private Ables joked that he was going to shoot him and lowered his rifle to the position of bayonet charge, his rifle discharged, the ball striking Payne in the back and passing through his body, coming out four inches below the navel. Payne died the following night at about 11:40 p.m. after hours of intense suffering.[8]

[COMMUNICATED.]

Editor Savannah Morning News:

I take great pleasure in stating that I have had several of my men treated at Bartow Hospital, one of whom was my own son. I visited my men constantly, and feel it my duty to say I never saw better nursing or kinder attention. No charges were ever made, either for my men or their nurses; I offered to pay for things that had to be purchased, but it was refused. Although one of my men died there, I am convinced that it was not for want of medical skill or proper nursing. Everything that could be done was done, both by the Doctor and Madame Cazier, who is the best nurse I ever saw. I cordially recommend Bartow Hospital to my fellow-soldiers as one of the best places provided for a sick soldier.

WM. WRIGHT,
Capt. of the Bartow Avengers, Wright's Legion,
near Savannah.

Savannah Daily Morning News, January 6th, 1862, Page 2

Fort Pulaski guarded the Savannah River near the City of Savannah. Lieutenant Lester recounted the Federal attack and seizure of the fort, thought by the Confederates to be impregnable, *"It was on the morning of the 10th of April, 1862, as well as I recollect, that we saw the first of the enemy's shells burst of Fort Pulaski. We did not know that this fort was a mere dirt dauber's nest, but thought it impregnable- a second Gibraltar – and believed that Major Olmsted and his men had nothing else to do but fight it out – "hold the fort," and achieve imperishable renown. Imagine our surprise, when charging and mortification, the next morning, when we heard that Fort Pulaski had been taken by the Federals. The enemy had stolen an island near the fort, planted their masked batteries in point blank range of the fort, and with the latest improved guns, had nothing to do but cut away their masks in front, knock down the old brick fort and silent the ancient artillery in it which had been mounted there for the last fifty years. After the fall of Fort Pulaski, it was believed by some that Savannah would be their next point of attack; but they wisely forbore to make the onset. They would have found some trouble and danger if they had made the attempt."*[9]

The Murphey Guards Company was mustered from DeKalb County and received so many recruits the company became too large. On April 1st, 1862, this company was divided and a new company was formed, called the "McCullough Rifles." John G. Rankin was elected Captain, and John W. McCurdy, George R. Wells and John Baxter were elected Lieutenants. Many members of this new company were formerly of Company A.

The Confederate Congress passed the first Conscription Act on April 16th, 1862, which drafted white men between eighteen and thirty-five for three years service. All soldiers already serving had their service extended to three years, or the duration of the war. New recruits soon filled the ranks of the 38th Georgia, since many preferred to volunteer, rather than be labeled "conscripts."

There had been talk for months of detaching the two artillery companies of the 38th Georgia, the Jo Thompson Artillery and Chestatee Artillery, from the regiment. Colonel George W. Lee was reluctant to agree to detach these companies from his command, because it would only leave eight companies of infantry. General Mercer assured him the two artillery companies would be replaced by infantry companies before being transferred.

Chronicle & Sentinel.

OFFICE ON BROAD STREET.
OPPOSITE AUGUSTA BANK

AUGUSTA, GA.,

TUESDAY MORN'G, March 10, 1862.

A FEDERAL OFFICER KILLED.—Our latest intelligence from Skidaway is to Saturday noon. We learn that on Saturday morning a member of Col. Wright's Legion was on picket duty at Waring's Point, and, in order to have a better look out, ascended to the top of a Laurel tree, while he left another picket a short distance behind him. In a short time after ascending to his position in the tree, he beheld a Federal boat approaching, and just as the boat turned a bend in the river the occupants of it discovered the pickets. They immediately beat a retreat, but as they turn d their boat the Confederate picket on the ground fired, when all the federals dropped in the bottom of the boat. Soon after the firing had taken place the Federals rose in the boat to resume their oars, but while the officer in command was urging his men forward, the picket in the tree fired with his rifle, and the Federal officer, who was standing in the stern of the boat, was seen to fa l and did not rise again. In confirmation of this event the Federal vessels in the neighborhood signalized to each other, and came closer together in the course of half an hour. The Georgian fired at a distance of 600 yards, and his shot was fatal to an invader of our soil. Many more of them will find hospitable graves beneath our tropical sun.—*Sav. News*, 10*th*.

On May 2nd, 1862, the "Irwin Invincibles," under command of Captain Henry L. Jones, from Henry County, Alabama, was transferred to the 38th Georgia. The Irwin Invincibles were mustered in during December 1861, as the 3rd Co. E, 25th Regiment Ga. Infantry. They were ordered from West Virginia to Georgia by the Secretary of War and arrived on Jan. 12th, 1862. They went into camp near Savannah, Ga. and were assigned to the 25th Regiment Ga. Inf., before being assigned to the 38th Georgia during May 1862. This company was later known as the "Henry Light Infantry." Another new company, "The Dawson Famers," was mustered into the service on May 15th, 1862, and assigned to the 38th Georgia Regiment. These men were mainly from Dawson and Forsyth Counties of Central Georgia. The way was now cleared to detach the two artillery companies from the 38th Georgia.

Chapter 3
On To Virginia!

Brigadier General Alexander Robert Lawton, commanding the Georgia Military District, selected the six best drilled regiments in Savannah to form an "elite brigade" of soldiers numbering 7,000 men strong. The brigade became known as Lawton's Brigade and was also known as the "Georgia Brigade." The regiments comprising the brigade were the 13th Georgia, 26th Georgia, 31st Georgia, Wright's Legion was re-designated as the 38th Georgia, 60th Georgia, and 61st Georgia.

These six regiments would endure vast hardships and fight side by side until the end of the war. Five of these six, including the 38th Georgia, were armed with Enfield Rifles with sword bayonets; along with other much need accouterments. The 31st Georgia Regiment was the only regiment of the brigade not issued Enfield rifles. They were issued smooth bore muskets, firing buck and ball cartridges. Captain C. R. Hanleiter of the Jo Thompson Artillery wrote of a flag presentation for the 38th Georgia, on June 6th, 1862:

RECRUITS WANTED.

I HAVE yet room in my company for twenty or thirty more recruits. I have very flattering inducements to offer in behalf of our Legion. We have good and kind field officers and the best war gun in the service, (the Enfield Rifle.) Recruits wishing to join my com pany will apply to Lieut. G. E. Goodwin, at Hunnicutt,Taylor & Jones'Drug Store,Atlanta, or to him at Harris Goodwin's residence, nine miles north of Atlanta. Transportation will be given to the Wright Legion, either by way of Augusta or Macon to Savannah. Men wishing to join as recruits will do well to study their interest, as few regiments have guns to compete with ours.

WM. WRIGHT,
Capt. Bartow Avengers, W. L.

feb18w3t.feb26dtill6mar

Southern Confederacy Newspaper, Feb 27th, 1862, Page 2

"Before 10, I was informed by Colonel Lee that Mrs. General Lawton had notified him that she would, in the course of the day, present the Legion with a Battle Flag; that a reply to address, or letter of presentation would devolve on him, and as he was totally inadequate to the task, he requested me to write out something that would answer the purpose. I endeavored to "crawfish", assuring him of my incompetency; but he would take no excuse, and I accordingly wrote out a short letter, which he fathered and expressed himself very pleased with. About 12 o'clock, PM, the Regiment was formed in battle line, and Captain Elliott, one of General Lawton's aides appeared with the flag, which he presented together with a note from the fair donor. The flag was received by Colonel Lee, who in turn, reading my letter, signed by himself, in reply and then ordering Captain Mathews (Captain James D. Mathews, Commanding the Tom Cobb Infantry) company to the front gave the flag to their charge. Captain Mathews replied in a handsome address, after which his company was marched backwards (it should have been an "about face") into line, and the regiment dismissed to assemble again at the sound of the

drums. About 2 o'clock the signal was given and the line was again formed and put in motion. On reaching General Lawton's residence, fronting the park, the column was countermarched, halted and fronted while the band played one or two of their best pieces. While the band was playing Mrs. Lawton appeared on the piazza, attended by Colonel Lee and Captain Elliott, then the regiment "presented arms" and the band played Dixie. This ceremony over, the column was again countermarched, heading for the Central Railroad Depot, and a salute was fired by a detachment each from my own and Captain Bomar's companies. Having loaned my horse to Lieutenant Shaw, I took my position on the corner of the Park and Whitaker street, where I remained shaking hands with the Officers and many of the men until they had all filed past me, when I returned to camp to look after the "odds and ends" as I had been directed to do."[1]

Captain Hanleiter wrote, *"Finally, Captain Bomar's company (The Chestatee Artillery) and my own were detached from the Legion, and the infantry companies, ten in number, were ordered to Richmond, as apart of General A. R. Lawton's new brigade. My command was immediately ordered to Beaulien, at the junction of the Vernon with Burnside rivers, about twelve miles southwest from Savannah, where we relieved a regiment (Colonel Evans', I believe), and did duty as heavy and light artillery and infantry until the night of the evacuation of Savannah, I being in command of the Post and both batteries during the entire period."*[2]

The Jo Thompson Artillery, Captain Hanleiter's company, would never rejoin the 38th Georgia and would serve as artillery until the end of the war. Lieutenant Augustus Shaw, of the Jo Thompson Artillery, was appointed as the 38th Georgia Adjutant officer and joined the infantry companies for the journey to Virginia.

The 38th Georgia Regiment boarded trains and departed for Virginia on June 6th, 1862. They had served around Savannah since November of 1861, nearly seven months. The Regiment was comprised of thirteen companies, but the two artillery companies were detached, leaving eleven infantry companies. The regiment was reorganized and each company received a new alphabetical designation:

1. Company A, "The Murphey Guards," of DeKalb county. Company mustered into the service Sept. 26th, 1861. It was divided Apr. 1st, 1862 and about half this company became Co. D.
2. Company B, "Milton Guards," Milton County. Company was mustered into the service Oct. 6th, 1861.
3. Company C, "Ben Hill Guards," Emanuel County. Company mustered into the service Oct. 1st, 1861.
4. Company D, "McCullough Rifles," DeKalb County. Captain, John G. Rankin (Old Reliable); Lieutenants, John W. McCurdy, George R. Wells and John Baxter. Most members of this company were from Company A

and were mustered into the service on Sept. 26th, 1861. Company A was divided to form Co. D on Apr. 1st, 1862.
5. Company E, "Tom Cobb Infantry," Oglethorpe County. Company Mustered into the service Sept. 29th, 1861.
6. Company F, "Thornton Line Volunteers," Hart County. Company mustered into the service Oct. 15th, 1861.
7. Company G, "Battey Guards," of Jefferson County. Company mustered into the service on Oct. 1st, 1861.
8. Company H, "Goshen Blues," Elbert County. Company mustered into the service on Oct. 15th, 1861.
9. Company I, "Irwin Invincibles," later known as the "Henry Light Infantry", Henry County, Ala. Captain, Henry L. Jones; Lieutenants, Andrew B. Irwin, James E. Jones and Reuben M. Campbell. Assigned to the 38th Georgia during May 1862. On approximately March 1, 1863, this company was again transferred and became 2d Co. A, 60th Regiment Ga. Inf. Finally, this company was transferred and became Co. K, 61st Regiment Ala. Inf. April 11, 1864.
10. Company K, "Bartow Avengers," DeKalb County. This company was mustered into the service Sept. 26th, 1861.
11. Company L, "Jo Thompson Artillery," This company was mustered into the service on Sept. 26th, 1861. Detached June 10, 1862 and assigned to the Army Of Tennessee C. S. A. Once detached, this company never returned to serve with the 38th Georgia Infantry Regiment.
12. Company M, "Chestatee Artillery," Forsyth county. Captain, Thomas H. Bomar; lieutenants, John Hendrix, McDaniel and Hendrix. It was mustered into Confederate State service as a Light Artillery Company October 9, 1861. Detached June 10, 1862 to serve as artillery. Rejoined the 38th Georgia Regiment Infantry during May of 1864 and served as infantry for the remainder of the war.
13. Company N, "Dawson Farmers," Dawson county. Captain, W. M. Blackburn; lieutenants, Hill, Marshburn and John W. Goswick. Mustered into Confederate service on May 15th, 1862.

The 38th Georgia Regiment arrived in Petersburg, Virginia on Monday evening, June 9th. Private Francis C. McCleskey of Company B, the Milton Guards, wrote a letter to his mother the following day, *"I would love to be at home mighty well. I don't think I even would get tired plowing or hoeing corn. I tell you plowing corn is a great deal better than being in the war, confined like we are hear. There is one thousand men here in our Legion all in one house and we can't get out they keep Guards at the door and won't let no private pass out."*[3]

The Georgia Brigade was sent to the Shenandoah Valley to reinforce General Stonewall Jackson's Army, at Port Republic, Virginia. The men of the 38th Georgia had learned little of the hard life of a soldier since joining the Confederate Army, but they quickly realized they had entered a new phase of "soldiering."

Private George Washington Nichols of the 61st Georgia wrote an account of the Georgia Brigades' trip from Petersburg to the Shenandoah Valley, then back to Richmond:

"Sent to the valley and then marched back to Richmond we drew about two pounds of boiled bacon and about a dozen hard tacks apiece, went into a large house out of the rain and stayed all night, and left next morning for Richmond. We stayed in Richmond about two days until the brigade arrived. We were then ordered to the great Shenandoah valley to join the famous "Stonewall" Jackson, and was assigned to Jackson's old division. We got on the Southside Railroad and went by way of Lynchburg, then to Charlottesville, where we could see the Blue Ridge Mountains. We then got on the Virginia Central Railroad and soon crossed the Blue Ridge Mountains, through a great tunnel, which was a mile and a quarter long, into the valley, and went on to Staunton. We remained here but a short while, and marched toward Port Republic. We arrived at this place on the 10th of June 1862. The battle of Port Republic was fought on the 9th of June, where the famous Stonewall Jackson routed the Union army commanded by General Shields. Here we saw a great many dead Union soldiers before they were buried.

We stayed here about two days, crossed the mountain and started on a long force march. We did not know where we were going. We soon found that our faces were turned towards Richmond. We had to march very hard, sometimes almost night and day, across mountains, creeks and rivers. We had to march from Port Republic to Richmond, except that we went by rail about fifty miles. On this march we rested one Sunday and had religious service near our camp, where the famous General Stonewall Jackson met to worship God. It was the first time some of us had ever seen him. We started very early next morning and marched very hard till late in the afternoon. We stopped to camp and cook rations. Our tents were all left behind. The clouds began to collect and thunder very heavily, and the rain began to pour down in torrents, with a heavy gale of wind. It rained for very nearly two hours, and we all got as wet as we could be. We started about day next morning on a forced march, with full creeks and branches to cross. The roads were so cut up with the wagons and artillery until we could hardly get along. Some of the boys would bog so deep into the mud till when they got out their shoes would remain often ten and twelve inches below the surface. Every man had to carry his own haversack, knapsack, gun and cartridge-box. Some of the boys had white sheets, and I believe a few had feather pillows. Jackson's old soldiers, who had been following Jackson in his campaigns, made sport of us. They would ask us what command we were

En Route for Old Stonewall.—Furloughed Men Notice.—We have just received a telegraphic dispatch from Col. G. W. Lee, commanding "Wright's Legion" dated at Petersburg: Va., the 10th inst., in which he advises us that his command is en route to reinforce "Stonewall" Jackson, and in which he also requests all of his furloughed men of the fact, and he *requires them to join him immediately*.—Furloughed men, therefore of "Wright's Legion" will govern themselves accordingly. What *soldier* will not rejoice to know that he is called to do service under "Old Stonewall"?

It gratifies us to know that the military authorities at Richmond are reinforcing "Old Stonewall," and that a portion of these reinforcements will be the command of Col Lee.—We know him and most of his men. Better *mettle* never went to meet the enemy, and a more gallant officer never led men. Success and triumph attend them!

June 17th, 1862 [*Atlanta Intelligencer.*

wagoning for, and what train that was. Some of "our boys" cursed out the war, others shed tears (for there were a lot of young boys in the brigade), and said but little, while others, I suppose, prayed. We were being initiated and taking the first degree in war. We had been mustered into the Confederate service eight months, and had learned but little about the rough life of a soldier."[4]

Chapter 4
The Battle of Gaines Mill - June 27th, 1862

General Stonewall Jackson secretly marched his army from the Shenandoah Valley to join General Robert E. Lee's army before Richmond. Together they would join in attacking Union General McClellan's massive army poised before Richmond. General Lee accurately assessed the true, timid nature of General McClellan, but McClellan totally misread Robert E. Lee. McClellan told President Lincoln, *"I prefer Lee to Johnston. The former is too cautious and weak under grave responsibility -- personally brave and energetic to a fault, he yet is wanting in moral firmness when pressed by heavy responsibility and is likely to be timid and irresolute in action."* General Lee strengthened the defenses of the earthworks surrounding Richmond to free up more soldiers to use in his planned offense. The Confederates attacked the Union army at Beaver Dam Creek, just outside Richmond on June 26th. This was the first of the Seven's Days battles. Though the Confederates were repulsed, the Union army retreated the next day, June 27th, to the vicinity of Old Cold Harbor and Gaines Mill. The Federals formed a new line reinforced with 96 pieces of artillery and awaited General Lee's next move.

Brigadier General Alexander R. Lawton
Library of Congress

Lee's plan called for Jackson to attack the right flank and rear of the Union army, which was arrayed with its back to the Chickahominy River. Jackson marched his Corp towards Cold Harbor, but abruptly halted his march several miles short and went into camp for some unknown reason. The battle raged just a few short miles to the front, but Jackson didn't budge until late in the day, finally moving his Army forward to attack around 4 p.m.

Lawton's Georgia Brigade brought up the rear of Jackson's army and was the last to engage the enemy. The Brigade remained about two miles from the battlefield at Gaines Mill for several hours before being ordered forward at about 5 p.m. As the brigade began to advance the order was passed down to the 38th Georgia to accelerate their pace, they promptly obeyed. The men were ordered to cast away haversacks, blankets, and other un-necessary encumberments. They alternated between marching and double quick time up a narrow, dirt road towards the action. Lawton reported, *"I then marched rapidly on, retarded much by the artillery and ambulances, which blocked up the narrow road. On reaching the edge of a cornfield, about 1 ½ miles from the nearest point of the battle-ground, I was informed that General Ewell was sorely pressed by the enemy and re-enforcements were promptly needed. I then marched forward at double-quick, and the men reached the wood on the south of the battlefield almost exhausted."*

The 38th Georgia carried 700 men into action.[1] A soldier of the 38th Georgia, identified only by his initials, L. A. M. recalled, *"We were halted and stacked arms and enjoyed a short rest in a pine thicket just out of range of the shot and shell. We saw Lee and Jackson on a little knoll with their field glasses scanning the front and the field generally. With them was General Jones, commanding our division (Jackson's) also General Lawton, commanding our brigade, and General Ewell. They seemed to be consulting and seemed anxious and uneasy. Suddenly they seemed to have come to some decision, for they separated, and Captain Lawton, A. A. G. of the brigade, called for the Thirty-First and Thirty-Eighth Georgia regiments of Lawton's brigade, which were in front of the column. The command "attention" was given and then "take arms; forward, file left," and we marched some distance through the thicket, coming closer every minute to the sounds of the battle. Suddenly came the command "halt; front, on the centre dress," and then that command which next to the long roll, causes a soldier's heart to creep up in his mouth and his courage to ooze out at his finger tips, "load."*[2]

The line of battle of the brigade was far too long for General Lawton to effectively control, so he sent his brother, Captain Edward P. Lawton, Assistant Adjutant-General to the left to accompany and command the 38th and 31st Georgia Regiments.

General R. S. Ewell – CSA
Library of Congress

"After having loaded, our field officers, whom we had seen talking to General Lawton, informed us that the two regiments (38th & 31st Ga.) had been selected to charge two batteries of artillery and a body of infantry in our front and with which troops our skirmishers of another brigade were engaged and they were expecting to see the signal to advance any minute. Scarcely had they told us when the command 'forward' was given and the two color bearers moved forward at once with the line of battle 2,000 strong following at quick time."[3]

Having no staff officers to direct the brigades' movement into line of battle, General Lawton saw two Confederate regiments standing in an open field and was informed these men had just been driven back from a parcel of woods to their front by the enemy. He ordered the brigade to move by the flank between the interval of these two regiments and fill the gap they had just abandoned.[4]

The 31st and 38th Georgia were on the left flank of the brigade and separated from the other four regiments. The other regiments entered a patch of woods, but the terrain covered by the 38th Georgia was an open field and they were the most exposed regiment of the entire brigade. One 38th Georgia soldier surveyed the terrain the two regiments would soon cover to engage the enemy, *"The Thirty-first Georgia, had much better ground to fight over as the ground was more uneven and had some trees and they could not be as easily seen, but in front of the Thirty-eighth Georgia a mouse could hardly hide."*[5] Once the main portion of the brigade entered the woods, they unleashed a withering volley from 3,500 Enfield rifles upon the Federals, loudly announcing their arrival on the battlefield. General Lawton saw it would be impossible to maintain control over his entire line of battle, so he sent members of his staff to the right and left wings, while he remained in the center of the brigade to provide direction. *"Onward the line advanced through the wood, firing at every step, and guided only by volleys from the enemy toward the thickest of the fight,"* Lawton reported.[6] Lawton and his twenty-year old aide, Captain Edward Cheves, entered the woods on horseback. Captain Cheves's horse was soon shot down and both fell to the ground, but young Cheves quickly jumped to his feet, uninjured. Cheves announced he was unhurt, and continued on foot with the brigade. Cheves veered to the left where the 38th and 31st were heavily engaged. General Lawton noted, *"In the midst of the wood I met with Major-General Ewell, then hotly engaged, who, as he saw this long line advancing under fire, waved his sword and cried out, "Hurrah for Georgia!" To this there was a cheering response from my command, which then moved forward more rapidly than ever."*[7]

The position of the main portion of the brigade provided some cover and concealment in the woods, but the 38th and 31st were not as fortunate. Lawton related, *"In emerging from the wood these two regiments found themselves in the hottest part of the field, where our friends were pressing on the enemy toward the left, and joined them in the contest at that point under a murderous fire."*[8]

Captain William Battey, Co. G., of the Battey Guards, reported, "Thus we marched under a most terrific fire to within about 180 yards of a body of 4,000 or 5,000 regulars."[9] Porter's Division of United States Regulars were professional infantry soldiers, being part of the regular army, not state volunteers. They were among the finest and most reliable Union troops on the field that day.

Brigadier General Fitz John Porter
Library of Congress

The U.S. Regulars had been engaged since 11 a.m that morning, repelling numerous Confederate infantry assaults throughout the day. A 38th soldier penned, "*As we emerged from the thicket and came in full view of the enemy, the order to double quick was given, and the "Rebel yell" ringing over the field, we began to charge."* The 3rd U.S. Artillery, also regular Army troops, supported U.S. infantry Regulars and were deadly shots. They had mowed down hundreds of Confederate troops since the battle had commenced that morning and now trained their guns on the 31st and 38th Georgia. The 3rd U.S. Artillery realized the battle was far from over and as the two Georgia regiments came into view, they opened a furious fire on the two exposed regiments. *"It seemed as if they were expecting us, for it was as if every cannon in the Yankee army was turned loose upon us. Shells burst in front of us, and scattering, carried death and wounds in their track. Solid shot cut men in twain, solid shot shrieked overhead and scared us just as bad, solid shot struck the ground just in front of us and scattered dirt and gravel in our faces and would ricochet and go howling to our rear. "*[10]

Post War Sketch of Lt. Col Lewis J. Parr
Sunny South, Jan. 10th, 1891

As previously referenced, the Georgia Brigade was armed mainly with Enfield rifles imported from England, the sole exception was the 31st Georgia, who were armed with smooth bore rifled muskets that fired "buck and ball," cartridges. The load consisted of a .65 caliber round lead ball combined with three .22 caliber buckshot pellets. The muskets were effective at about 200 yards and deadly at close range. The disadvantage was a regiment with rifles could decimate a regiment with muskets before the muskets could return an effective fire. Then commenced the whistle of the Minié balls from the infantry, which, while not so demoralizing, was far more deadly. "It was terrific, it was murderous, and made the stoutest heart tremble." [11] Lieutenant Colonel Parr, commander of the 38th, was hit by a shell fragment or cannon ball, shearing off his left arm close to the shoulder; he was also shot through the right hand by a rifle ball. Major James D. Mathews of Co. E, the Tom Cobb Infantry, was shot through the legs and thought mortally wounded, but survived the ordeal. Lieutenant Lester, of Company E, vividly recalled an incident involving the wounding of Mathews, *"After Major Mathews was wounded, it seemed he would die. He called for water, and George W. Smith (of Co. E) went to a small branch, a few steps in the rear, to get him a drink. As he arose to go back with the water, there stood Joseph J. Lumpkin, who is over six feet high, firing away fearlessly on the enemy. How he escaped immediate death is a wonder. "* [12]

Captain Edward Cheves, General Lawton's young aide, was killed in action "while gallantly pursuing the line of his duty he fell pierced through the heart by a rifle ball," General Lawton lamented. [13]

The 38th's introduction to combat was horrific. *"Before the Thirty-eighth had advanced 150 yards the lieutenant colonel and major were shot down, three or four of our most prominent captains had been killed and a dozen lieutenants killed and wounded and 250 privates were killed and wounded. The regiment was left without a commander, some companies without an officer."* [14] The men of the 38th Georgia had drilled for months in anticipation of this moment, but always under the leadership of capable officers. Without these officers to lead the regiment and encourage the soldiers to remain steady, the regiment was in danger of floundering and being shot to pieces.

The loss of the key officers and punishing enemy fire had a demoralizing effect on the regiment. *"The terrific fire that we were under produced a demoralization that was fearful to witness. Men huddled up and became entangled with one another. Some lay down and seemed to spread out like adders, while some just turned their backs and went at a Nancy Hanks* [15] *gait to the rear. What officers were left were nobly doing their duty, but did not seem to know exactly what to do."*

"Our color bearer, Sergeant James W. Wright, of Co. E, who had been selected for conspicuous gallantry, wildly and frantically waved the colors, but otherwise stood still," asserted a member of the 38th, *"No one seemed to be in command. In fact there was no commander. Some of us could see plainly that this panic would get worse if not checked, but all seemed powerless to do anything, and what would have terminated in a most disgraceful affair was suddenly checked by the only man who seemed to know what to do, did that thing at once."* [16] Colonel Clement A. Evans, commanding the 31st Georgia Regiment, rallied the 38th and prepared them to charge the enemy. [17] An unidentified 38th soldier wrote of Colonel Evans, *"I saw him run down our line and seize our colors. With a voice that was firm and clear he was heard above the din and roar he shouted: 'Georgians, is this the way to meet the enemy? Thirty-eighth rally to your colors! On the centre dress.' His words, his fearlessness and cool bearing and his commands acted like magic. Everyone caught the spirit, and though men were falling every second, yet almost as soon as I am writing it, our line was restored. The sight of that Colonel, with sword in one hand and colors in the others was irresistible, and I don't think I ever witnessed so grand and sublime a sight. With the colors in one hand as the line was being reformed, I hear him say 'Captains Battey and McLeod, give your companies to your lieutenants and take command of the regiment.'"* [18] Captain Lawton then ordered Captain Battery to have the regiment lie down and continue firing, Captain Battey recalled, *"I commanded my men to 'Fire and load lying,' which order they promptly executed until nearly all the cartridges were expended."* [19] General Charles S. Winder, commanding the veteran "Stonewall Brigade," arrived on the field and filed into line to the left flank of the 31st and 38th Georgia. The Stonewall brigade was somewhat scattered from moving through the woods and swamp. General Winder halted his brigade and began cobbling together several scattered regiments to form a solid line of battle. General Winder reported, *"I ordered all troops whom I found in front to join this command, making the line continuous. Lieutenant-Colonel Gary, Hampton's Legion: Colonel. B. T. Johnson,*

First Maryland Regt.; Twelfth Alabama Regt.; Fifty-second Virginia, Lieutenant Colonel. J. H. Skinner; Thirty-eighth Georgia, Captain and Assistant Adjutant-General Lawton, commanding, joined this line." [20]

Jackson realized the sound of steady, continuous rifle fire from his own troops and the enemy, meant the battle lines were stagnant and darkness would soon end the battle in a stalemate. Jackson tersely sent orders to his division commanders, *"Tell them, this affair must hang in suspense no longer; let them sweep the field with the bayonet!"*[21]

Winder ordered his entire line to charge, even before Jackson's couriers could deliver his orders. It was between sunset and dark when Captain Lawton and Captain Battey ordered the 38th forward in a "gallant" charge against the enemy. *"Our color bearer was again himself, and grasping the colors...and assured him he would carry them wherever he said. With his sword pointing to the front Colonel Evans said: 'Carry them yonder, and quickly follow him boys.'"* The 38th raised the "Rebel yell," and charged the enemy, moving forward in, *"splendid style over the field, the enemy retiring long before it was possible to use the bayonet. We routed their infantry-Sykes' regulars, the finest troops in the army."* [22] The 38th reformed their line of battle just after dark and prepared to execute a second charge, along with the Fifth Virginia and First Maryland Regiments. Three Federal Regular Infantry regiments held the line to their front, protecting a five gun battery. The three Confederate regiments charged and routed the Federals, capturing their battery. Captain E. P. Lawton said of the 38th Georgia's second charge, *"It was perfectly successful. We drove them from the field (the Federals), and have never heard of them since, except the prisoners we took."* [23]

A. A. SHAW, ADJT.
Post war sketch of Adjutant Augustus A. Shaw
The Sunny South, January 10th, 1891

"We blackened the ground with their dead and wounded. Not only this, but the charge had been at this point to break their lines if possible and as we did fresh troops took our place and doubled them up all along by flanking and cross firing them." [24] Confusion reigned over the battlefield as the Federal lines suddenly collapsed, daylight was quickly fading. The Federal line had been broken at many points almost at once. The U.S. Regulars retreated towards the bridges over the Chickahominy River. Yet, help was on the way for the hotly pressed Federals. Two Union brigades, General French's and General Meahger's, had crossed the river to support Porter's line and were pushing their way through their retreating comrades. They forced their way through at the point of the bayonet. These two fresh brigades formed the rearguard covering the retreating Federals and began firing on the advancing Confederates.

The 38th began receiving deadly fire from distant sharpshooters during this lull in the battle near night fall. The fading light, dust, and smoke from the battle greatly reduced visibility. *"The fire of the enemy's sharpshooters was so deadly that our Adjutant (Lieutenant Augustus Shaw) took it into his head that there our own men firing into us, and went out to stop it."* [25] Shaw boldly walked forward across the fields into the no-man's land and disappeared.

Lieutenant George Lester, further recalled, *"The day after the battle we performed the sad duty of burying our dead, far from their native homes. They filed a soldier's grave – buried without coffin or shroud."* [26] Private Daniels of the Ben Hill Guards, wrote, *"After that hard fight in which we were successful, some of us were passing over the battle field where the dead and dying were thick upon the ground, when all of a sudden we came upon Gus Shaw (Adjutant Augustus Shaw) dead, poor fellow, with his body shot literally to pieces. We gathered up his remains as best we could, and wrapped them in a blanket, and because he was such a favorite among the men, held a special burial service at his grave. There was hardly a surviving man who did not shed tears when they heard Gus Shaw was killed, but those were times when one couldn't weep long, even for his own kin."* [27]

The Thirty-eighth Georgia at Coal Harbor.

CAMP LAWTON, NEAR RICHMOND, VA., July 13th, 1862.

Editor Savannah Republican:

As there has been a mis-statement in regard to the Thirty-eighth regiment in your paper of the fight at Coal Harbor, on Friday, 27th June, I think it nothing but justice to them to state the facts as to their position and what they did in that engagement.

I did not see the article in your paper, but others informed me that you had published that we made a gallant charge on the batteries and failed; but a Texas regiment came up and took it. Such is not the fact. After receiving the fire of six regiments of regulars, for nearly two hours, at a distance of one hundred and eighty yards, in an open old field, and the enemy behind a fence, with some protection from the depth of the road in the lane, we were ordered to charge by Capt. Lawton, which we did with a Maryland regiment and the Fifth Virginia. We captured the batteries and put the enemy to flight. We never moved from the position we occupied in line of battle, until we were ordered to charge, and we never wavered, but accomplished all that could be done—put the enemy to flight and slept on the ground that night. Col. Parr and Major Mathews were wounded early in the engagement, and the command devolved on me, and with Capt. Lawton's and Capt. McLeod's assistance our work was well done. You may judge how severe the conflict was, for out of six hundred who went into the fight, one hundred and fifty-six were killed or wounded.

Two regiments had occupied the same ground that we did, but had to retire, and, I presume one of them was taken for the 38th, and your correspondent made the mistake in that way. I know you are too proud of Georgia to do injustice to her sons.

Yours respectfully,
WM. H. BATTEY,
Capt. Comd'g 38th Reg't G. V.

Savannah Republican, July 18th, 1862, Page 2

After the 38th Georgia completed the sad chore of burying their own comrades, the Tom Cobb Infantry was detailed to bury the remains of young Captain Cheves, General Lawton's aide who fell near their line of battle. "The last thing we did was make a detail to bury a young man from Savannah. He served on General Lawton's staff. It was sad to look upon one so young and handsome, cut off in the bloom of youth." [28] Captain Cheves was eulogized by General Lawton, "Though a mere youth, he had exhibited a degree of zeal, intelligencer and gallantry worthy of all praise, and not one who fell on that bloody field has brought more sorrow to the hearts of those who knew him best." [29]

Lieutenant Lester wrote, "One of our company believed that he (Shaw) was killed, and stated the spot where he was buried. Some of the company seemed dubious on the subject but our comrade, who was a positive man said, "Don't I know Adjutant Shaw?" The soldier stated that he raised the cap off his face as Shaw was lying dead on the battlefield, and was certain it was Adjutant Shaw.

"About a week later, by some means we got a Yankee newspaper through the lines. In this paper was the report of the battle of Cold Harbor (Gaines Mill.) Among other things in this account was the capture of Adjutant Shaw, of the 38th Georgia Regt," [30] Adjutant Shaw was beloved by the men of the 38th Georgia and described as a very intelligent officer. Private Daniels of Company C, described Shaw as, "a finer fellow never lived, he was as clever as the day is long." [31] Shaw entered no-man's land between the lines and walked closer and closer to the suspected "Confederate" sharpshooters, as though he was looking for someone fallen on the field. He quickly realized he was mistaken when he was challenged by Federal sharpshooters of General French's Brigade, 57th New York Volunteer Infantry Regiment. French's Brigade and the Irish Brigade hurried to support the collapsing Union lines about 6 p.m., but arrived too late to do anything but provide a rear guard for the retreating Federals. [32] The Federals halted Shaw with, *"Who comes here?"* Shaw replied, *"A friend on the opposite side."* The Federals promptly took Shaw prisoner. Captain Gilbert Frederick of the 57th NY recalled that Shaw said, *"That his regiment had been in the fight all day and he was out looking for the body of his Colonel. He further said that three lines of the rebels had broken at the advance of our brigade, and that there would be bloody work upon the morrow."* [33]

Shaw spent about a month in a Federal prison before he was exchanged. He was exchanged and returned to the 38th during early August of 1862. Lieutenant Lester later recalled, *"Sometime after this the Adjutant came walking into camp, and he was saluted with "Don't I know Adjutant Shaw! It was clearly a case of mistaken identity."* [34] The 38th Georgia soldiers held a grave side service and buried the man they supposed was Adjutant Augustus Shaw. They probably marked his grave with a wooden board from an ammunition or ration box. The grave was later exhumed and the remains transported to Oakwood Cemetery in Richmond for reburial, just a few miles from the battlefield. Today, Oakwood Cemetery records show A. Shaw, 38th Georgia Regiment, Company I, buried in Division C, Row Q, Grave #75.[35] The unknown Confederate soldier, mistakenly identified as Adjutant Augustus Shaw, still wears his identity, over 150 years after his death.

Confederate prisoners.—We find the following in the army correspondence of the Philadelphia *Inquirer*.:

The following are among the officers of the rebels now in our possession:

Colonel Lamar, of the Seventh Georgia Regiment, mortally wounded.

Lieutenant-Colonel Towers, Seventh Georgia Regiment, mortally wounded.

Captain S. B, McChesney, Third Louisiana.

Adjutant G. B. Sloan, first South Carolina.

Adjutant A. Shaw,. Thirty-eighth Georgia.

Lieutenant H. Shaw, Third Louisiana.

In addition to these there are about three hundred privates. We took more than 2,000 privates on Sunday and Monday that we were unable to bring away.

Rome Weekly Courier, Rome, Georgia, July 25th, 1862, Page 4

The 38th Georgia had thirty-eight soldiers killed in action when the battle ended. Another 118 were wounded, 22 of the wounded would die of their wounds, bringing the total number of fatalities to sixty men, the highest number of dead from one battle in the entire war. [36] Most of the mortally wounded died within a few days of the battle. Others lingered for months and at least one, Private James Lewis Powell of Elbert County

Georgia, for over a year, before succumbing to their wounds. Powell died at the home of his mother in Elbert County on October 9th, 1863. He survived for one year and three months after being wounded.

Captain George W. McCleskey, commanding Company B, The Milton Guards, was wounded in the chest during the fierce battle on June 27th. He was carried to the 1st Georgia Hospital in Richmond, where he died from his wound at half past midnight, on Thursday morning, July 17th, 1862.[37] His son, Private Francis McCleskey, also of the 38th, was at his father's side when he perished. He wrote a letter to his mother, Ms. Angeline McCleskey, telling of his father's death.

Richmond Virginia, July 17th, 1862
1st Georgia Hospital

Dear Mother, resolving time permits me to drop you a few lines it is with a sad heart that I write you these lines. Mother, Father departed this life last night at half past 12 O'clock. I know this is heart breaking news to you, but dear Mother there is great consolation he is gone he is in heaven. He said his way was clear, it filled my heart with joy to know he has gone to rest. Dear Mother you must not take it too hard. Contend your self to meet him in a better world. He is done with this war he was a great and good Soldier he died in defense of you and your dear little Children. He done all that he could do, his Soul is now at rest in where wars shall never rage. Mother he died very easy. He was praying all day yesterday though he new any of the boys that would come in. Jackson and Thomas came in to see him late yesterday morning and he new them and talked with him he new me all the time until he died. He didn't complain of being in any paine at all. Mother it your request to bring him home I would do it if I could but there is no hope at all. They wont transport him at all, and they say (the doctors) that just to have him and take him home this fall. It is so warm that the doctors say it is impossible to carry him home. Mother we will have him put away directly and if you want him brought home we will bring him this fall or as soon as we can. Uncle Guss is here with us he came with Jackson and Thomas. Father asked him to stay with us a day or two but Guss asked one of the doctors here and he done all he could to get Father home. The doctors new Uncles Davey Mc (Davy McCleskey) he said he would advise me to bury him here.

George (Francis's 16 year old brother) is tolerably bad off but I think he is better this morning. I hope he will up in a few days he has fell off mightily. Mother I am going to send George home as he gets able provided I can get him off, I have no idea that he can Stand what a Soldier has to under go he as not been mustered into the service and I am not going to let him be if I can help it. Mother you must keep in as good spirits as you can you know we all to die some time and always bear in mind that father died in a good way defending his country and family, it does my soul good to know how's pleasant he looked when he was laying on his death bed, it seemed that he had nothing to bother him. Mind he did not struggle for his breath. He wanted to cough just before his breath left him, but he was too weak.

*Mother I will write to you again in a few days and I want you to write me as soon as you get this letter. Tell the Newton boys' family that they were here yesterday but did not have time to write them was both well they said they would write as soon as they could. Tell their wives not be uneasy about them *** *** ***** could not get off to come and so tell his folks he is well and would write every chance he had. Mother be sure and write soon. These lines leave me well hoping and praying they will find you all well.*

F. C. McCleskey [38]

Killed or Mortally Wounded at Gaines Mill, Virginia - June 27th, 1862

Company A – "The Murphy Guards," DeKalb County, Georgia

1. Private Alexander C. Austin - Killed 6/27/1862.
2. Private Joseph Maxine Thompson - (or James M. Thompson) Wounded at Gaines Mill, Va., 6/27/1862. Died at Camp Winder, Richmond, Va., 12/30/1862.

Company B, "The Milton Guards," Milton County, Georgia (Currently Fulton County)

3. Private Robert D. Gibbs - Killed at Gaines Mill, Va., 6/27/1862.
4. 2nd Corporal William F. Griffin - Killed 6/27/1862.
5. Private Martin James - Severely wounded in the thigh 6/27/1862 and died 7/5/1862.
6. Captain George Washington McCleskey - Wounded 6/27/1862. Died from wounds 7/17/1862.
7. Private George W. Vinson - Killed 6/27/1862.

Company C, "The Ben Hill Guards," Bulloch & Emanuel Counties, Georgia

8. Private John Boyd - Killed 6/27/1862.
9. Private Silas S. Corbin - Killed 6/27/1862.
10. Private John A. Kersey - Killed 6/27/1862.
11. Private P. Hugh McLean - Killed 6/27/1862.
12. Private George W. Smith - Killed 6/27/1862.

Company D, "The McCullough Rifles," DeKelab & Fulton Counties

13. Private Robert C. Hambrick - Wounded 6/27/1862. Died of wounds 7/14/1862.

Company E, "The Tom Cobb Infantry," Oglethorpe County, Georgia

14. Private James S. Callahan - Mortally wounded 6/27/1862, died 6/29/1862.
15. Private John Henry Day - Mortally wounded 6/27/1862, died from wounds 8/1/1862.
16. Private James R. Fleman - (or Fleeman) Mortally wounded 6/27/1862, died 7/21/1862.
17. Private John Wesley Fleeman - Killed 6/27/1862.
18. Private George W. Nash - Wounded 6/27/1862, died 7/11/1862.
19. Private Samuel Turner - Killed 6/27/1862.

20. Private James M. Glenn - Killed 6/27/1862.
21. Private Woodson B. Harris - Wounded 6/27/1862, died at Richmond of wounds, 8/4/1862.
22. Private William Glenn Howard - Killed 6/27/1862.
23. Private John Peter Huff - Killed 6/27/1862.
24. Private John W. Jackson - Killed 6/27/1862.
25. Private Sion Thomas Smith - Wounded 6/27/1862, died in Richmond, Va. hospital 8/1/1862.
26. Private Elijah Pratt - Killed 6/27/1862.
27. Private Doctor Simon Smith - Wounded 6/27/1862, died 11/19/1862.
28. Private Benjamin Thornton Hunt - Wounded 6/27/1862. Died from wounds 7/30/1862.

Company F, "Thornton's Line Volunteers," Hart and Elbert Counties, Georgia

29. Musician James M. Hutcherson (or Hutchinson) - Killed 6/27/1862.
30. Private James Lewis Powell - Wounded at either Gaines Mill or Savage Station. The record is not clear. Sent home to Elbert County, Ga. and remained there until his death on 10/9/1863.

Company G, "The Battey Guards," Jefferson County, Georgia

31. 2nd Sergeant James Gunn - Wounded 6/27/1862 and died 7/1/1862.
32. Private William L. Gunn - Killed 6/27/1862.
33. Private Samuel R. Patterson (or James R. Patterson) - Wounded 6/27/1862, died 7/5/1862.
34. Private John F. Perdue - Wounded 6/27/1862 and died at Richmond, Va. 7/5/1862.
35. Private Michael W. Pool - Killed 6/27/1862.
36. Private George A. Rodgers - Killed 6/27/1862.
37. Private James W. Rodgers (or Rogers) - Wounded 6/27/1862. Died of wounds 7/5/1862.
38. Private James M. Smith - Killed 6/27/1862.

Company H, "The Goshen Blues," Elbert County, Georgia

39. Private James M. V. Anderson - Killed 6/27/1862.
40. Private Gabriel H. Booth - Killed 6/27/1862.
41. 1st Corporal George W. Butler - Killed 6/27/1862.
42. Private Thomas Sanford Carruth (or Carrouth) - Wounded near Richmond, Va. June 1862. Died of wounds in Elbert Co., Ga. 8/7/1862.
43. Private James Rousey - Killed 6/27/1862.
44. Private William H. Vaughn (or Vaughan) - Killed 6/27/1862.

Company I, "Irwin Invincibles," Henry County, Alabama

45. Private Thomas M. Martin - Killed 6/27/1862.
46. Private Crawford Hicks (Or Hix) - Killed 6/27/1862.
47. Captain Henry L. Jones - Killed 6/27/1862.
48. Private Irwin Register - Killed 6/27/1862.

49. Private George W. Bond - Wounded 6/27/1862. Died of wounds July 13 or 16, 1862.

Company K, "DeKalb & Fulton Bartow Avengers," DeKalb and Fulton County, Georgia

50. Color Corporal John M. Dorris - (or J. M. Dowis) Killed 6/27/1862.
51. Private Horton H. Hornbuckle - Wounded 6/27/1862. Died at Charlottesville, Va., 2/6/1863.
52. Private Jesse L. Henry - Killed 6/27/1862.
53. 1st Sergeant John S. Johnston - Killed 6/27/1862.
54. Private John W. Phillips - Killed 6/27/1862.
55. Private Josephus S. Richardson - Killed 6/27/1862.
56. Private Aaron Jordan Wilson - Wounded 6/27/1862. Died of wounds 6/30/1862.
57. Private Robert F. Jones - Killed 6/27/1862.
58. Private Charles Harris Goodwin - Wounded 6/27/1862. Died of wounds in DeKalb Co., Ga. July 25, or 8/4/1862.

Company N, "The Dawson Farmers," Dawson County, Georgia

59. Private William M. Forbes - Killed 6/27/1862.
60. Private Timothy S. Loggins - Killed 6/27/1862.

* Sources of casualty information. [39]

Chapter 5
The 2nd Manassas Campaign, July - Aug 1862

Ordnance Sergeant William C. Mathews, of the Battey Guards, wrote of the 38th Georgia's reorganization after the Seven Days Battles, *"After the battles around Richmond and a few days rest on the outskirts of the city, we were ordered forward to Gordonsville and Orange County Court House at Liberty Mills. At this place while enjoying a much needed rest, the regiment had a sort of reorganization as follows:*

Colonel George W. Lee having resigned, and Captain John Y Flowers, of Company A, who had been made major, also having resigned. Lieutenant Colonel, Lewis J. Parr was made Colonel, Maj. James D. Mathews Lieutenant Colonel, and Captain William H. Battey, Company G, was made major. Lieutenant Poole was made captain of Company A, and private James S. Jett elected 3rd lieutenant of same company. Lieutenant Andrew J. McMakin was made captain of Company B, and private Simpson A. Haygood elected 3rd lieutenant of same company.

Lieutenant George H. Lester was made captain of Company E, and Lieutenant John J. Daniel resigned, Sergeants Jabez M. Brittain and Robert T. Dorough were elected 2nd and 3rd lieutenants of same company.

Lieutenant's John H. H. Teasley and Benager T. Brown of Company F, having died, privates John G. Curry and Thomas D. Thornton were elected 2nd and 3rd lieutenants of same company. Lieutenant John W. Brinson was made captain of Company G, and private William C. Mathews elected 3rd lieutenant of same company.

Lieutenant James C. Hall, Co. H, Goshen Blues, from Elbert Co., Ga.
Courtesy of Mr. Chandler Eavenson, descendsant.

Lieutenant James C. Hall of Company H, having resigned, (note: he died 8/16/1862 at hospital), Sergeant Henry R. Deadwyler was elected 3rd lieutenant of same company. Lieutenant Andrew B. Irwin of Company I was made captain (died in hospital July 1862) and private John D. Grantham was elected 3rd lieutenant of same company. Lieutenant Julius J. Gober having died and Captain William Wright resigned, 2nd lieut. Gustin "Gus" Goodwin was made captain and privates Enoch H. Morris and William R. Henry elected 2nd and 3rd lieutenants of same company. Lieutenant Shaw was ordered back to his company (Jo Thompson Artillery) at Savannah, and Lieutenant Levin W. Farmer was detailed to act adjutant." [1]

Sergeant George Washington Gaddy, of Company A, from DeKalb County, Georgia, sat down and wrote a letter to his wife Emily and their children, while the regiment was at Gordonsville, Virginia. He noted the regiment was greatly reduced in strength from the battles losses at Gaines Mill and from widespread sickness.

Virginia, Gordonsville Station July 22, 1862

Dear wife and Children it is by the goodness of God that I one time more have the pleasure of droping you a few lines to inform you of my presant condition I can inform you that I am not very well at this time but I am still with the reigement we are about eighty miles from Richmond City at this time we have bin marching nearly all the time and we have bin in this cuntry we don't stay at any one place more than two or three days at a time. I hope that these few lines will come safe to hand and find you all well and doing well.

Sgt. George Washington Gaddy, Co. A, of DeKalb Co., Georgia.
Courtesy of Mr. Peter Carmouche Jr., descendant.

I want to here from you all very bad at this time but I want to see you all a heap worse. I have not herd any thing from you sense I left home and I am very uneasy a bout you all and this is five or so letters that I have sent to you and I have not heard any anser from you yet I want you to write as often as you can for I shal not get all the letters that you send to me so if you don't write often I will not here from you hardly at all when you write to me direct yore letters to the thirty eighth Georgia reigement Ga. voulunteers in lawtons brigade in care of Capten Pool Company D for ware not known here as the wright legion.

I wish I could tell you where we are a going but I cant tell any thing a bout where we will go far I dont no one day where we will go the next for we are following Stone wall Jackson and no one knows where he is a going but himself we are all nearly worn out now and if we have to follow him much longer there will not be many of us left for we have to march so hard that we cant stand it and we dont get much to eat we have flower and beef and hardly any salt to go with it. We don't have any tents to ly under and if it was not the hopes of our freedom I dont thank that any body would suffer it any longer for it is all most impossible to buy anything to eat in this part of the cutry for it not here to buy.

We hardly have any men in the company that is able for duty at this time. I dont think the company will ever will get to gether again for they are so badly scattered. Emily you may tell M.R. Miller that C. Miller (Private Charles Miller, died in Staunton, VA, Hospital, Aug. 12th, 1862) and Crison (Private Henry Bryce Miller, died Staunton, VA, Aug. 24th, 1862) is both sick at this time and are in the horsepitile to gether. Thomas Wilson (Private Thomas M. Wilson, died at Stauton, VA. Aug. 5th, 1862) also and was sent yesterday to Staunton and several others with them.

Emily I want you to do the best you can for yore self and the little ones I left be hind and all that I left thare and if I never return home to you and them again I want you to sell one hundred acres of the old place if you can and try to pay what I owe and keep the rest for a home for you and the children and if it all has to be to be sold I want you to have the same portion but I want you to keep it all together if you can this may the last time that I will be permitted to write to you in this life for death is in the camp as well as the battlefield. but still and put my trust in God that I may be preserved and that we may be permitted to all meet again in this life and I want you pray for me that if we never meet on earth again that we may all meete in heaven whare thare will be no more parting nor no more weeping nor morning for the so goodly of those loveing ones may God bless you all. I still remain your loving husban until death.

From G. W. Gaddy to Emily Gaddy and Children[2]

George Washington Gaddy never saw his wife and children again; he was killed less than two months later at the battle of Antietam (Sharpsburg), Maryland, on September 17th, 1862. His remains are likely buried in an unmarked grave at Washington Confederate Cemetery, Hagerstown, Maryland.

"On the 7th of August the brigade was transferred from Jackson's old division to General Ewell's division, and we started on the march after Pope's army. The regiment was detailed to support an artillery battery at the battle Cedar Run (Cedar Mountain, August 9th, 1862) and although under a heavy artillery fire, were not engaged otherwise. Our loss was six or eight wounded. On the night previous to this little tussle, there considerable signaling, with torch lights. This was quite a novel scene to us, and entirely a new way of communication."[3] Records show two 38th Georgia soldiers wounded at Cedar Mountain by artillery fire, both were from the Dawson Farmers Company. Private Zebulon C. Payne was wounded and disabled. Private John W. Kellis was wounded and died of his wounds in 1862.[4]

A Confederate soldier watched the Federals bury their dead on the day after the battle of Cedar Mountain, he remarked, *"The Yanks came over under a flag of truce, asking permission to bury their dead, which was granted; and their burial parties were at work on the field under the friendly fluttering's of a white flag, packing away their comrades for dress parade when Gabriel sounds the great*

Reveille. Ah, my silent friends! You came down here to invade our homes and teach us how to wear the chains of subordination and reverence a violated constitution. In the name of Dixie we bid you welcome to your dreamless couch under the sod that drank your blood, and may God have mercy on your poor souls and forgive you for all the despicable depredations that you have committed since you crossed the Potomac." [5]

Capt. Robert P. Eberhart, Co. H, "Goshen Blues," of Elbert Co.,Ga.
The Sunny South, Jan 10th, 1891

"Early the following day (Sunday) after the fight we were on the march. After this battle, Maj. William H. Battey, who had been in command of the regiment since the battle of Coal Harbor (Gaines Mill, June 27th 1862), was taken violently ill, and turned over command to Robert P. Eberhart, senior captain present." [6]

After the Confederate victory at Cedar Mountain, Jackson waited on the field for two days, inviting the Federals to resume the battle. The Federals refused to attack. Then Jackson received news that Pope's army had arrived at nearby Culpeper Court House with a considerable force. Jackson ordered a retreat back to Gordonsville. Lieutenant Lester recalled, *"We commenced to retrace our steps to Gordonsville, where we arrived foot sore and tired; pitched our tents on the same ground which we had so recently left."* [7]

The overall Federal strategy during the peninsula campaign called for McClellan to attack Richmond from the East, while General Pope's army advanced on Richmond from the North. After the Seven Days battle around Richmond, Pope was still far from Richmond, but McClellan had been pushed back to Harrison Landing on the James River. McClellan was demoralized and began evacuating his troops down river and back to Washington. McClellan's army was no longer an immediate threat to Richmond. McClellan planned to move his army by ships back to Washington and then join forces with Pope's army. With McClellan suppressed, Lee and Jackson began to plan to destroy Pope's army before he could be re-enforced by McClellan. General Lee realized speed was of the essence in pressing the attack on Pope. He must strike Pope before he could unite with McCellan's army.

Pope's army was posted in a strong line on the North bank of the Rappahannock River. General Lee's plan called for General Longstreet to demonstrate in front of Pope's line, feigning an attack to keep Pope fixed in place. Jackson would then move his Corps around Pope's right flank and strike out for his rear to cut his communications with Washington and confuse Pope. Lee, moving with Longstreet's Corps, would then cross the Rappahannock River and join Jackson in attacking Pope, before he could be reinforced by McClellan. This was a risky plan and Lee realized he was violating one of the basic maxims of war, never divide your army in the face of a superior force.

Newly promoted Lieutenant William C. Mathews wrote, *"The 'foot cavalry' started after Gen. Pope, passing through Culpeper Court House and on to the Rappahannock, where we were unmercifully shelled, losing two killed and about twenty wounded. We turned up the river, and at a rocky ford called Henson's Mill, we effected a crossing."* [8]

Lieutenant George Lester detailed the advance and several interesting incidents of the Georgia Brigade,

"On the 20th of August we were on the tramp again, marching towards the Rapidan River which we crossed at Summerville ford, near Brandy Station, continuing our march to the Rappahannock River where we found the enemy on the north bank. By the usual sign they soon gave us to understand that we would not be allowed to cross at this place without a struggle. After the descent of a few shells into our ranks, we changed direction, not however, till Lieutenant Robert T. Dorough received a slight wound in the shoulder with a piece of shell. Marching up the river for a few miles, we effected a crossing without resistance. After crossing we found all manner of clothing, blankets, and stationary – indeed almost everything needed by a soldier. Some of the boys loaded themselves down with these things, cast away by the fleeing enemy only to drop them down one by one until all were left strewn upon the road as before; for as Bill Arp [9] *says 'we were in a strain' about this time – Stonewall Jackson was on a forced march, with the head of the column, toward Manassas Junction, where was stored the soldier's glory.*

On the morning of the 24th of August we re-crossed the Rappahannock and stacked our arms just behind a steep little hill near the river, where we had planted a battery. In a very short time, the different mess had their firers underway, and was preparing the dough for the oven. However just before we were ready to commence the baking process the enemy's battery on the opposite hill saw fit to send a challenge to our battery on the hill above us, which was instantly accepted, and the duel commenced in good earnest, which warmed up to that extent that we commenced looking around for a safer place and preparing to move. The sick were dispatched up the branch, under the shelter of the hills and trees. These feeble boys were interested with the dough that we had been prepared for baking. The Mess of Willis B. Jackson turned theirs over to Willis for safety, he being on the sick list. He started out ahead with all the speed he could,

Major General John Pope – USA
Library of Congress

with the aforesaid dough stuck under his arm for safe transportation, but the shells came too fast and thick for Willis. He raised his arm and let her slide to the great grief of his hungry mess mates who were following far in the rear. When the mess came up to where their fondest hopes had been cast into the mud, which contained the last drop of grease the possessed, the very grease, too, that came from the piece of meat Bill Smith had carried next to his heart till the grease dripped off his heels. It almost made them shed tears!"[10]

At early dawn August 26th, Old Stonewall turned his face and directed his steps toward the Bull Run Mountain, passing through Thoroughfare Gap, pushed on day and night, to reach the goal not far ahead. This was a very trying time to the weary, foot-sore fellows; but to the Tom Cobb Infantry, the sight of Jackson was enough to inspire them a with the determination that there should be no such word as fail. Proceeding direct to Bristow's Station, on the Orange railroad, we captured several trains. [11]

Private George W. Nichols of the 61st Georgia Regiment, of Lawton's Brigade, wrote how the brigade wrecked and captured two trains, at about dark on August 26th. "*On Lawton's Brigade's part of the line there came two heavy supply trains loaded with bacon, hard-tacks and almost everything that we needed to eat. Our artillery had not caught up and we had not had time to tear up any railroad track, so the engineer pulled his throttle wide open and let his engine fly by while he and the fireman lay down in the tender. Many shots were fired but to no effect. Doubtless they would have gone on to Pope's army, for we were then away in the rear, but Lieutenant John Brannen, of our company, threw a heavy oak cross tie across the track in a curve and ditched the front engine. The other engine crushed into the rear of the first train and made a terrible wreck of both trains with some of the cars completely smashed to pieces. We were not scarce of supplies then, for every one took all he wanted. Some of the men carried off a whole side of bacon.*" [12]

Part of Lawton's Brigade clashed with the Federals about four miles from the massive Federal supply depot. Lieutenant Lester recalled the novel experience of how the Tom Cobb Infantry watched the battle, without participating in the fight.

Trains possibly wrecked by Lawton's Brigade, Bristoe Station, Aug. 26th, 1862.
Photographic History of the Civil War, Vol. VI. Published 1911

"This march of ours had been a race with Gen. Pope "Headquarters in the saddle," since we crossed the Rappahannock the last time, for Manassas Junction. And now we had beat him there, and we were between him and his forces at the Junction. Jackson

determined to hold the Federal forces in check until his bare-foot soldiers could draw from Uncle Sam's Quarter-master's Department, shoes, blankets, and four day's rations, which they sorely needed and so richly deserved. According to the previous plan (I suppose) the Tom Cobb Infantry were detailed for picket duty, and sent out about three-quarters of a mile from the railroad and posted on a small elevation within 800 yards of a river (Broadrun), over which was a bridge. This bridge was the nearest point of crossing to our picket post, and was, according to my recollection, west from where we were posted, and to our right to the south from whence the enemy was expected.

One regiment of one brigade (Lawton's) the 60th Ga., (I think) and two regiments of Louisianans were stationed in our front and about two or three hundred yards from us. On the next hill, just on just on the corner of an open field of considerable size, about 2 o'clock, a brigade of Federals attacked these three regiments of ours, while there was still another regiment of Federals behind a stone fence to their left front (of our regiment's left front). This brigade of Federals marched down and charged across this open field in splendid order, but met with such a deadly fire from that little band that they were nearly annihilated. The few that were left fled in confusion. Another came and bit with the same fate; still another came this time without such a great loss, as the regiment behind the rock fence were pouring a deadly fire into our regiments engaged. Their volleys seemed not to be so effective as on the previous charges, but enough to demoralize them to that extent that they turned their backs to flee, but were rallied after considerable effort on the part of their brave leader, and a hand to hand fight ensued, when butts of guns and bayonets were freely used. The flag was taken from the 60th Ga. Regiment in the melee by the Federals, but our boys rushed at them and retook their flag.

Our forces having held the field against all odds as long as it was necessary for Stonewall Jackson's purpose, retired in good order across the river, having crossed bayonets with the enemy -a thing of rare occurrence and rejoined their commands. During the time of this deadly struggle just described, the Tom Cobb Infantry's were at their picket post upon the hill where they were placed in the morning, watching, with admiration, the heroic courage displayed by their comrades. This was the first time they had ever been as eye witness to such a scene without being themselves engaged. It was a grand sight to see the serried hosts of the enemy, with gleaming bayonets, charge down on our troops, and watch the wave and surge of the battle -three small regiments, being charged by twice their numbers, with bayonets fixed, glittering in the rays of the sunshine like so much glass. One could imagine only destruction to any who would dare to oppose them. But the Georgians and Louisianans stood there in their front like a stone wall 'till they came near enough to see the whites of their eyes, when they sent forth a volley that thinned their ranks so, that the rest fled in confusion." [13]

The Tom Cobb Infantry was nearly cut off and captured during their retreat from the picket post. Lieutenant Lester continued the account. *"We were permitted to remain on post until the entire command had crossed to the opposite side of the river. By this time we were nearly encompassed by the enemy, in front and rear. Receiving orders to join our regiment just at this state of affairs - being cut off from the river bridge and closely pressed by the enemy in the rear - we hurried away as fact as our legs could carry us, even man taking care of himself - throwing away blankets, and everything except guns that tended to impede our flight. Fording the river about three hundred yards below the bridge, under a heavy fire from the enemy, who had by this time reached the river barks a short distance above us, we all landed safely on the other side, worse scared than hurt - for none of us were touched. We saw the old familiar flag of the 38th, that had been completely riddled in the battle of Cold Harbor, floating in the breeze but a few hundred yards distant indicating the location of our command, which we has- tended on to join, congratulating ourselves that we had run the gauntlet with complete success. We were now pushed forward to Manassas Junction, where awaited us everything that a soldier could desire; arrived at the Junction late in the afternoon, loaded ourselves down with the good things, so carefully stored there by Uncle Sam. After supplying ourselves with all we could carry we set fire to the balance and went on our way rejoicing."* [14]

The Confederates fondly described the feast at the massive Federal supply depot. *"What a prize it was, and with what zest did the half-starved Confederates set out to plunder the vast depot in which were housed the provisions for an army of 60,000 men! Jackson looked on indulgently as his men clothed and fed themselves from the spoil of the enemy. One precaution he took. "The first order that General Jackson issued," writes Major Mason, "was to knock out the heads of hundreds of barrels of whiskey, wine, brand, etc., intended for the army. I shall never forget the scene when this was done. Streams of spirits ran like water through the sands of Manassas, and the soldiers on hands and knees drank it greedily from the ground as it ran."* [15]

A Confederate Chaplain watched as the starving Confederates looted the supply depot, he wrote, *"I will not attempt to describe the scene I here witnessed for I am sure it beggars description. Just imagine about 6000 men hungry and almost naked, let loose on some million dollars worth of biscuit, cheese, ham, bacon, messpork, coffee, sugar, tea, fruit, brandy, wine, whiskey, oysters, coats, pants, shirts, caps, boots, shoes, socks, blankets, tents, etc.. Here you would see a crowd enter a car with their old confederate grays and in a few moments come out dressed in Yankee*

Confederates plundering the Federal supply depot
Battles and Leaders of the Civil War, Vol 2

uniforms; some as cavalry; some as artillerists; others dressed in the splendid uniform of Federal officers. I have often read of the sacking of cities by a victorious army but never did I hear of a railroad train being sacked. I viewed this scene for almost two hours with the most intense anxiety. I saw the whole army become what appeared to me an ungovernable mob." [16]

The remaining Federal supplies were burned and Lawton's Brigade and the 38th Georgia continued their march towards the old battlefield at Bull Run. Lieutenant George Lester wrote of a humorous incident on the march, *"Our march was now in the direction of Bull Run by the way of Centreville and across the Natural Bridge. Proceeding within about five miles from the Junction, we struck camp for the night. Sometime after midnight we were aroused by a native of this (Oglethorpe) County, but a member of the 13th Ga. Regiment. Now, John Dunn, the name of this distinguished individual, who was a sort of "Ichabod Crane" being very tired, laid himself down for a happy snooze "Sleepy Hollow" near the Junction. Those who know John will recollect that he was very tall, gawky, and by no means a Solomon.*

John had scarcely gotten into the land of dreams when, (as he told us,) there occurred a tremendous earthquake, which came near engulfing him as he stood on the edge of the awful chasm. The quaking and shaking of the earth at this dreadful moment almost frightened "Ichabod" out of his wits - this was no place for him - and his legs, true to their nature, carried him away at break-neck speed, never stopping to take even a last look, but he safely landed in the ' bivouac of his company, escaping the "perils by land." He related that Manassas Junction, with a mile or two around, had gone done to fathomless depths in this terrible earthquake, and that he was so near the edge of the abyss he lost his hat and blanket, which went down with the balance. This earthquake turned out to be the blowing up of a car load of powder, causing a tremendous report which jarred the county for miles around. Notwithstanding this information, "Ichabod"' was never known to nap it outside of his company's protection again. [17] *Just before the dawn on the following day (August a 28th,) orders came to fall in, and we were soon wending our way toward Centreville and over the stone bridge; proceeding directly across the battle field of the first Manassas battle (on 21st July, 1861,) took our position on the north side of Marietta and Alexander turnpike road and rested when we were made aware of enemy's presence again by the firing of pickets all along the lines."* [18]

Jackson's corps arrived on August 28th at the previous battlefield where the Battle of Manassas was fought. He posted his corps behind an unfinished railroad embankment, a readymade breastwork, and waited. General Pope's army was widely scattered and frantically searching for Jackson's army. Pope didn't find Jackson, but Jackson found Pope. It was reported to Jackson late in the afternoon, that large numbers of Federals were marching down the Warrenton Pike near Groveton, or Brawner's Farm, heading north towards Centreville, their line over a mile long. Soon the picket line began firing and the boom of cannon fire sounded, Lawton's Brigade was about to be ordered into action.

Lieutenant William C. Mathews recalled the state of the regiment just before the battle of 2nd Manassas, "The long, rapid marches, with scarcely anything to eat, had sadly reduced the numbers of the 38th Georgia Regiment, as well as all commands and the regiment only carried into the Second Manassas 260 men" [19]

Gen. John Gibbon – USA
Library of Congress

Jackson ordered a battery of artillery to open fire on the Federals. The Federals marching down the road were General John Gibbon's Midwestern Brigade, consisting of the 2nd, 6th, 7th Wisconsin, 19th Indiana and 24th Michigan. Only the 2nd Wisconsin had seen combat prior to this day, but these were some of the finest troops in the Union army. They would soon earn the sobriquet "The Iron Brigade," because they stood like iron during battle. This was the first, but not the last, time the Georgia Brigade would face these hardy men from the Midwest.

Jackson rode out alone to observe the Federals marching on the turnpike. He rode back to an assembly of officers. Jackson ordered his commanders behind the railroad cut, *"Bring out your men, Gentlemen."* The battle of Second Manassas was about to begin. [20]

Jackson attacked General Rufus King's Division as they marched along the Warrenton Turnpike. The battle opened with a volley of artillery shells fired by Jackson's horse artillery which was positioned on a ridge overlooking the pike. The barrage hit the center brigade in the division, commanded by General John Gibbon. Gibbon ordered the 2nd Wisconsin Regiment to drive away what he supposed was only "horse" artillery. Lieutenant Lester wrote, "We were not kept long in suspense, but were soon formed into line of battle, and advanced slowly - till we had cleared the woods and crossed the railroad; found ourselves in an open field with slight ascent. No sooner than we were in open ground than the order came to double quick, which we did for about two or three hundred yards to the top of the hill, led by Gen. Ewell in person and on foot, where we haltered for a few minutes. We were now under heavy fire from the enemy's infantry, as well as a heavy shelling from their batteries. While here Captain Charles A. Hawkins was wounded slightly on the head, Sergeant Walter S. Robertson assisted him off of the field. The company was now left without a commissioned officer, but with Gen. Ewell in their front, urging them on, they would have gone into the very jaws of death. Again the command was given to charge. By this time the whole face of the earth around was being completely scoured by grape and canister from the enemy guns." [21]

Private Samuel House Braswell, of Company A, from DeKalb County, Georgia, was shot through the right wrist by a Minié ball, the ball breaking the radius bone in his arm. Confederate surgeons removed two inches of shattered bone from his wrist and performed a resection. When the wound healed, it contracted his wrist, resulting in a permanent twist to his right wrist. He was permanently disabled, yet lived for 52 more years, dying in 1914.[22]

Pvt. Samuel House Braswell, Co. A, of DeKalb Co., Ga., wounded at 2nd Manassas.
Courtesy of Mr. Roy Hunter

The Second Wisconsin deployed in line of battle and started moving over a ridge beside the turnpike, towards the Confederate artillery. They crested the ridge and received a volley of fire from the Stonewall Brigade, which was positioned a short distance from the ridge. The fourth regiment in Gibbon's line was the 19th Indiana, they had never seen combat prior to this moment. They were ordered to support the 2nd Wisconsin by forming on the left flank of the 2nd. As soon as the 19th Indiana lined up to the left of the 2nd Wisconsin, a Confederate battery unlimbered on their left flank and opened fire. The 19th Indiana turned to face the flank artillery attack and its infantry support. They shot it out with the Confederate infantry for about one half hour, but in that short time, 259 of the 423 Wisconsin men engaged were shot down. [23]

General Gibbon quickly fed his other regiments into the fight, forming them to the right and left of the 2nd Wisconsin. The 7th Wisconsin formed to the right of the 2nd Wisconsin and the 19th Indiana formed to their left flank. The 6th Wisconsin formed to the right of the 7th Wisconsin, but there was a 250 yard gap between these two regiments. Gibbon realized this gap was a fatal flaw in his line and appealed to General Abner Doubleday for support. Doubleday quickly sent the 76th New York into the gap to the right of the 7th Wisconsin and the 56th Pennsylvania to the right of the 76th New York, closing the gap between the 6th and 7th Wisconsin. Both Union and Confederates brigades now had a solid line of battle, only 80 yards apart. Here they stood firing into each other, both refusing to yield an inch. Darkness was quickly approaching when Jackson ordered Lawton's brigade to charge the Federal line in front of the 7th and 2nd Wisconsin. Through some mismanagement, only the 38th and 26th Georgia, to the right of the 38th, charged the enemy line. It was about 7:45 PM and twilight made it hard to distinguish the Federal lines to their front.

A Union Lieutenant recalled the deployment of the 76th New York and the attack by the 38th and 26th Georgia. *"But a few moments elapsed after entering the wood, before sharp and continuous musketry firing was heard very near, and up the hill hidden by the woods. A strange mounted officer came riding down through the woods, shouting "Come on! Come on! Quick! Quick!" The Seventy-sixth was*

Private Booker Woodson Hopper (left) and his brother Private Daniel Hopper, of Company E, from Oglethorpe County, Georgia. Daniel was killed in action August 28th, 1862 at 2nd Manassas. *Photo courtesy of the William D. Hopper family, descendants.*

immediately in motion—over fences, through the bushes, around the trees, over logs—the bullets and shells tearing through the woods like a hailstorm through a wheat field, on rushed the Regiment. Several of the men were killed and wounded before leaving the wood. After going about twenty rods, the Regiment emerged into an open field.

Here was battle in real earnest. Just in front and a little to the left were the gallant boys of the "Iron Brigade," (composed of three Wisconsin and one Michigan regiments), fighting and falling in a manner terrible to behold. Just at this juncture, as the rebels were preparing in great numbers in the woods beyond, for a charge upon our lines, the 76th New York and 56th Pennsylvania were ordered into line to fill a gap between the Sixth and Seventh Wisconsin. By this timely movement, the noble "Iron Brigade" was saved from total annihilation.

During a lull in the action, a body of men was seen moving on the extreme left flank. As they came forward they shouted, 'Don't shoot your own men!' At that distance it seemed doubtful whether they were friends or enemies, and it was not without much hesitation that the Colonel gave the order, "By the left oblique! Aim! Fire!" No rebel of that column, who escaped death, will ever forget that volley. It seemed like one gun." [24]

The 38th Georgia survivors recalled that volley, Lieutenant Lester wrote, *"It was now quite dark, and the light caused by the shells and other flying missiles of death, striking against rocks, which almost covered the road produced a scene grand and beautiful beyond, though not altogether pleasant. As we said, the General Ewell led the order to charge was given and on we went, General Ewell leading the regiment. The enemy reserved his fire until we were within fifty yards of their lines, when they sent such a volley into our ranks it seemed that the very earth itself belched forth fire. We were ordered to lay down and fire which we did. The engagement lasted till about 9 o'clock, with terrible effort. The ground being literally strewn with the dead, dying and wounded."* [25]

Pope launched a series of assaults against Jackson's corps throughout the next day, but all were repulsed. Pope was convinced he had isolated Jackson and would crush him before Longstreet arrived, but Lee and Longstreet's corps arrived on the battlefield about noon on August 29th. Longstreet posted his corps on Jackson's right flank. Lee urged Longstreet to attack, but Longstreet asserted the time wasn't right.

Pope renewed his attacks the following day, August 31st, against Jackson's position, he was under the mistaken impression Jackson was retreating. He ordered his massed legions forward in pursuit of the retreating Confederates, but Jackson wasn't retreating, his troops still held the line behind the railway cut. It seemed Jackson would be overwhelmed by the massive Union force arrayed on the open plains of Manassas. Longstreet's artillery batteries then opened fire on the Union left flank, cutting down scores of Federal soldiers and breaking up the Union infantry attack against Jackson. Longstreet ordered his entire wing forward and 28,000 Confederates descended on the Union left flank, the largest infantry assault of the entire war. The left flank of the Union army was crushed and they retreated back across Bull Run Creek. The Federal rearguard fought valiantly and narrowly prevented another disaster on the scale of the "First Bull Run." The Confederates had won a great victory and the Federal army retreated to Centreville.

Captain William McLeod prepared an official causality report for the battle of Second Manassas on January 12th, 1863. This report lists 29 killed and 80 men wounded on August 28th, 1862. Ten of the wounded men died of their wounds. All the 38th Georgia casualties were recorded as occurring on August 28th, with no additional casualties reported for the two subsequent days of battle. [26]

Killed or Mortally Wounded at 2nd Manassas - Aug. 28th, 1862

Company A, "The Murphy Guards," DeKalb County, Georgia

1. Private John T. House - Wounded 8/28/1862. Died at home in Cobb Co., Ga. 11/28/1862.

Company B, "The Milton Guards," Milton County, Georgia (Currently Fulton County)

2. Private William Washington Dinsmore (or Densmore) - Wounded 8/28/1862. Died 8/30/1862.
3. Private William Thomas Jameson - Wounded 8/28/1862. Died 9/15/1862.
4. Private John King - Wounded 8/28/1862. Died 9/6/1862 at Gordonsville, Va.

Company C, "The Ben Hill Guards," Bulloch & Emanuel Counties, Georgia

5. 4th Corporal Abraham Boyd - Killed 8/28/1862.
6. Private James W. Dudley - Killed 8/28/1862.
7. Private William J. Grant - Killed 8/28/1862.
8. Private John McLean - Killed 8/28/1862.

Company E, "The Tom Cobb Infantry," Oglethorpe County, Georgia

9. Private Daniel H. Hopper - Killed 8/28/1862.
10. * Private John Baughn Jackson - Killed 8/28/1862.
11. Private Joseph Warren Coile - Killed 8/28/1862.
12. Private Larkin R. Scisson - Killed 8/28/1862.

Company F, "Thornton's Line Volunteers," Hart and Elbert Counties, Georgia

13. Private John W. Black - Killed 8/28/1862.
14. Private Thomas M. Eavenson - Wounded 8/28/1862. Died 8/25/1864, or at Harrisonburg, Va. 9/28/1864.
15. Private Benjamin "Martin" Maxwell - Killed 8/28/1862.
16. Private Thomas C. Stephenson - Killed 8/28/1862.

Company G, "The Battey Guards," Jefferson County, Georgia

17. Private Robert L. Gunn - Killed 8/28/1862.

18. Private Andrew J. Hancock - Reported as wounded and missing at 2d Manassas, Va., 8/28/1862. Confederate Service Record states he was killed at 2nd Manassas, 8/28/1862.
19. Private Jordan Smith - Killed 8/28/1862.
20. Private Weems F. Smith - Killed 8/28/1862.
21. Private Robert Stewart - Killed 8/28/1862.
22. Private James Alfred Ginn - Killed 8/28/1862.
23. Private George T. Willoughby - Wounded at 2nd Manassas, died of wound at Middleburg, Va., 9/16/1862.

Company H, "The Goshen Blues," Elbert County, Georgia

24. Private James Alfred Ginn - Killed 8/28/1862.
25. Private William P. Moon - Wounded 8/28/1862. Died of wounds 9/2/1862.
26. Private John W. Parham - Killed 8/28/1862.

Company I (New Company I) Henry County, Alabama

27. Private George W. Applewhite - Killed 8/28/1862.
28. Private John Fleming - Killed 8/28/1862.
29. Private John Jacob Gregory - Killed 8/28/1862.
30. 3rd Sergeant James C. Hall - Killed 8/28/1862.
31. Private Benjamin N. Williams - Wounded 8/28/1862. Died of wounds 9/11/1862.

Company K, "DeKalb & Fulton Bartow Avengers," DeKalb and Fulton County, Georgia

32. Private John H. Akers - Wounded 8/28/1862. Died from wounds 9/4/1862.
33. Private John W. Chandler - Killed 8/30/1862.
34. 1st Corporal Francis Marion Gasaway (or Gazaway) - Killed 8/28/1862.
35. 1st Lieutenant Gustin E. Goodwin - Wounded 8/28/1862. Died of wounds 9/7/1862.
36. 2nd Lieutenant Enoch H. C. Morris - Killed 8/28/1862.

Company N, "The Dawson Farmers," Dawson County, Georgia

37. Private Aaron A. Brooks - Killed 8/28/1862.
38. Private Elias Dufrees (or Dufraes, or Dupree) - Wounded 8/28/1862. Died 10/27/1862.
39. 4th Sergeant David James Wallis - Wounded 8/28/1862. Died of wounds 10/8/1862.[27]

*Note: Private John Baughn Jackson was not listed in Henderson's Roster of Georgia Confederate Soldiers and was not found in Confederate Service Records, but was reported by Lt Lester, in the History of the Tom Cobb Infantry, as being killed in action.

The Union army was beaten, but not destroyed. General Lee was not satisfied with the result, so he sent General Stonewall Jackson after the Union army. Marching by a circuitous route, Jackson hoped to intercept and block the retreat of the Union army. Jackson caught up with the Union Army at Chantilly, Virginia, which was also known as the battle of Ox Hill. Lieutenant Lester of the Tom Cobb Infantry recounted the pursuit,

"Our command marched northward and crossed Bull Run at Ludley, on the 30th, and came the next day into the Little River Turn-pike, which leads eastward, intersects the, Warrenton road At Fairfax Court-house. Here Stonewall Jackson attacked the enemy on Ox Hill and defeated them, though not without a severe struggle, which lasted about an hour, during which time we were drenched in a heavy shower of rain. Fortunately, we had none of our company hurt in this engagement." The other companies of the 38th Georgia Regiment were not as fortunate; Private Robert Stewart of the Battey Guards was killed. Private Joshua K. Lewis of the Ben Hill Guards, and Private Andrew J. Brown, of the Henry Light Infantry, were both wounded. [28]

Lieutenant Lester described the 38th Georgia's movements after the battle of Second Manassas, *"This affair closed on the evening of the 1st of September, 1862, which closed the fighting of the second battle of Manassas. We took up our march on the 3rd of September, still being the advanced force of the army, and passed through Leesburg on the 4th, and camped that night at the spring that may be justly ranked as the largest spring in Virginia. On the morning of the 5th we crossed the Potomac at White's Ford and advanced northward to Frederick City and occupied the Baltimore and Ohio railroad and remained there till September 10th, when we resumed our march in the direction of Williamsport, where we re-crossed the Potomac into Virginia and arrived at Martinsburg the 12th of September and captured many valuable stores from the enemy; proceeding thence to Harper's Ferry and captured the place, without much resistance on the part of the enemy. Jackson's army captured a large amount supplies and munitions of war, together with eleven thousand prisoners and seventy-three pieces of artillery. After the capture we immediately resumed our march to rejoin General Lee at Sharpsburg."* [29]

Chapter 6
The Bloodiest Day – Antietam, September 17th, 1862

Lieutenant William C. Mathews of the Battey Guards wrote, "After the victory at Manassas the march was resumed towards Washington City and on the 1st of September we had the fight at Chantilly or Ox Hill. After a rest of a day or two, we marched towards the Potomac by way of Leesburg. Crossing the Potomac at Edwards Ferry, we entered Maryland and marched to Monocacy Junction on the Baltimore & Ohio R. R., forty miles from Baltimore. After a day or two of rest and foraging, passing through Fredrick town and Middletown and Boonsboro, we crossed the Potomac at Williamsport and marched to the town of Martinsburg, Va. Down the line of the B & O R.R. we then marched to Harper's Ferry, which place after a little fight in which our loss was slight, we captured with its garrison of 11,000 men, supplies and munitions of war. Hardly had we congratulated on the splendid achievement, when we were ordered to leave, and after marching all night long and crossing the Potomac after midnight at Shepherdstown at daylight we were in line of battle to the right of Sharpsburg." [1]

The 38th Georgia stacked rifles in the West Woods, just behind this church on the night of Sept 16, 1862. – Library of Congress.

Lawton's Brigade stacked rifles in the West Woods, just behind the small white Dunker Church on September 16th. However, the simple lack of food for General John Bell Hood's Division would thrust Lawton's Brigade onto some of the bloodiest ground of the war. *"On the evening of September 16th, the (Georgia) brigade, commanded by Colonel M. Douglass, was ordered to march toward the enemy's line. After marching some distance, the brigade filed into a piece of woods, and there remained, with arms stacked, until about 10 p.m."* [2]

General Hood described arriving on the battlefield on the afternoon of September 16th, his men had not eaten a meal in three days and were nearly starved, "During the afternoon of this day I was ordered, after great fatigue and hunger endured by my soldiers, to take position near the Hagerstown pike, in an open field in front of the Dunkard Church. General Hooker's corps crossed the Antietam, swung around with its right on the pike, and about an hour before sunset, encountered my division. I had stationed one or two batteries on a hill in a meadow, near the edge of a cornfield just by the pike, the Texas Brigade on the left and that on Law on the right. We opened fire and a spirited action ensued, which lasted till a late hour in the night. When the firing had in a great measure ceased we were so close to the enemy that we could distinctly hear him massing his heavy bodies in our immediate front. The extreme suffering of my troops for want of food inducted me to ride back to General Lee, and request him to send two or more brigades to our relief, at least for the night, in order that the soldiers might have a chance to cook their meager rations. He suggested I go see General Jackson and endeavor to obtain assistance from him. I finally discovered him alone, lying upon the ground, asleep by the root of a tree. I roused him and made known the half starved condition of my troops; he immediately ordered Lawton's (Georgia Brigade), Trimble's, and Hay's brigades to our relief. He exacted of me, however a promise that I would come to the support of these forces the moment I was called upon. I quickly rode off in search of my wagons, that the men might prepare and cook their flour, as we were still without meat; unfortunately the night was far advanced, and although every effort was made amid the darkness to get the wagons forward." [3]

Lawton was commanding Ewell's division, as Ewell was wounded at 2nd Manassas. His division, in Jackson's Second Corps, consisted of three Confederate brigades, which included his own Georgia Brigade, Hay's Louisiana Brigade, and Trimble's Brigade. Colonel Marcellus Douglass, recently commanding the 13th Georgia Regiment, was placed in command of Lawton's Brigade. About 10 o'clock that night, Jackson ordered Lawton to move his division forward to relieve Hood's men in line of battle. Major J. H. Lowe of the 31st Georgia reported, "About 10 p.m., the (Georgia) brigade was ordered to relieve Brigadier-General Hood, whose command was in line of battle near the enemy's lines. The brigade marched up and formed line of battle, and the Thirty-first Georgia Regiment, in Lawton's brigade, commanded by Lieutenant Colonel [J. T.] Crowder was ordered out as skirmishers, which order was promptly obeyed. During the night sharp skirmishing ensued. [4]

A Yankee officer spent a miserable night about a mile to the front of Lawton's Brigade in the North Woods, *"Once or twice during the night, heavy volleys of musketry crashed in the dark woods on our left (the East Woods.) There was a drizzling rain, and with the certain prospect of deadly conflict on the morrow, the night was dismal. Nothing can be more solemn that a period of silent waiting for summons to battle, known to be impending."* [5]

Lawton's entire Brigade numbered only 1,150 soldiers, its ranks greatly reduced by sickness, stragglers, and battle losses at Gaines Mill and 2nd Manassas. The same was true for most of Lee's army; brigades were the size of regiments and regiments the size of fully manned companies. The Confederates were outnumbered by the enemy by more than two to one. General Lee's entire army mustered only about 41,000 soldiers, while Union General George McClellan fielded 87,000 Union soldiers. Lieutenant Mathews of the Battey Guards recalled, "Into this battle the (38th Georgia) regiment carried only 123 men."[6] General Jubal Early wrote, "These two brigades (Lawton's and Trimble's) were posted in the positions occupied by General Hood's brigades, Trimble's brigade under Colonel Walker, being on the right, next to General D. H. Hill's division, and Lawton's brigade on the left of it. In this position they lay on their arms during the night, with occasional skirmishing in front between the pickets." [7]

The skirmish line of the 31st covered the entire front of the other five regiments of the Georgia Brigade. This line was to serve as a "trip wire" to prevent the Federals from surprising the brigade during the night. "He (Lieutenant Colonel Crowder) strung out his men about fifty feet apart wherever he could find protection behind a tree or other objects, but these did not afford any shelter from the shells and solid shot coming from the other side of the creek," recalled Private Isaac Bradwell of the 31st Georgia.[8] About 3 a.m. in the morning, the first shots of the battle were fired, "During the night sharp skirmishing ensued." [9] The other five regiments of the brigade, including the 38th, were posted in a line stretching from nearly the Hagerstown Pike on the left flank, towards the East Woods. Their line was parallel to farmer David R. Miller's 32 acre cornfield to their front.

Lawton's Brigade was too small to cover their entire front with a solid line of battle, with only just over 1,000 men in the ranks. Private Bradwell wrote, "Our Brigade was stretched out in a very thin line, with wide intervals between the regiments to occupy as much space as possible. They were formed in the open fields to the west of the road…running north from Sharpsburg, and in front of them not more than a hundred feet was a low rail fence. From this fence to a forest to north was a field of high corn standing very thick on the land." [10]

"The ground held by the brigade was somewhat lower than the cornfield, and, in nearly its entire length, was covered by low stone ledges, and small protuberances, which afforded some protection and, in places, a rail fence was thrown down and piled as a breastwork. In other places there was no protection, either of rock-ledge, inequality of the ground, or fence rails," In rear of Lawton's Brigade, "on the plateau nearly opposite and about 225 yards from the church, were four batteries of Colonel S.D. Lee's artillery battalion--the Ashland (Va.) Artillery, Captain P. Woolfolk, Jr.; Bedford (Va.) Artillery, Captain T. C. Jordan; Brooks' (S.C.) Artillery, Lieutenant William Elliot; and Parker's (Va.) Battery, Captain W.W. Parker. There was also in the vicinity of the church and on the ridge south and west of it some guns of Cutts' artillery battalion." [11]

"Never did a day open more beautiful," reported Union Major Rufus Dawes, who was camped a short distance to the front of Lawton's Brigade. Major Lowe of the 31st Georgia reported, "At dawn, when the enemy could be seen, heavy skirmishing commenced and continued for an hour. The skirmishers, after their ammunition was nearly exhausted, were ordered to retire or fall back with their brigade." [12] General Stonewall Jackson wrote, "At the first dawn of day skirmishing commenced in front, and in a short time the Federal batteries, so posted on the opposite side of the Antietam as to enfilade my line, opened a severe and damaging fire. This was vigorously replied to by the batteries of Poague, Carpenter, Brockenbrough, Raine, Caskie, and Wooding. About sunrise the Federal infantry advanced in heavy force to the edge of the wood on the eastern side of the turnpike, driving in our skirmishers. Batteries were opened in front from the wood with shell

and canister, and our troops became exposed for near an hour to a terrific storm of shell, canister, and musketry." [13]

Lieutenant W. C. Mathews wrote of the disaster that befell the 38th shortly after sunrise, when their commanding officer was cut down early in the fight, *"We had the misfortune to lose our brave Major W. H. Battey, one of the best men in the service. He had left his sick bed to overtake us. (He) was not hardly able to sit on his horse, but his will and courage carried him onto the battlefield, and early in the fight, he was cut in two by a solid shot."* [14] Three young Grogan brothers, Josiah, Gideon and Joseph, from Fulton County, Georgia, were all shot down on the firing line. The two youngest brothers, Joseph, 17 and Gideon, 19, were killed in action. Josiah, the oldest, age 23, was stuck by a shell fragment on the lower left side of his backbone, breaking two lower ribs loose from his backbone and "mashing them in." He survived the battle, recovered from his wound, and faithfully rejoined the regiment, serving to the bitter end at Appomattox Court House in April 1865.[15] Josiah lived over 50 years after the war ended, dying an old man in 1917. He undoubtedly never forgot the horrific day his two younger brothers were shot down and killed at Antietam. Twenty-two year old Private Thomas J. Jordan, of the Battey Guards, was struck down and mortally wounded, dying three weeks later in a hospital in Richmond, Virginia. [16]

General Joseph Hooker led his Federal First Corps, numbering 14,000 men, across the fields heading south, on a collision course with Lawton's Brigade. "Hooker came out of the North Wood. His battle flags were bright and he had drums and brazen horns. Loud and in time, regular as a beat in music, came the Huzzah! Huzzah! Of his fourteen thousand men." General Hooker spied what may have been part of Lawton's Brigade picket line concealed in the cornfield, "My object was to gain the high ground nearly three-quarters of a mile in advance of me, and which commanded the position taken by the enemy on his retreat from South Mountain. We had not proceeded far before I discovered that a heavy force of the enemy had taken possession of a corn-field (I have since learned about a thirty-acre field) in my immediate front, and from the sun's rays falling on their bayonets projecting above the corn could see that the field was filled with the enemy, with arms in their hands, standing apparently at 'support arms.' Instructions were immediately given for the assemblage of all of my spare batteries, near at hand, of which I think there were five or six, to spring into battery, on the right of his field, and to open with canister at once.

In the time I am writing every stalk of corn in the northern and greater part of the field was cut as closely as could have been done with a knife and the slain lay in rows precisely as they stood in their ranks a few moments before. It was never my fortune to witness a more bloody, dismal battle-field. Those that escaped fled in the opposite direction from our advance, and sought refuge behind the trees, fences, and stone ledges nearly on a line with the Dunker Church, as there was no resisting this torrent of death-dealing missives." [16]

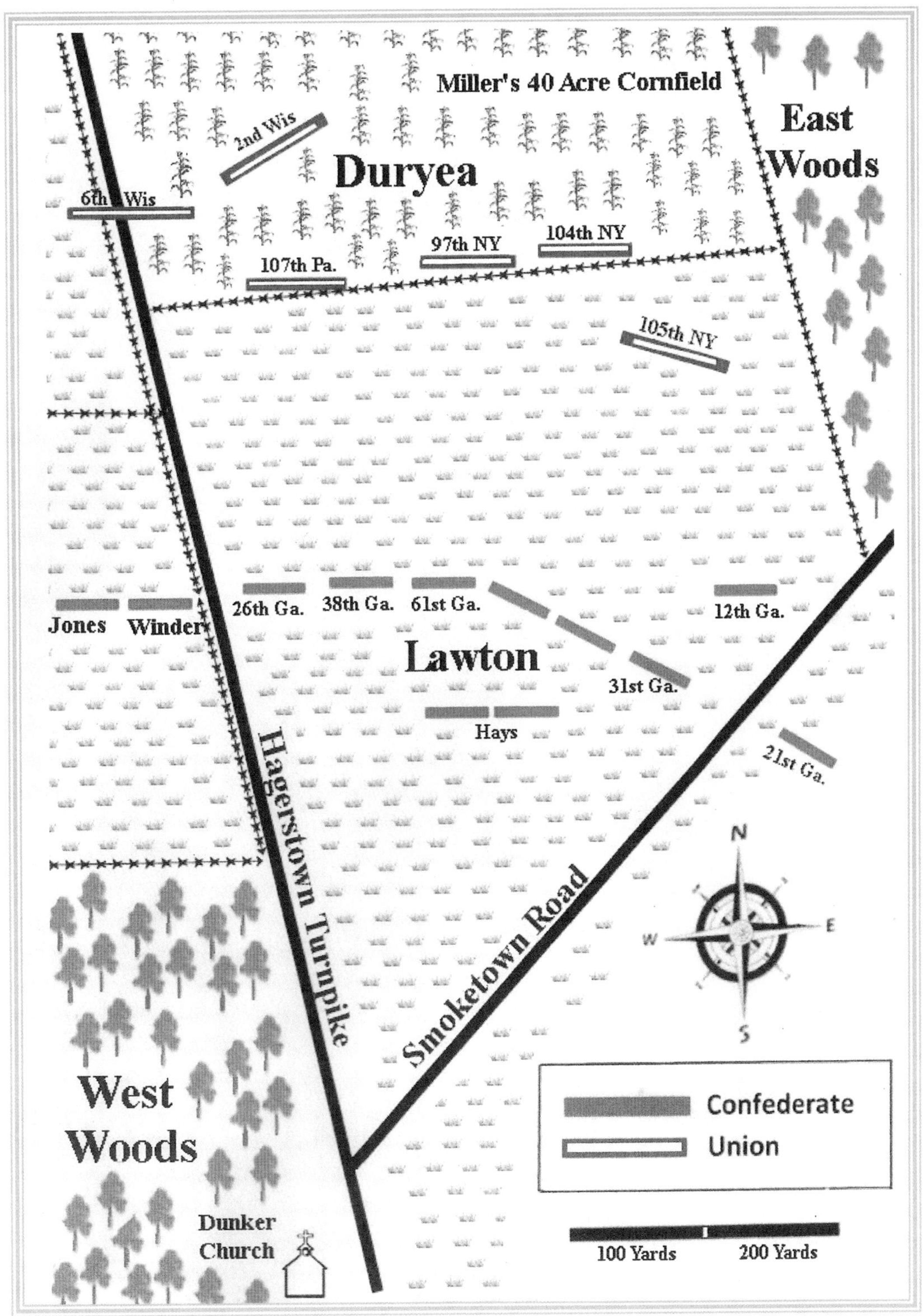

Map by D. Gary Nichols

Private Isaac Bradwell of the 31st Georgia described the moment, *"The artillery fire opened with great fury and it must have been the signal for the infantry to advance. They moved forward at the same time in the woods, with several lines supporting one another, fringed in front by skirmishers to develop our position. As these came up the fire of our pickets drove them back on their main advance line. So accurate was the aim of our men that they, too, were halted until the second line came up and opened on them. Colonel Crowder was shot and disabled, and so many of the regiment killed or wounded by the artillery and infantry fire that the rest were forced to flee through the corn to their friends in the main line. When they reached the fence and were getting over it- in more haste than dignity- they were guyed for coming over it so great a hurry. Many of our men were laughing and saying, "What's the matter? What are you running for?" to which came the reply: "You'll soon see!""*[17]

What they saw was Union Brigadier General Duryea's Brigade advancing. This was the first Federal brigade to break through the corn and come face to face with Lawton's Brigade. Duryea's brigade contained four regiments, the 107th Pennsylvania and the 97th, 104th, and 105th New York and numbered about 1,000 soldiers, close to an even match in numbers against the Confederates. But, Lawton's Brigade, with Trimble's Brigade to its right, wouldn't be fighting Duryea's Brigade alone, the flags of other Union brigades could be seen approaching from the north and still other Union brigades approached from the northeast. The size of the massive force advancing towards Lawton's Brigade alarmed the brigade commander, according to Private Bradwell, *"The number of regimental standards floating in the morning air indicated the immense numbers of the advancing enemy. Colonel Douglass, fearing the result of an attack by so large a force on his weak brigade, ran from regiment to regiment exhorting the men not to fire until the enemy reached the fence and began to get over it – to shoot low and make every bullet count."* [18]

Lawton's Brigade had been instructed by Colonel Marcellus Douglass, *"Watch for the Union line to reach the edge of the corn and for each man to fire down his "own corn row."* A Georgia Brigade soldier recalled the moment the battle opened in earnest as Duryea's Federal Brigade emerged from the corn. *"The enemy approached within a distance of 100 yards. The enemy also did not fire until they attained a position near us. They took their position deliberately - the officer in command, dressed the men and acted in every respect as though on parade. His words of command could be distinctly heard. Col Douglass then ordered his command to fire, and at short range. He then ordered the second volley; both which were discharged before we received the enemy's fire. At each discharge wide gaps were made in the enemy's ranks, and the groans of their wounded were horrible. Our men bore their sufferings in silence. These were the first volleys of the great fight and after that every man fired at will."* [19] *"At first no attention was paid by either line to the rail fence in their respective fronts, but each stood and fired on the other, neither party endeavoring to advance, soon, however, the severity of the fire dictated more caution and most of the men, on both sides, laid down and sought cover."* [20]

Private Josiah Tuck Hewell of Company H, the Goshen Blues, a thirty seven year old farmer from Elbert County, Georgia was lying down firing at the enemy. He was struck by a Minié ball that passed through his left side, then through the head of his penis, continuing and passing through the calf of his right leg, shattering the bones in his leg and leaving a large exit wound. He instantly became a cripple and endured constant pain for the remaining 34 years of his life, with small pieces of bones periodically working their way to the surface of his leg, leading to his death by blood poisoning in 1896.[21]

"The struggle continued between Duryea's right (the 107th Pennsylvania and 97th New York) and Lawton's left (the 26th, 38th and 61st Georgia). At first the 26th Georgia was 120 yards east of the Hagerstown road, but it obliqued to the left until it gained high ground, about 50 yards from the road and directed a right oblique fire upon the right of the 107th Pennsylvania; the left of the Pennsylvania regiment and the entire 97th New York, was under the fire of the 38th and 61st Georgia. On the west side of the road was Jackson's old division, not yet engaged, but a few of its skirmishers, at the fence of the road, were firing at the flank of the 107th Pennsylvania. The 38th made a desperate effort to gain the cover of a ledge in its front, near the corn, but was disastrously repulsed; the 61st was content with holding on, suffering terribly from a crossfire. Neither side gained any advantage of ground, but Lawton's men lost more heavily, as they were fired at from both front and right, the fire from Seymour's men in the East Woods enfilading the three left regiments, at the same time Duryea's right partially enfiladed the right regiments. In addition both Lawton's and Trimble's brigades were then, and had been since daybreak, under a distressing artillery fire from the Union batteries in front and from the long range guns beyond the Antietam which, with the infantry fire, General Early reported, "Subjected the two brigades (Lawton's and Trimble's) to a terrible fire." [22]

1st Lieutenant John Wilson McCurdy commanded Company D, The McCullough Rifles, but as casualties mounted, he took command of the regiment until he was shot in the right leg, four inches above the ankle. He would spend the next two years on crutches and was never fit to rejoin the regiment. [23]

The right wing of Lawton's Brigade was holding its own, "The 105th and 104th New York, on reaching the south edge of the corn pushed out into the open field 160 and 120 yards respectively and were opened upon with such vigor by Lawton's right and the 12th Georgia, and S.D. Lee's guns, that they soon fell back to the corn, the former carrying with it, its mortally wounded commander, Lieutenant Colonel Howard Carroll." For about 30 minutes, the two brigades hammered away at each other, but then Duryea's brigade began to fall back through the corn. [24]

Duyree reported the right wing of Lawton's Brigade devastated two of his regiments, *"The conflict continued until there were only about 100 men of the One hundred and fourth and One hundred and fifth New York Regiments left on the right of the brigade. At this point the cannoneers of one of the batteries were compelled to abandon their guns. The remnants of the two regiments above named rallied behind a large rock and continued to pour in a deadly fire until re-enforcements came up and covered the guns. The enemy's dead upon the field were almost in as perfect line as if on dress parade."* [25]

Lawton realized his two brigades were being cut to pieces and ordered up Hay's Brigade for support, "Hay's Louisiana brigade came on the field. It had bivouacked in the woods northwest of the church. Soon after daylight Lawton ordered Hays to move quickly and fill the interval of 120 yards between the 26th Georgia and the Hagerstown road. Hays crossed the road, 120 yards north of the Dunkard Church, and was advancing due north to close the interval, when he was directed by Lawton to bear to the right and take position immediately in rear of his brigade. This was done and he remained in this position until Colonel Douglass commanding Lawton's Brigade, requested him to come to his assistance. With his 550 men he advanced under a deadly fire from Matthews' and Thompson's guns, and was still advancing when Hartsuff's Brigade came down through the cornfield and East Woods and opposed him." [26] Immediately upon Duryea's retirement, Lawton's skirmishers pushed into the corn in pursuit, and the entire line, supported by Hays' Brigade, was ordered forward, when it was discovered that the advance of Doubleday's Division, on either side of the Hagerstown road, threatened to turn the left flank, upon which the left of the brigade obliqued toward the road and became engaged with the 6th and 2nd Wisconsin of Doubleday's advance and, at the same time, Jackson's old division, west of the road became engaged. The battle now raged near and along the Hagerstown road and in fields west of the East Woods. [27]

A soldier of Company E, The Tom Cobb Infantry, remembered the moment well, *"All the company were shot down, wounded or detailed, but Wm. J. Smith, and just previous to the last charge of the regiment, Captain Hawkins was lying wounded, and saw that there was none left of his company but William. K. Smith; said to him, "Smith, give it to them." Smith replied, "Captain, you had better believe I will," and on went in the charge, and William. J. Smith fell a martyr of the lost cause; and no braver man or more generous man ever fell on either side."* [28] As the fresh Federal troops advanced through the cornfield in another assault, Lawton's Brigade faced their old antagonist, the Federal Iron Brigade. "Up to this time the 2nd Wisconsin had not given or received a shot, nor had it seen an enemy, but, as it reached the south edge of the corn, the men saw before them, the left of Lawton's Brigade, about 200 yards distant. There was no time for extended observations, for, as the 2nd Wisconsin and the seven companies of the 6th came into view, the 26th, 38th, and 61st Georgia rose from the ground and simultaneously both lines opened fire." [29]

Major Rufus R. Dawes of the 6th Wisconsin, wrote, *"At the front edge of the corn-field was a low Virginia rail fence. Before the corn were open fields, beyond which was a strip of woods surrounding a little church, the Dunkard church. As we appeared at the edge of the corn, a long line of men in butternut and gray rose up from the ground. Simultaneously, the hostile battle lines opened a tremendous fire upon each other. Men, I cannot say fell; they were knocked out of the ranks by the dozens."* [30]

"There was but a short halt at the south edge of the corn, the Wisconsin men went over the fence bounding it; Kellogg came up the road with his three companies, and all went forward, firing and shouting, driving back and to the left the three Georgia regiments, to the foot of the high ground, where, only a skirmish line now, and under cover, they held on. This encounter near the road was at the time Hays came up farther to the Confederate right and made the charge in which most of the officers in both his own and Lawton's Brigade were killed or wounded."[31]

Two 38th Georgia Regiment color bearers were gunned down during the ferocious battle, Lieutenant Lester wrote, *"In the blood arid carnage of this battle, the glorious old battle flag, that was presented to us by Mrs. General Lawton, was always borne on, uplifted by strong arms. Sergeant Jas. W. Wright carried it through one charge and had his knee smashed by a ball. The flag went down, but was caught up by Malcolm Scoggins and borne forward until he was shot down. It was then seized by Lieutenant Wright, (Lieutenant Miller Armistead Wright) of the "Ben Hill Guards."* [32]

Colonel Lowe recalled, *"Re-enforcements then came, (Hay's Brigade) and with them we made a charge in the most gallant manner." During that time (before the charge) the brigade lost its commander, and nearly every regiment lost its regimental commander; also the greater portion of the different companies lost their company commanders. After the charge the brigade fell back, and, in taking off the wounded, a great many were lost for a short time from their regiments."* [33]

Lawton's Brigade was cut to pieces, nearly half the brigade had fallen before the enemy guns. A Lawton's Brigade member recalled the moment, *"There is only a man every ten feet or more to resist the last and greatest effort of the enemy. Heavy reinforcements have been sent into the woods. These come forward in such numbers that the few Confederates defending the position are beaten back step by step to the reserve line held by General Walker and his Virginians. The eighth ball pierces the body of Colonel Douglass, (Georgia Brigade Commander) and he falls helpless in the arms of his soldiers. He begs them to let him die on the battlefield with his men, declaring he would rather die there than in the arms of his wife at home."* [34]

Nineteen year old Private Nathan B. Bagwell, of the Company B, Milton Guards, was shot in left leg below the knee and disabled. He is captured by the enemy and his left leg was amputated below the knee. [35] Private Martin Hall of Company B was shot in left forehead, the bullet depressing the right plate of his skull. He survived, but never returned to the army. He had constant fits until he died of his wound in 1872.[36] Private John W. Christian of Company H, the Goshen Blues was shot through the right hand, requiring the amputation of the second finger and rendering his hand useless for the rest of his life. [37] Private George Washington Gaddy, Company A, The Murphy Guards, was killed in action. [38]

Colonel Douglass, whose brigade had been hotly engaged during the whole time, was killed, and about half of the men had been killed and wounded. Hays' brigade, which had been advanced to Colonel Douglass' support, had also suffered terribly, having more than half killed and wounded, both of General Hays' staff officers being disabled...these three brigades, which were reduced to mere fragments, their ammunition being exhausted, retired to the rear. [39]

Major Dawes of the 6th Wisconsin Volunteers watched the Confederate line, including Lawton's Brigade, crumble, *"We jumped over the fence, and pushed on, loading, firing, and shouting as we advanced, another line of our men came up through the corn. We all joined together, jumped over the fence, and again pushed out into the open field. There is rattling fusillade and loud cheers. 'Forward' is the word. The men are loading and firing with demoniacal fury and shouting and laughing hysterically and the whole field is covered with rebels fleeing for life, into the woods. Great numbers of them are shot while climbing over the high post and rail fences along the turnpike."* [40]

Lieutenant John Baxter of Company D, The McCullough Rifles, recalled, *"The hottest of the battle lasted about half an hour. Company D went in with 17 men and came out with two unhurt. There was no field officer on the ground that I could see when we were ordered out. Sergeant E. T. Harris (Edmond T. Harris) was wounded by my side and caught on my shoulder and asked me to take him out as we retired from the field."* [41]

"Dead horse of Confederate Colonel, both killed at Battle Antietam" -Library of Congress
Horse of Colonel Henry Strong, 6th Louisana "Tigers" of Hays Brigade. Hay's Brigade was part of Lawton's Division and this horse, along with Colonel Strong were killed during a charge on the Federals, that included the 38th Georgia.

Jackson realized the time had come to call for Hood's Division, "Thinned in their ranks and exhausted of their ammunition, Jackson's division and the brigades of Lawton, Hays, and Trimble retired to the rear." [43] Many of General Hood's men had still not finished preparing their meal at 7 o'clock in the morning, General Hood said, "dawn of the morning of the 17th broke upon us before many of the men had time to do more than prepare the dough. Soon thereafter, an officer of Lawton's staff dashed up to me saying, "General Lawton sends his compliments with the request that you come at once to his support." 'To Arms' was instantly sounded and quite a large number of my brave soldiers were again obliged to march to the front, leaving their uncooked rations in camp. As we passed about sunrise, across the pike through the gap in the fence in front of the Dunkard Church, General Lawton, who had been wounded, was borne to the rear upon a litter, and the only Confederate troops left on that part of the field, some forty men who had rallied around the gallant Harry Hays.

The Union attackers sensed victory, but then General Hood's Confederate Division of 2,300 soldiers, swarmed from the West Woods and formed a line of battle. Screaming like banshees, they immediately opened fire on the advancing Federals. General Hood reported, "I at once marched out on the field in line of battle and soon became engaged with an immense force of the enemy, consisting of not less than two corps of their army."

Major Rufus Dawes of the, 6th Wisconsin Volunteers never forgot the moment Hood launched his counterattack, *"A long and steady line of rebel gray, unbroken by the fugitives who fly before us, comes sweeping down through the woods around the church. They raise the yell and fire. It is like a scythe running through our line. 'Now save, who can.' It is a race for life that each man runs for the cornfield."*

Private Lawrence Daffan, Co. G, Fourth Texas Infantry, was part of Hood's Texas Brigade, preparing a meal in the West Woods when they were called upon to attack. He recounted coming up on the rear of Lawton's Brigade as the Texas Brigade advanced, *"We were ordered 'forward.' We emerged from the timber into the stubble field; some of it I think had recently been plowed. As we emerged from the timber, a panorama, fearful and wonderful, broke upon us. It was a line of battle in front of us. Immediately in front of us was Lawton's Georgia Brigade. After we left the timber we were under fire, but not in a position to return the fire. As we neared Lawton's Brigade, the order came for the Texas Brigade to charge. Whenever a halt was made by a command under fire, every man lay flat on the ground, and this was done very quickly. Lawton's Brigade had been on this line fighting some time before we reached them. Lawton's Brigade attempted to charge, and did charge; their charge was a failure, because their numbers had been decimated; they had no strength. Then the Texas Brigade as ordered to charge; the enemy was on the opposite side of this stubble field in the cornfield. As we passed where Lawton's Brigade had stood, there was a complete line of dead Georgians as far as I could see."*

General Hood's Division swept the Federals from the field, fighting surged back and forth across the infamous cornfield, the cornfield changed hands too many times to count. Hood's Division had pushed the Federals back, but at a tremendous cost. Union counterattacks pounded Hood's Division until it shattered, with the remnants, including those of Lawton's Brigade, streaming back to the West Woods near the Dunker Church. That evening, General Lee asked Hood, "Where is your splendid Division?" Hood replied, "They are lying upon the field where you sent them, sir." Colonel John B. Gordon, who later commanded Lawton's Brigade, noted, "I looked at the sun. It moved very slowly; in fact, it seemed to stand still." Major John Lowe, of the 31st, discovered he was the ranking officer remaining in Lawton's Brigade, "Finding that I was senior officer present, I reformed the brigade and reported to Brigadier-General Early, and was ordered to take position on the right of the division in line of battle."

The 38th's role in the battle of Antietam ended by 7:30 a.m., but the battle would rage throughout the day, ending just after dark. Lieutenant Lester of the Tom Cobb Infantry wrote, *"The battle raged all day without ceasing. Never did men display more courage than the boys here."* It remains the bloodiest single day in American history, with an estimated 23,100 men killed, wounded, or missing. At least 3,500 men were dead on the field.

Lawton, commanding Ewell's Division, was seriously wounded and would never return as the Brigade commander. The Brigade suffered horrific losses near the southern edge of Miller's infamous, cornfield. Of the 1,150 soldiers present, 567 were reported killed, wounded, or missing in the two hours of fierce battle, about half the brigade. Particularly hard hit were the senior officers of the brigade. Lawton's Brigade commander, Colonel Marcellus Douglass was killed in action, along with Major Archibald P. McRae, commanding the 61st Georgia. Captain William Battey, of Jefferson County, Georgia, commanding the 38th Georgia, was killed in action. Lieutenant Colonel J. T. Crowder, commanding the 31st Georgia was wounded. The brigade, along with much of Lee's army, was scattered and that night, a Confederate staff officer recalled, "half of lee's army were hunting the other half."

The 38th had carried 123 men into battle and suffered 23 killed or mortally wounded, 46 other wounded, and 1 missing. When the battle ended the evening of September 17th, only 28 soldiers and three officers gathered around the colors. Lieutenant W. C. Mathews lamented, "We had withstood the heavy assaults of Hooker and Mansfield and nearly annihilated their corps, outnumbering ours three to one, but death had reaped a rich harvest of brave souls and only a handful of the command was left. Lieutenants Vaughn,

DEATH OF CAPTAIN BATTEY.—It is with sincere sorrow that we chronicle the death of Capt. Wm. H. Battey, of the Jefferson County Battey Volunteers, Wright's Legion. Despatches received from Richmond announce that he fell in the recent battle near Sharpsburg, Md. He commanded the Legion in one of the late battles before Richmond, and distinguished himself for gallantry and good judgment.

Capt. Battey was our friend and classmate in early life. We knew him well, and can tender a heartfelt sympathy to his bereaved widow and children. We begin to awake to the magnitude of war's calamities in contemplating the bloody remains of such men as he. Heaven, in its mercy, bring the horrible strife to an end.—*Savannah Republican, Sept. 27, 1862, P. 1*

Hawkins, Dorough, Wright, and McCurdy were wounded severely. The only officers left when the fight ended were Lieutenant Baxter, Wells, and Mathews."

The battle ended in a stalemate, but Lee refused to abandon the field and remained in position all day on September 18th, inviting an attack by the Federals; but the attack never came. Private Bradwell of the 31st recalled, "*Our brigade took no part in this day's fight after this and that evening at sundown could muster on forty-eight men who could stand in line. The next morning they were deployed one hundred yards apart, facing the enemy, who showed no disposition to fight. Each side had had enough, and they stood there all day long watching each other like gladiators too weak from loss of blood to renew the fight.*" Lieutenant W. C. Mathews of the Battey Guards wrote, "*After spending the day after the fight (Sept. 18th) burying our dead, we crossed the Potomac, the remnant of the 38th Georgia.*" On the night of the September 18th and into the early morning hours of September 19th, the Confederate army retreated across the Potomac River at Shepherdstown ford. Lawton's Brigade served as part of the rearguard.

Killed or Mortally Wounded - Antietam (or Sharpsburg), Sept 17th, 1862

Company A, "The Murphy Guards," DeKalb County, Georgia

1. Private William F. Austin - Killed 9/17/1862.
2. 5th Sergeant George Washington Gaddy - Killed 9/17/1862.
3. Private David Young Miller - Killed 9/17/1862.
4. 3rd Corporal John A. Sellers - Killed 9/17/1862.

Company B, "The Milton Guards," Milton County, Georgia (Currently Fulton County)

5. Private Mack V. Bruce - Killed 9/17/1862.
6. Private Lewis Burgess - Killed 9/17/1862.
7. Private John W. Nix - Killed 9/17/1862.

Company C, "The Ben Hill Guards," Bulloch & Emanuel Counties, Georgia

8. 1st Corporal Mathew M. Coleman - Wounded & captured 9/17/1862. Federal records state he died at White House Hospital near Sharpsburg & also at Line's Farm Hospital.

Company D, "The McCullough Rifles," DeKalb & Fulton Counties

9. Private Charles Thomas Akins (or Aiken) - Killed 9/17/1862.
10. Private Francis Marion Baxter - Killed 9/17/1862.
11. 3rd Corporal Josiah Wheeler Gresham - Killed 9/17/1862.
12. Private John T. Tucker - Killed 9/17/1862.

Company E, "The Tom Cobb Infantry," Oglethorpe County, Georgia

13. Private Robert H, (or Robert A.) Mathis - Killed 9/17/1862.
14. Private William S. Shearer - Wounded 9/17/1862. Died of wounds 11/20/1862.
15. 1st Sergeant William James "Willie" Smith - Killed 9/17/1862.

Company F, "Thornton's Line Volunteers," Hart and Elbert Counties, Georgia

None.

Company G, "The Battey Guards," Jefferson County, Georgia

16. Captain William Henry Battey - Killed 9/17/1862.
17. Private Thomas J. Jordan - Wounded 9/17/1862. Died at Richmond, Va., 10/17/1862.

Company H, "The Goshen Blues," Elbert County, Georgia

None.

Company I (New Company I) Henry County, Alabama

18. Private John A. Hawkins - Wounded and captured at Antietam, Md., 9/17/1862. Died of wounds in U.S.A. General Hospital at Frederick City, Md., 10/28/1862.

Company K, "DeKalb & Fulton Bartow Avengers," DeKalb and Fulton County, Georgia

19. 2nd Corporal James E. Chandler - Killed 9/17/1862.
20. Private Gideon F. Grogan - Killed 9/17/1862.
21. Private Joseph D. Grogan - Killed 9/17/1862.
22. Private Joshua "Josh" T. Hammond - Killed 9/17/1862.
23. Private Ellis W. Wiggins - Killed 9/17/1862.

Company N, "The Dawson Farmers," Dawson County, Georgia

None.

Chapter 7
The Battle of Shepherdstown, or Boteler's Ford
September 19th & 20th, 1862

Once the Confederate Army safely crossed the Potomac River, Lee posted a rearguard at Boteler's Ford, near Shepherdstown, Virginia, to discourage pursuit by the Union Army. Confederate General William N. Pendleton was posted on the heights overlooking the river crossing, along with 44 guns of the reserve artillery. Pendleton was ordered to take command of two under strength infantry brigades, General Lewis Armistead's Virginia Brigade, commanded by Colonel James G. Hodges and General Lawton's Georgia Brigade, commanded by Colonel John H. Lamar of the 61st Georgia. Lieutenant George Riley Wells commanded the 38th Georgia Regiment and he posted a picket line along the edge of the Potomac River.

Brigadier General William Pendleton, CSA
Library of Congress

Lawton's Brigade was ordered by Pendleton, *"to picket the ford and screening themselves as well as possible, to act as sharpshooters on the bank. I...instructed the colonels commanding to keep their force at the ford strong, vigilant, and as well sheltered as occasion allowed, and to have the residue well in hand, back of adjacent hills, for protection, till needed. My directions were also given them not to fire merely in reply to shots from the other side, but only to repel any attempt at crossing, and to guard the ford."* [1]

Lawton's Brigade was just a shadow of its former self after suffering serious losses at Gaines Mill, 2nd Manassas, and Antietam, "This brigade was very much reduced, having suffered terribly on the 17th, and a considerable number of the men being just returned from the hospitals, were without arms." [2] The estimated strength of the entire Georgia Brigade was only 400 soldiers and Armistead's Brigade numbered roughly the same.

General Jackson wrote, *"Early on the morning of the 19th we recrossed the Potomac River into Virginia near Shepherdstown...In the evening the command moved on the road leading to Martinsburg, except Lawton's (Colonel Lamar, of the Sixty-first Georgia, commanding), which was left on the Potomac Heights. On the same day the enemy appeared in considerable force on the northern side of the*

Potomac, and commenced planting heavy batteries on its heights." [3] Pendleton reported, *"About 8 a.m. of the 19th, the enemy appeared on the distant heights opposite, and found our army entirely and safely across the ford, and*

Author's depiction of 38th Georgia Regt flag lost at Sheperdstown.

on the Virginia side of the Potomac. They soon brought up and opened with artillery, much exceeding ours in weight. Still, our rifles did excellent service in keeping at bay for hours the entire hostile host, artillery, cavalry, and infantry, which, in various positions, appeared; care being taken not to waste ammunition in mere long-range exchanges of shot. Our troops that had been briefly resting in the valleys were now ordered farther inland, to be out of reach of the shells.

During most of the forenoon the enemy's fire was furious, and, under cover of it, in spite of persistent vigor on the part of our batteries, a heavy body of sharpshooters gained the canal bank on the northern and hostile side of the river. This proved to us an evil not slightly trying, since it, exposed our nearer cannoneers to be picked off, when serving their guns, by the enemy's effective infantry rifles. The enemy's fire, which had for a season relaxed, became fiercer than before, and so directed as to rake most of the hollows, as well as the hills, we occupied.

At the same time their infantry at the canal breastwork was much increased, and the crack of their sharpshooters became a continuous roll of musketry. Colonels Lamar and Hodges both reported to me that the pressure on their small force, the whole of which remaining I had ordered to the river, and the sum total when all were there was, they informed me, scarcely 300, was becoming too great to be borne. I directed them to hold on an hour longer; sunset was at hand, and I had communicated with Colonel Munford, who promised at dark to be with us; that by that time I would have the batteries withdrawn; they should, after due notice, retire next the batteries, and the cavalry should fall in between them and the enemy, so that all would get rightly out." [4]

CAPT. G. R. WELLS, Stone Mountain.
The Sunny South, January 10th, 1891

A general advance of the Federals was not expected by Lawton's Brigade, but the Federals rapidly forced a crossing at the ford. Captain John B. Isler of the First U.S. Sharpshooters was ordered to cross the river. Captain Isler recalled, *"It was about 5.30 p. m. when one of my officers announced to me that General Porter had ordered us to find a ford to cross the river and repulse the enemy at any hazard. This order was conveyed to us by Colonel Barnes. I then ordered to cease firing and advance. Owing to the extent of my line of skirmishers, only part of my command received the notice, and with it (about 60 men) I advanced. When in the canal which runs parallel with the river Potomac, the Fourth Michigan Volunteer Regiment marched down and acted as our support while we forded the river. The river being unknown to us, we found a ford with some difficulty. The enemy's musketry was very sharp during the crossing over, but occasioned only a loss of 4 men, as he was evidently retreating before us. While we were crossing, the Fourth Regiment Michigan Volunteers rendered us efficient support by firing volleys over our heads. My men as they crossed also fired several volleys. After we had crossed, the Fourth Michigan followed us, and when on the other side we jointly advanced up the bluff in front."* [5.]

George Sykes (Library of Congress)

This advance caught Lawton's Brigade and the 38th by surprise. Ordnance Sergeant William C. Mathews of Company G, the Battey Guards, recalled, the 38th was, *"Put on picket duty at the ford, but was surprised and driven off, losing one man killed and two wounded, and I am sorry to say we lost in this little fight our battle flag that we had carried from Savannah, Georgia. Our color bearer was killed and in the confusion of the retreat it was forgotten."* [6]

The 38th Georgia's abandoned flag was picked up by the U.S. Sharpshooters of Griffin's Brigade, Morell's Division. [7] The flag was described as, *"A silk flag of large size. Its color was originally pink, but now faded by exposure to the weather. It had diagonal bars of blue, with the white stars, and is bordered with rich yellow fringe. It must have been very handsome when new."* [8]

It was almost dark, but the enemy had forded the river and broke the Confederate line. Several batteries of artillery had begun to pull out and head to the rear. General Pendleton recounted the pandemonium that soon followed, *"Deep dusk had now arrived; certain batteries, as allowed, were on their way inland, while others, as directed, were well using ammunition still on hand. My own position was taken near the point of chief importance, directly back from the ford, so that I might the better know of and control each requisite operation. The members of my staff vigorously seconded my endeavors, under furious fire, in carrying orders and supervising their fulfillment and everything appeared likely, under favoring Providence, to result in effecting the withdrawal planned.*

This prospect was, however, suddenly changed. A number of infantry-men rushed rapidly by the point I occupied. Arresting them, I learned that they were of the sharpshooters who held guard at the ford; that their body had all given way and that some of the enemy were already on our side of the river. Worn as were these men, their state of disorder, akin to panic, was not, justly, to be met with harshness. They were, however, encouraged to be steady and useful in checking disorder, and affording such tokens as they might, in the settling dark, to make the enemy cautious. No other means had I of keeping back an advance." [9]

The 38th also lost two ambulances in addition to their flag. A member of the Union Second District Columbia Regiment reported, *"We were ordered by General Griffin to cross the river to Shepherdstown on Thursday. We did so, driving the rebel pickets before us, capturing one lieutenant; two cavalrymen, battery, blacksmith forge, and two ambulances belonging to the Thirty eighth Georgia."* [10] Captain Isler of the U.S. Sharpshooters pushed his command to the top of the bluff and expected to there spend the night. He posted a picket line, but soon received orders to retire back across the river. Captain Isler wrote, "We expected to stay for the night…but, after about one hour, I was ordered to withdraw my command and to recross the river, where we camped for the night." [11]

The battle had ended for the day and the Georgia Brigade played no further part in the battle the next day. It was nearly dark when Pendleton abandoned the battlefield and sought help. He rode to the rear and several hours later found the main body of the Confederate army camped several miles from Shepherdstown. He was in a panic; he roused General Lee from his slumber and delivered the astonishing news that all the artillery of the Confederate reserve artillery (44 guns) had been captured by the enemy. Lee asked, "All?" Pendleton replied, "Yes General, I fear all." Lee decided to wait till daylight before launching a counterattack, but when Jackson heard of the disaster, he immediately ordered General A. P. Hill's Light Division to return to the ford.

The Federal's had captured only four or five of Pendleton's guns, a few caissons and the flag of the 38th Georgia Regiment. Federal General John Fitz-Porter reported, *"Darkness concealed the movements of the enemy and enabled them to remove a portion of their artillery before our attacking party scaled the heights. The result of the day's action was the capture of 5 pieces, 2 caissons, 2 caisson bodies, 2 forges, and some 400 stand of arms; also 1 battle-flag. Our loss was small in numbers. Some excellent officers and men were killed and wounded. The party was recalled during the night, and the whole command bivouacked within reach of the fords."* [12]

A. P. Hill (Library of Congress)

The next morning, the Federals again crossed the Potomac River with about 2,500 men. They had no idea a large Confederate force under A. P. Hill was approaching their position at the ford. Hill arrived and immediately ordered his Light Division to attack. Most of the Federals retreated safely across the river, but some stayed to fight. Hill attacked the remaining Federals on three sides and shoved a large number over the bluff and into the river, capturing 200 prisoners. Many had to swim for their lives as the Confederates fired down at them from the bluff above, many drowned.

The 118th Pennsylvania, known as The Corn Exchange Regiment, was among the last cross the river and lost 282 men out of 800 present. Hill reported of the attack by his division, *"A daring charge was made, and the enemy driven pell-mell into the river. Then commenced the most terrible slaughter that this war has yet witnessed. The broad surface of the Potomac was blue with the floating bodies of our foe. But few escaped to tell the tale. By their own account they lost 3,000 men, killed and drowned. From one brigade alone, some 200 prisoners were taken. This was a wholesome lesson to the enemy and taught them to know that it may be dangerous sometimes to press a retreating army."* [13]

Private Alexander Atkinson of the Ben Hill Guards was reported wounded at the battle of Boteler's Ford. He died in a Confederate hospital in Staunton, Virginia, December 15th, 1862.[14] General Early wrote of the Georgia Brigade giving way at the ford and attributed their retreat to the mismanagement of the Brigade by General Pendleton, the commanding general, *"On the afternoon of the 19th the enemy commenced crossing a small force at Boteler's Ford, and Lawton's brigade gave way, abandoning its position. This brigade was very much reduced, having suffered terribly on the 17th, and a considerable number of the men being just returned from the hospitals, were without arms, and, without knowing the particulars of the affair, I am satisfied its conduct on this occasion was owing to the mismanagement of the officer in command of it."* [15]

It was long believed no regimental flags of the 38th had survived the war. However, there is a surviving flag in the Museum of the Confederacy, in Richmond, Virginia, attributed to the 31st Georgia, listed as being "abandoned" at the battle of Shepherdstown Ford, September 19th, 1862. The evidence supporting this attribution to the 31st was circumstantial, at best. Then the author discovered a long forgotten passage, written in 1891, in an obscure Georgia newspaper titled "The Sunny South," that solved this mystery and proved this flag belonged to the 38th Georgia. It was the few lines written by Sergeant William C. Mathews, formerly of the Battey Guards, 38th Georgia Regiment. Mathews wrote, *"Put on picket duty at the ford, but was surprised and driven off, losing one man killed and two wounded, and I am sorry to say we lost in this little fight our battle flag that we had carried from Savannah, Georgia. Our color bearer was killed and in the confusion of the retreat it was forgotten."* [16]

These few, long forgotten lines, provide the only known first-hand account discussing the loss of the Confederate flag found abandoned on the battlefield at Shepherdstown. After exchanging several mails, during 2012, with renowned Confederate flag expert, Mr. Craig Biggs and the Museum of the Confederacy, the Museum of the Confederacy agreed to change their records and this flag is now attributed as belonging to the 38th Georgia Regiment. This flag is cataloged at Museum of the Confederacy, in Richmond, Virginia. It as flag # WD 106.

Chapter 8
The Battle of Fredericksburg, Sept. – Dec. 1862

Lieutenant Lester of The Tom Cobb Infantry, recalled life after the bloody battle of Antietam:

"After this dreadful encounter at Sharpsburg, and the recrossing the Potomac, the Tom Cobb Infantry were allowed the privilege of enjoying a season of rest, which was greatly needed, for they were completely worn down, ragged and famished, this being the first rest we had been allowed since leaving Savannah in June. We were encamped for some of time at Bunker's Hill and spent the time most pleasantly in exterminating vermin that infested our clothing (I hope this will not be offensive to ears polite). We gathered apples, of which there were in abundance. These we utilized in every manner that the imagination could devise-roasted, baked, fried, in pies, dumplings, and now and then in the shape (if boiled juice. At this place many of our boys reported for duty who had been wounded and sick; so that when we came to leave we had a pretty full company. During this season of rest, by way of variety, we were used in tearing up and destroying the great Baltimore and Ohio railroad.

During these operations on this railroad we were marched through the town of Martinsburg, which town was considered sprinkled with Union citizens. As we marched by a fine dwelling on the main street a bevy of ladies were standing on the colonnade and each displayed a miniature U. S. flag pinned on their bosoms. Our boys observed this and forgetting their usual politeness and chivalrous bearing, called out, "look out, ladies, the Georgia boys are the devil to charge Union colors on Yankee breastworks."

About the middle of November the Tom Cobb Infantry, with Stonewall Jackson's corps, were put on the tramp again, heading for Richmond. Soon we reached the Shenandoah River, which stream we waded, the water coming to our waists and was extremely cold; then crossed the Blue Ridge at Swift-Run-Gap. This was a memorable march to all present, in consequence of the apple jack captured on the route. Nearly the whole of the army, men and officers, were on a big drunk. After crossing the mountains we soon arrived at Orange Court House; then by way of Madison Court House to a point called Port Royal, some 25 miles below Fredericksburg, immediately on the Rappahannock River. Here we pitched our tents for a few days again, and enjoyed a season of rest. Our duty being picket service, our picket posts being on one side of the river and the Federals on the other.

Miller Wright, in company with his father, Hon. A. R. Wright, returned from Richmond yesterday morning.—He was wounded in the foot at Sharpsburg—the ball entering at heel and passing through the foot. He was one of only three that escaped in his company, and was in command of his regiment, when wounded.

Rome Tri-Weekly Courier, Oct. 2, 1862 -- page 2

At this camp the following incident occurred: One of our boys received a box from home full of good things, and among others a gallon jug of Mack Young's best peach brandy (if any of the accursed stuff could be called best,) upon which two of our sober-sided fellows got most gloriously drunk. This jug of Mack's brandy created a wonderful change in Sober-sides. He mounted a stump and orated thus: No government has prospered since General Scott whipped the British at Lundy's lane, except the Confederate government. What is this government but the brave men by which I am surrounded, whose valor has excited the administration of the world? The Confederate government is wiser than all other governments; it is an example to all the enlighted nations of the earth; it piles its commissioners stores in brick houses, five stories high and ten miles long, and feeds its soldiers on blackberries and hardtack - a government element whose cavalry feeds on butter-milk, its Surgeons feed on whiskey. You can say, "tents" Tents indeed! What does our government want with tents an old custom of the effete past? Give us a mud-hole to sleep in, cover us over with a brush heap. This is a tent worthy of our valor. We ' cover ourselves with glory, and need no other covering. Sir, did you say, "clothing!" That is what our government is created for. It is to take charge of our clothing while we are fighting and when we have whipped our enemies, O' that we had enemies! We have not enough. We want foeman worthy of our steel. With our government to back us we can whip the word in arms. These just and noble sentiments were sprung as by inspiration from this jug of brandy. How much longer he might have orated is not known, but the throats of the boys became unusually dry about this time which suddenly brought these eloquent remarks to a halt. We are happy to state that the orator survived." [1]

THIRTY DOLLARS REWARD.

THE above reward will be paid for the apprehension or delivery of John C Richardson, of company N, 38th Georgia Regiment, who deserted his company on or about the 27th day of June last. Said John C. Richardson was enlisted by Capt Elisha Blackburn at Atlanta, Georgia, on the 15th day of May, 1862, to serve for three years or the war; is five feet ten inches high; dark hair; hazel eyes; dark complexion; born in the State of North Carolina; age 32 years, and by occupation when enlisted a school teacher. Thirty Dollars will be paid if lodged in any safe jail so that I may get him. If delivered to his company the reward and all expenses will be paid.

W. A. HILL,
nov20-8t Capt Com Co N, 38th Ga Regt.

Southern Confederacy, Nov. 22, 1862 -- page 2

The number of soldiers prepared for duty in the Army of Northern Virginia greatly increased after the battle of Antietam. The ranks of Lee's army and Lawton's Brigade swelled, as many of the sick and wounded returned to the army. The 38th Georgia's numbers swelled to about 450 soldiers, a considerable increase from the 32 men who rallied around the colors at the close of the battle of Antietam. Though the number of men prepared for duty in Lawton's Brigade increased, the condition of the brigade was deplorable. In a letter written just before the battle of Fredericksburg, an officer of Lawton's Brigade wrote,

"Lawton's Brigade is composed of the 13th, 26th, 31st, 38th, 60th and 61st Georgia Regiments, and I venture to assert that a more gallant set of men were never embodied under one command. At the last report from our Brigade we had seven hundred and five (705) men without shoes and there are numbers without a

single blanket to shelter them from the cold. This is not fiction, but a statement of the truth. Georgians! Think of this, think of such a number of these men, who have aided in making the name of Georgia illustrious, marching 20-25 miles a day, with nothing to shelter their feet from contact with the snow, frost and rocks, and without a blanket to shelter them from the chilling blast at night, and this, too, without a murmur at their hard fate."[2]

Union Major General George McClellan was relieved of command of the Army of the Potomac during the fall of 1862, by President Abraham Lincoln, due to his reluctance to aggressively pursue Lee's army. Lincoln appointed Major General Ambrose Burnside as the new commander. Burnside initially refused the appointed, stating he was not competent to command the army, but he reluctantly accepted. Lincoln probably found his humble demeanor refreshing, especially after the pompous behavior of McClellan. The problem was, Burnside was not simply being humble, he was truly incompetent. Lincoln's error would cost the Union army thousands of lives before he realized the truth.

Burnside's Union army numbered about 120,000 men on the eve of the battle of Fredericksburg, while Lee's army totaled 85,000. This was the largest number of men fielded for any single battle during the entire war. Burnside's army was organized into three grand divisions, with each division containing infantry, cavalry, and artillery. Lee's army was organized into two corps, with Longstreet commanding the First Corps and Jackson commanding the Second Corps. General Jubal Early succeeded Lawton in command of Ewell's old division, since Lawton had been seriously wounded at Antietam and was still recovering. Lawton's Brigade numbered about 1,500 men and was temporarily commanded by Colonel Edmund N. Atkinson of the 26th Georgia. The 38th was commanded by Captain William McLeod, the senior Captain of the regiment.

Burnside planned a rapid crossing of his army over the Rappahannock River at Fredericksburg, followed by a quick march towards Richmond. Poor staff work sabotaged his plan; the pontoon bridges needed to cross the Rappahannock River were delayed in arriving. By the time they arrived, Lee's army was firmly positioned on the on the hills to his front, overlooking the city of Fredericksburg.

Burnside ordered the construction of three pontoon bridges to cross the Rappahannock, on December 11th, 1862. Confederate marksmen of Barksdale's Louisiana Brigade shot down the Union bridge builders as they began the bridge construction across from the city. Federal artillery opened fire on the town in an effort to drive away the Confederate marksmen, but when the smoke cleared the Confederates again opened fire on the bridge builders. The Federals launched men in boats to capture the town. Once the city was captured, the bridge building resumed in earnest. Two other bridges were constructed, one close to the city, another further downstream from the city of Fredericksburg, near Hamilton's Crossing. The Federal Army crossed the river over these three bridges in heavy force on December 12th.

General Franklin's Grand Division crossed the lower bridge near Hamilton's Crossing and spread out, arranging his line parallel to hills overlooking the plain.

Franklin's 65,000 soldiers faced Jackson's 37,000 Confederates.

Early's Division, including Lawton's Brigade, was posted about 12 miles down river, at the town of Port Royal. Jackson ordered Early to move his division to Hamilton Crossing, located about five miles downriver from the city of Fredericksburg, during the afternoon of Friday, December 12th. A Georgia army newspaper correspondent accompanied the 38th during their march from Port Royal to the battlefield, he wrote,

"Friday night (Dec 12th) with two days rations in the haversacks, the 38th Georgia of Lawton's Brigade took up line of march in the direction of Fredericksburg. Buoyant at the prospect of a fight all hands were in fine humor, and the witty jokes and merry laugh of these veterans of war ran along the lines as if they were on their way to a frolic instead of a fight. Marching until 11 o'clock P. M., we slept on our arms two miles of Hamilton Station, the scene of the next day's action.

Scarcely had bright Aurora gilded the eastern sky when the long roll and the gruff voice of the Orderly aroused us from our sweet dreams of home and loves ones, before Old Sol had thrown his silver rays over us, we are in motion. The morning was obscured by a thick fog which last until 10 o'clock A. M. Heavy booming of cannon mingled with the rattling report of the small arms of the skirmishers announced the commencement of the conflict. Coolly, and at common time the brigade moved on towards the battlefield. Ever again a shell would burst near us or plough up the ground around, making a big fuss but doing no damage. Arriving at the battlefield, the brigade was posted in a thick grove the left of the Station (Hamilton's Station) the 38th Georgia, Captain McLeod commanding occupied the centre of the brigade. Quietly and firmly resolved to conquer or died, they awaited the coming of the enemy. They had not long to wait." [3]

Major General Ambrose Burnside, USA
Library of Congress

The following morning was Saturday, December 13th and a heavy blanket of fog covered the woods and plains surrounding the brigade. Lieutenant Lester wrote of the movement, "Our season of rest and security was now destined to close. We were immediately placed in line of battle in a piece of woods, near Hamilton's crossing. Our brigade now occupied the extreme right of the army." [4]

Longstreet's First Corp formed the left wing of the Confederate line, posted on the hills overlooking the plains between the hills and the City of Fredericksburg. Jackson's 2nd Corp would form the right wing of the line, posted on a small range of timbered hills overlooking a large open plain stretching to the Rappahannock River. This position was on the extreme right of Lee's line. A railroad bed was located just at the foot of the hills, in front of the Confederate line. Early described the terrain overlooked by the Confederate line, "The river flats or bottoms immediately below Fredericksburg widen out considerably and continue to widen until they are one and a half to two miles in width." [5] Lawton's Brigade was posted as a reserve, several hundred yards behind the main Confederate line in a thick stand of timber. They couldn't see the Federals massed on the plains in front of the range of hills. At 10 O'clock a.m., the fog burned off and the Confederates on the front lines saw the Union army arrayed on the plains before them. Just after 11 a.m., the Federals began a massive artillery bombardment of the Confederate line to soften it for an infantry attack. The Confederate artillery reserved their fire, awaiting the Federal infantry assault they knew would soon follow.

The Federal bombardment lasted about an hour, but the Confederates suffered little damage, being under the cover of the hills and trees, though a few were killed and wounded. At about noon, General George Meade's and General John Gibbon's infantry divisions were ordered to attack Jackson's line. Most of the Union infantry was in the open when Jackson ordered his artillery to open fire. The South Carolina Pee Dee Artillery was posted on Prospect Hill, nearly directly in front of Lawton's Brigade. A Pee Dee artilleryman wrote, *"We were posted on a chain of hills. Just in the edge of the woods before us was a wide level plain extending to the river, some three or five miles wide. I could see fully half the whole Yankee army, reserves and all. It was a grand sight seeing them come in position this morning; but it seemed that that host would eat us up any how. I felt uneasy until I saw General Lee, and right behind him the 'Old Stonewall,' riding up and down our lines, looking at the foe as coolly and calmly as if they were only going to have a general muster. The Yankee batteries came into position beautifully, and commenced shelling the woods we were in. It was hard to take it, but we had strict orders not to fire. Their infantry advanced in beautiful order. When one thousand yards distant we poured a perfect storm of shell into them from fifty or one hundred guns, but on they came. Our infantry was too much for them they had to leave. Oh! it did me good to see the rascals run; but here comes a fresh line. Far as the eye can reach the line extends. They have the fate of their predecessors, but another new line advances. I had been uneasy; perhaps scared before, but now had death or defeat been offered me I would have taken the former. Some of our bravest were down. Pegram's men (a Virginia battery stationed by our side on the right) had left their guns. Captain Pegram wrapped his battle flag around him, walking up and down among his deserted guns. It was a time to test a man's courage."* [6]

The Federal attack stalled, but they soon renewed the attack. About 1 p.m. Meade's division rolled straight ahead, directly to his front was General Maxcy Gregg's South Carolina Brigade. Gregg's Brigade was posted several hundred yards inside the woods. A marsh fronted Gregg's position and some thought it impenetrable; it wasn't. Gregg was under the impression that other Confederates troops were posted in front of his brigade, but there was only a picket line. When this picket line was struck by Meade's Federal division, the pickets fled to the left and right, but none fell directly back to warn Gregg of the approaching Federals. Gregg had ordered his brigade to stack rifles, as they were the second line and not to fire to the front. He said he feared mistakenly firing on friends, as frequently happened in the confusion of battle. The Union soldiers poured through the trees and closed on Gregg's Brigade and opened fire. Many of the South Carolinian's realized that these were Federal soldiers firing on them, but Gregg galloped about on his horse yelling for his men to cease fire and not to fire on friends. His soldierly appearance drew a flurry of rifle fire from the Federals and he was shot through the spine and mortally wounded.

The South Carolina Brigade was overwhelmed, shattered and swept from the field. Some of the Federals swept to their left, rolling up the flank of Archer's Brigade, others swept to their right, dispersing Lane's North Carolina Brigade. A huge gap yawned in the Confederate line.

The Confederate artillery massed on Prospect Hill was in danger of being

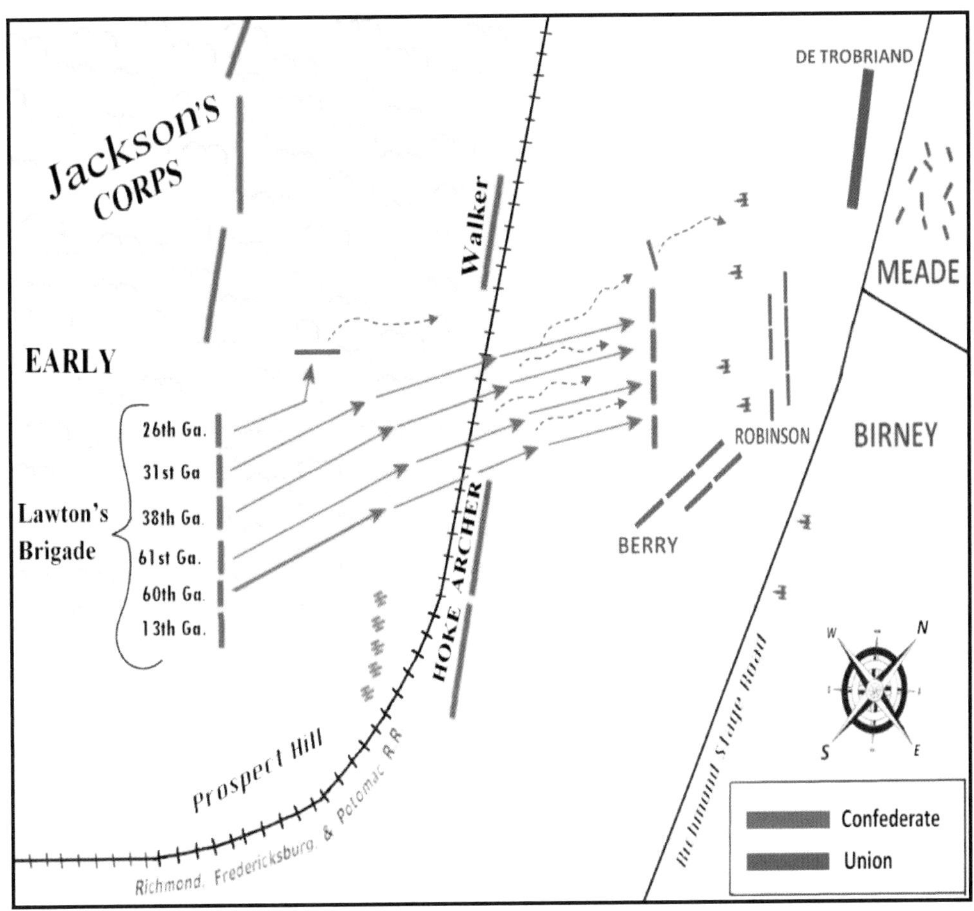

captured. Jackson's line was in danger of being sliced in half, placing the Federals squarely on the flank of Lee's main line. The entire Confederate army was in danger of being pushed off the hills overlooking Fredericksburg.

At this critical moment Early ordered Lawton's Brigade forward to seal the breech. Colonel Atkinson, commanding Lawton's Brigade, ordered the brigade forward. The 13th Georgia, located on the right flank of the brigade, was separated by a small hill from the rest of the brigade and failed to hear the order to advance. Lieutenant Lester of the Tom Cobb Infantry recalled the afternoon, *"We lay in line of battle the monotony relieved every few minutes by the shells tearing through the woods. About the hour mentioned the order was given, "Attention---load—forward!" We knew now, by the rapid picket firing, that warm work was just ahead."* [7]

The brigade moved forward only a short distance when the left regiment, the 26th Georgia, brushed against the enemy, Colonel Clement Evans, commanding the 31st Georgia Regiment, reported, *"The brigade moving forward about 250 yards, Captain Grace, commanding the Twenty-sixth Georgia, on the left, encountered the enemy, being apprised of their proximity to him by a volley poured into his ranks, which for a moment checked his advance. But quickly recovering, the regiment delivered its fire, reloaded, and, advancing, drove the enemy before them through the woods. Having encountered the enemy so soon, they became for the time separated from the brigade."* [8]

The remaining four regiments, consisting of the 60th Georgia, Colonel Stiles; 61st Georgia, Colonel Lamar; 38th Georgia, Captain McLeod, and 31st Georgia, Colonel Evans, pushed straight ahead. They came upon the enemy in a minute of time after they were first encountered by Captain Grace, receiving their fire without producing scarcely a perceptible check; fired in return, and, with loud cheers, dashed forward.

Colonel Clement Evans of the 31st Georgia reported of Lawton's Brigade's charge, "We soon came upon the lines, posted directly behind a brush fence in these woods. Here we charged, formed again the lines and routed the enemy. They soon formed again at the edge of the wood behind another brush fence. Another charge upon this line soon moved them from this position. The battle became a perfect stampede upon the part of the Yankees." [9] Captain McLeod, commanding the 38th, was knocked out of action by a Federal shell early in the action, but the regiment continued the advance without him.

Lieutenant Colonel Stiles, commanding the 60th, the regiment charging beside the 38th, told of an amusing incident during the charge of Lawton's Brigade. Stiles related, "Men were double-quicking to the point of peril and he (Colonel Stiles) was running from one end to the other of his brigade line to see that all parts were kept properly 'dressed up,' when he observed one of the conscripts who had lately been sent to his regiment--a large, fine-looking fellow--drop out and crouch behind a tree. Colonel Stiles was described as, "a tall, wiry, muscular man, was accustomed to carrying a long, heavy sword, and having it at the time in his hand, as he passed he struck the fellow a sound whack across his shoulders with the flat of the weapon, simultaneously saying, "Up there, you coward!" To his astonishment the man dropped his musket, clasped his hands and keeled over backwards, devoutly ejaculating, "Lord, receive my spirit!"

The bizarre announcement startled Colonel Stiles, he explained, "The entire dénouement was so unexpected and grotesque and his haste so imperative, that he scarcely knew how he managed to do it, but he did turn and deliver a violent kick upon the fellow's ribs, at the same time shouting, 'Get up, sir! The Lord wouldn't receive the spirit of such an infernal coward;' whereupon, to his further amazement, the man sprang up in the most joyful fashion, fairly shouting, 'Ain't I killed? The Lord be praised!' and grabbing his musket he sailed in like a hero, as he ever afterwards was. The narrator added that he firmly believed that, but for the kick, his conscript would have completed the thing and died in good order." [10]

Colonel Evans wrote, *"From this time the contest consisted of but a series of temporary halts made by the enemy, only to be driven away from their positions. At the railroad the enemy made their most determined resistance, and for a few minutes poured a heavy fire into our line. Seeing that a charge was the most effectual plan to dislodge them, the order was given, and so rapidly accomplished that many of the enemy were captured, and, a few, in their attempts to get away, received the application of the bayonet. As an incident of the battle, I desire to state that one of the enemy, after surrendering, leveled his gun to fire at our passing line, but a bayonet thrust from the hands of Captain W. D. Wood, of the Thirty-first Georgia, prevented the intend barbarism."* [11] Lawton's Brigade and the 38th were punishing the Federals, but incurred significant casualties as well. Sergeant William Absalom Booth, of the Goshen Blues was wounded and disabled. He was never able to rejoin the regiment afterwards. [12] Sergeant William T. Thornton of Company F, Thornton's Line Volunteers, was shot through the right eye and mortally wounded, dying from his wounds a month and a half later. [13] Private Miles B. Vaughn of Company H, The Goshen Blues, was shot through the head and killed instantly [14]. Sergeant William R. Henry of Company K, The Bartow Avengers, was shot through the ankle joint. His leg was amputated five inches below the knee. [15]

"From their first stand at the woods to the river was a wide, level bottom, some half mile across. Over this field the Yankees fled in perfect confusion, many hundreds of them losing their lives and being captured." [16] Private Jasper L. Harbin of Company H, the Goshen Blues, a youth of seventeen years, captured four Union soldiers single handed. He captured a lieutenant and three privates with an empty gun and delivered them safely to General Early." [17]

The South Carolina Pee Dee Artillery continued to pound the Federal infantry arrayed on the plains below Prospect Hill. The artilleryman cheered as Lawton's Brigade pushed the Federals out of the woods and chased them across the railroad tracks, *"Our cannon flamed and roared, and the roar of musketry was terrific. The foe halts, wavers and flies. We double charging our gun, pour the canister among them. As they get out of range of that we send them an occasional shell to help them on. 'Cease firing!' What means that yell to the right. No one answers, nor do we need an answer, for our gallant boys are seen pouring from the woods, double quicking on the charge. On they go...nearly up to the Yankee batteries. How my heart did beat then. My hat couldn't stay on my head. I would have hollered if I had been killed for it the next minute, simply because I couldn't help it."* [18]

One Confederate soldier quipped, *"The gallantry of some of the officers of the 38th Georgia was truly remarkable. Two in particular as related to my position, as bearing the most conspicuous part were Captain Hawkins of Co. E, and Captain Brinson, of Co. G. The later displayed an almost fearless courage throughout the engagement, being frequently in advance of the Regiment waving his cap and happily cheering the men on to victory."* [19]

Captain James L. Jones, was killed, *"Captain Jones, of the Irwin Invincibles, Co. I, while gallantly leading his company in the charge, was shot through the head by the enemy's sharpshooters, and expired immediately, while waving his sword and urging the Invincibles on to battle. He had been promoted to the Captaincy of his company but a few months before, since Captain Andrew B. Irwin, who died at Richmond. His death casts a gloom over the whole regiment."* [20]
An unidentified Confederate soldier became separated from his own regiment and joined the 38th in the charge, *"Charging them like an avalanche with frightful slaughter for near a mile, taking in the route a large number of prisoners. Coming to a ditch in an open field, the enemy rather recovered and tried to make another stand, being under cover of their batteries planted just beyond on an eminence. Here the brigade halted, and the writer himself was of opinion it had ventured the pursuit as far as prudence would dictate, but elated with victory-the spirit being fully aroused, and tempted by the old "Stars and Stripes" still unfurled on the breeze just ahead -- the Thirty-Eighth followed by portions of two other regiments, the Thirty-First and Sixty-First, if I mistake not, sprang the fence and charged the ditch "taking in out of the weather" almost an entire brigade, besides bringing to the dust nearly all who attempted to escape."* [21]

A Confederate army correspondent accompanying the 38th also witnessed the action; *"The Brigade reserved their fire until they were within one hundred yards of the enemy. A regular volley was poured into them that thinned their ranks and scattered them like chaff. A charge was ordered and with the wildest yells and gruff 'hurrahs' Lawton's Brigade made at them, shooting them down and taking prisoners as they went. The route was complete. Thrice they rallied partially their broke and scattered columns. But the fierce looks and worst of all, the glittering steel of the 'Empire State Boys,' (Georgia was known as the "Empire State" of the south) whose homes and property they were seeking to devastate, they could not stand, and after a feeble effort they would again take to their heels."* [22]

Colonel Evans of the 31st reported, *"At this part of the railroad a short neck of woods juts out into the plain, so that on our right and left were the open fields, while before the line lay this neck of thickly matted woods. Under its shelter the enemy fled, pursued by these four regiments with so much precipitation that both parties entered the ditches beyond almost together. At the railroad and in these ditches a large number of prisoners were captured and sent to the rear, among whom was one colonel and several officers of minor grade. Colonel Atkinson, in command of the brigade, participating fully in the enthusiasm of the charge, was wounded in the arm above the elbow soon after entering the field, and fell into the hands of the enemy."* [23] *"It was believed that the (38th Georgia) regiment captured as many prisoners as it had men in the ranks."*

Lieutenant Lester told of a humorous incident that occurred during the battle, *"In charging the enemy at Fredericksburg, cross the flat-lands mentioned, which was drained by ditches, one of our company fell in a ditch, a Yankee in another ditch, close at hand. The contending forces had swept past and left each in this condition. The predicament was exceedingly embarrassing and unpleasant. If our man attempted to retire the Yankee would shoot him; if Yank made an attempt to leave Reb would kill him. They did not enter into terms of capitulation, as each side refused to surrender. There they sat and sat, eyeing each other like two ferocious Thomas cats, each knowing it was death to attempt to leave. They remained each in his ditch and watched each other until dark when our man vacated under cover of darkness and marched back to our company, where he amused the crowd by narrating his adventure."* [24]

Colonel Evans took command of the brigade after Colonel Atkinson was wounded, he wrote *"I cannot forbear to mention in terms of unqualified praise the heroism of Captain E. P. Lawton, assistant adjutant general of the brigade, from the beginning of the advance until near the close of the fight, when he received a dangerous wound and was unavoidably left in the open plain where he fell. Cheering on the men, leading this regiment, or restoring the line of another, encouraging officers, he was everywhere along the whole line the bravest among the brave. Just as the four regiments emerged from the neck of woods referred to, his horse was shot under him, and in falling so far disabled him, that thousands less ardent or determined would have felt justified in leaving the field, but, limping on, he rejoined the line again in their advance toward the battery."* [25]

Private William Absalom Booth, Company H, The Goshen Blues
Elbert County, Georgia
Photo courtesy of Ms. Amy C. Parker
Descendant

Captain Lawton led the 38th in pursuit of the fleeing Federals, *"Forward boys, forward, was the only command heard above the din and roar of battle, and forward they went. The 38th without any commander, Major McLeod having been knocked down early in the action, was in one hundred yards of a battery of eight guns that was belching forth grape shot and doing a deal of damage to the Brigade. Having thus disposed of the infantry, an order was given to charge the battery, which the men made at with a desperation unparalleled in the annals of history, approaching sufficiently near to put to flight the cowardly brigade placed to its support, to make the gunners abandon their pieces, and shoot down at the guns several of the horses. One more effort and they have killed the horses attached to it, the gunners that are alive have scampered away, and the 38th exhausted and without ammunition, have their hands upon it, with three loud cheers....Thus they held the battery some length of time, but ammunition being exhausted and the enemy reinforcing heavily on the right, with none coming to their aid, they had to retire reluctantly leaving behind the 'glittering prize.'"* [26]

Captain Edward Payson Lawton was shot down and mortally wounded in the final charge of the 38th Georgia on the Union battery. He was the brother of General Alexander R. Lawton and was beloved by both the men and officers of the brigade for his heroics in battle. He had helped lead the 38th Georgia Regiment in the battle of Gaines Mill, after the other officers were shot down. He was born in 1832 and educated at Brown University, Rhode Island. He was captured by the Federals when Lawton's Brigade was forced to retreat and was taken to an Alexandria, Virginia hospital where he died on December 26th, after receiving all possible medical attention. His wife traveled to Alexandria and returned south with his body. His widow watched with gratitude as his remains were provided a formal Federal military escort and military honors, as his body was transferred to Confederate officers under a flag of truce at Fredericksburg.

The four regiments of Lawton's Brigade were in a moment of crisis. They were far in advance of the main Confederate line of battle, isolated. and nearly out of ammunition. An army correspondent observed, "At length the Federal line assumed a crescent shape, and bent around and enveloped the brigade. In front the enemy had been scattered like chaff, but on the right and left of the brigade, they still maintained ground and poured into both flanks a terrible fire of grape and canister. The difficulty was the brigade was not properly supported on either side. It had moved too rapidly for the rest of the line. In other words it fought too fast, and got too far ahead." [27] Their right and left flanks were in the air, when the Federals began their second grand advance. A Federal line of battle emerged fronting Lawton's Brigade, and overlapping their line far to the left and right.

Private George Nichols was in the 61st, charging on the right flank of the 38th in the attack. He wrote of the crisis, *"In the mean time the Yankees being reinforced. Succeeding in rallying their fleeing columns and fell upon the brigade with more fury than they did before, killing and taking prisoners. Here we also lost some brave, noble boys. The Union line was much longer than ours, so the enemy at our right begun to swing around us and advance, which caused considerable confusion in our ranks. They were so close to us that they began to order us to halt and throw down our guns and surrender."* [28] At least fourteen members of the 38th were captured during the attack. They may have surrendered when the regiment was being overwhelmed, or were possibly captured when the Federals advanced a picket line over the drainage ditches between the lines. These ditches were filled with both Union and Confederate soldiers seeking shelter from the flying bullets and artillery shells. Colonel Atkinson, the acting Brigade commander for Lawton's Brigade, was wounded in the shoulder during the charge on the battery and sought shelter in one of these drainage ditches between the lines. Atkinson was captured when the Federal's advanced a picket line over the ditch later in the afternoon. He was the only Brigade commander on either side to be captured during the battle.

Lawton's Brigade was forced to beat a hasty retreat back to the wooded hills. Private Isaac Bradwell, 31st Georgia, charged with Lawton's Brigade, on the left flank of the 38th Georgia. Bradwell recalled General Early's welcome to the retreating soldiers of Lawton's Brigade, "When we got back to the woods we were completely demoralized and without semblance of organization. Presently General Early came riding about among us in the midst of bursting shells and whizzing grapeshot. The old fellow was furious and hailed every man he saw, asking if he belonged to that "blankety – blankety Georgia brigade." I have always considered this the most disgraceful affair we ever took part in our service in the war, but the blame should rest with our commander rather than on us. The color bearer of the 31st Georgia Regiment threw their colors in muddy ditch during the retreat, but luckily the banner was retrieved after the battle." [29]

The retreat of Lawton's Brigade all but ended the battle on the right wing of the Confederate line near Prospect Hill. The battle on the Confederate left raged from noon until after dark. The Federals launched massive waves of Union soldiers over the open plains towards Marye's Heights, only for Confederate artillery and infantry to cut them to pieces. Thousands fell before Confederate guns and not a single Union soldier reached the stonewall at the bottom of Marye's Heights. The Federal army remained in position on December 14th, but did not attack. The Federals finally retreated back across the Rappahannock River on the night of December 15th - 16th and pulled up their bridges behind them.

Colonel Evans of the 31st assumed command of Lawton's Brigade during the retreat, he said, "It is gratifying to me to be able to record that officers and men generally behaved with a courage characteristic of the Southern soldier, continuing for the brigade a well-deserved reputation. The report of casualties will testify how severe the fire was through which these brave men passed in driving the enemy before them. In the heat of the contest these four regiments may have gone too far, but brave men in that important struggle feel that they scarcely went far enough." [30]
A Confederate Army Correspondent for the Central Georgian newspaper accompanied the 38th Georgia Regiment during the battle and sat down to write his report, just two days after the battle, on December 15th, 1862:

Army Correspondent
Near the Battlefield

Dear Georgian – The long expected battle of Fredericksburg has at last been fought. Once more the two grand armies of the North and South have met in battle to contest the supremacy of their arms. Once more the "Dixie Boys" as usual, have triumphed and driven Old Abe's horde of thieves to the shelter of their gunboats, (Note: The Federals had no gunboats protecting their army, but had heavy artillery posted on the ridge line on the north side of the river) under whose friendly protection they are today hiding their cowardly carcasses, and the name of Fredericksburg will be inscribed on the folds of our regimental flags, by the side of other battles where our arms triumphed...Too much praise cannot be bestowed upon the officers and men of the brigade (Lawton's Georgia Brigade) for their gallant and meritorious conduct in action. General Early commanding the Division remarked that the only fault he had to find was that the brigade ventured too far without support. Captain Brinson (John W. Brinson) drew notice to himself from the brigade commander for his coolness and good order in which he carried the Battey Guards in action. Captain Hawkins (Charles A. Hawkins) and Lieutenant Durrough (Robert T. Dorough) of the Tom Cobb Infantry were conspicuous for their bravery, as were their company. The former was wounded early in the fight but would not leave the field until the regiment was relieved. But it is needless to draw distinctions where all acted so well their part... Guillaume [31]

Massive battles created chaos accounting for soldiers in their aftermath. Twenty-one 38th Georgia soldiers were listed as missing in action when the battle ended. The Federals held the field where the 38th charged across the plain to attack the Federal artillery battery, until the night of December 15th - 16th, when they retreated back across the river. Many of the 38th killed and wounded during the charge were necessarily left behind as the regiment retreated back to the safety of their own lines.

Union General Marsena Patrick, Provost Marshal General of the Army of the Potomac, received a message on the morning of December 17th. He was told messages had been exchanged under a flag of truce with the Confederates across the river agreeing to a prisoner exchange. He wrote, "After learning that their flag agreed to exchange prisoners at 12 m. today I came up and made full arrangements, then went back and got the Reb prisoners to the River, 459 in number, 215 having already gone to Fort Monroe." [32]

Sergeant W. C. Mathews of the Battey guards wrote of the 38th Georgia losses after the battle, "In this battle out of 450 men, we lost 37 killed and 92 wounded." [33] Twenty-one 38th soldiers were reported as "missing" after the battle and it was still unclear as of December 19th, (the date a casualty report was written) how many were killed, wounded, and captured. Surviving service and pension records document 14 of the 21 missing were captured on December 13th and twelve were exchanged on the battlefield December 17th. These men had already returned to the regiment, two days before the casualty report was written, yet they were still appeared on the list of missing on December 19th.

Two of the 14 captured soldiers were wounded and had already been sent north to Union hospitals, before the exchange occurred on December 17th. Private Benjamin H. Hickman, Co. F, from Elbert County, Georgia, was wounded, captured and died at a hospital in Washington, DC. He is the only member of the 38th Georgia buried at Arlington National Cemetery. Private Isaac Starling, of the Co. I, the Alabama Company, also died at a D.C. hospital from a gunshot wound to the side of his head. Soldiers captured after the summer of 1863 would endure much longer terms of imprisonment, due to breakdown of the prisoner exchange agreements between the Confederacy and the Federal government.

Reported Missing in Action - December 13th, 1862

Company A – "The Murphy Guards," DeKalb County, Georgia
1. Corporal John W. Ball - Captured 12/13/1862, exchanged 12/17/1862.
2. Private David Alexander Chesnut - Captured 12/13/1862, exchanged 12/17/1862.
3. Sergeant Andrew J. Miller - Captured 12/13/1862, exchanged 12/17/1862.
4. Private William T. Marlow - Captured 12/13/1862, exchanged 12/17/1862.

Company B, "The Milton Guards," Milton County, Georgia (Currently Fulton County)
5. Private Wiley B. Herring - Possibly captured 12/13/1862, exchanged 12/17/1862 at Fredericksburg.
6. Private Thomas Jefferson "Jeff" Woodliff - Confirmed as killed in action and his remains recovered, buried at Confederate Cemetery, Spotsylvania Court House, Va.
7. Private John M. Moore - Confederate Service Record reads "Missing in Action, "Supposed Killed" at Fredericksburg 12/13/1862. Burial site unknown.

Company C, "The Ben Hill Guards," Bulloch & Emanuel Counties, Georgia
8. Corporal Isaac Washington Coleman - Captured 12/13/1862, exchanged 12/17/1862.
9. 2nd Lieutenant Jacob "Jake" Perry Pughsley - Captured 12/13/1862, exchanged 12/17/1862.

Company D, "The McCullough Rifles," DeKalb & Fulton Counties
10. Private Pressley "Press" Lanier - Captured 12/13/1862, exchanged 12/17/1862.
11. Private Moses H. Newsom - Captured 12/13/1862, exchanged 12/17/1862.

Company E, "The Tom Cobb Infantry," Oglethorpe County, Georgia
None.

Company F, "Thornton's Line Volunteers," Hart and Elbert Counties, Georgia
12. Private Benjamin H. Hickman - Wounded and captured at Fredericksburg, Va., 12/13/1862. Died of wounds in Douglas Hospital at Washington, D. C. 2/8/1863. Buried at Arlington National Cemetery, Va. Marker reads B. H. Hickman, GA, CSA.

Company G, "The Battey Guards," Jefferson County, Georgia
13. Private William F. Atkinson - Captured 12/13/1862, exchanged 12/17/1862.
14. 2nd Lieutenant William J. Farmer - Confederate Service Record states "taken prisoner at Fredericksburg, not heard from since." No additional information.
15. Sergeant Thomas J. Stewart - Killed at Fredericksburg, Va. 12/13/1862. Buried at Mount Jackson Cemetery in Shenandoah Co., Va.

Company H, "The Goshen Blues," Elbert County, Georgia
None.
Company I, "Irwin's Invincibles," Henry County, Alabama
16. Private Isaac Starling - Wounded and captured at battle of Fredericksburg, 12/13/1862. Died at Old Capital Prison, hospital, D.C.
17. Private Thomas Hawkins - Listed on casualty roster as "missing" 12/19/1862. Possibly captured 12/13/1862, exchanged 12/17/1862.
Company K, "DeKalb & Fulton Bartow Avengers," DeKalb and Fulton County, Georgia
18. Private Isaac Washington Autry - Private Captured 12/13/1862, exchanged 12/17/1862.
Company N, "The Dawson Farmers," Dawson County, Georgia
19. 1st Lieutenant John W. Goswick - Captured 12/13/1862, exchanged 12/17/1862.
20. 1st Sergeant John Wesley Mashburn - Listed on casualty roster as "missing" 12/19/1862. Possibly captured 12/13/1862, exchanged 12/17/1862 at Fredericksburg.
21. Private Van Buren Martin - Captured 12/13/1862, exchanged 12/17/1862.

Two of the 21 missing soldiers were later confirmed as killed in action, Private Thomas Jefferson Woodliff, of the Milton Guards and Sergeant Thomas J. Stewart, of the Battey Guards, were confirmed as killed in action after the Federals retreated. Two other missing soldiers simply vanished, with little explanation in their Confederate service records. 2nd Lieutenant William J. Farmer, of the Battey Guards, record reads "taken prisoner at Fredericksburg, not heard from since." Private John M. Moore of the Milton Guards, service record notes, "Missing in Action, "Supposed Killed" at Fredericksburg 12/13/1862.

Killed or Mortally Wounded - Fredericksburg, December 13th, 1862

Company A – "The Murphy Guards," DeKalb County, Georgia
1. Private James Poss - Killed at Fredericksburg, Va., 12/13/1862.
2. Private Thomas Jefferson "Jeff" Woodliff - Killed at Fredericksburg, Va. 12/13/1862.

Company B, "The Milton Guards," Milton County, Georgia (Currently Fulton County)
3. Private Martin Carter - Killed at Fredericksburg, Va., 12/13/1862.
4. Private Thomas Orr - Wounded in battle of Fredericksburg, 12/13/1862, send to Henningsen Hospital at Richmond, Va., died of wound Nov. 1 or Dec. 5, 1862.
5. Private John M. Martin - Killed at Fredericksburg, Va., 12/13/1862.
6. Private John M. Moore - Listed as "Missing in Action, "Supposed Killed" at Fredericksburg 12/13/1862. Not listed in Henderson's Roster of Georgia Confederate Soldiers, found in Confederate Service Records.

7. Private William A. Baker - Mortally wounded at Fredericksburg, 12/13/1862, leg amputated close to body, died.
8. Private Robert C. Hampton - Reported as mortally wounded in the battle of Fredericksburg, 12/13/1862, in a newspaper list of casualties, dated 1/7/1863.

Company C, "The Ben Hill Guards," Bulloch & Emanuel Counties, Georgia

9. Private Thomas L. Phillips - Killed at Fredericksburg, Va. 12/13/1862.
10. Private William S. Mason - Killed at Fredericksburg, Va. 12/13/1862.

Company D, DeKelab & Fulton Counties

11. Private Hiram P. Adams - Wounded at Fredericksburg, Va. 12/13/1862. Died of wounds in General Hospital # 24 (Moore Hospital,) at Richmond, Va. 1/24/1863.

Company E, "The Tom Cobb Infantry," Oglethorpe County, Georgia

None.

Company F, "Thornton's Line Volunteers," Hart and Elbert Counties, Georgia

12. Private McAlpin Anslem Bentley - Wounded at Fredericksburg, Va. 12/13/1862. Died of wound 6/6/1864.
13. Private James C. Nelms - Wounded at Fredericksburg, Va. 12/13/1862. Died of wounds, date not given.
14. 5th Corporal William Thomas Fortson - Wounded at Fredericksburg, Va. 12/13/1862. Died from wounds 12/15/1862.
15. Private Benjamin H. Hickman - Wounded and captured at Fredericksburg, Va. 12/13/1862. Died of wounds in Douglas Hospital at Washington, D. C. 2/8/1863.
16. 5th Sergeant William T. Thornton - Wounded through right eye at Fredericksburg, Va. 12/13/1862. Died from wounds 1/29/1863.

Company G, "The Battey Guards," Jefferson County, Georgia

17. 2nd Lieutenant William J. Farmer - Killed Fredericksburg, Va. 12/13/1862.
18. 3rd Sergeant Thomas J. Stewart - Killed at Fredericksburg, Va. 12/13/1862.
19. Private Miles B. Vaughan - Killed at Fredericksburg, Va. 12/13/1862.

Company H, "The Goshen Blues," Elbert County, Georgia

20. Private John A. Smith - Wounded at Fredericksburg, Va. 12/13/1862. Died from wounds at Mount Jackson, Va. 9/4/1864.
21. Private Wiley T. Roberts - Killed at Fredericksburg, Va. 12/13/1862.
22. 2nd Sergeant Jonas Reese (or Reep) - Killed at Fredericksburg, Va. 12/13/1862.

Company I, "Irwin Invincibles," Henry County, Alabama

23. Captain James E. Jones - Killed in the battle of Fredericksburg, shot through the head, 12/13/1862.
24. Private John W. Smith - Wounded at Fredericksburg, Va. 12/13/1862. Died of wounds at Richmond, Va. 2/16/1863.

25. Private Isaac Starling - Wounded at Fredericksburg, Va. 12/13/1862. Died of wounds 1/1/1863.

Company K, "DeKalb & Fulton Bartow Avengers," DeKalb and Fulton County, Georgia

26. Private Benjamin L. Wilson - Killed at Fredericksburg, Va. 12/13/1862.
27. Private James R. Mitchell - Killed at Fredericksburg, Va. 12/13/1862.

Company N, "The Dawson Farmers," Dawson County, Georgia

28. Captain James E. Jones - Killed in the battle of Fredericksburg, shot through the head, 12/13/1862.

Chapter 9
Chancellorsville Campaign, Jan. - May 1863

Just after the battle of Fredericksburg the (Georgia) brigade was commanded by the gallant Colonel Clement A. Evans and was in winter quarters on the Rappahannock River near Port Royal. A snow storm came and snow fell to a depth of a foot. A member of the Tom Cobb Infantry recalled, “After the battle of Fredericksburg our toils and dangers ended for the present, for here quarters at Hamilton's Crossing, and remaining until spring, enjoying ourselves by eating, sleeping and joking, with an occasional drill. Many of our boys received boxes of provisions, etc., from their homes, which only a soldier can appreciate properly." [1]

[COMMUNICATED.]

REPORT OF THE BATTEY GUARDS' SOCIETY.

JEFFERSON COUNTY, GA., DEC. 1st, 1862.

Mr. Editor: I herewith transmit a report of the "Battey Guards' Sewing Society," which was organized in August, 1861, by electing Mrs. Wylanty Smith, President, and other officers. It may not be improper to premise the report by saying that, while this Society was organized in honor, and for the benefit of "The Battey Guards," of this county, our individual contributions have not been confined to this noble company, nor to the companies of this county merely, but have have been dispensed in various ways for the good of our noble defenders, wherever found or wherever from. And, we may add, that our report, as here exhibited, falls short of our actual contributions for the "Battey Guards," as much that was accomplished while the Society was in its formative state, and even after its organization, has never been recorded. Hence, the following report is very imperfect:

We have sent the "Battey Guards" 50 pair pants, 50 shirts, 60 pair socks, 80 pair shoes, 50 pair drawers, 19 coats, 12 quilts, and 12 blankets.

The example of our first President, Mrs. Wylanty Smith, who died in the midst of her usefulness, the heroism of many of the "Battey Guards" who have fallen in our defence, and the death of their brave and gallant Captain, Wm. H. Battey, who fell at Sharpsburg, but stimulates us to a more firm resolve, to work and pray for our noble boys and for our final success. We have received, in cash, $800.

Respectfully yours,
SALLIE HUDSON, Sec.

Daily Constitutionalist, (Augusta GA), Sunday, Jan. 4th, 1863, Page 3

Lawton's Brigade and the 38th participated in a little known battle shortly after the battle of Fredericksburg, a snowball battle involving between 6,000 and 7,000 soldiers. "These contests were unique in many respects. In the first place here was sport, or friendly combat, on the grandest scale, perhaps, known in modern times. Entire Brigades lined up against each other for the fight. And not the masses of men only, but the organized military bodies--the line and field officers, the bands and the banners, the generals and their staffs, mounted as for genuine battle. There was the formal demand for the surrender of the camp, and the refusal, the charge, and the repulse; the front, the flank, the rear attack. And there was intense earnestness in the struggle-- sometimes limbs were broken and eyes, at least temporarily, put out, and the camp equipment of the vanquished was regarded as fair booty to the victors." [2]

Captain Tip Harrison of the 31st Georgia wrote of the snowball battle. *"Early one morning Hoke's brigade of North Carolinians appeared in front of the camp of Lawton's Georgia brigade and mockingly demanded the surrender of the camp. Colonel Clement A. Evans was in our brigade and quickly ordered us out to repel the threatened attack of Hoke's "Tar Heels."*

The Thirty-first and Thirty-eighth Georgia regiments were occupying that part of the camp where Hoke appeared with his six regiments of sturdy, jolly North Carolinians. Our two regiments formed in line quickly to defend the camp, and hastily summoned the other four regiments, the Thirteenth, Twenty-sixth, Sixtieth, and Sixty-first Georgia to our assistance.

Hoke, like a good general that he was, did not wait until his enemy was fully prepared to meet him, but ordered forward his Tar Heels. They came at us with the usual "rebel yell," and with haversacks and hands filled with hard snowballs. They flanked us, drove us, ran over us, and pressed on to the center of our camp. There they met the other regiments, which had formed to check their progress, and it was soon apparent to those of us who had been routed at the first onslaught that the Tar Heels had spent their strength. We quickly reformed and dragging our skulking crowd from their huts, charged the North Carolinians in the rear.

It was then their turn to break, and in a twinkling they were broken, routed and driven out of our camp. We captured their mounted officers, their flags and made prisoners of hundreds of their men. It was three miles to their camp and rout ended only when the broken fragments of Hoke's once grand line found refuge in their huts.

Colonel Evans was not satisfied with this splendid repulse, so he ordered out his brigade that same day, and in the afternoon we were marched over and challenged the Tar Heels for another battle on their own grounds." [3] Many soldiers were keen for a re-attack, since they planned to bring away anything portable that could be captured, but a good many of the old soldiers refused to go on the raid. [4]

Captain Harrison described the attack of the Georgia Brigade on General Hoke's encampment and how the victory in the morning turned to defeat in the afternoon,

"I do not care to recall all the unpleasant incidents of that second engagement with our Tar Heel friends. Suffice to say were beaten, and for three miles we ran the men of Hoke's brigade, but as old Abe Dudney remarked once, we ran them by "working in the lead."

One of my company, known throughout the command as 'Bull' Averett, in his retreat towards his camp, was closely pursued by three or four Tar Heels. They pressed him hard until he reached a deep gully in the woods, across which he attempted to jump, but his feet struck the bank and he fell back into the ditch. It was deep and its sides steep. 'Bull' did not know how long it was and started down its rough winding way. His pursuers 'lined up' along the banks and pelted him unmercifully. 'I surrender,' said he. 'We are not taking any prisoners.' replied the exultant Tar Heels. 'Here,' said one, 'You came to capture our camp and steal our cooking utensils, Take this biscuit and put it in your oven.' And with all his might he struck my luckless comrade on the side of his head with his hardened snowball. Finally, when he found he could expect no mercy at their hands, 'Bull' seized the root of an overhanging big tree and dragged himself out of the ditch and escaped, bruised and thoroughly demoralized." [5]

Lieutenant Colonel Stiles, commanding the 60th, was roughly handled and had, *"several very heavy bruises on his face...at one point in the conflict (he) had been dragged from his horse and captured by Hoke's men, but later had been recaptured by his own command, and on both occasions had been pretty roughly handled. One would have supposed these veteran troops had seen too much of the real thing to seek amusement in playing at battle."* [6]

"The glory of our victory of the morning was eclipsed by the utter defeat we sustained that afternoon. Many of the boys said that Colonel Evans deserved no better fate because he did not know when to quit fighting," [7] said Captain Harrison. The two battles taken together made the day's fighting a draw, and on that each side rested." [8]

On January 9th, 1863, a letter written by an unidentified 38th Georgia soldier appeared in the Georgia, Augusta Chronicle Newspaper. This well educated writer, probably an officer, felt Lawton's Brigade and the 38th Georgia hadn't received due recognition of its' battle achievements thus far during the war.

Camp of Ewell's Division,
Lawton's Brigade,
Thirty-Eighth Georgia Regiment, Dec. 18th, 1862

Mr. Editor - It is in no spirit of controversy that I ask the use of a short space in your columns, for no one is more opposed to originating or fomenting State or sectional jealousies than myself. The correspondents of most of the Virginia papers have favorite regiments or brigades, from the fact they have some relative or friends connected therewith, and they aim at the promotion of these friends and relatives. Whilst this is perfectly natural, it does injustice to regiments from other States serving in Virginia; because those who have no newspaper correspondents are passed over in silence, no matter how gallant a part they may have taken in the engagement. But very few papers south of Richmond have regular correspondents near the army; they gather what intelligence they publish from Virginia papers; and thus many a gallant fellow's name, like his deeds, passes into oblivion. I think it is due their memory, as well as a reward to those who survive them, that their names should be published; as it not only serves to stimulate the living to strive to win themselves a name in the history of this great Revolution, but as a simple act of justice.

Lawton's Brigade came cheerfully to Virginia to share with their brothers who had proceeded them, the dangers and fatigues of what promised and has proven to be one of the most active and brilliant campaigns on record, It has participated in every engagement in which Jackson's corps has engaged, and whatever the honor of history may award Stonewall Jackson as a victorious General, Lawton's Brigade claims to have contributed their share towards raising him to that military eminence - through so far they have failed to receive credit for it; and whilst they have born in silence this injustice, they are pained none the less. The bones of many a gallant Georgian lay bleaching on the mountains and in the vallies of Virginia unhonored even with the silent darkness of the soldier's grave, and their memory will never receive so much as a passing notice from a public journal -- thus showing great neglect on the part of the government it was their pride to serve, and affording but poor encouragement to the army in general. Though not a Regiment of our Brigade has emblazoned on their battle flags Coal Harbor, Manassas, Sharpsburg, or Fredericksburg, like many that I have seen, they have the proud consciousness of knowing that they contributed as much as those that been heralded in almost every Virginia paper towards the general success.

But I can speak more definitely of our Regiment, as the ensanguined field of Coal Harbor, and where they sustained in every respect the high reputation for courage and endurance that Georgia soldiers previously bore in this engagement. The total loss, including all the Field Officers and many Company Officers, was one hundred and eighty-three. At Manassas, our loss was ninety-three. At the memorable field of Sharpsburg, where Georgia lost some of her most gifted and gallant sons --among whom was Col. Marcellus Douglass, our Brigade Commander, and our own regimental commander, Captain Wm. H. Battey -- our loss was one hundred and forty-three. And now, at Fredericksburg, where another draft was made upon the treasure of Georgia's blood, our loss was one hundred and forty-five -- making a total, not including those that died from sickness, of a loss of five hundred and sixty-four out of a regiment that numbered eleven hundred men when they landed in Petersburg, Va. If this does not entitle the Thirty-Eighth Georgia Regiment to a place in the front of the picture, we have struggled in vain. But we are willing to abide the time of the historian for our reward, and we only ask the publication of this statement that our silence may not be construed into assent. I am most respectfully,
One of the Thirty-Eighth [9]

In addition to the feelings of lack of recognition from the newspapers, other tensions within the Regiment strained the commissioned officers. There was great dissension among the 38th Georgia officers after the battle of Fredericksburg. Captain William L. McLeod, commander of the Ben Hill Guards Company, was the senior captain of the regiment, present for duty, and acting commander of the regiment. He had briefly led the regiment during the battle of Fredericksburg, until he was knocked out of the ranks by a shell early in the charge. Captain John W. Brinson of Company G, the Battey Guards, tendered his resignation one week after the battle, due to ill feelings between him and Captain McLeod. He wrote on December 20th, 1862,

"The Commanding officer of the 38th Georgia Regiment (Captain William McLeod) in a recent report brought to the attention of the brigade Commander under charges of "straggling and frequent neglect of duty" thereby implying my incompetence as a company commander...there now exist and have for the last two

months existed between myself and that officer that the above charges were prompted from a disposition to do me injustice not only on this but on all possible occasions, that I am determined not to, if possible to serve under him as captain of a company, and consequently as well as on account of a sense of incompetency in many respects, I beg to tender you my unconditional resignation as captain of Company G, 38th Georgia Vols."

Signed Your Obd. Servt, Captain John Brinson [10]

The officers of the regiment had a meeting to discuss promotions to fill the vacant positions of Colonel, Lieutenant Colonel, and Major after the battle of Fredericksburg, on January 19th, 1863. At least eighteen officers attended the meeting, the senior officer being Captain Charles Hawkins, of Company E. They knew Captain William McLeod was eligible to serve as the Lieutenant Colonel, or Major, of the regiment, but they recommended he not be promoted. The eighteen officers signed this letter, to be sent to acting brigade commander, Colonel Clement A. Evans:

Camp of 38th Regt Geo Vol
January 19th 1863

At a Meeting held pursuant to an order from Col. C. A. Evans Cmdg Brigade - We the undersigned tender these as our objections to the promotion of Captain W. L. McLeod to Lieut Col. of the 38th Geo Regt.

1st He claims to command an artillery Company has drawn pay as such, and applied for a transfer on the ground that he never regularly belonged to the regiment.

2nd His extreme youth being only 20 yrs of age.

3rd His want of judgment & Stability

4th Injustice to men & officers being subject to extreme prejudices which in our opinion unfits him to command brave & true men.

5th Gambling with his men & officers.

6th Advising men & officers to resort to illegal measures to get home.

7th Cowardice in action.

Chas A. Hawkins Capt & Brigade Inspector
Lt. R G Dorough Comdg Co (E)
Lt. G G Maddox Comg Co (B)

Lt. S A Hagood Comp (B)
Lt. K R Cross " (B)
J. Cole Vaughn Sr Comdg Co (G)
L. W. Farmer Lt. " "
J. O. Maxwell Lt Cmdg Co "F"
T. D. Thornton Lt (F)
John W. McCurdy Lt Co (D)
R.M. Campbell Lt. Comdg Co (I)
John Baxter Lieut. Co. D
A. C. Bell Lt Commanding Co. N
Lt. A. H. Hill Com N
G. W. Stubbs Lt. Com Co K
John Oglesby Lt Com (H)
John O. Andrew Lt Co. (H)
Henry R. Deadwyler Lieut Co H

We therefore beg leave to recommend the following named officers as suitable for the field officers of the 38th Reg. Ga. Vol.

Major Jas D. Mathews for Colonel.
Capt. R P Eberheart for Lt. Colonel.
Capt John C. Thornton Major

Captain McLeod quickly received word of the regimental officers banding together to deny him promotion and responded the very next day:

HD 38th Geo. Regt.
Jany 20th 1863

Genl'

In accordance with an order from Col Evans, comdg Brigade - The officers held a meeting, and recommended for promotion to fill vacancies junior officers over me, and object to my promotion for grave charges alleged in their memorial. Therefore I most respectfully demand a court of inquiry, and insist upon my promotion by seniority and a right. I deny any charges alleged. And feel satisfied the officers charging me are moved by selfish motives, a number having just come home know nothing about the facts – or pretended facts.

And I enter my solemn protest against the promotion of officers I commanded over me. It being directly opposed to the law authorizing promotions – And tendering to insubordination. The law directing if the officers entitled to promotion are incompetent to rise – they shall be cited to the incompetent Board, not officers of the Regiment. If the report of the meeting be sent forward – I ask that – that my protest may accompany this report. But allow me to suggest that the promotion rest until after the court of inquiry.

I am sir,

Your Obd. Servt.
Wm. L. McLeod

Colonel Evans immediately placed Captain McLeod under arrest, stating this was the first he had heard of these serious accusations. Colonel Evans sent the below report to his superiors:

Hdqrs Lawton's Brigade
Jan 20, 1863

Respectfully forwarded.

The 38th Ga. Reg. has had no field officer in command since the 27th of June 1862. Col. Wm Lee having resigned, Lt. Col. Parr and Maj. Matthews both severely wounded. Lt. Col. Parr resigned in December last, and Maj. Matthews is still at home, perhaps permanently disabled. Capt. Wm McLeod the Senior Captain, was absent from the last of June 1862, to the last of October 1862 and since then has had command of the Regiment

On account of the absence of all field officers, the uncertainty of their return, the increasing discontent of men and officers, the regiment has been under no efficient organization or discipline since its arrival in Virginia.

The promotion of good to competent officers by the rule of Seniority is preferable to filling vacancies by election, but the rule rightfully subjects the officers claiming promotion to all objections for competency and when this power is fairly and firmly exercised, good officers will be promoted and those who are incompetent will be kept back. I have therefore in this case as in all others notified (by Circular) the Company officers of their right to file objections to the promotion of any officer, which notice has called out the 'objections to the promotion of Capt McLeod' which accompanies this report. I send also the "protest of Capt. McLeod."

Should the transfer of Capt. McLeods Company be ordered the question of his promotion will be disposed of without reference to the other objections; -- Capt. McLeod is only twenty years of age, and is deficient in that Stability, Consistent, uniform discipline, and prudent management of a regiment so necessary to be exercised by the officer charged with that important trust.

His "gambling with men & officers" & "advising them to resort to illegal measures to get home" brought to my notice for the first time, by these "objections." I find an enquiry to have sufficient cause of probability to authorize me to order his arrest and to direct charges to be preferred in proper form.

If Maj. Mathews is not permanently disabled he is worthy to be promoted Colonel -- To Captain Eberheart's promotion there is no objection --The ill health of Capt. Thornton which has prevented him from doing duty for several months is an objection to his promotion.

The vacancies in the Several Company Offices may be filled by promotion according to rank after a report upon their competency made by the Board of Examiners appointed by myself for this Brigade.

C.A. Evans, Col. Cmg, Brigade

The record is unclear as to whether Captain McLeod was actually promoted to Major or Lieutenant Colonel. He was however, released from arrest and placed in command of the regiment the following month. He would have little time to enjoy his new position as regimental commander, for he was killed on the first day of battle at Gettysburg, July 1st, 1863. The record states that when killed at Gettysburg he was serving as "acting Colonel" of the regiment. [11]

Lieutenant Lester of the Tom Cobb Infantry wrote, *"Our only duty during this winter was picket duty on the river. Many little incidents occurred doing our picket services that were interesting to us. Our Yankees manifested all their aptitude for trade and commerce. Quite a little commercial trade sprung up between our boys and the Yankees. We would make little rafts and start them afloat across the river, laden with tobacco, almost our only article of commerce, and receive in exchange many little Yankee tricks and newspapers. Many of these rafts would capsize and the cargo prove to be an utter loss, while others would reach the land safely. An interesting little thing occurred that many of the boys will remember with pleasure: One of our picket posts was situated immediately on the river bank. Directly across the river Yankees had established a post. Just in the rear of the Yankee's post, upon a bluff, was a large residence. One Sunday evening a lot of young ladies came out from the house and waved their handkerchiefs. The Yanks ordered them to return to the house. Instead Of doing so they struck up a song, "The Bonnie Blue flag," much to our delight and the annoyance of the Yankees."* [12]

The Alabama Company, Company I, the "Irwin Invincibles was transferred to the 60th on March 1st, 1863, since the 60th only had nine companies and the 38th had eleven companies. Lieutenant Mathews of the Battey Guards wrote of this transfer nearly 30 years after the war, "Company I, our Alabama company, of whom we thought so much, was transferred, much against their consent and ours, forcibly to the 60th Georgia, that regiment only having nine companies. This robbery was done by order of Gen. Early, and which we consider an injustice to this day." [13]

Lieutenant John Baxter of Company D, the McCullough Rifles, sat down and wrote a letter home to his cousin, Jane Hardman, on Sunday, March 8th, 1863. He describes the health of the regiment, living conditions and seeing Yankee observation balloons across the Rappahannock River.

Camp of the 38 Ga. Reg, Fredericksburg Va., March 8, 1863

Dear Cousin, With pleasure this Sabbath morning write you a few lines in answer to your kind letter received the 4th of this inst. and was truly glad to hear from you and hear you was all well. This leaves me and Samuel (Corporal Samuel Wylie Cochran of Co. D) well and hope will reach and find you and family well. Cousin I have nothing to write that will interest you at this time. The health of the Regiment is tolerable good there is some cases of the fever and 2 cases of the small pox, but it does prove fatal here like in some places we have some very wet weather here but we are tolerable well fixed here for the weather we have good tents with chimneys to them. The boys is all in good spirits they engage themselves finely and generally look well but want to come home very bad and wants the war to end very bad and I know that I do. Jane it will be no trouble for girls to marry when this war ends for they all say that if they get home they will marry if they can get any body to have them. But if old Lincon calls his three milion of men and we have them to whip them it will thin a heap of us but I hope I never will have to fight any more.

I am satisfied that there is danger in it for their balls is blind they had as soon hit a man in the face as anywhere else. The boys say that the old bombs comes on in had as soon take one as another. Cousin we can see the Yankeys go up in their balloons every 2 or 3 days we are in a bout 3 miles of them. Some think that there will be a fight here before long but they got so badly whipped here before that they will not try it anymore. Cousin I would like the best in the world to see you and let me hear from you as often as possible. Tell uncle that he is excusable for not writing to me as he does not write letters but you can write for him. I have not heard from home in some time tell I received your letter and I am very uneasy. I fear that Almeda (John Baxter's 8 year old daughter) is very bad off or Emaliza (his wife). Oh Jane I want to see them very bad. I will close for this time. You must excuse my bad writing and composing and give my love and respects to uncle and aunt and all the family and a good portion to your self.

To Miss E. J. Hardman
John Baxter
Please write as soon as you get this.[14]

"It was at Hamilton's cross that the noble and gallant Gen. John B. Gordon took command of the brigade. Gen. Gordon was a gallant Georgian and among the best brigade commanders in the Army of Northern Virginia. He had been shot at least five times during the battle of Antietam, holding the position of the Confederate center, thereafter known as 'Bloody Lane.' He quickly won the respect and admiration of the entire Georgia Brigade. Though only temporally assigned command of the Brigade, the men and officers of brigade petitioned for his permanent assignment to the brigade." [15]

General John B. Gordon, CSA
US National Archives

General Gordon recalled of his convalescing and return to the army, *"It was nearly seven months after the battle of Antietam, or Sharpsburg, before I was able to return to my duties at the front. Even then the wound through my face had not healed; but Nature, at last, did her perfect work, and thus deprived the army surgeons of a proposed operation. Although my enforced absence from the army was prolonged and tedious, it was not without its incidents and interest.*

Some of the simple-hearted people who lived in remote districts had quaint conceptions of the size of an army. One of these, a matron about fifty years of age, came a considerable distance to see me and to inquire about her son. She opened the conversation by asking:
"Do you know William?"
"What William, madam?"
" My son William."
I replied: "Really, I do not know whether I have ever met your son William or not. Can you tell me what regiment or brigade or division or corps he belongs to?"
She answered: "No, I can't, but I know he belongs to Gin'al Lee's company."
I think the dear old soul left with the impression that I was something of a fraud because I did not know every man in "Gin'al Lee's company "--especially William.[16]

General Gordon was ordered to take command of the Georgia Brigade, "On my return to the army, I was assigned to the command of perhaps the largest brigade in the Confederate army, composed of six regiments (13th, 26th, 31st, 38th, 60th, and 61st Ga.) from my own State, Georgia. No more superb material ever filled the ranks of any command in any army." [17] The Georgia Brigade was no longer known as "Lawton's Brigade," but became known as "Gordon's Brigade."

The armies began to prepare for the renewal of hostilities during April. 1st Sergeant Francis C. McCleskey of the Milton Guards wrote of Federal spy balloons watching their camps, *"We are still near Fredericksburg I can't tell how long we will remain here but I hope all Summer or until the War ends. The Yankees seem to be uneasy on something. I can't tell what. They are constantly up in there Balloons looking over at our army. And you may depend that we have got a good Army here."*

Francis also spoke of retrieving the remains of his father, Captain George W. McCleskey, who died of wounds at Richmond on July 17th, 1862 and was buried there at Oakwood Cemetery, *"You said that C. H. Maddox has been to see Dr. Shelton about carrying Pap home. I would love to have him taken home, but if you can't get him home Satisfy your Self. We taken Jeff Woodliff (Private Thomas Jefferson Woodliff, of Company B, killed December 13th, 1862 at the battle of Fredericksburg) up last week he was considerably decayed. We put him in a coffin and buried him the best we could. Though that was bad. We had to bury him in the same clothes and he did not look natural."* [18]

Winter had ended and a new "fighting season" was beginning. The Georgia Brigade, now Gordon's Brigade and 38th Georgia were called to action during the 2nd Battle of Fredericksburg, which was part of the battle of Chancellorsville. Union General Burnside, had been relieved of command following the Federal disaster at Fredericksburg. The new leader of the Army of the Potomac, Major General Joseph "Fighting Joe" Hooker, created a brilliant plan. First, Major General John Sedgwick's VI Corps of 25,000 would cross the river at Fredericksburg and threaten to attack, holding the Confederate army in place. At the same time, Hooker would swiftly march 12 miles upriver with the majority of his army, about 100,000 men, cross the river, push though the Wilderness, and break into the open ground. This movement would flank Lee's army out of Fredericksburg and force him to give battle on favorable ground, where his army would be destroyed.

Sedgwick reported, "On Tuesday, the 28th ultimo, in compliance with the orders of the commanding general, received that morning, the Sixth Corps moved to the vicinity of Franklin's crossing, near the mouth of Deep Run; the First Corps, Major-General Reynolds, to a position about 1 mile farther down the river, and the Third Corps, Major-General Sickles, took position slightly to the rear and between the positions of the First and Sixth Corps. All the troops encamped that night behind the heights, without fires, and concealed from the observation of the enemy. During the night the pontoons were carried to the river by hand. At the upper crossing, and shortly before daylight, (Wednesday, April 29th) Brooks' division, of the Sixth Corps, crossed in the boats, Russell's brigade taking the lead, and receiving the fire of the enemy's pickets and reserves. The enemy's rifle-pits were immediately occupied, and three bridges were rapidly laid, under the direction of Brigadier General Benham. At Reynolds' crossing, 1 mile farther down, the passage was delayed by a severe fire from the enemy's sharpshooters, but was at length gallantly accomplished, General Wadsworth crossing with a portion of his division in the boats, and driving the enemy from their rifle-pits." [19]

On Wednesday, April 29th, Sedgwick's corps crossed the Rappahannock River, several miles below the town and deployed to attack Lee's lines, posted on the ridge running parallel to the river. When Sedgwick failed to attack, Lee quickly concluded this was a diversion and Hooker's main force planned to cross the river about 12 miles upstream near Chancellorsville. Lee ordered Early to hold Sedgwick in check at Fredericksburg. Lee then led the rest of the Army of Northern Virginia into the Wilderness on the night of Thursday, April 30th, to meet Joe Hooker's army.

Sergeant Francis L. Hudgins of Company K, the Bartow Avengers, vividly related the events leading to the battle of Chancellorsville and the 2nd battle of Fredericksburg,

"On the morning of April 29th, the 38th Georgia Regiment was awakened from slumber by the beating of the long roll. We fell into line, hastily formed and reported to General Gordon, who thanked us for promptness, being the first regiment in the brigade to respond.

The brigade being formed, we moved out and took position along the railroad northwest of Hamilton's Crossing, near Dead Horse Hill, (today known as Prospect Hill) where Pelham's battery was in position on the 13th of December previous. The 13th Georgia Regiment was on skirmish at Deep Bottom, and the incessant crack of the rifle and yells told us that it was heavily engaged and, to all appearances was holding its own. Capt. W. L. McLeod, commanding the 38th, detailed me to return to our old camp, hurry up the cooking detail, and bring them and all others who were not excused by the surgeon to the firing line, as all others would be needed to successfully oppose "Fighting Joe" Hooker's terrible advance." [20]

Private Isaac Bradwell of the 31st Georgia Regiment related, "The 13th (Georgia Regiment) was one of the best regiments in the brigade, but it could offer little resistance to Sedgwick's army, and after some fighting it fell back and gave the enemy an opportunity to put in their pontoon bridges without opposition." [21]

Sergeant Hudgins wrote of 38th Georgia skirmishers stealing Yankee haversacks while the Yankees called roll. *"On the way back we passed Hill's Division coming up, and all moving to the left in the direction of Chancellorsville. After we returned with our detail, General Gordon came walking along our line and said: "Boys we are Georgians here, and do you know what Georgia expects of you? Every man to do his duty." Sharp skirmishing continued all day. A portion of Hay's Louisiana and part of the 38th Georgia Brigade reinforced skirmishers along the telegraph road in front of Deep Bottom, where General Sedgwick was endeavoring to throw his pontoon bridge across the Rappahannock River. After crossing the river the enemy formed their line near the brick houses and stayed there all night. After dark some of our skirmishers crawled down to the brick houses; and while the Yankees were calling roll they slipped into the houses from the back way and brought out all the haversacks belonging to the Yankees which they had time to get. They then returned to their positions along the telegraph road and said: "Boys, the enemy is down there by the thousands, and these well filled haversacks are the best proof of it."* [22]

Gen. John Sedgwick, USA
Library of Congress

Private Bradwell of the 31st Georgia Regiment recalled the pontoon bridges were quickly completed and, "As soon as this was done, Sedgwick's whole force came across, formed a line parallel with the river, and fortified their positions with excellent breastworks. Behind these they remained very quiet and at first did not seem deposed to make trouble." [23]

Up to this point, Hooker had flawlessly conducted this campaign and had stolen a march on Lee's Army. Three Union corps had crossed the Rappahannock and were well into the Wilderness. General George Meade almost gushed with glee when he met General Slocum on the afternoon of April 30th, near Chancellorsville, "This is splendid, Slocum; hurrah for old Joe; we are on Lee's flank, and he does not know it. You take the Plank Road toward Fredericksburg, and I'll take the Pike, or vice versa, as you prefer, and we'll get out of this Wilderness." Meade's high hopes were at once crushed by Slocum's reply, "My orders are to assume command on arriving at this point, and to take up a line of battle here, and not to move forward without orders." [24] Hooker had lost his nerve and this halt would prove fatal to the Army of the Potomac; Lee's army was on the way to meet him in the Wilderness. The first Confederate brigades arrived around midnight and took up positions on the roads blocking the Union Army's advance.

On Friday, May 1st, Hooker issued a congratulatory order to his army, "It is with heartfelt satisfaction the commanding general announces to the army that the operations of the last three days have determined that our enemy must either ingloriously fly, or come out from behind his defenses and give us battle on our own ground, where certain destruction awaits him." [25] Hooker briefly launched an offense attack against Lee on Friday morning, but soon demurred. He had surrendered the initiative and Lee wasted little time in seizing the advantage. During Saturday morning, May 2nd, Hooker realized Sedgwick's ruse at Fredericksburg had failed to come to fruition and he ordered the 1st Corp to join his army at Chancellorsville. Later that same day, Lee divided his army yet again, sending Jackson on a flanking march that surprised and crushed the Union right flank and sent the Federal army reeling. It was a brilliant victory, but came at a great cost; Jackson was mortally wounded that night while reconnoitering the Federal lines, shot down by his own soldiers.

That evening, fearing further disaster, Hooker ordered Sedgwick to push across the river at Fredericksburg and march to his relief at Chancellorsville. He was calling for Sedgwick's 25,000 soldiers to come to the rescue his 100,000 men. Sedgwick received orders to attack Fredericksburg on the night of Saturday, May 2nd, "That night at 11 o'clock I received an order, dated 10.10 p.m., directing me to cross the Rappahannock at Fredericksburg immediately upon receipt of the order, and move in the direction of Chancellorsville until I connected with the major-general commanding; to attack and destroy any force on the road, and be in the vicinity of the general at daylight." [26]

On Sunday, May 3rd, Lee and General Jeb Stewart, now in command of Jackson's Corps, pressed their attacks and united both wings of the Confederate army. Hooker thought Sedgwick had pulled back across the river at Fredericksburg, but he remained on the south side with Early's Confederates.

Sunday morning, May 3rd, Sedgwick and his 25,000 troops faced Early's 12,000 Confederates, aligned over a seven mile front. Lieutenant William C. Mathews of Company G, the Battey Guards, described the array of Confederate forces at the beginning of the battle, "Our Division (Early's) was composed of the following brigades: Gordon's Georgia Brigade, Extra Billy Smith's Virginia Brigade, Hoke's North Carolina Brigade, and Hay's Louisiana Brigade, and Barksdale's Mississippi Brigade, assigned temporarily. These were posted in the following order: Gordon on the right at the (Hamilton's) Crossing, Barksdale on the left in front of Fredericksburg, Hays in the center, Smith left center, and Hoke right center. [27] Gen. Early was also supported by 56 guns of the Confederate reserve artillery.

Early's main force was positioned near Prospect Hill, the site of the Union break through during the first battle of Fredericksburg, December 13, 1862, but the main Federal attack would soon occur at the Marye's Heights, about 5 miles north. This position proved impregnable during the first battle of Fredericksburg, then eight thousand Confederates supported by heavy artillery had nearly annihilated over 40,000 attacking Federals. Now, General Early had less than 1,000 soldiers defending Marye's Heights, the 18th and 21st Mississippi Regiments supported by one battery of artillery, the Washington Artillery.

Lieutenant William C. Mathews of Company G, the Battey Guards, wrote of the attack on Sunday morning, May 3rd,

"The next day we took the road in the direction of the heavy booming that was announcing the most terrific battle of the year. We had left only the Louisiana Brigade composed of the 5th, 6th, 7th, 8th, and 9th Regiments to guard Deep Run Crossing. But we had not gone more than four miles before we were suddenly about faced and ordered in a hurry back to Deep Run. The Yankees had crossed in immense force and were marching on Hamilton's Crossing and attempting by a flank movement to storm the Marye's Heights.

We had so much ground to cover that our line of the left presented to their well filled front a little more than their line of advancing skirmishers. With the rising of the sun their columns commenced advancing from the river. It was soon evident they intended to press the gallant Mississippians from the coveted heights.

They commenced pouring a heavy fire on our thin line of skirmishers which they soon drove in. A considerable demonstration was made on our right, but only a feint. From the top of the trees we could see them advancing, their heaviest column on our left, and we could also see them give back in confusion when a heavy fire from the Mississippians would make many of them bite the dust.

Our Brigade was the only idle spectators of the scene, and I can say that I never witnessed a more grand and terrific one. Twice did they charge them and nearly succeeded in capturing their batteries, but were driven back by fresh troops that they had in reserve. The best troops have to yield to vastly superior numbers and this was our case. We could not assist the Mississippians, who overpowered by the Yankees suddenly gave way and with a wild, loud, and deafening yell the Yankees raised the Stars and Stripes in front of the city on the very spot that they captured 9 pieces of the Washington Artillery, this happened about 3 o'clock in the afternoon. We quickly and with good order formed a line of battle at right angles with our former one and by the time that it was formed it was late at night." [28]

The route to the rear of Lee's army at Chancellorsville was clear. Sedgwick pushed his VI Corps forward on the Plank Road, about five miles from Fredericksburg. At Salem Church he encountered stiff resistance from Wilcox's Alabama Brigade. Wilcox was soon joined by General Lafayette McLaws, along with four Confederate brigades, totaling about 10,000, these troops ensured the "further advance of the enemy was effectively opposed."[29]

Sergeant Hudgins of Bartow's Avengers scouted the enemy lines the next morning, "That night General Gordon moved up in front of Fredericksburg and formed his brigade north of the telegraph road, near Lee's Hill, just south of Welborn's mill on Hazel Run. The morning following General Gordon sent me out as a scout to locate the position of the enemy and report as early as possible. I found the enemy in a strong position along Hazel Run, and as I returned I met the 31st Georgia deployed as skirmishers already advancing. I reported to General Gordon, who was with them, and he said: "Join the 31st on the left." [30]

Confederate Dead of Barksdale's Mississippi Brigade at the Stonewall, May 3rd, 1863 – US National Archives

Lieutenant Mathews watched as Gordon addressed the Georgia Brigade, "We lay on arms and after a most refreshing sleep were awakened about sunrise and skirmishers immediately advanced in the direction of the heights. It was now certain that we were going to take them by storm. General Gordon rode in front of our brigade." [31]

Gordon recalled the moment, "As I was officially a comparative stranger to the men of this brigade, I said in a few sentences to them that we should know each other better when the battle of the day was over; that I trusted we should go together into that fort, and that if there were a man in the brigade who did not wish to go with us, I would excuse him if he would step to the front and make himself known. Of course, there was no man found who desired to be excused, and I then announced that every man in that splendid brigade of Georgians had thus declared his purpose to go into the fortress. They answered this announcement by a prolonged and thrilling shout, and moved briskly to the attack." [32]

Lieutenant Mathews continued, "We commenced the advance then and for two miles and we saw not a Yankee, but on ascending a hill near the old plank road we got sight of them in line of battle behind the road that afforded some protection to them. I thought that the tug of war had come, for we had to charge down the hill and across a deep mill pond and then up a long slant to dislodge them. They had commenced shelling us also." [33]

Gordon had received orders from Early to advance his Brigade, Gordon wrote, "When we were under full headway and under fire from the heights, I received an order to halt, with the explanation that the other troops were to unite in the assault; but the order had come too late. My men were already under heavy fire and were nearing the fort. They were rushing upon it with tremendous impetuosity. I replied to the order that it was too late to halt then, and that a few minutes more would decide the result of the charge." [34] The brigade raised the "rebel yell" and charged the hill. A Union soldier on the hill listened and turned to his Colonel and said, "Do you hear that yell? We had just as well get away, for the damned rebels will take this hill as sure as hell." [35]

Lieutenant Mathews described the 38th charging down the hill and through the mill pond, *"After a little halt, during which times bayonets were fixed, we commenced the charge. We went down the hill like an avalanche and into that mill pond where the water on the right of the regiment reached to our waist. We were soon across and under the brow of the hill, reformed, and started again. Then the bullets commenced their music but before we could get near the road the Yankees were going like a parcel of sheep through the woods, having wounded but 3 of our regiment, while we had killed several and took some prisoners, that were so scared they could not move and also took a portable blacksmith shop, a number of ambulances and several horses. Being a foot I made for one of the wagons and soon cut a great big sorrel horse from the harness and mounted him. I had hardly secured my prize before we made for the hill where the Washington Artillery had been taken the day before, and there we run another force into the city, without a scratch."* [36]

38th Georgia Regiment.

EDITOR INTELLIGENCER:—The following is a list of casualties in Companies A, D and K, 38th Georgia regiment, in the late battle near Fredericksburg:

Company A, Lieut. J. S. Jett commanding—Wounded: John Elson, seriously in head; S. H. Braswell, flesh wound in thigh; William Riddling, seriously in foot.

Company D, Capt. J. G. Rankin commanding—Wounded: H Estes, slightly in leg; Sergt J. A. Lafoy, slightly in heel.

Company K, Lieut. Stubbs commanding—Killed: B. L. Wilson. Wounded: Serg't H. M. Burdett, seriously in foot; J. B. Boyd, slightly in arm.

Savannah Republican, May 14th, 1863, Page 1

Sergeant Hudgins of Company K recalled the charge, "We crossed Hazel Run under a hot fire from the 20th and 21st New York Regiments, which were deployed to oppose our advance. We drove them up Marye's Hill, and I passed through the Marye house yard. Colonel Zimmerman, commanding, was mortally, wounded and fell into our hands and died the next day. We had now retaken the position that General Barksdale had lost; and we held it until late in the evening." [37] Gordon's brilliant charge nearly led to his court martial, "General Early playfully but earnestly remarked, after the fort was taken, that success had saved me from being court-martialed for disobedience to orders." [38]

Lieutenant Mathews of the Battey Guards continued,

"These heights we occupied until late in the evening, the Yankee sharpshooters popping at us from the town the whole time. We had to keep ourselves in a ditch all the time and when one put his head out there would be a dozen bullets whizzing all around him. About an hour before sunset orders came for us to leave that place and advance to the support of the Louisianans who were preparing to storm a hill some two miles distant and rather in our rear, Lieut. Col McLeod ordered me to get the six companies out of the ditch and carry them to a spot some 1/4 miles distant and there halt until he brought up the other four companies who were posted on the other side of a ravine. I gave the order to my six

companies to be ready to leave when I gave the command for them. I then ordered them to about face and to forward at a double quick, for I knew as soon as we got out of the ditch that it would be a hot place. We had scarcely got out of the ditch before they fired upon us, and "JEWHILIKENS," I never heard anything in my life till then, at the first fire over twenty men were shot down. We retired at a full run and in pretty good order, and before we had gotten out of range they gave us another fire, but we had gotten out of range nearly and it did little damage. Lieut. Col. McLeod coming up with the other four companies, we formed our line and advanced at a run to the right of the La. Brigade.

But before we had succeeded in joining him a great many of our brigade had fallen from exertion in the long double quick. By the time that we had united, the Batteries from the other side of the river opened a heavy enfilading fire upon us and the batteries of the enemy in front of us. The air seemed with thousands of death dealing missiles. I saw one solid shot hit between the ranks of the 13th Ga. and cut off half a dozen knapsacks from the backs of the men and never killed one. And strange to say, we lost but few men at this critical period in the whole brigade. Our line was now west of Fredericksburg, going north in the direction of the river.

Sometime after dark, as we had now driven everything from our front, we halted, the lines were put in shape, and skirmishers thrown out. Having reached a wood we were out of danger of the enfilading fire from the river batteries but were exposed to a still more severe one from the grape and canister that was hurled from the battery at the front. Just at dark a charge was ordered and another trying period in the day was enacted, darkness nearly on us, and we were going like a storm through the thick wood with nothing but the flash of the cannon and small arms to guide us. We had gotten to within 300 yards of the battery, I suppose, when everything was comparatively hushed. The battery had limbered up and left in a hurry and the only thing that we could hear (for we could not see) was the rumbling of the cannon wheels as the battery was changing its base towards the river, and had it been daylight we could have gotten it for certain. We halted on the ground where the battery had been worked and stood all night completely exhausted by our day's work." [39]

Early's command and the Georgia Brigade was now pressing the rear of Sedgwick's VI Corps. Sedgwick was now surrounded on three sides, with his back to the river. The battle lines at Chancellorsville were stagnant, but Lee saw an opportunity to destroy Sedgwick's isolated VI Corps. Lee arrived before noon along with Anderson's Division, a force roughly equaling the strength of Sedgwick's Corps. Lee was frustrated by his inability to coordinate a massive attack on May 4th, and that night and early into the next morning, Sedgwick escaped the trap by quietly marching his VI Corps over a pontoon bridge on the Rappahannock River to safety.

Private George W. Nichols of the 61st recalled, "It was now after sundown, very near dark, and we could not follow up our success. We lay quiet on the field of battle all night. Next morning- the Yankees were all back across the river and the battle of Chancellorsville and Marye's heights was over." [40] Hooker withdrew the remainder of his army north of the Rappahannock the night of the 5th and 6th of May.

Lieutenant Mathews recalled bringing in Yankee prisoners and massive amount of plunder left during the hasty retreat.

"By daybreak the next morning, we commenced bringing in prisoners, and plundering their knapsacks that were thrown all over the field. I got about a peck of coffee mixed with sugar, a nice portfolio with paper and pens and anything that could name from a paper of pins up. I never was so bothered in all my life. A fortune in my grasp and not way carry it off. I was vexed but contented myself with a few small trinkets, and my coffee and sugar, and canteens.

I have my horse yet, and am going to keep him, though I will have to buy him from the government who claims all captured property in that line. The loss in our regiment is all follows: 1 killed, 20 severely wounded, 9 slightly. Besides this there are a dozen that got bruised by shells of little account. I have a bruise as black as ink about my knee that must have been done by a spent ball, but I never felt it until the next day. It now hurts me every now and then, but I pay no attention to it. Our regiment was very fortunate in losing so few to be exposed to so severe a fire of bullets, grape, etc. None at all hurt in the Batty Guards. I omitted to mention that the La. Brigade captured two batteries of 13 pieces on our left. They tried to get out of the way but the Louisianans being closer to them than we were to our battery, overrun them and took every man. Fighting men these Louisianans. Since the fight our regiment has been sent down to Guinea Station to guard prisoners and we start to Richmond tomorrow with 1500 a foot as they can't furnish transportation to so many on the railroad, on account of hauling rations. We have already sent some 6,000 to Richmond and we carry the remainder in the morning. I have been busy all day in taking down their names, etc." [41]

Sergeant Hudgins, of Company K, wrote, *"Here I shall give General Hooker's official dispatch of the battle of Chancellorsville from a Confederate standpoint, which we sang on all occasions afterwards when in a singing mood:*

"I thought I would cross the river
And whip out General Lee;
But, blast his Rebel liver,
He turned the trick on me!
Chorus.
Sixteen cents a dozen
A dollar and a quarter a day.
Joe Hooker is a fine man-
That's what the Yankees say.
Brave Sedgwick was commanded
To move upon the right,
And, when the army landed,
To move on Marye's Height.

I, 'Fighting Joe' Hooker
Upon their left bore down,
Aiming with quick destruction
To flirt them at every bound.
A courier now comes hastening
And brings the glorious news
That Jackson is skedaddling
Back to Richmond goes.
His words are scarcely spoken
When cannon booming near
Gave out the startling token
That Stonewall's in our rear.
Confusion now confounded
Within our ranks prevailed;
For, leaving dead and wounded,
For foreign parts we sailed" [42]

The 38th was ordered to Guinea Station just after the battle of Chancellorsville to escort Federal prisoners captured in the battle, to Belle Island Prison, at Richmond, Virginia. A Yankee officer taken prisoner during the battle was a keen observer of the condition of the Confederate Army. He may have been observing the soldiers of the 38th Georgia and remarked, *"Their artillery horses are poor, starved frames of beasts, tied to their carriages and caissons with odds and ends of rope and strips of rawhide; their supply and ammunition trains look like a congregation of all the crippled California emigrant trains that ever escaped off the desert out of the clutches of the rampaging Comanche Indians. The men are ill-dressed, ill-equipped and ill-provided, a set of ragamuffins that a man is ashamed to be seen among, even when he is a prisoner and can't help it. And yet they have beaten us fairly, beaten us all to pieces, beaten us so easily that we are objects of contempt even to their commonest private soldiers, with no shirts to hang out the holes of their pantaloons, and cartridge-boxes tied around their waists with strands of rope."* [43] Such was the condition of the Confederate Army during the spring of 1863.

Sergeant Francis L. Hudgins of Company K, detailed the memorable story of marching the prisoners from Guinea Station to Belle Island prison camp at Richmond, *"The 38th Georgia Regiment was ordered to Guinea Station just after the battle to guard the prisoners who were to be sent to Richmond. When we arrived, General Jackson, who had been wounded, was in the Chandler house at Guinea Station, where he died on the 10th of May, 1863. Before his death, we started with two thousand prisoners to Richmond, having to walk the entire way and guard them. Our journey was tedious and slow. The dirt road which we traveled was along the railroad, and the Rappahannock River ran parallel to the road, just a few miles to our left.*

The first day the prisoners planned an escape, which was to overwhelm the guards at night, seize all the arms, kill the reserves, go down to the river, signal the vessels in the stream, which would take them aboard, and they would be free.

The first night we camped near the village of Bowling Green. After we struck camp, Capt. W. L. McLeod, commanding the 38th Georgia went into the village to an entertainment, leaving Capt. John G. Rankin in command. Captain McLeod has been told of the proposed plan of escape, but he did not believe a word of it.

After he rode away, Captain Rankin had the men fall in and take arms, doubled the guard, and ordered them to be extra vigilant. He then marched the reserve out of sight. There were no fires, and Captain Rankin said: 'Boys, we'll sleep with one eye open to-night.' Doubling the guard and marching the reserve out of sight of course greatly excited the prisoners, and they soon began saying: "Hello Johnnie, where is your reserve?" These questions were repeated several times, until forbearance ceased to be a virtue, and one fellow said: 'Your scheme to escape from us to-night has leaked out; and if you make the attempt, we are prepared for you, and there will be none you left to tell the tale.' The night wore away, and morning found us all alive. There was no further effort on the part of the prisoners to effect their escape. But for the prompt action of Captain Rankin, we might have been swept out of existence as a regimental organization. The 38th Georgia finally arrived at Richmond with their prisoners, 'Arriving there with them the very day that Stonewall Jackson's remains reached Richmond, and as we marched the prisoners down one of the main streets, we meet the funeral cortege of Jackson coming up the same street. His horse was following the procession without the gallant Stonewall, who had let us to victory on so many hard-fought battle fields.' After camping a few days on the island in the James River, the 38th Georgia Regiment returned to the front. [44]

Chapter 10
The Battle of 2nd Winchester, June , 1863

Private Bradwell of the 31st Georgia recounted Gordon's Brigade departing Fredericksburg and marching to Winchester, Va., "After the Chancellorsville battle, our brigade made camps in an oak grove near Hamilton's Crossing. The situation was elevated, and we could see the enemy's couriers riding to and fro carrying orders and the men drilling near the camps every day." [1] "Here in this camp our regimental chaplains held divine service day and night. Our beloved General Gordon was often among the worshipers. He had become almost an idol in the brigade with officers and men, often leading in the prayer and exhortation service. A great many professed religion, joined the church and were baptized. We drew plenty of clothing and shoes. Every gun was examined and if they were not all right we had to get one that was. Our cartridge boxes were filled, and we knew something was up. General Lee had been reinforced until his army was eighty thousand strong," wrote Private G. W. Nichols, of the 61st Georgia. [2] Private Bradwell recalled Gordon's Brigade's departure from Fredericksburg, "On June 1, we made great heaps of logs on our side of the camps towards the enemy, and when night came we set fire to these and quietly marched away. No doubt the enemy on the other side of the river was puzzled to know what these brilliant fires on the hill meant. Ewell did this to keep Hooker's balloon spies from seeing us moving. Generals Ewell and Hill were both promoted to corps commanders, and Jackson's old corps was divided between them. Early's division was assigned to Ewell's corps.

"We traveled only at night for some time, camping only a short time before day, lest our movements should be observed by the enemy's scouts and signal men. When had been gone several days and were entirely out of sight, we made rapid marches by the way of Culpepper Courthouse and Front Royal, where we crossed the Blue Ridge into the Valley. The weather was hot and the roads dry and dusty. This dust, worked up by the wagons trains and artillery, settled on us until we were as brown as the dust itself, General Gordon, riding along by us, said in loud voice: 'Boys, if your mothers could see you now, they wouldn't know you.' Some of us were limping along on blistered feet, and the General greatly endeared himself to us by his conduct on this occasion. Getting down from his horse, he mounted a private soldier in the saddle, while he fell into the ranks with a gun on his shoulder and trudged along with us."

"General Lee had planned to invade Pennsylvania and Winchester was on his route. This place must first be captured and the way made clear of all enemies before the grand advance could be made. After Jackson left the valley in June, 1862, only a small cavalry force remained there and the Federal General Milroy made Winchester his headquarters and held it with a force of six thousand five hundred infantry and cavalry. They had little trouble to keep in check the few Confederate scouts operating there, and a practical period of peace prevailed for twelve months. The officers and men had sent north for their wives and sweethearts, and were

boarding them at the hotels, boarding houses, and in the homes of the citizens of the city. To the west of the town in an eminence commanding the city and surrounding country side in every direction except on the west, where there were some good positions for artillery. Milroy's main fort stood on this hill. To protect this from any attack from that direction, he had one of these fortified also. He had every reason to believe himself secure, as his position was well-nigh impregnable. In the midst of the great fort stood a tall flag pole, from the top of which floated in the breeze a United States flag thirty feet long. From this secure place Milroy exercised his authority over the defenseless Southern people in a manner so arbitrary as the secure for himself the ill will of everyone. But now the moment had arrived, and the ax was about to fall with a mighty stroke and break up this happy state of affairs with General Milroy and his Army." [3]

Maj. Gen. Robert H. Milroy, National Archives

Union General Milroy wrote in a letter to his wife during January, 1863, *"In this city (Winchester) of about 6,000 inhabitants ... my will is absolute law - none dare contradict or dispute my slightest word or wish. The secesh here have heard many terrible stories about me before I came and supposed me to be a perfect Nero for cruelty and blood, and many of them both male and female tremble when they come into my presence to ask for small privileges, but the favors I grant them are slight and few for I confess I feel a strong disposition to play the tyrant among these traitors."* [4] At Front Royal, the three divisions constituting Ewell's corps took different roads. While Early's division marched direct to Winchester, Johnson's and Rhodes divisions took routes leading to the north of that place so as to cut Milroy off from any means of escape to the fords of the Potomac. The 61st Georgia Regiment was detached from the Georgia Brigade and ordered to guard Early's wagon train, as there was rumored to be a regiment of Federal cavalry in the vicinity." [5]

Lieutenant Mathews, of the Battey Guards, told the story of the battle of Winchester, *"About 10 miles from the town we took a left hand road and came into the Valley Pike at Newton. About 3.12 miles from Winchester, we halted and formed line of battle and advanced a chain of skirmishers. Our line was formed for battle in the fields and forests, for Milroy had gotten wind of our approach and had come out of his fortified position at Winchester to meet us, supposing we were only a straggling band of Confederates. We then commenced advancing slowly and steadily with now and then a stray shot or bomb shell whistling over us. About a half-mile distant we came to Kernstown, the scene of Jackson's terrible conflict a little over a year ago. Here we met with a slight resistance, but very slight, as far it concerned us, for we never wavered a moment.*

We swept everything before us like a rolling torrent. The Yankees broke and ran like sheep. About a mile farther, we came in sight of their battle line and battery on the hill, about half mile distant. A charge was ordered and at them we went at a bull run. They only fired one or two rounds of grape from their battery before they limbered up and cut out before we got anywhere about it. The infantry had a little more spunk, or else relying on their position thought themselves a match for us and remained. At the distance of about 200 yds they poured a heavy volley into us that throwed our men into some confusion, but under tried leaders, they wavered but a moment only, and advanced again with redoubled energy. They poured another volley into us at the distance of 100 yards, but no attention was paid to it. The enemy was posted behind a stonewall in their final stand." [6]

Captain Charles A. Hawkins and brother Sergeant John M. Hawkins, of Company E, The Tom Cobb Infantry, from Oglethorpe County, Georgia. Captain Hawkins was mortally wounded on June 13th, 1863 at the battle of Second Winchester. Courtesy David Wynn Vaughan Collection

Captain Charles A. Hawkins of Tom Cobb Infantry was shot through the body and mortally wounded while gallantly leading his company in this final charge. *"When within seventy-five yards of the wall, as the enemy poured their last volley into the ranks of that heroic old Brigade before breaking and running. He fell in front of his company, the Tom Cobb Infantry, leading the desperate charge and exclaiming as he fells, 'Boys they have killed me, but go on!' He was stuck by a Minié ball entering his left side, passing entirely through his body."* [7] Captain Hawkins died the following evening and was buried at Stonewall Cemetery in Winchester, not far from where he fell. Lieutenant William Mathews lamented his loss, "Among the killed was the brave, the gallant, and the dashing Captain Charley Hawkins, who fell the head of his company, mourned by the whole regiment. [8] Lieutenant Mathews continued the battle account, *"The Yankees seeing the last one them would be killed, broke and run in every direction, throwing away their guns and everything else. We halted at the top of the hill and shot at them as they ran, and poor devils, we slaughtered them like dogs. I never saw such running and screaming in all my life. We were now about 1.12 miles of the city on the opposite side of which was then strong fortifications and stockade fort. About 1.4 mile from the city, on this side was also another fortified hill, on both of which hills they had very strong forces of infantry, cavalry and artillery. They immediately commenced such as shelling as I have not seen in any of the battles, from both of those hills, and we were ordered to fall back to the foot of the hill so as to be out of danger of their missiles of destruction.*

A little before sunset I witnessed one of the grandest sights of the war. Our videttes on the hill reported a cavalry advance, and we were ordered to the top of the hill to give them a reception. They had formed their line by the time that we got to the top and numbered as far as I see about 1500. We had not long to wait. The bugle was sounded and as its notes died away here they come, with sabers clashing and horses snorting. We were ordered not to fire until they were in a hundred yards of us, the 60th Ga. on our right not being made to hold their fire, poured a volley into them at the distance of about 350 yards that spoiled the fun. Though not withstanding they emptied a great many saddles and get them up so badly that their horses commenced throwing them and a most awful stampede ensured. Everyone commenced shooting then and you ought to have seen that skedaddle. Horses and men were running in every direction. Some of the horses came into our line. The shelling being rather too hot for us, we were again ordered back to the foot of the hill, and darkness coming soon afterwards, but a stop to our days work, and we lay down to take the rest we needed so much.

"Tired nature's sweet restorer" soon came to our relief, and the toils of battle were forgotten in our sweet dreams of home and loved ones, then perhaps clustering around the family altar, sending up petitions in behalf of absent ones engaged for a bloody struggle for independence. The heavy thunder and lightning announced the coming of a storm, but the silent sleepers of the "Stonewall band" heard not the pattering drops of rain on their uncovered hears, only a slight jerk of caps over heads and we slept as soundly as we ever did at home. The 'Stonewall

Corps' has long since become accustomed to bullets and hardships and now dread neither.

The next morning I and many others were nearly covered with water and as wet as drowned rats. The water had collected all around us and we were nearly floating. A volunteer force of two officers and 30 men were called for from our regiment and the 26th Ga. to go on a scout and ascertain whether there any Yankees on the hill this side of Winchester or not, and also to act as skirmishers, for the day and night. Thinking there might be a prospect for some fun and plunder; I volunteered to go from our regiment and was placed in command of the detachment from the 38th Georgia and 26th Ga., consisting of 15 men from each regiment and ordered forward by General Gordon. Arriving at the foot of the hill, I halted and seeing no battery on the hill, but very few men employed as skirmishers, I reported the fact to General Gordon, and he ordered me to charge them instantly. Mustering up all my courage that I could possibly command, I gave the command "Forward, double quick," and at them we went. They gave us a pretty stiff little fight, but we waded right through them, killing two and wounding 4 or 5 out of about 50. They only wounded two men of the 26th Ga., that was with me."“We kept on after them and run them clean into town. Fearing that they might lead me into an ambush, I halted 200 yds. of the town and took a position behind a stonewall. Captain Hope of the "Maryland Line" came up shortly afterwards and joined his chain of skirmishers to my right. He being the ranking officer assumed command of both detachments, and shortly afterwards ordered us to advance into town. My detachment was on the left and his on the right of the turnpike. We started at a pretty fast run, the bullets whistling all around from the *Yankees posted behind the house, and soon were into town fighting from one house to the other. The ladies commenced screaming and children crying, and our boys and Yankees whooping and hollering combined to make a sight that was "some to hinder."*

We continued to advance and by the time that we had gained two squares we were joined by 6 or 8 little boys from 10 to 14 years of age armed with shotguns and every available weapon that could be procured. Every now and then we would tumble a Yankee, but they either from bad marksmanship or, excessive fright, barely touched one of our men. We could have killed more had it not been for the ladies in the street, getting in the way. Some of them scolded our men for shooting at the Yankees (real Unionists) and would place themselves in the way. While the majority of them would say "go to it gray coats, kill the nasty rascals, etc." The boys that were with us fought like wild cats. They were acquainted with the locality of the place and could get at the Yankees better than we could. Arriving opposite the "Our House" a young lady and a little boy came running out of a house and showed us where an old Union man had a whole lot of Yankee plunder and told us for God's sake to kill him as he had pointing out all the "Secesh" to Milroy. I took four men and picking up a hatchet that was lying on the side wall, I made for his house calling him out I told him to show where he hid the plunder. He was nearly scared to death, and denied being a Union man. I told him the least said would be better for him. Whereupon he carried me to a little back room and opened the door and told me to walk in, which I did, I thought I had suddenly got in a variety store, there was so many things of every description. I told the boys to make haste and help themselves to what they wanted. I only took a canteen that was hanging on a peg, and we left and went on down town. Just before we go to the Taylor hotel, we saw two ladies standing in front of a house, who turned out to be Milroy's wife and daughter. Milroy's daughter told one of my men that there wasn't a general in our army that could whip her pa, and there was but one and he was dead, thank God, (meaning Jackson.) Wonder if she don't think her pa is whipped now.

By the time that we got to the Taylor Hotel, the Yankees commenced being reinforced and we had to get out there faster that we go got them out. We got out with a whole hide though. There was but four men wounded in this street fight but they belonged to the "Maryland Line." We fell back to our stonewall and there held our position, the Yankees not returning outside of town.

All this time, there had not been a single shot fired from our batteries, why it was I cannot tell. We now captured every hill except two, the one beyond town and another to the left of the Romney road, protected by the former. We were also gradually closing on them from every side. General Hay's Louisiana Brigade was now detailed to storm the hill on the left of the Romney Road, and a battery of 12 pieces was put in position on the hill that I had taken in the morning to engage the attention of the batteries of the high hill beyond town. The plan worked to perfection. The Louisianans run right over the hill on the Romeny road capturing 11 pieces of artillery with little or no loss. One more hill was now to capture and the garrison, town, and everything would be ours. That was the hill where the stockade fort is, and was considered a Gibraltar. Our Brigade was formed in line of battle and night coming we lay to sleep on our arms anxiously awaiting the approach of day as our brigade had been selected to storm the fort, and we all knew that it was not a light task and fraught with immense danger. But Georgians never shrink in time of need. We resolved to add one more feat to our wide reputation for fighting. A little before day we commenced the advance. Silently and cautiously our long line of six well filled regiments moved onward. We are in 300 yds of the hill and fort, and not a gun has been fired though the stars and stripes are streaming from the top of the fort. Suddenly the stillness was broken by the heavy sound of musketry in the direction of Jordan's Springs. We knew too well what that meant. They had evacuated the fort and town, and were trying to break through our lines. We pressed suddenly forward and in a few minutes were on top of the hill, having met no resistance at all." [9]

Private Bradwell of the 31st recalled entering the fort, *"As we entered it (the fort) General Gordon came galloping in from somewhere on a large black United States army horse which we called "Old Milroy," supposing him to belong to the commander of the fort. The General rode up to the flag pole in the center of the fort and, haling down the colors, detached them from the rope and placed one end of them on his saddle. Remounting, he put the spurs to his horse and sailed out of the sally port ahead of us, while the "Star Spangled Banner" floated out thirty feet behind in the morning air."* [10]

Lieutenant Mathews watched as the, *"The stars and stripes were hauled down and a Confederate flag raised in its place. We halted but a few minutes and then moved in the direction of Jordan's Springs, taking prisoners and negroes every ten steps. Soon we were in their immense wagon train and every man commenced such as scene of pillage that I never witnessed before. Anything in the world that you could imagine was in this train. The first thing that I got hold of was a horse and the next thing I commenced loading him with plunder. We were not allowed to pillage long before we were ordered forward. We took prisoners the whole way from Winchester, until we came to Johnson's Division. Nearly every man in Battey Guards mounted himself, but have since had give up their stock.*

We got every prisoner that was in the town except for old Milroy, and about a thousand that cut their way out owing to some mismanagement on the part of Johnson's Division. The Yankees came upon them while they had their arms stacked, and before they could get into lines broke through and got away. Old General Johnson (12th Georgia notoriety) got after Milroy as I was told and run him 4 miles. Our loss was quite small in the Regiment and Brigade, 6 killed and 13 wounded in the 38th Georgia, besides a few slight wounds that we never make any account of. Rube Atwell (Private Reuben Atwell) was killed in the Battey Guards. I saw him stop in the charge and write something in his blank book and in a few minutes afterwards was shot through the heart and died instantly. He was at least 20 paces ahead of the company and loading and shooting as fast as he could. Pat Smith (Sergeant Patrick Henry Smith) went and got his things and these words were found written in his book, 'If I fall in the defense of my country tell my father that I want to meet him to meet me in heaven.' This must have been what he wrote in the book when he stopped for there was nothing else written in it but his name.

Sergeant Patrick Henry Smith, Company G, The Battey Guards, Jefferson County, Georgia
Source: The Sunny South, Jan 10th, 1891

Having secured everything, we rested a day in the vicinity of Winchester, and yesterday morning started on the march again. Last night we camped at a little town called Darksville and this morning came on through Martinsburg and 1 mile this side and camped. We are now waiting further orders, we may leave in a day or two or we may leave in an hour. Unlike most of our marching and fighting, we have had plenty to eat all the time this trip. There is not such a manifestation of Southern feeling among the people as there was last year, though we are warmly greeted in many places. We get any quantity of milk and butter for very small prices. Butter 25 cts, milk 25 cts gallon, eggs 10 cts, and other things in proportion. [11]

Captain Charles Hawkins' death notice appeared in the Augusta Chronicle Newspaper, July 30th, 1863:

Captain Charles A. Hawkins of the Thirty Eighth Georgia Regiment.

The gallant and meritorious young officer fell mortally wounded in the assault of Gen Ewell's Corps upon the fortifications of the enemy at Winchester on the 13th of June, 1863. He died on the evening of the 14th, surviving almost twenty four hours. He was struck by a minie ball, it entering the left side and passed entirely through his body. The enemy was posted in heavy force behind a brick wall or fence. Lawton's Brigade which the Thirty Eighth was attached, command by Gen. Gordon was ordered to charge and drive them from their strong position, which order was promptly obeyed. In front of the enemy was an open space of several hundred yards through which our men made the charge. It was here that Capt. Hawkins led, within seventy-five yards of the wall, as the enemy poured their last volley into the ranks of that heroic old Brigade before breaking and running.

He fell in front of his company, the Tom Cobb Infantry, leading its' desperate charge and exclaiming as he fell, "Boys they have killed me, but go on!" Can a people ever be conquered where dying defenders exclaim as he? My countrymen, "Go on!" "Go on!" Comes a voice from the spirit land. At the battle Sharpsburg Capt. Hawkins' entire company were killed or wounded, he himself receiving a painful wound. Also, at Manassas and Fredericksburg and other fields of less note. He displayed the best qualities of the soldier and officer. His bravery was daring, dashing, fearless; the very best type of southern chivalry. Well educated, he was generous, brave, he was regarded with pride and affection by his men. An officer of his regiment speaks of him: "Kind, generous, brave (almost to a fault) our regiment has lost its best officer and we have lost a friend whom we loved as a brother."

He was the son of judge Thomas H. Hawkins of Oglethorpe county, who has lost in the death of his promising son the great pride of his affections. He left home to enter the service on the 4th of September 1861, in which he continued with unabated ardor and devotion up until the time of his death. He died in the 22nd Year of his age, entirely reconciled to the disposition of providence, which led him from the battle field to the land of peace and rest. He professed religion some years and joined the Baptist Church, of which he was a member at the time he fell. He returned home on furlough and married an accomplished young lady. With her he stayed one night after marriage and left for his command on the Rappahannock, to see his wife no more. Alas! How soon do Earthly joys pass away! How quickly do our new born happiness fade as the dew on the grass, as the fragrance of the flower, or the first sliver of light of the early morn, they vanish away.

The announcement of a great national calamity as the fall of the immortal Stonewall Jackson, affects the popular hearts as a crashing thunderbolt; but the news of the death of a brave, generous, noble southern patriot who falls fighting in the lines for country alone, touches the heart with peculiar sadness. The name of Capt. Hawkins, with the names of thousands of other gallant dead of this revolution, will perhaps be unrecorded, lost in the great aggregate of the unknown, who make up the roll of the Confederate armies; but while a brave and generous people remember with gratitude and pride their defenders, his name will be honored by all who knew him in the generation in which he lived and for which he died. – Americus [12]

Four soldiers were killed or mortally wounded, June 13th, Private William G. Bagwell of Company B, the Milton Guards, Private James W. Herrington of Company C, the Ben Hill Guards, Captain Charles Hawkins of Company E, the Tom Cobb Infantry, and Private Reuben "Rube" Atwell, Company G, the Battey Guards. One soldier was recorded as killed on June 14th, Private Joseph W. Hooks of Company C, the Ben Hill Guards, from Emanuel County, Georgia.

Chapter 11
The Road to Gettysburg

The roads north were clear after Lee deposed of Milroy's Union army at the battle of Second Winchester. Sergeant Hudgins of Company K, recalled the crossing of the Potomac during the campaign, "We crossed the Potomac at Shepherdstown, Va., on the 22nd of June with flags fluttering and bands playing 'Maryland, My Maryland.' We now passed the historic battlefield at Sharpsburg, where the battle was fought on the 17th of September previous, and crossed Antietam Creek at the celebrated bridge and mill. We then passed Boonesborough, Waynesboro, and old Thad Steven's iron works, which were burned. Gordon's Brigade was now in the advance of the Confederate army in the great Cumberland Valley and was the first Confederate force to enter Gettysburg, little thinking then that within ten days we would witness the greatest battle of the war. We continued our advance, passing York, Captain W. H. Harrison, commanding the advance company entering the place." [1]

Lieutenant Lester wrote of the invasion of Maryland and Pennsylvania, *"We entered the state of Pennsylvania on the 23rd and occupied Chambersburg on the same day. Here we struck the famous Pennsylvania militia mounted on their fat horses, but at the first fire of a cannon they wheeled and fled, followed by the derisive shouts and jeers of Ewell's ragged infantry. Here our regiment was detailed and sent to York, a considerable town. At this place General Early made a levy on the population for hats to cover his bare-headed men.*

Here the following incident occurred: A Negro named Ned, a servant of one of the boys, (while our men were engaged in cutting down U. S. flags, many of which were displayed from poles,) concluded to forage a little on the enemy and entered a Dutchman's confectionery and proceeded very deliberately to empty a barrel of crackers upon his blanket and to carry them out in the street to boys, to the great consternation of Dutchy and amusement to the boys. From here we followed the militia to Wrightsville, on the Susquehanna River.

Here we captured all everything to gladden the heart of a soldier, among other things a train load of whiskey, and I regret to say that another big drunk was the consequence. The militia having crossed the river here and burned the bridge, the pursuit was abandoned. By this time the Federals had gotten between us and our main army; so we went around by Hillsboro, and we joined the main army near Gettysburg. On this march we burned the irons works of that "dear departed" old man, the Honorable Thaddeus Stevens, which aroused the vengeful ire of that fangless old viper, who hurled the hot thunder-bolts of his vengeance at the South in the United States Congress during the balance of his life." [2]

"We continued our advance to Wrightsville, on the Susquehanna, where the Pennsylvania militia was drawn up to oppose our advance. We drove them pell-mell through the streets of Wrightsville and across the river, they burning the bridge to prevent capture. Here we carried the Confederate flag farther North than any other troops during the war. The bridge being burned, we could go no further; so we retraced our steps in the direction of Gettysburg by way of York.

On the evening of June 30th, Companies A and K of the 38th Georgia Regiment, commanded by Captains Miller (Capt. William A. C. Miller) and Stubbs, (Capt. George W. Stubbs) were placed on the picket line. On the morning after the first day of July, General Gordon marched the brigade away and left us standing,

LIEUT. COL. W. L. McLEOD

and we still have not been relieved to this day. Miller and Stubbs held a council and very wisely decided to abandon our position and try to overtake the brigade. Bill Jenkins, (William H. Jenkins) kettle drummer of Company A, beat the long roll, and we quickly formed and marched with quick step until we overtook our command. This incident has never been explained to me," wrote, Sergeant Francis Hudgins. [3]

Gordon described the Brigade's arrival at Gettysburg and their role in the first days' battle, "Returning from the banks of the Susquehanna, and meeting at Gettysburg, July 1, 1863, the advance of Lee's forces, my command was thrown quickly and squarely on the right flank of the Union army. A more timely arrival never occurred. The battle had been raging for four or five hours. The Confederate General Archer, with a large portion of his brigade, had been captured. Heth and Scales, Confederate generals, had been wounded. The ranking Union commander on the field, General Reynolds, had been killed, and Hancock was assigned to command. The battle, upon the issue of which hung, perhaps, the fate of the Confederacy, was in full blast. The Union forces, at first driven back, now reinforced, were again advancing and pressing back Lee's left and threatening to envelop it. The Confederates were stubbornly contesting every foot of ground, but the Southern left was slowly yielding. A few moments more and the day's battle might have been ended by the complete turning of Lee's flank. I was ordered to move at once to the aid of the heavily pressed Confederates." [4]

The 26th Georgia was detached from the Georgia Brigade to support the Division's artillery. Gordon led the brigade forward in the assault on a small hill known as Blocher's Knoll, today known as Barlow's Knoll.

Sergeant Hudgins wrote of the charge of the Georgia Brigade and 38th Georgia Regiment at Gettysburg, commencing about 3 p. m.

"*When we arrived on the field, A. P. Hill was in position and hotly engaged with Buford's Calvary and General Reynolds, commanding the 1st Corps, then the advance of the United States army; General Rhodes division, of the Stonewall Corps, being in advance, joined A. P. Hill on the left. We now came up and joined*

Rhodes near the Stearns House, the Federal line being just across a small stream called Willoughby Run, just west of the Almshouse. General Gordon gave his ringing order, "Forward at right shoulder, shift arms!" It was a grand sight to see the Federal infantry on the bank of the stream awaiting motionless our approach.

Many a brave fellow on each side of the stream knew full well that in a few short seconds his soul would appear before the God of battle. When within about seventy-five yards, General Barlow, commanding the Federal line, opened fire. We raised the Rebel yell and continued our advance. When we reached the creek, the gallant Captain W. L. McLeod, commanding the 38th Georgia Regiment, was killed, and Lieutenant Oglesby, (Lt. John Oglesby Jr.) of Elberton, and Lieutenant Mathews, (Lt William C. Mathews) of Tennille, Ga., were severely wounded. We were now nearly together. Sergeant Major Phil Alexander ordered a Federal soldier to throw down his arms. He started to comply, and Alexander ordered another to the same thing. Glancing at the first man, who was raising his gun to fire, Alexander turned on his right heel and struck him with a heavy sword, splitting his head wide open. The gun went off at the same instant, sending a Minié ball through Alexander's hip." [5]

Sergeant Hudgins was shot through the body and severely wounded during the assault. *"It was there that I received my closest call. A Minié ball entered my left breast, going down over and out under the fifth rib. The gallant Lieutenant Baxter (Lt John Baxter of Co. A), the bravest of the brave, walked up to me where I stood, the blood spurting at every breath, looked me straight in the face, and said: "I think you are about gone up, old fellow." I thought then, and I haven't changed my mind since, that was the poorest consolation I ever had offered me in my life. The ball that passed through me made thirteen holes in my blanket. Dr. Taylor, assistant surgeon, came up, and I asked him what he thought of my chance. He examined me rather hurriedly and said: "I don't see why you should not get well." I said: "Doctor, I'll take your advice." Then I walked out to the field hospital and from there watched the progress of the great battle."* [6]

Sergeant Francis C. McCleskey, the son of Captain George Washington McCleskey, the commander of Company B, the Milton Guards, was killed instantly during the charge. His comrades reported, *"He was honorably discharging his duty when the fatal ball pierced his brain. He was shot in the head and instantly killed by a Minié ball while bravely driving the enemy before him."* [7] Private William Joseph Powell, of Co. F, from Elbert County, Georgia, was shot and killed during the charge. [8] The Federal line of battle was shattered and the regiments were running towards the town of Gettysburg. The ground was covered with the Federal dead and wounded, General Gordon said, "*I had no means of ascertaining the number of the enemy's wounded by the fire of this brigade, but if these were in the usual proportion to his killed, nearly 300 of whom were buried on the ground where my brigade fought, his loss in killed and wounded must have exceeded the number of men I carried into action."* [9]

Gordon was ordered by Early to halt his brigade near the Alms House to realign his line of battle. [10] Sergeant Jeptha Campbell of Co, H, the Goshen Blues, told the story of his company throwing out a skirmish line,

"The 38th Georgia was ordered to throw out a line of skirmishers and take in all the Federals who wished to surrender. " In front of that building (Alm's House) was a large three-story brick barn, in which a large number of Federal troops had taken refuge. A call was made for men to volunteer as skirmishers, as the position was a most dangerous one and required soldiers of steady nerves and clear heads. Jeptha Campbell, Green Seymour, William Kirby, John King, Jasper Harbin and Phil W. Alexander, all Elbert (County, Ga.) boys and members of company H, stepped forward and announced their readiness to undertake the dangerous service. They were told to go on picket duty and take all prisoners who wished to surrender, penetrating the enemy's lines as far as possible. The brick barn described was about 250 yards from the Confederate line of battle, and directly in the path of the gallant party of skirmishers. Campbell and Kirby went directly to the barn, and entered its door. They were surprised to find the building filled with Yankee soldiers, and it didn't take the boys long to discover that they were in a mighty tight spot.

Private Jonathan Greene Seymour
Company H, The Goshen Blues
Elbert County Georgia
Photo courtesy of Ms. Linda Seymour Beyers
Descendant of J. Greene Seymour

Kirby, who knew not the meaning of the word fear, began to curse and abuse the Yankees for everything he could think of. Jep Campbell says he decided on a different and more conciliatory approach and explaining to the Union soldiers that Kirby was drunk and not responsible for what he said, called on them to surrender, stating that the barn was surrounded by Confederates, and the building would be riddled by shells unless they gave themselves up. The soldiers, who were badly demoralized by the defeat of the day, believed all that Campbell told them, and consented to lay down their arms and surrender. So as not to let them know that the enemy who surrounded them were only two in number, Campbell made them march from the rear door of the barn by pairs, laying down their arms as they did so. Thus 280 Federal soldiers surrendered to and at the order of only two Elbert County boys, and were marched back and turned over to the command." [11]

"The other skirmishers of the Goshen Blues regrouped and pressed forward into the streets of Gettysburg, even though the town was still under Union control. Campbell, Seymour, King, Kirby and Harbin all then united and went over into the town of Gettysburg. They mingled freely with the Yankee troops, who were too demoralized to molest them. They visited the Federal hospitals and saw the wounded men treated, walking around as freely in the Federal camp as the Yankees themselves." [12]

Gordon reported it was not, "*possible for me take any account of the prisoners sent to the rear, but the division inspector credits this brigade with about 1,800. I carried into action about 1,200 men. The loss of the brigade in killed and wounded was 350, of whom 40 were killed.*" [13] Lieutenant Mathews of the Battey Guards recalled the dead and wounded, "*Our loss in this first day's fight was quite severe. The regiment carried into action 400 men and had 28 killed and 76 wounded. Among the killed was Lieutenant Colonel McLeod, a brave and fearless young officer, and who had been in command of the regiment on this march. He was only 21 years old and was a splendid young officer. Lieutenant Goodwin, (William Franklin Goodwin) Company K, Lieutenant Oglesby, Company H, were both killed while gallantly leading their respective companies, and Lieutenant William C. Mathews, Company G, was severely wounded in the arm.*" The regiment was not actively engaged during the 2nd and 3rd days of the battle, but were exposed to a heavy fire both days, losing several men wounded. [14] The Confederates nearly won the day on July 2nd, but the attack of Pickett's charge was soundly defeated on July 3rd, forcing Lee to retreat back to Virginia.

On the 4th of July, Lee's army began the long retreat back to Virginia. Many of the seriously wounded members of the regiment and brigade were left behind in the make shift hospitals and fell into the hands of the enemy. Others, less severely wounded walked, or were loaded into wagons, for the long trek back to Virginia. Sergeant Hudgins recalled, "On the 4th of July, Tom Raines, of the 13th Georgia, and I. N. Nash, (Isaac Newton

September 5th, 1863
The Constitutionalist
Augusta, Georgia

W. L. McLEOD—At the battle of Gettysburg, on the evening of July 1st, while gallantly cheering his men and grasping at the colors, he received his mortal wound. A minnie ball struck him just above the right temple, passing through his brain. He survived the wound about five hours. Yes, it was as the lingering beams cast by the low descending sun playing as beautifully upon his calm, natural countenance, that his gallant spirit winged its (?) (?) regions of light, with his familiar voice echoing in the breeze—Mother—Mother. Though not 22 years of age, he occupied a post of high honor and responsibility. At the commencement of the war he was a student at Oglethorpe University of Georgia. He quit his studies, went home and raised a company and entered the service in the Wright's Legion as Captain of Artillery. In June 1862, the Legion, all except the artillery companies, was ordered to Virginia. He preferred an active field and accompanied the Legion to Virginia in time to engage in the fights in front of Richmond. At Cold Harbor all the field officers fell, and he assumed command and led the regiment through the rest of the fight. Colonel Lee resigned and by virtue of seniority he was Major of Regiment. The other officers being unsuitable for duty, he was in command of regiment from October until he fell on the bloody field of Gettysburg.

He was recommended for Lieutenant Colonel after the Winchester fight. Major Hale, of General Early's staff, in speaking of him after the Gettysburg fight, said, "He behaved so gallantly; but poor Mc. is no more—he fell as all gallant men, coolly and gallantly doing his duty."

"He was a gentleman of the highest tone," and the recipient of constant attention from his associates, especially those that admired the noble and brave. His bearing up to the time of his wound, displayed the greatest coolness and bravery.

Bereaved mother, heart-rended father and mourning friends, though your fireside has been robbed of the idol of your hearts, what a consolation to know that he fell so nobly doing his duty. It is indeed hard to know that we will never again behold his bright eyes and lovely countenance. Let us not think of him as in the cold tomb; but in the bright and sweet resting place of the weary traveler. We will no more hear his merry laugh on the march and in camp and his shrill encouraging voice amid the storm of battle. Yes, on the bloody field of Gettysburg, where he (?) this last battle, repose the remains of this "young hero." Thus he laid his life on the altar of his country. We loved him and will delight to ever keep fresh in our minds his heroic death. May his spirit rest with God of Peace.

A Friend

Photo Courtesy of the David Wynn Vaughan Collection
Private A. W. Vaughan, Company H, Goshen Blues, of Elbert County, Georgia, wounded and captured at Gettysburg.

Nash) of the 38th, and myself, all badly wounded were placed in one ambulance and then began the return to Dixie." [15] Gordon's Brigade was assigned the "post of honor" during the retreat, serving as the rear guard of Lee's army. Captain Mathews recalled, "On the sad retreat from Gettysburg the regiment under command of Captain John G. Rankin, Company D, was on the rear guard until the Potomac was crossed, and in all that sad retreat performed arduous and trying duties, being constantly on duty and under arms night and day." [16]

While the Georgia Brigade remained in the rear covering the retreat, the wagon trains carrying Sergeant Hudgins, Private Nash, and other wounded 38th soldiers, slowly moved far ahead of the main column. Lee ordered General Imboden and his cavalry brigade to escort the Confederate ambulance train back to Virginia. The ambulance train departed Gettysburg on July 4th, at four o'clock in the afternoon. The walking wounded trailed the column, which was 17 miles long. General Imboden recalled the hellish retreat of the massive train of the wounded,

"After dark I set out from Cashtown to gain the head of the column during the night. My orders had been peremptory that there should be no halt for any cause whatever. If an accident should happen to any vehicle, it was immediately to be put out of the road and abandoned. The column moved rapidly, considering the rough roads and the darkness, and from almost every wagon for many miles issued heart-rending wails of agony. For four hours I hurried forward on my way to the front, and in all that time I was never out of hearing of the groans and cries of the wounded and dying. Scarcely one in a hundred had received adequate surgical aid, owing to the demands on the hard-working surgeons from still worse cases that had to be left behind. Many of the wounded in the wagons had been without food for thirty-six hours. Their torn and bloody clothing, matted and hardened, was rasping the tender, inflamed, and still oozing wounds. Very few of the wagons had even a layer of straw in them, and all were without springs. The road was rough and rocky from the heavy washings of the preceding day. The jolting was enough to have killed strong men, if long exposed to it. From nearly every wagon as the teams trotted on, urged by whip and shout, came such cries and shrieks as these:

> *"'O God! Why can't I die!"*
> *" My God! Will no one have mercy and kill me!'*
> *"Stop! Oh! For God's sake, stop just for one minute; take me out and leave me to die on the roadside."*
> *"I am dying! I am dying! My poor wife, my dear children, what will become of you?"*

Some were simply moaning; some were praying, and others uttering the most fearful oaths and execrations that despair and agony could wring from them; while a majority, with a stoicism sustained by sublime devotion to the cause they fought for, endured without complaint unspeakable tortures, and even spoke words of cheer and comfort to their unhappy comrades of less will or more acute nerves. Occasionally a wagon would be passed from which only low, deep moans could be heard.

No help could be rendered to any of the sufferers. No heed could be given to any of their appeals. Mercy and duty to the many forbade the loss of a moment in the vain effort then and there to comply with the prayers of the few. On, we must move on. The storm continued, and the darkness was appalling. There was no time even to fill a canteen with water for a dying man; for, except the drivers and the guards, all were wounded and utterly helpless in that vast procession of misery. During this one night I realized more of the horrors of war than I had in all the two preceding years." [17]

Federal cavalry harassed these wagon trains and a member of the Tom Cobb Infantry remembered a keen loss to the regiment, "On our retreat we had lost, by capture, one of the favorite of the company and also of the regiment. This was Lieutenant A. W. Webb Robinson, who was captured in an ambulance, being very sick. He was taken to a Federal prison, where he languished until the close of the war.

A bigger soul never was encased in a small body than Webb Robinson's. As Commissary Clerk under Captain Quinn, he rendered many an extra favor to the Tom Cobb Infantry in the way of rations in straitened times. Such was his popularity with the company that they prevailed on him to give up his position and accept a Lieutenancy in the company, where he was soon captured as before stated." [18]

Gordon reported, "In the afternoon of July 5th, on the retreat from Gettysburg, my brigade, acting as rear guard, was pressed by the enemy near Fairfield, Pa. I was ordered by Major-General Early to hold him in check until the wagon and division trains could be moved forward. Detaching one regiment (the Twenty-sixth Georgia), I deployed it, and after a spirited skirmish succeeded in driving back the enemy's advance guard and in withdrawing this regiment through the woods, with the loss of 8 or 10 killed and wounded."[19]

Captain Tip Harrison, of the Georgia Brigade, wrote a detailed account of Colonel Atkinson's 26th defending the retreat, *"Returning from Gettysburg Pa. in July 1863, Gordon's brigade for a part of the time 'brought up the rear,' and it became necessary at times to turn upon the Federals who were following our column. Col Atkinson with his fighting regiment of Wiregrass boys, was sent back to face the somewhat exultant Yankees who were annoying us with their artillery from nearly every hilltop behind us, and with strong lines of skirmishers in advance. We had just entered a typical body of timber, rectangular in shape, covering possibly 50 acres of land. Through this large grove of heavy trees the road over which our men were traveling ran. Col Atkinson deployed his regiment across the road just before passing the body of woodland. The rest of our corps went on and left the 26th and its' colonel to do the proper thing-to check the enemy until our trains might gain time and get beyond the reach of the annoying Yankee shells. Col, Atkinson waited quietly in these woods until the advanced line of Federals came close up to where he had his men lying down waiting for them. With a yell (at his command) the regiment arose and charged the enemy (numbering more than twice his own command) and drove them pell-mell back through the woods and across the open space beyond. He halted his men at the edge of the grove and continued firing as long as there was a Yankee to be seen. Then quickly facing one half of the men to the left-the other to the right, they opened a gap in the center and turning back he filed both wings of his command to the rear, and quickly vacated the woods, and re-entering the road soon overtook the brigade. The Yankees, supposing from the noise the 26th made and the execution they did, that this big body of woods 'was just swarmed with rebels,' hurried up to the support of their retreating line several pieces of artillery and a body of infantry. The guns were run into battery and for a half hour they vigorously shelled the woods, while the Confederates were leisurely retreating over the hills three or four miles from the scene. We never heard or saw the Yankee column after this well planned and brilliantly executed movement. The only damage done by this furious bombardment was to the beautiful trees in this park-like body of woods, which doubtless had been tenderly preserved for years by the thrifty Pennsylvania landlords in the neighborhood."* [20]

Sergeant Hudgins endured an agonizing ride in a wagon back to the Potomac River wrote, *"At Hagerstown, Md., the Federal cavalry shelled our ambulance train. We turned back and met Hood's Division, with the 10th Georgia in front. They soon brushed the cavalry out of the way, and we proceeded on our way to Williamsport, where Nash and I crossed the Potomac and were taken on to Winchester, where Nash's hand was amputated, and I was sent to Richmond, arriving on the 20th of July."* [21]

The Confederate army faced destruction along the banks of the Potomac River. The pontoon bridge over the Potomac River near Williamsport, Maryland, had been destroyed by Federal cavalry to slow Lee's passage back to Virginia. The river was swollen by rains and impossible to ford. The Army of Northern Virginia was trapped with its' back to the river, but immediately began constructing massive field fortifications on July 11th while awaiting the arrival of the Federal Army. Meade's army arrived at Williamsport on July 12th and began probing Lee's lines. Meade prepared for an attack on July 13th and very heavy skirmishing occurred along the lines as he arranged the Union army for attack. By this time, the river had dropped precipitously and the Confederates had completed a new bridge across the Potomac. They began crossing that night and into the early morning hours. The narrow pontoon bridge was far too small to quickly move Lee's entire army across the river and was mainly reserved for wagons and artillery. Most of the infantry waded across the river. Gordon vividly recalled crossing the Potomac with his brigade, he wrote:

"The fording of the Potomac in the dim starlight of that 13th of July night, and early morning of the 14th, was a spectacular phase of war so quaint and impressive as to leave itself lastingly daguerreotyped on the memory. To the giants in the army the passage was comparatively easy, but the short-legged soldiers were a source of anxiety to the officers and of constant amusement to their long-legged comrades. With their knapsacks high up on their shoulders, their cartridge-boxes above the knapsacks, and their guns lifted still higher to keep them dry, these little heroes of the army battled with the current from shore to shore. Borne down below the line of march by the swiftly rolling water, slipping and sliding in the mud and slime, and stumbling over the boulders at the bottom, the marvel is that none were

drowned. The irrepressible spirit for fun-making, for jests and good-natured gibes, was not wanting to add to the grotesque character of the passage.

Let the reader imagine himself, if he can, struggling to hold his feet under him, with the water up to his armpits, and some tall, stalwart man just behind him shouting, 'Pull ahead, Johnny; General Meade will help you along directly by turning loose a battery of Parrott guns on you.' Or another, in his front, calling to him: 'Run here, little boy, and get on my back, and I'll carry you over safely.' Or still another, with mock solemnity, proposing to change the name of the corps to 'Lee's Waders,' and this answered by a counter-proposition to petition the Secretary of War to imitate old Frederick the Great and organize a corps of 'Six-footers' to do this sort of work for the whole army. Or still another offering congratulations on this opportunity for being washed, 'The first we have had, boys, for weeks, and General Lee knows we need it.'" [22]

Once the Georgia Brigade reached the opposite shore, they were back on Virginian soil. Just four days after the 38th returned to Virginia, the regiment camped near Darksville, in Berkeley County, Virginia (now West Virginia.) On July 18th a group of young officers sat down to write a letter of condolence to the mother of Sergeant Francis McCleskey, who was killed in action during battle at Gettysburg on July 1st.

Camp of 38th Georgia Regt
Near Darksville, Va.
July 18th, 1863

Mrs. Angeline McCleskey,

Dear Madame, As you are doubtless aware before this time it has again become our sad and painful duty to record the Death of another member of your family who was a brother in arms with us. At about 4 o'clock on the evening of the 1st of July inst. (after several days hard marching) we were again called on to participate in another terrible and bloody conflict at Gettysburg, Pa. in this fierce struggle. Frances was called on to lay his life's blood on his Country's altar which he did nobly and bravely. He was honorably discharging his duty when the fatal ball pierced his brain. he was shot in the head and instantly killed by a Minié ball while bravely driving the enemy before him.

While we are called on to mourn his absence and death we mourn not as those who have no hope. We are cheered by the hope that his noble and magnanimous spirit is with that of his Dear Father bask in sunshine and happiness around the eternal throne of God in Heaven where no wars, no strife, and where evenings never come, and where the poor, tired care worn, scared, and sunburned Soldier may forever rest from his labors and where the sound of the war drum, the sharp clash of Rifles, and sullen roar of Artillery is heard no more throughout great Eternity.

Frances was always very kind and at all times obliging to all his companions and in fact to all who came in his way. No one has ever fell that has cast such deep and lasting gloom and sorrow on all the members of this company since the Death of his Father (Francis's father was Captain George W. McCleskey, company commander of Company B, mortally wounded at Gaines Mill) as the death of Frances has caused. he was as beloved and caressed by all who knew him. The Boys have often said that they knew but little difference between their love for him and their own Brothers. he was ever in a good humor with everything and everybody. he possessed quite an agreeable and accommodating disposition. was always attentive to duty never complaining but cheerfully performing whatsoever fell to his lot and his duties were very heavy yet he was fully competent for the task. he hadn't an Enemy in Camp as far as we knew. he was a friend to all high minded honorable Men. And all such were friends to him. none speak of Frances but to praise and do him honor.

But alas he is gone from us to return no more. yet we are consoled by the thought that though he can not come to us again we can (by the help of the great and All wise Disposer of Events) go to him. let this thought console and comfort you in your afflictions. We know that your loss has been very heavy and that your afflictions have been very severe, but he that has taken from you a Dear Husband and two dutiful sons has promised that our trials shall not be above that which we are able to bear We feel and acknowledge that our capacities are too limited to give to Frances that full and rich need of praise which is due him.

We humbly ask of you that we may be allowed to share your grief. But at the same time allow us to offer words of consolation and comfort. We feel and know that it is very hard to give up Dear Friends and loved ones, but we know that it is written, that it is appointed unto all men once to die. And we try to exclaim from the depths of our souls "Thy will O Lord, not ours be done in us in all things and at all times." Will you all pray for us that are left that we may be so imbued with a spirit of Grace that under all circumstances we may be fully prepared to meet death at any time. It now becomes our duty to say something concerning Frances effects. he had on his person when he died $18.25 cents and fifty cents in stamps but they were wet and spoiled. he also had one small note of $10.00 on one of the Boys here and his Blank Book & Pocket knife which we will send home by the first reliable person passing that way. Some of the Boys were owing him small amounts and he owing some small amounts to some of them. If you see proper you can write to some one here to settle those small amounts for you and it will be done. If not we will send all to you and you can make such disposition of them as you choose. There is due him by the Government 2 months wages which at $20 pr. Month will make $40.00. In order that you may get it you will be compeled to authorise someone by power of Attorney to collect it for you at the Second Auditors Office in Richmond. We forgot to state in the proper place that we buried Frances as well as the circumstances would permit us. he was buried on the field near by where he was killed. Hoping this may prove satisfactory and prove a source of comfort rather than anguish.[23]

We remain your Most Obedient Servants

Lieutenant Kinchen R. Cross
Lieutenant Simpson A. Hagood"
Lieutenant Augustus C. Bell
Sergeant Thomas B. Newton

To: Mrs. Angeline McCleskey
Freemansville, Georgia

Sergeant Hudgins of Company K miraculously survived his severe wound and the tortuous journey in a wagon back to Virginia, he said *"I was sent to Richmond, arriving on the 20th of July. I stayed at Richmond until the 12th of August and reported to General Gordon for duty at Orange Courthouse. I ask you the reader to think of being shot through the body, hauled out of Pennsylvania across Maryland, West Virginia, and then to Richmond, and getting well in forty-two days."* [24]

Killed or Morally Wounded at Gettysburg, July 1st - 3rd, 1863

Company A – "The Murphy Guards," DeKalb County, Georgia

1. 2nd Corporal David A. Green - Killed at Gettysburg, Pa., 7/1/1863.

Company B, "The Milton Guards," Milton County, Georgia (Currently Fulton County)

2. Private Allen Bayless Day - Killed at Gettysburg, Pa., 7/1/1863.
3. 1st Sergeant Francis C. McCleskey - Killed at Gettysburg, Pa., 7/1/1863.
4. Captain William L. McLeod - Killed at Gettysburg, Pa., 7/1/1863.
5. Private David Daniel Douglas - Wounded and captured at Gettysburg, Pa. 7/2/1863. Admitted to Hammond U.S.A. General Hospital at Point Lookout, Md. 1/12/1864. Died from gunshot fracture of right leg, 3/10/1864.

Company C, "The Ben Hill Guards," Bulloch & Emanuel Counties, Georgia

6. Private Benjamin H. Boyd - Wounded at Gettysburg, Pa., 7/3/1863. Captured at Boonesboro, Md. 7/5/1863. Died of wounds in General Hospital at Frederick, Md. 7/28/1863.

Company E, "The Tom Cobb Infantry," Oglethorpe County, Georgia

7. Private Richard H. H. Boggs - Wounded in lungs at Gettysburg, Pa., 7/1/1863 and died there of wounds 7/20/1863.
8. Private John M. Jackson - Wounded and left at Gettysburg, Pa., 7/2/1863. Died of pneumonia in Chambersburg, Pa., hospital 7/8/1863.

Company F, "Thornton's Line Volunteers," Hart and Elbert Counties, Georgia

9. Private Eastin L. Fortson - Killed at Gettysburg, Pa., 7/1/1863.
10. 5th Corporal William Joseph Powell - Killed at Gettysburg, Pa., 7/1/1863.

11. Private William J. Shiflet - Killed at Gettysburg, Pa., 7/1/1863.

Company G, "The Battey Guards," Jefferson County, Georgia

12. Private James A. Jordan - Killed at Gettysburg, Pa., 7/1/1863.

13. Private John B. Willoughby - Wounded in abdomen at Gettysburg, Pa., 7/2/1863 and died there of wounds in General Hospital 8/3/1863.

Company H, "The Goshen Blues," Elbert County, Georgia

14. 1st Lieutenant John Oglesby Jr. - Mortally wounded, 7/1/1863, at Gettysburg and died the next day at Gettysburg, Pa.

15. Private Jacob E. Sanders - Severely wounded in abdomen at Gettysburg, Pa., 7/1/1863 and died there from wounds, 7/6/1863.

16. Private Alexander Vaughan, (or Vaughn) - Wounded at Gettysburg, Pa.. 7/1/1863 and died there of wounds, 7/17/1863.

Company K, "DeKalb & Fulton Bartow Avengers," DeKalb and Fulton County, Georgia

17. 4th Corporal James E. Ball - Severely wounded in chest and captured at Gettysburg, Pa., 7/2/1863. Died of wounds, in hospital at, or near, Gettysburg, Pa., 7/10/1863.

18. 2nd Lieutenant William Franklin Goodwin - Killed at Gettysburg, Pa. 7/2/1863.

19. Private J. H. Wilson. - Killed at Gettysburg, Pa., 7/2/1863.

Chapter 12
The Battle of Wilderness, Day One – May 5th, 1864

Lieutenant William Mathews wrote of the events following the battle of Gettysburg, *"After the regiment returned to Virginia they camped a while at Madison Court House and then near Brandy Station, on the Rappahannock. In the winter of 1863-64 our quarters were on Clark's Mountain, near Orange Court House and many of the regiment received furloughs."* [1]

Gordon recalled the winter of 1863-64, spent along the banks of the Rapidan in Virginia:

"My camp and quarters were near Clark's Mountain, from the top of which General Lee so often surveyed with his glasses the white-tented city of the Union army spread out before us on the undulating plain below. A more peaceful scene could scarcely be conceived than that which broke upon our view day after day as the rays of the morning sun fell upon the quiet, wide-spreading Union camp, with its thousands of smoke columns rising like miniature geysers, its fluttering flags marking, at regular intervals, the different divisions, its stillness unbroken save by an occasional drum-beat and the clear ringing notes of bugles sounding the familiar calls. On the southern side of the Rapidan the scenes were, if possible, still less warlike. In every Confederate camp chaplains and visiting ministers erected religious altars, around which the ragged soldiers knelt and worshipped the Heavenly Father into whose keeping they committed themselves and their cause, and through whose all-wise guidance they expected ultimate victory. The religious revivals that ensued form a most remarkable and impressive chapter of war history. Not only on the Sabbath day, but during the week, night after night for long periods, these services continued, increasing in attendance and interest until they brought under religious influence the great body of the army. Along the mountain-sides and in the forests, where the Southern camps were pitched, the rocks and woods rang with appeals for holiness and consecration, with praises for past mercies and earnest prayers for future protection and deliverance. Thousands of these brave followers of Southern banners became consistent and devoted soldiers of the cross. General Lee, who was a deeply pious man, manifested a constant and profound interest in the progress of this religious work among his soldiers. He usually attended his own church when services were held there, but his interest was confined to no particular denomination. He encouraged all and helped all." [2]

A soldier of the 38th, know only by the initials J. C. C., sat down two months after the battle of Gettysburg and wrote a letter to the Augusta Chronicle Newspaper. He talked of the great religious revival that swept through the Army of Northern Virginia and brigade inspections. He also wrote of what he clearly foresaw as the likely battle sites of the coming Spring and the indomitable spirit of the men of the 38th Georgia:

Camp Gordon's Brigade, Thirty-Eighth Georgia Regiment, August 28th, 1863

Knowing that a brief synopsis of passing events of this vicinity will not be unacceptable to the numerous readers of your valuable journal, while familiar faces occasionally visits regimental camps and is as heartily welcomed by the patriot from Georgia, I have concluded to write you.

A Christian spirit pervades the brigade to an unusual degree. The religious revivals are attended with flattering results - conversions and baptisms being of daily occurrence. Nightly meetings are held in the respective regimental camp grounds and these contribute much towards the good work. Long may it continue. What a pleasing contrast is there presented between Federal and Confederate soldiers. With one a foreign mercenary on an errand of plunder, with no higher idea of liberty that the excuse to satiate a vitiated appetite, and gratify his corrupt, lustful nature. The other, called to the field by a sense of honor and the higher obligation of duty, leaves all he holds dear and bares his bosom to the dangers of the battle field willingly, rather than see his home despoiled by a brutal foe without honor or mercy.

We have learned, with feelings of mortification there are those in Georgia who despair of the success of our cause. Shame on all such. So low a weakness can emanate only from a cowardly nature. We, who have fought at the torn and bloody fields of Cold Harbor, Malvern Hill, Second Manassas, Second Fairfax, Sharpsburg, Fredericksburg, Mary's Heights, Winchester, Gettysburg, and other minor engagements, and added to the galaxy of glorious names, Douglass, McClesky, Battey, Lawton, Hawkins, Brenan and hosts of others too numerous to mention, are understanding that our efforts should be fruitless, and the blood of our heroes be spilt in vain. When first the tocsin of war vibrated through the borders of our young republic, we rallied behind, the banners that we have so often borne to victory, and beneath whose tattered folds we still intend to battle until that success which is inevitable, crowns our efforts with peace and liberty. We are ever ready and faithful to the last," and rather than purchase peace at the price of bondage and chains, we are willing to make our beautiful land the of the grave of liberty. I am glad to be able to say that the call of the President has been heeded. The roads leading to the different brigade encampments are daily thronged with convalescents returning to duty. Politics are not discussed in the army. The general opinion among the men is that Gov Brown will make as good a Governor as anybody else. I doubt if sufficient interest is felt in the result to cause a ballot in camps.

We have daily drills and semi-weekly inspections. We are soon to be inspected by our glorious old chief, Robt. E. Lee, and all is astir in preparation. Fourth Georgia brigade is Excelsior. They acknowledge no superious in drill, disipline or equipment. The chivalrous soldier and christian gentlemen John B. Gordon, who is our commander, is beloved by all as a friend, confided in as a father, and respected as an officer. He is one among the few who can of whom it can be truthfully written, "He owed his promotion to merit alone," for his scarred face bears the evidence of the carnage of the battle field of Sharpsburg. His generous nature and affable deportment has won for him hosts of admirers.

There are some speculations in our camps concerning the furloughing of officers and men, it is a object upon which there is much difference of opinion. If your correspondent is allowed to express his views in the matter, it is briefly this: Both armies are recruiting and marshalling their hosts of another conflict, which will occur between Culpeper and Fredericksburg, or at one of the places named; for I cannot beleive Gen. Lee will attack them in their current entrenchments. We can better afford to await the advance of Meade than do this.

Never was an army more bouyant in spirit or under better discipline than ours. The commsuriat is well supplied. We are also being furnished with an excellent stock of clothing and shoes. The only trouble we now have to encounter, is want of forage for army horses. The railroad transportation is insufficient. Something must be done immediately in this connection or great losses will ensue to the Government during the winter. The poor stock cannot weather, as exposed as the necessarily will be to the rigors of this climate. I hope this matter will meet with consideration due it from the proper authorities. J.C.C.[3]

Research revealed the probably identity of the writer, known only as J.C.C., is likely Sergeant Major John Cottle Chew, of Company C, the Ben Hill Guards. John C. Chew was a very well educated man and was later elected as a Legislator for Burke County, Georgia, in the 1880's. This letter was compared to post war writings of John C. Chew and is remarkably similar to his prose writing style.[4]

Another letter to the editor was submitted to the Augusta Chronicle at the same time, purporting to be a member of Gordon's Brigade, known only as "Sergeant." This unidentified soldier is possibly another member of the 38th Georgia Regiment. The writer tells of brigade inspections and the abundance of food staples, but notes the scarceness of vegetables.

Gordon's Brigade, Va., August 28th, 1863

We are yet at the same camps we have been for some time past. Everything about it looks neat and clean Our arms are in splendid condition bright as a silver dollar. We are expecting an inspection of our arms every day, and every one is trying to see if he cannot have the best gun when he goes on inspection.

Our Brigade was reviewed yesterday evening by Gen. Gordon, our Brigadier. Those who witness it say it was a sight worth looking at. There is very litle sickness in our camp. The Yankees, I suppose have fallen back across the Rappahannock River. From appearances one would judge that they are very anxious to engage with us again. We hear every day of thosands of them deserting.

We have been picket at Rapidan Station for about two weeks. The citizens living in that neighborhood treated us very kind during our stay. Everything is very quiet here at this time. It is time we were having some rest, for we have been marching and fighting nearly all this summer.

We have been lately amply supplied with clothing and shoes by our Quartermasters. Provisions are very high and scarce here, but we get a plenty of beef, bacon, flour, and corn meal. We do not get any vegetables. Apples, peaches and other fruits are scarce. Watermelons sell from four to six dollar each. The weather is getting tolerable cool and we will soon need blankets. If we get into a fight with the Yankees, those of us who get out safe will have a chance to supply ourselves of what we want.

Sergeant. [5]

Lieutenant Simpson A. Hagood, of Company B, the Milton Guards, wrote a letter to his wife Adaline, on Sunday, March 20th, 1864. He told of being on picket duty and being called to the residence of a local family whose eighteen year old son had died of consumption.

Camp of 38th Geo Regt Vols, Near Orange CH Va, March 20 1864

Dear Adaline,

I write you this beautiful Sunday morning that you may understand that I am alive and well yet and getting plenty to eat.

*Your letter of the 11th inst was at Camp waiting for me yesterday evening when I returned from Picket where I had been for three days. The morning we went on Picket I went to the Commander of the Picket line and obtained leave to stay at Camp and not go on Picket. I had not been to my Tent more than five minutes when our Colonel came down to the Tent and assigned me to the Adjutant of Co. "A" of our Regiment for the three days ***[6] Picket *** The people here say that this has been the dryest warmest winter and Spring that they have ever saw. they seem to think that this will not be a very good Crop year on account of the dry warm winter. They think we will have a great deal of hot weather in the summer and fall. Be that as it may it will have but little to do with the movement of the Army after it leaves winter quarters. The Army never stops for rain in Summer unless it wants to cross a River that is swollen so that the men cant ford it and not then if the Pontoon bridges are at hand. I still think this army will have no fighting of any consequence to do until the Northern Army has done something and if Johnston drives the Yankees back or they are driven back by Longstreet, Polk, Smith *** and Picket and Early which I believe will be and they are going to fight before we do. Then I dont think we will have much fighting to do at this point.*

The Yankee Army at this place have nearly all (comparatively speaking) been sent to those other Armies.

If you have not already heard you will doubtless soon hear great windy tales by disaffected persons of the States of Louisiana and Arkansas having by the vote of a popular majority of the people gone back into the "Union". Pay no attention to those tales. Dont for a moment believe them for the people of these States are as far from retracing their steps as the people of Virginia or Georgia. They are as loyal to the South as they ever were and are as fully determined to fight the War to a successful issue as any people can.

*The reason why I caution you all about these reports is this. I see through the papers abstracts from Northern papers stating that those two States had gone back and had elected Officers from Governor *** to Magistrates of Union Men and* *was frameing *** for giving into the *** held in that portions of those *** Yankees hold and the Ballot Box *** lay Yankee Soldiers and those only *** who would cast their suffrage as the United States Officers were pleased to dictate. Consequently, the Citizens didn't vote in these elections. The soldiers elected those black hearted traitors. The consequence is that those good and Loyal people will have to endure anything that Yankee hatred and Yankee cupidity is pleased to heap upon them. You may all rest assured that their brave and Lion like hearts arent changed or crushed but for the time being is awed into silence by the weight and presence of Superior numbers of vandals. These elections were carried at the point of the Bayonet like they carried the elections last fall in Kentucky.*

There never were better or more brave soldiers than the Louisianians and Arkansians of this Army and they say all the letters they can get through the lines about them to hold on. The hearts of the people at home are with them but they are kept from helping them because of being cut off from them by the Yankee army along the Mississippi River. Believe not their Big tales about reunion.

Last Monday evening I received a note from a Lady living about two miles from Camp where one of our boys are doing guard duty for her. her name is Crawford she had a son 18 years old who died of Consumption on Sunday night. The note she sent me was an invitation to come and stay with them that night and sit up with the Corps. Henry Wayne (who is on guard for her) had recommended me to her. I went found them all in deep grief. I sat up with the Corps with three others until 2 O'Clock in the morning when we went to bed and slept until breakfast. I arose washed and ate then helped to place the young man in the coffin and Hearse. They started to the burying ground with him. I received a very pressing invitation from the Old Lady better to accompany them to the burial ground but my leave of absence would not let me (now buryings or funerals in this Country only this was by special invitation). I was very sorry for the family which consists of a Widow Lady and a daughter of about 18 years. she also had one son in the Army.

When the Yankees were in possession of her home last Spring there was a large Camp near by and they destroyed nearly all her property. she professed to be very thankful for the visit and when I left she pressed me warmly by the hand and asked that I call and see them again. she also asked that God would bless me with life, health and prosperity through the War. Her daughter brought her Album to me and asked that I would place some Verse in it before I left. This rather stunned me for I knew I was no Poet and moreover I couldn't bear the idea of borrowing a verse from some others work. I came very near declining to but thought it would never do to back out. I began to call my little store of knowledge for something that would be appropriate. The old Lady had told me the night before of the manner in which her son died. I made his last words my theme:

A Brother Dear has passed away
From time and all sublunary things
He has joined the Hosts of endless day
In Heaven he rests and sweetly sings.

He can never return again
To Earth and dull mortality
Your great loss is his eternal gain
He lives in bright felicity.

Remember his last parting words
Sister Dear in Heaven meet me
Always Love and serve the Lord
Christ will surely sweetly guide thee.

We shall there no more be ___
But in everlasting happiness
The sorrowing sighing broken heart
Spend Eternity in rapturous bliss

Then Virginia dry those briny tears
Let comfort take the place of grief
Religion will dispel all your cares
And bring instant joyous relief

The young lady declared the verses beautiful and showered a thousand thanks on me for it. It is dinner time. I must close for the present. I will write again in a few days. Baylis Bruces Wife arrived in Camp a few days ago on a visit. It looks like all the wives of our Men are coming. Tom Newton is now gone to meet his wife at Orange CH. If we dont all get home next fall I want you to come to see me as soon as we get into winter quarters again but it is too late in the season now for you to come. when I direct a letter to you it is my desire that you break open the letter with your own fingers. my letters are generally to all but I want you to break them yourself. myself and Levi is well.

Same as ever.

SA Hagood[7]

Lieutenant William C. Mathews noted the reorganization of the 38th Georgia during the winter of 1863-64, *"During this winter the following changes occurred in the organization: Captain Phillip E. Davant of the 3rd Georgia Battalion was elected Lieutenant Colonel. Captain Bomar of Company M, whose company joined us here, was made Major. Lieutenant Dourough was made Captain of Company E; Lieutenant Deadwyler was made Captain of Company H; Lieutenant McAfee of Company I (The Dawson Farmers) -formerly Company N; Lieutenant John C. Hendrix of Company M, (Chestatee Artillery, now serving as infantry in the 38th Georgia) was made Captain of that company; Lieutenant J. M. Brittian, Company E, having been made a chaplian vice Captain G. W. Mashburn resigned; Private A. W. Robinson was elected Lieutenant of the same company, and Private Ben Morris was elected Lieutenant of Company C and E. L, Law was appointed Adjutant. The 38th Georgia was in a few engagements, at Bristow and Mine Run, on the Rapidan. In theses engagements the regiment lost about three men killed and eight or nine wounded."*[8]

The men of the 38th and the Georgia Brigade knew the coming of Spring meant hostilities would soon resume. Lieutenant Lester reported, "About the first of May it became evident something was going to be done. As is well known, General Grant was now placed in command of the Northern army and began to maneuver to cross the Rapidan."[9]

Sergeant Hudgins wrote, *"On the 4th of May, 1864, the 38th Georgia Regiment was on Clark's Mountain, opposite Summerville and Robinson's Ford's, on the Rapidan River, where we had been picketing the previous winter. General U. S. Grant had succeeded to the command of the Federal army and had personally assumed command of the Army of the Potomac, with the intention of crossing the Rapidan River and crushing the Confederate army under General R. E. Lee in a single blow. Grant began crossing his army at Ely's and Germanna Fords on the 4th; and General Ewell put his corps, the 2nd, or 'Stonewall's,' in motion, going by the way of Locust Grove down the old turnpike, to the Wilderness, Johnson and Rhodes and Downs leading, Early's Division of that corps bringing up the rear. We camped that night on the old turnpike east of Locust Grove."* [10]

General Gordon noted the events leading to the battle, *"Lee, in the meantime, was hurrying his columns along the narrow roads and throwing out skirmish-lines, backed by such troops as he could bring forward quickly in order to check Grant's advance and to ascertain whether the heaviest assault was to be made upon the Confederate centre or upon the right or left flank. Field-glasses and scouts and cavalry were equally and almost wholly useless in that dense woodland. The tangle of underbrush and curtain of green leaves enabled General Grant to concentrate his forces at any point, while their movements were entirely concealed. Overlapping the Confederate lines on both flanks, he lost no time in pushing to the front with characteristic vigor "A more beautiful day never dawned on Clark's Mountain and the valley of the Rapidan than May 5, 1864. There was not a cloud in the sky, and the broad expanse of meadow-lands on the north side of the little river and the steep wooded hills on the other seemed 'appareled in celestial light' as the sun rose upon them. At an early hour, however, the enchantment of the scene was rudely broken by bugles and kettledrums calling Lee's veterans to strike tents and 'fall into line. My command (The Georgia Brigade) brought up the*

Gen. Grant USA - 1864
Library of Congress

rear of the extreme left of Lee's line, which was led by Ewell's corps. Long before I reached the point of collision, the steady roll of small arms left no doubt as to the character of the conflict in our front. Dispatching staff officers to the rear to close up the ranks in compact column, so as to be ready for any emergency, we hurried with quickened step toward the point of heaviest fighting. Alternate confidence and apprehension were awakened as the shouts of one army or the other reached our ears. So distinct in character were these shouts that they were easily discernible. At one point the weird Confederate 'yell' told us plainly that Ewell's men were advancing. At another the huzzas, in mighty concert, of the Union troops warned us that they had repelled the Confederate charge; and as these ominous huzzas grew in volume we know that Grant's lines were moving forward. Just as the head of my column came within range of the whizzing Miniés, the Confederate yells grew fainter, and at last ceased; and the Union shout rose above the din of battle." [11]

Gordon's brigade awoke early on the morning of May 5th, Sergeant Hudgins wrote, "We resumed our march in the direction of the Wilderness, and quite soon we heard the rattle of musketry, and we knew the great battle had begun. Warren, commanding the 5th Army Corps, had assailed Johnson and Rhodes furiously; and when Gordon's Brigade, of Early's Division, arrived on the field, the ordnance and ambulance trains were all moving to the rear."

General Gordon continued, *"I was already prepared by this infallible admonition for the sight of Ewell's shattered forces retreating in disorder. The oft-repeated but spasmodic efforts of first one army and then the other to break through the opposing ranks had at last been ended by the sudden rush of Grant's compact veterans from the dense cover in such numbers that Ewell's attenuated lines were driven in confusion to the rear. These retreating divisions, like broken and receding waves, rolled back against the head of my column while we were still rapidly advancing along the narrow road. The repulse had been so sudden and the confusion so great that practically no resistance was now being made to the Union advance; and the elated Federals were so near me that little time was left to bring my men from column into line in order to resist the movement or repel it by countercharge. At this moment of dire extremity I saw General Ewell, who was still a superb horseman, notwithstanding the loss of his leg, riding in furious gallop toward me, his thoroughbred charger bounding like a deer through the dense underbrush. With a quick jerk of his bridle-rein just as his wooden leg was about to come into unwelcome collision with my knee, he checked his horse and rapped out his few words with characteristic impetuosity. He did not stop to explain the situation; there was no need of explanation. The disalignment, the confusion, the rapid retreat of our troops, and the raining of Union bullets as they whizzed and rattled through the scrub-oaks and pines, rendered explanations superfluous, even had there been time to make them."* [12]

Lieutenant George Lester of the Tom Cobb Infantry recalled the moment, "The battle was opened by an attack on Jones' Va. brigade. The Yankees were met with a galling fire but pressed on and drove General Jones back, thus making a gap in the line. General John B. Gordon's (our) brigade was ordered to his relief." [13]

Sergeant Hudgins watched Gordon as the brigade prepared to form a line of battle. They encountered Rhodes Brigade and Gordon's old regiment, the 6th Alabama Regiment, "We had now reached Battle's Brigade, of Rhodes's Division and General Gordon said: "Steady, 6th Alabama!" Those instantly recognized their former captain and colonel and instantly replied: "We will." Warren was making his main attack, and those gallant Alabamians were holding their own." [14]

Ewell greeted Gordon as he rode at the head of the George Brigade, *"The rapid words he did utter were electric and charged with tremendous significance. 'General Gordon, the fate of the day depends on you, sir,' he said. 'These men will save it, sir.' Quickly wheeling a single regiment into line, I ordered it forward in a countercharge, while I hurried the other troops into position. The sheer audacity and dash of that regimental charge checked, as I had hoped it would, the Union advance for a few moments, giving me the essential time to throw the other troops across the Union front. Some of my men were killed and wounded before the first regiment was placed in position."* [15]

As the Georgia Brigade was forming a line of battle, the 38th was at the rear of the brigade and was the last to file into line. Gordon then ordered: "Forward into line on the right; right oblique and load as you march!" Gordon rode in front of the brigade and began steeling the men's nerves for battle. Private Bradwell, 31st Georgia watched as, *"Gordon rode along the front of the entire brigade, seeming as one inspired with burning word of eloquence. With hat in hand he passed along, his face radiant as he spoke to his men in these words, 'Soldiers, we have always driven the enemy before us, but this day we are going to scatter them like the leaves of the forest..,' he raised the fighting spirit in his men to the highest pitch."* [16]

Private Nichols of the 60th also witnessed the event, *"General Gordon turned to us with a deep determined look - to move them or die. General Gordon addressed us in about these words: 'Boys, there are Yankees in front and lots of them, and they must be moved or the day is lost, and we must move them. Now all who are faint hearted, fall out, you shall not be hurt for it; for we do not want any to go but heroes – we want brave Georgians.' The brigade cheered loudly and prepared to charge. Gordon rode to the center of the brigade's line of battle and gave the command "Forward!"*[17]

"Just as we were ordered forward, Irvin Spivey, of the Twenty-sixth Georgia Regiment, hallooed. He could halloo the queerest that I ever heard any one. It was a kind of a scream or low, like a terrible bull, with a kind of a neigh mixed along with it, and it was nearly as loud as a steam whistle. We called him "The Twenty-sixth Georgia's Bull," and the Yankees called him "Gordon's Bull." He would always halloo this way when we charged the enemy, and we were informed that the Yankees understood it as a signal for them to move back," wrote Private George Nichols. [18]

Gordon's command of "Forward!" fell, "With a deafening yell which must have been heard miles away, that glorious brigade rushed upon the hitherto advancing enemy, and by the shock of their furious onset shattered into fragments all that portion of the compact Union line which confronted my troops." [19]

The Georgia Brigade met their old antagonist, the famous "Iron Brigade" of the Union Army. The Georgia Brigade and 38th fought the "Iron Brigade" at 2nd Manassas and Antietam, but for the first time in the Iron Brigade's history, their ranks were shattered and they were swept from the field, with the Georgians close on their heels. Gordon told of the unique situation facing the Georgia Brigade, due to the success of the attack, *"At that moment was presented one of the strangest conditions ever witnessed upon a battle-field. My command covered only a small portion of the long lines in blue, and not a single regiment of those stalwart Federals yielded except those which had been struck by the Southern advance. On both sides of the swath cut by this sweep of the Confederate scythe, the steady veterans of Grant were unshaken and still poured their incessant volleys into the retreating Confederate ranks. My command had cut its way through the Union centre, and at that moment it was in the remarkably strange position of being on identically the same general line with the enemy, the Confederates facing in one direction, the Federals in the other. Looking down that line from Grant's right toward his left, there would first have been seen a long stretch of blue uniforms, then a short stretch of gray, then another still longer of blue, in one continuous line. The situation was both unique and alarming. I know of no case like it in military*

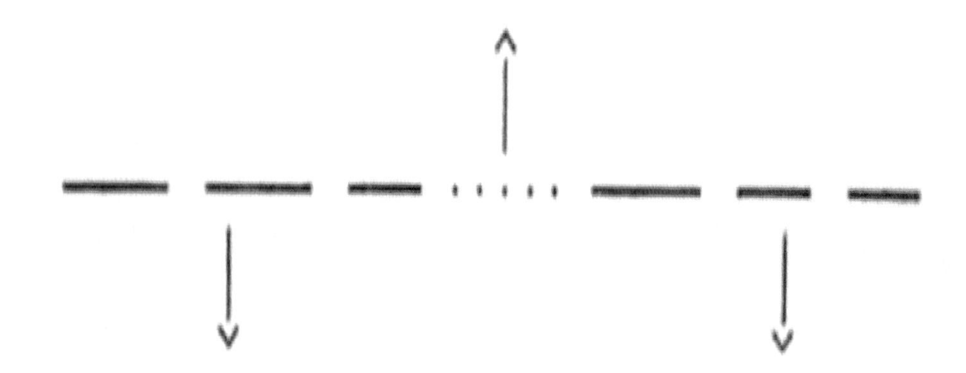

history; nor has there come to my knowledge from military text-books or the accounts of the world's battles any precedent for the movement which extricated my command from its perilous environment and changed the threatened capture or annihilation of my troops into victory. The solid and dotted portions of the line, here given, correctly represent the position of my troops in relation to the Federals at this particular juncture: the Union forces are indicated by a solid line, the Confederates (my command) by a dotted line, and the arrows indicate the direction in which the forces were facing.

The advance was made with such spirit that the enemy was broken and scattered along the front of my brigade, but still held his ground or continued his advance on my right and left. It will be seen that further movement to Grant's rear was not to be considered; for his unbroken lines on each side of me would promptly close up the gap which my men had cut through his centre, thus rendering the capture of my entire command inevitable. To attempt to retire by the route by which we had advanced was almost, if not equally, as hazardous; for those same unbroken and now unopposed ranks on each side of me, as soon as such retrograde movement began, would instantly rush from both directions upon my retreating command and quickly crush it. In such a crisis, when moments count for hours, when the fate of a command hangs upon instantaneous decision, the responsibility of the commander is almost overwhelming; but the very extremity of the danger electrifies his brain to abnormal activity.

In such peril he does more thinking in one second than he would ordinarily do in a day. No man ever realized more fully than I did at that dreadful moment the truth of the adage: 'Necessity is the mother of invention.' As soon as my troops had broken through the Union ranks, I directed my staff to halt the command; and before the Union veterans could recover from the shock, my regiments were moving at double-quick from the centre into file right and left, thus placing them in two parallel lines, back to back, in a position at a right angle to the one held a moment before. I left a thin line (Thirty-first and Thirty-eighth Georgia Regiments) to protect my front, and changed front to the right with three regiments (Thirteenth, Sixtieth, and Sixty-first Georgia), and moved directly upon the flank of the line on my right... At the same time I caused the regiment on the left (Twenty-sixth Georgia) to make a similar movement to the left, which was also successful. [20]

This quickly executed maneuver placed one half of my command squarely upon the right flank of one portion of the enemy's unbroken line, and the other half facing in exactly the opposite direction, squarely upon the left flank of the enemy's line. This position is correctly represented by the solid (Federal) and dotted (Confederate) lines here shown.

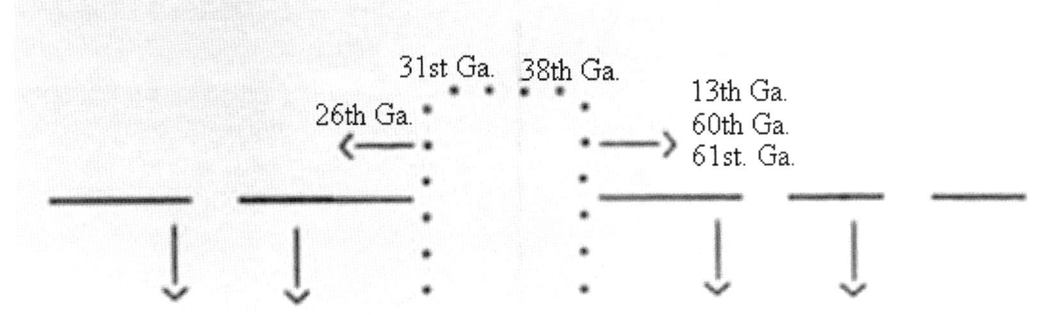

By this time portions of Battle's brigade rallied, and with other troops of Rhodes' division came forward and assisted in driving the enemy back and establishing the line, which was afterward held. [21]

Private Nichols of the 61st described the attack and amazing capture of an entire Union regiment, *"We forwarded and soon struck the Yankees. They began to fire at us and we at them. I never heard such a yell as we raised. We could scarcely hear a gun fire, and could hardly tell when our own guns fired, only by the jar it gave us. We soon routed the first Yankee line. We all pushed right on and on with the yell, until we had driven the first line into the reserve line. The two lines did not stand but one or two volleys before both began to waver and retreat in confusion. We soon had them into the third line, and on into the fourth, and on until we seemed to have five or six lines in one confused mass, with many of them lying down and surrendering-, or coming back with their hands up to show that they were surrendered. We would send them to the rear. Our officers could hardly get a man to go to the rear with them. We killed a great many of them, and drove them off of Lee's position and on for nearly two miles.*

The brigade forwarded in thick woods in the wilderness. Every man seemed anxious to go ahead, and it seemed that everyone had an iron will - determined to move those Yankees from General Lee's chosen position or die in the attempt. Our regiment's (61st Georgia Regt.) position was on the extreme right, and we kept getting further to the right until Company D was not more than a skirmish line. We found that the enemy had retreated so far from our front till they were all out of our sight. We came to an open field, or open place in the woods, and found that they had divided and some had retreated to the right and some to the left of this open field. So Major J. D. Van Valkenburg, of our regiment, was left with our company and a part of his old company to watch that side and prevent a flank move by the enemy. General Gordon had driven the enemy from Lee's position and General Lee sent Gordon orders to fall back. Colonel Lamar sent Lieutenant Eugene Jeffers with orders to Major Van to fall back to the line. "We had just gotten started when we saw a regiment of Yankees between us and our line. I felt bad, for we did not have over forty men and there were about five hundred Yankees.

They appeared badly confused to see Confederate soldiers coming up in their rear, when they were not expecting any. We stopped and Major Van advanced, for his quick military eye took in the position that we were placed in. He walked up with a quick step and with drawn sword and ordered their commander to surrender. He refused and ordered Major Van to surrender. He refused and hallowed back to tell General Gordon to send up a brigade, for we had an obstinate regiment cut off and they refused to surrender. Captain Kennedy and the rest of us, in a low voice, began to command forward. Major LeGrand B. Spence, the commander of the Yankee regiment ordered his regiment to stack or ground arms. He expected a volley from a brigade.

We hurried up and Major Van commanded: "Officers to the front." All the officers came to the head of the regiment and Major Van led off and ordered the enemy to follow him. Captain Kennedy took their flag and carried it out and we, forty men, formed a thin guard around them and marched them out. I tell this truthfully, for I was an eye-witness. The enemy told us after we got past our lines that they had been sent in to reinforce their line and that they came up through that open field referred to above, and that they never saw any of their retreating men and that when they saw us and Major Van's actions, they thought that Gordon's Brigade had them cut off. They said we had captured every man in the regiment but one, who was sent back to inform their commander that they could not find any men only their dead on the battlefield. When we got them out and they found that they had been captured by Major Van's stratagem they were the worst set of mortified officers and men I have ever seen. Major Van turned them over to the provost guard and took a receipt for about forty regimental and company officers and four hundred and seventy-four noncommissioned officers and privates. If I am not mistaken it was the Seventh Pennsylvania Regiment of reserves that we captured."

On General Gordon's return (so we were informed) General Lee met him with a big smile, jerked off his glove and gave him a hearty hand shake, calling him Major General Gordon. So you see General Gordon was promoted from brigade to major general on his return from one of the most successful charges of the war, where Gordon's Brigade of six Georgia regiments of not much, if any, over two thousand men, had captured about twenty-five hundred prisoners, had killed and wounded about as many more and had routed five or six times our own number." [22]

Lieutenant Lester of the Tom Cobb Infantry noted, "Our loss in this charge was small in numbers but considerable in quality. We had one of our best soldiers disabled - Willis B. Jackson was wounded by a Minié ball in the neck." [23]

"The ground taken by Gordon's Brigade was held by them the remainder of the day, and A. P. Hill with the 3rd Corps came up the Plank Road and joined Ewell's right. The musketry fire in Hill's front was heavy all afternoon and ceased only at nightfall. After dark Hill's men on the right commenced cheering, and it ran all along his front and along Ewell's line and was repeated, thus bidding defiance to Grant and his army, who were in our front in this vast jungle, the Wilderness," wrote Sergeant Hudgins of the Bartow Avengers. [24]

General Gordon recalled the evening after the battle. The Georgia Brigade was ordered to move to the left flank of General Lee's army, *"Both armies rested for the night near the points where the first collisions of the day had occurred. It would be more accurate to say they remained for the night; for there was little rest to the weary men of either army. Both sides labored all night in the dark and dense woodland, throwing up such breastworks as were possible--a most timely preparation for the next day's conflicts. My own command was ordered during the night to the extreme left of Lee's lines, under the apprehension that Grant's right overlapped and endangered our left flank.*

Thus ended the 5th of May, which had witnessed the first desperate encounter between Grant and Lee. The fighting had not involved the whole of either army, but it was fierce and bloody. It would be unjust to claim that either of the famous leaders had achieved a signal victory. Both sides had left their dead scattered through the bullet-riddled underbrush. The Confederates drew comfort from the fact that in the shifting fortunes of the day theirs was the last advance, that the battle had ended near where it had begun, and that the Union advance had been successfully repulsed." [25]

Private Bradwell of the 31st Georgia recounted the evening after the fight, *"Our brigade (Gordon's) spent the night after this great fight in noisy rejoicing over its splendid achievements, passing the word in loud shouting to the next brigade to the right that they had whipped everything. This was communicated to the next until the news went from brigade to brigade to Lee's extreme right. The word came back to us, by the same means from every part of the line, of the same import. So great and continuous was the noisy demonstration that the enemy decided we were receiving reinforcements from the coast; but this was only bluff on our part to affect the morale of our foe. When day dawned each side had constructed breastworks out of logs and everything lying about on the ground. Ours consisted of logs and dirt dug up with bayonets and cast up with tin plates and our naked hands. Those of the enemy contained also the dead bodies of their own men, besides army blankets, knapsacks, and anything they could find in the darkness of the night. These two lines ran parallel to each other for miles through this wilderness and about one hundred or one hundred and fifty yards apart. As to the dead being used in the construction of the enemy's defenses, I do not pretend to say this from what I heard, but I make this assertion from what I actually saw on the third morning of the fighting after we had routed them from their works."* [26]

Confederate service records show four 38th Georgia soldiers were killed, or mortally wounded, on May 5th, at the Wilderness. Private Asa Chandler Brown of the Thornton Line Volunteers and Private A. J. Andrews of the Goshen Blues were killed in action. Corporal Beverly W. Hall, of the Goshen Blues, from Elbert County, Georgia, was mortally wounded and died June 9th, 1864. 1st Sergeant Joseph A. Maddox, of Company K, from DeKalb County, Georgia, was mortally wounded and died of pneumonia at the Gordonsville, Virginia, General Receiving Hospital (Exchange Hotel) on December 17th, 1864.[27]

Chapter 13
The Wilderness, Day Two - May 6th, 1864

General Gordon was ordered to move his brigade to the extreme left of the Confederate line of battle late on Thursday evening, May 5th. "The night of the 5th of May was far spent when my command reached its destination on the extreme Confederate left. The men were directed to sleep on their arms during the remaining hours of darkness. Scouts were at once sent to the front to feel their way through the thickets and ascertain, if possible, where the extreme right of Grant's line rested. At early dawn these trusted men reported that they had found it that it rested in the woods only a short distance in our front, that it was wholly unprotected, and that the Confederate lines stretched a considerable distance beyond the Union right, overlapping it. I was so impressed with the importance of this report and with the necessity of verifying its accuracy that I sent others to make the examination, with additional instructions to proceed to the rear of Grant's right and ascertain if the exposed flank were supported by troops held in reserve behind it. The former report was not only confirmed as to the exposed position of that flank, but the astounding information was brought that there was not a supporting force within several miles of it." [1]

The night of May 5th, and into the early morning hours of May 6th, Gordon's scouts probed the Union lines and discovered their left flank was "in the air." Gordon even rode forward to confirm their reports and when he determined they were correct, he urged Early and Ewell to launch a flank attack on the their exposed position. General Gordon wrote:

"In a woodland so dense that an enemy could scarcely be seen at a distance of one hundred yards, that Union officer had left the right flank of General Grant's army without even a picket-line to protect it or a vedette to give the alarm in case of unexpected assault. During the night, while the over-confident Union officer and his men slept in fancied security, my men stole silently through the thickets and planted a hostile line not only in his immediate front, but overlapping it by more than the full length of my command. All intelligent military critics will certainly agree that such an opportunity as was here presented for the overthrow of a great army has rarely occurred in the conduct of a war.

As soon as all the facts in regard to the situation were fully confirmed, I formed and submitted the plan which, if promptly adopted and vigorously followed, I then believed and still believe would have resulted in the crushing defeat of General Grant's army. Indeed, the plan of battle may almost be said to have formed itself, so naturally, so promptly and powerfully did it take hold of my thoughts.

When the plan for assault was fully matured, it was presented, with all its tremendous possibilities and with the full information which had been acquired by scouts and by my own personal and exhaustive examination. General Early at once opposed it. He said that Burnside's corps was immediately behind Sedgwick's right to protect it from any such flank attack; that if I should attempt such movement,

Burnside would assail my flank and rout or capture all my men. He was so firmly fixed in his belief that Burnside's corps was where he declared it to be that he was not perceptibly affected by the repeated reports of scouts, nor my own statement that I myself had ridden for miles in rear of Sedgwick's right, and that neither Burnside's corps nor any other troops were there. General Ewell, whose province it was to decide the controversy, hesitated. He was naturally reluctant to take issue with my superior officer in a matter about which he could have no personal knowledge, because of the fact that his headquarters as corps-commander were located at considerable distance from this immediate locality. In view of General Early's protest, he was unwilling to order the attack or to grant me permission to make it, even upon my proposing to assume all responsibility of disaster, should any occur.

Who was responsible for the delay of nine hours or more while that exposed Union flank was inviting our attack? Both General Early and I were at Ewell's headquarters when, at about 5:30 in the afternoon, General Lee rode up and asked: "Cannot something be done on this flank to relieve the pressure upon our right?" After listening for some time to the conference which followed this pointed inquiry, I felt it my duty to acquaint General Lee with the facts as to Sedgwick's exposed flank, and with the plan of battle which had been submitted and urged in the early hours of the morning and during the day. General Early again promptly and vigorously protested as he had previously done. He still steadfastly maintained that Burnside's corps was in the woods behind Sedgwick's right; that the movement was too hazardous and must result in disaster to us...The details of the whole plan were laid before him. There was no doubt with him as to its feasibility. His words were few, but his silence and grim looks while the reasons for that long delay were being given, and his prompt order to me to move at once to the attack, revealed his thoughts almost as plainly as words could have done. Late as it was, he agreed in the opinion that we could bring havoc to as much of the Union line as we could reach before darkness should check us. It was near sunset, and too late to reap more than a pittance of the harvest which had so long been inviting the Confederate sickle.

The solid and dotted lines here given sufficiently indicate the approximate positions occupied by the respective armies at the beginning of my flank attack. [2]

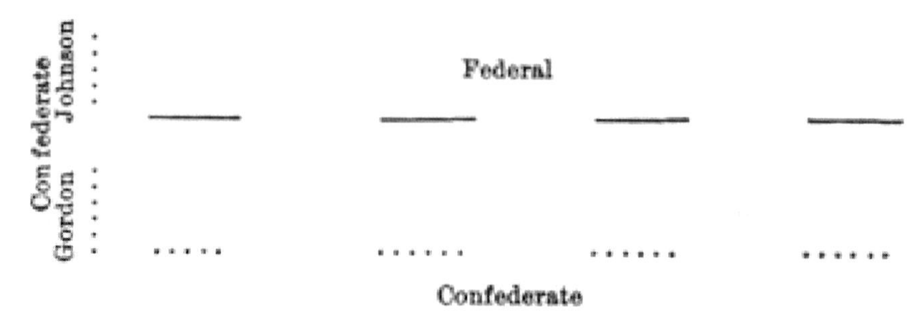

4th Sergeant Francis Hudgins, of Company K, from DeKalb County, Ga., describe the position the 38th Georgia occupied that Friday morning and the attack that soon followed, *"On the morning of the 6th Gordon's Brigade was moved across the turnpike road from the south to the north side, relieving Harris's Mississippi Brigade, which occupied the Confederate left, and had thrown up fairly good earthworks during the morning. Our front was comparatively quiet; and we were eating Yankee crackers and pickled pork and drinking genuine coffee sweetened with plenty of sugar captured the day before, when, about midday, our skirmishers were driven in, and an attack on our works was begun. This was soon repulsed, and the Federal line returned to its former position in that vast wilderness of pines."* [3]

Private Bradwell of the 31st wrote of the role of Gordon's Brigade's sharpshooter battalion on May 6th. The sharpshooter's battalion was composed of select men from every company in the Brigade, including the 38th.

"The battalion of sharpshooters, under Captain Keller, was kept in reserve and shifted from point to point whenever it was thought they might be needed to restore the line if it should be broken. This battalion was strung out in a very thin line, thirty or forty feet apart, wherever each man could find protection for himself in front of the other brigades composing the division (Early's). Opposite them the enemy had a heavy line of skirmishers, five men on each post. This battalion of Confederates was composed of select men from every company in Gordon's Brigade, the best marksmen and the most fearless, well trained for this special duty. They were armed with short Enfield rifles, and when strung out thirty or forty feet apart they could hold any advancing line in check. From what I saw in going over the ground the next morning, very few shots fired by them were ineffectual. In front of them at every post occupied by their opponents, as far as I went, were lying from one to five dead men. Prisoners told us that they were compelled to reenforce their skirmishers several times during the day with new men to take the places of those killed and wounded." [4]

Sergeant Hudgins carefully listened as Gordon spoke directly to the men of the 38th, to raise their battle fervor. *"We fell in just at sundown and moved by the left flank until we passed out of our own works; then the order, 'By the right flank, forward!' was given. General Gordon rode along our line and said: '38th, this is the 6th Corps we are going to attack, the same fellows we fought on Marye's Heights. Those that we didn't get then, we want now.'"* [5]

Private Bradwell recalled the moment the Brigade was ordered to advance, with the Sharpshooter Battalion leading the charge as skirmishers. The soldiers in the main line of battle were ordered to hold their fire, until the Sharpshooters had halted to rejoin the main line.

"The order to advance was now given, and the skirmishers ran up the hill and were on the enemy, then cooking their evening meal on thousands of small fires, secure, as they thought, behind their breastworks. Poor fellows! None of them suspected the bolt that was about to strike them. Suddenly, and only a few yards away, the long line of gray-clad soldiers appeared and opened on them seated in groups about the fires with their guns stacked back of their works. Never was lightning from the clouds more unexpected, and confusion reigned supreme. About this time the main line came on the scene, and so anxious were they to open fire that they disregarded the orders and poured a deadly volley into the confused enemy, endangering very much the lives of our sharpshooters, who fell on their faces, shouting back to us not to shoot until we had passed over them. No attention was paid to this,' and we were at their works," [6]

Gen. Alexander Shaler, USA
Library of Congress

Sergeant Hudgins advanced up the rear of the Yankee breastworks, "There wasn't a skirmisher on their flank, and the first intimation of our presence was when we opened fire. My company struck the Federal breastworks squarely on the end. I advanced up the rear of their works at least a mile, and at each step their confused mass became more dense." [7]

General Gordon described the unexpected attack on the Union right flank, "In less than ten minutes they struck the Union flank and with thrilling yells rushed upon it and along the Union works, shattering regiments and brigades, and throwing them into wildest confusion and panic. There was practically no resistance. There could be none. The Georgians, commanded by that intrepid leader, Clement A. Evans, were on the flank, and the North Carolinians, led by a brilliant young officer, Robert Johnson, were sweeping around to the rear, without a shot in their front. There was nothing for the brave Federals to do but to fly. There was no time given them to file out from their works and form a new line of resistance. This was attempted again and again; but in every instance the swiftly moving Confederates were upon them, pouring a consuming fire into their half-formed ranks and shivering one command after another in quick succession." [8]

Gen. Truman Seymour, USA
Library of Congress

Sergeant Hudgins recounted his "capture" of a Yankee frying pan and the numerous dead Federal soldiers behind their breastworks,

"Their bark fires in the rear of these works were well filled with coffeepots of steaming coffee and frying pans of pickled pork. One of these pans I have still in my possession, which I carry with me to all reunions as a recollection of the times that tried men's souls. I have been on many battlefields of the war; but here were more Federal dead than at any other place I ever saw before or since, possibly with the exception of Fair Oaks, in front of Richmond, in 1862. In our advance on the Federal lines we had doubled the 6th Corps, commanded by General Sedgwick, back on its center and had captured Generals Shaler and Seymour, who commanded that portion of the Federal right." [9] The gallant Union leaders, Generals Seymour and

Shaler, rode among their panic-stricken troops in the heroic endeavor to form them into a new line. Their brave efforts were worse than unavailing, for both of these superb officers, with large numbers of their brigades, were quickly gathered as prisoners of war in the Confederate net; and nearly the whole of Sedgwick's corps was disorganized, wrote General Gordon. [10]

Sergt. Francis L. Hudgins, Co. K
The Sunny South, January 10th, 189

Sergeant Hudgins recalled how General Gordon and his courier accidently rode their horses into a mass of Federal soldiers attempting to reform after the devastating attack, "After the engagement was over for the night, General Gordon with his courier, Bill Beasley, from LaGrange, Ga., rode into the Federal lines. Beasley rode alongside and said: 'General these are Yankees.' Gordon said, 'No'; but Beasley said: 'I am positive.' Both then turned their horses and dug the rowels in their flanks, while the enemy fired a volley at close range. Strange to say, neither man nor horse was touched." [11]

General Gordon wrote of how only darkness halted the advance of his brigade and saved Union General Sedgwick's VI Corps from destruction.

"It will be seen that my troops were compelled to halt at last, not by the enemy's resistance, but solely by the darkness and the cross-fire from Confederates. Had daylight lasted one half-hour longer, there would not have been left an organized company in Sedgwick's corps. Even as it was, all accounts agree that his whole command was shaken. As I rode abreast of the Georgians, who were moving swiftly and with slight resistance, the last scene which met my eye as the curtain of night shut off the view was the crumbling of the Union lines as they bravely but vainly endeavored to file out of their works and form a new line under the furious onset and withering fire of the Confederates. "The impressive feature of that memorable night was the silence that succeeded the din of battle. The awe inspired by the darkness and density of the woods, in which two great armies rested within hailing distance of each other, was deepened by the low moans of the wounded, and their calls for help, as the ambulance corps ministered to blue and gray alike." [12]

Gordon's flank attack stunned the Union army. An officer approached General U.S. Grant at his Headquarters in a panic, he cried, "General Grant, this is a crisis that cannot be looked upon too seriously. I know Lee's methods well by past experience; he will throw his whole army between us and the Rapidan, and cut us off completely from our communications." Grant snapped, "Oh, I am heartily tired of hearing about what Lee is going to do. Some of you always seem to think he is suddenly going to turn a double somersault, and land in our rear and on both of our flanks at the same time. Go back to your command, and try to think what we are going to do ourselves, instead of what Lee is going to do." [13]

Sergeant Hudgins said of the day after the battle, *"On the 7th (of May) the 38th Georgia was thrown out on the picket line. We captured many prisoners that day and were reserved at night. On the morning of the 8th we began the march by the right flank to Spotsylvania Courthouse, passing through the burning woods, where the smoke was so dense as nearly to suffocate us."* [14]

Two brutal days of combat left many uncertain of Grant's intentions, yet his tenacity was unequaled by any previous Federal commander. Many Union and Confederate soldiers alike suspected he would retreat like previous commanders, yet he did not retreat, but pushed his army forward to Spotsylvania Court House. General Lee anticipated Grant's next move and the Confederate army was there awaiting his arrival.

Sergeant Hudgins noted, *"The brilliant performance of the Georgia Brigade at the Wilderness earned General Gordon promotion to, "Major General, and C. A. Evans, Colonel of the 31st Ga., the beloved and gallant commander of the Georgia Division, to a brigadier generalship."* [15]

Killed or Mortally Wounded at Wilderness, May 6, 1864

Company D, DeKelab & Fulton Counties

1. Private Adolphus Rosenthal - Wounded in the foot 5/6/1864, died in Richmond, Virginia, hospital 5/27/1864.
2. Corporal Tilnon P. "Sally" Smith - Killed 5/6/1864.

Company E, "The Tom Cobb Infantry," Oglethorpe County, Georgia

3. Private James S. Boggs - Killed 5/6/1864.
4. Private George L. Mathis (Or Mathews) - Killed 5/6/1864.

Company G, "The Battey Guards," Jefferson County, Georgia

5. Private Howell L. Peebles (or Peeples) - Wounded 5/6/1864, died of wounds in hospital, May 31, or 6/15/1864.[16]

Chapter 14
The Battles of Spotsylvania Court House - May 10th, 1864

After the battle of the Wilderness Lee and Grant's armies raced for Spotsylvania Court House, a small hamlet in the Virginia country side. Lieutenant Lester recalled how his company was greatly reduced due to battle losses. "After the terrible carnage at the battle of the Wilderness, our command was engaged mostly in skirmishing between this place and Spotsylvania Court House. Our company was now reduced again to a very small number. The Confederate army arrived at Spotsylvania first and immediately began constructing earthworks. Our brigade commander, the noble hero, General John B. Gordon, was promoted to Major-General, Colonel . C. A. Evans, of the 31st Ga. Regiment, was promoted to Brigadier-General and took command of our brigade. A braver soldier never drew sword in defense of the "lost cause" than Clement A. Evans." [1]

Colonel Clement A. Evans was promoted to Brigadier General during May of 1864. He would remain as the Georgia Brigade commander for the duration of the war. The Brigade was now called "Evans' Brigade." Evans was later called on to serve a division commander, but another brigade commander was never formally assigned to command the Georgia Brigade. After the war, the brigade would be known as the Lawton-Gordon-Evans Brigade, named after all three of its illustrious commanders.

Gordon was appointed to serve as the division commander for three brigades and he reported, "The march to Spotsylvania Court-House was begun by my brigade, with Early's division, on the night of the 7th. On the morning of the 8th I was placed in command of this division, consisting of three brigades--Pegram's (Virginia), Johnston's (North Carolina), and Gordon's (Georgia)--and on the afternoon of the same day reached Spotsylvania Court-House. Orders from Lieutenant-General Ewell directed that I should use my division as a support to either Johnson's or Rodes' division, or to both, as circumstances should require. I had, therefore, placed my largest brigade (Gordon's, now Evans') in rear of Rhodes' right and Johnson's left, and directly in front of the McCool house. The other two brigades were held in reserve near the Harris house." [2]

The Georgia Brigade occupied a portion of the Confederate lines forming a huge salient, or bulge, in the line shaped like a mule shoe. This bulge, about half a mile wide and deep, was a weak point in the Confederate line.

Private Bradwell wrote of the morning of May 10th, "At daylight we found ourselves at a line of Confederate soldiers who were hastily constructing breastworks of poles, dirt, and everything else they could lay hands on. We threw ourselves down behind these for a rest, but were soon called to order and marched to the right to assist in fortifying our position. As far to the right as could be heard there was the incessant noise of the pick, the shovel, and the ax. There were no idlers in Lee's army that day, for the desultory skirmish in front warned every man to do his best and to be ready for bloody work we all knew to be impending." [3]

Colonel Theodore Lyman of General Meade's staff wrote of the Confederate prowess in constructing earthworks, "It is a rule that, when the Rebels halt, the first day gives them a good rifle-pit; the second, a regular infantry parapet with artillery in position; and the third a parapet with an abattis in front and entrenched batteries behind. Sometimes they put this three days' work into the first twenty-four hours." [4] The Georgia Brigade arrive at Spotsylvania Court House on the afternoon of May 8th. The Confederates used their time well and elaborate earthwork defenses were in place by May 10th.

About 6 p.m. on the evening of May 10th, Union Colonel Emory Upton led twelve handpicked Union regiments, 5,000 men with fixed bayonets, in an attack on the mule shoe salient. They formed in a wood not far from the edge to the mule shoe salient. Upton's men were formed in a compact formation and were ordered to charge in mass at the Confederate earthworks. The men were ordered not to fire, not to stop to help the wounded, and not to stop for anything. Once they breeched the Confederate line, they were to spread right and left widen the gap for other Union troops following behind them to exploit the breakthrough.

Colonel Emory Upton (Library of Congress)

Gordon recalled the attack, "On the afternoon of the 10th I received orders to move my division rapidly from the left of our lines to the support of Rodes' division, now being heavily assaulted by the enemy. When my division reached this position the enemy had carried the portion of work held by Doles' brigade, Rodes' division, and had reached a point more than 100 yards in rear of the line. My leading brigade (Johnston's North Carolina) was immediately formed, by direction of Lieutenant-General Ewell, across the head of the enemy's column and ordered to charge. In the mean time Gordon's brigade was also formed and ordered forward." [5]

The Georgia Brigade was resting in the rear of the point of attack in an open field. Private Bradwell of the 31st recalled the moment the brigade was called to action.

"General Gordon and Colonel Evans, next in command, had ridden away, I suppose to see how the fighting was progressing in front of where we were, there being no field officer present except Lieutenant Colonel Berry, of the 60th Georgia....a courier arrived riding as fast as his horse could run. When he arrived he hastily asked where General Gordon was. No one knew, and he rode off to find

him…But now we saw an officer of General Lee's staff coming towards us riding at

"Battle of Spotsylvania"
By Thure de Thulstrup, 1887, Library of Congress

the same pace. When he reached us he made the same inquiry and received the same answer, he spoke to Colonel Berry and told him that General Lee's line was broken and ordered him to take command of the brigade and follow him. He turned his horse's head and trotted back in front of us, saying repeatedly: "Come on, boys; come on." Ahead of us on the left was a straggling piece of woodland through which our works extended to the right a short line of works out in the field behind which were crouching our sharpshooters and other who had taken refuge there.

About one hundred and fifty yards in the rear of line ran an old road parallel to them. We entered the woods by this road, trotting along in fours, following Colonel Berry, when suddenly just as I stepped over the body of a dead Confederate soldier he wheeled to the left and shouted, 'Here they are men!' In an instant we turned, and, to our surprise, there stood in the twilight not more than a hundred feet away, a blue mass of Yankee soldiers apparently indifferent to our approach. We instantly brought our guns into position, and a line of fire flashed along the regiment as we closed on the enemy. Stupid under the influence of liquor, they retired slowly and sullenly, while our men beat them with clubbed guns back to the breastworks, which they defended obstinately for a while.

Our men beat them out of the works, and they retired to another line built by our men and afterwards abandoned when they first came to this place…I suppose they were thousands of them. In all my experience of the war I was never under a heavier fire. It seemed as if a hand or a head above the protection of the works would be pierced by a Minié ball. Word was passed down the line for the three regiments of our brigade to mount the works and charge the enemy out of their den." [6]

Evidence suggests the 38th, led by Lieutenant Colonel Phillip E. Davant, was one of the three regiments selected to make the charge. Bradwell continued, *"My! Thought I, how it possible for living men to face such a fire. Yet every man responded. Some fell back dead as they mounted the works, but the rest swept forward and in ten minutes had cleared the enemy from our immediate front."* [7] The three Georgia regiments charged and most of the Federals, not killed or captured, fled to the woods just beyond the salient. Lieutenant Colonel Davant urged the 38th forward as he leapt over the earthworks and charged the Federals alone, the rest of the 38th had halted at the earthworks. The retreating Federals promptly captured Davant and his Adjutant, John Law. Gordon recalled his capture, "As Hancock's troops were driven out of our lines, the commander of one of my regiments, Colonel Davant of the Thirty-eighth Georgia, became so enthused that he ran in pursuit ahead of his men, and passed some distance beyond the breastworks. A squad of Hancock's retreating men at once halted, and, in the quaint phraseology of the army, 'quietly took him in.' Davant, surprised to find himself in the hands of Hancock's bluecoats instead of in the company of his Confederate comrades, attempted to give notice to his men in the rear that he was captured. His adjutant, John Gordon Law, my first cousin, heard the colonel's call, and sprang forward through the thicket to aid him. Law was likewise captured, and was kept in prison to the end of the war." [8] The 38th learned a hard lesson at the battle of Fredericksburg that they never forgot; never chase the enemy far in advance of Confederate lines.

Gordon recounted his first cousin's, tale of his kind treatment immediately after his capture. *"He (Law) delights to tell of the great kindness shown him by the guard to whose care he was assigned. The soldier in blue who guarded Law was a private, and had no possible use for a sword-belt; but he wanted it, nevertheless. Instead of taking it forcibly, he paid for it, in greenbacks, the full price named by Law. In answer to Law's lament that he was going to prison without a change of clothing or any blankets, this generous Union boy offered to sell him his own blankets. Law replied to the suggestion:*

I have no money to pay you for your blankets, except Confederate bills and the greenbacks which you have just paid me for the sword-belt."

"Oh, well," said the Federal private, "you can pay me for the blankets in Confederate money, and if I should be captured it will answer my purpose. If I should not be captured I will not need the money. Give me your graybacks' and you keep my 'greenbacks' to help you along during your stay in Fort Warren." [9]

The Georgia Brigade pushed the enemy out of the Confederate lines and darkness enveloped the armies. Private Nichols recalled the bands playing in the darkness, "After the battle was over and the wounded very well cared for and night had closed around us, one of our bands began to play 'The Dead March' just in the rear of the 'death angle.' You could hardly ever hear a man speak, and it seemed that we all wanted to shed tears of real sorrow; some that had lost relatives or dear friends did have to wipe their eyes. When our band ceased playing, one of the

Union bands played ‘Nearer, My God, to Thee;’ then our band began to play ‘The Bonny Blue Flag,’ after which the Union band played ‘The Star Spangled Banner;’ then our band played ‘Dixie Land,’ and the Union band finally struck up ‘Home, Sweet Home;’ this probably brought tears rolling' down many powder-blackened cheeks in both armies. When the Union band played ‘Star Spangled Banner’ we could hear their soldiers huzza, and when our band struck up ‘Dixie’ it looked like it cheered every man, and we raised a yell; but oh I how different when ‘Home, Sweet Home,’ was played. It brought to our mind two of the sweetest, dearest words in the English language — Home and Mother." [10]

Chapter 15
Spotsylvania Court House – May 12th 1864

General Gordon was ordered to take charge of his division and act as a reserve and at the first instant of danger, to move his men to the contested point without waiting for orders. Gordon later reported, "Orders from Lieutenant-General Ewell directed that I should use my division as a support to either Johnson's or Rodes' division, or to both, as circumstances should require. I had, therefore, placed my largest brigade (Gordon's, now Evans') in rear of Rodes' right and Johnson's left, and directly in front of the McCool house. The other two brigades were held in reserve near the Harris house." [1] Grant was elated by the partial success of the attack of May 10th. If twelve regiments could crack the Confederate line, what could an entire Union Corps accomplish? He charged General Winfield Scott Hancock, the very best Union Corps commander, with leading his entire Second Corps, 20,000 soldiers, against the Confederate line. They were to be in place and attack just before daybreak on May 12th.

Major General Winfield Scott Hancock (National Archives)

Lee received intelligence indicating Grant was retreating with his army towards Fredericksburg on May 11th. He wanted his army ready to move at a moment's notice to follow and attack Grant, but the artillery in the Mule Shoe would slow the movement of his army. Lee ordered the majority of the artillery removed from the Mule Shoe salient. Without this artillery in place, the Mule Shoe was extremely vulnerable to attack. This was a critical error at the exact time and place Grant was preparing to attack.

Gen. Edward Johnson – CSA
Library of Congress

Late on the evening of May 11th, General Edward Johnson began to suspect the enemy was massing for an attack on the Mule Shoe. The pickets reported the rumblings sound of moving wagons and a hushed roar in the distant forests. General Johnson pleaded with Lee to return the artillery to the Mule Shoe and General Lee agreed. The orders were sent to the artillery commanders to return to the salient. The darkness and the rain slowed the movement of the orders to the artillery units. Once the orders were received, the artillery commanders began to return, but there was no sense of urgency in their movement. Gordon reported, "During the night of the 11th I received information from Major-General Johnson that the enemy was massing in his front, and under the general instructions I had received from corps headquarters I sent another brigade (Pegram's) to report to him." [2] It was raining that night and at 4 a.m. a heavy fog covered the ground, Hancock ordered his 20,000 men to prepare to attack. The men formed in a mass, compact formation and prepared to advance. Thirty minutes later, Hancock ordered his men forward and 20,000 men rolled forward toward the tip of the Mule Shoe salient. The soft wet ground muffled the sound of their feet. The rain, darkness, and their dark uniforms helped conceal their movement forward. The mass of men crested a ridge and before them lay a huge swale, covered in abatis and tangle foot obstructions. The Federal's frantically clawed their way through the obstructions as the Confederate line began to come to life.

The Stonewall Brigade was posted near the tip of the salient and stood ready to repel the Federal assault. They were ordered to fire, but instead of the deep sound of rolling rifle fire, the caps only exploded on their guns, their powder was wet. Hancock's men rolled over the breastworks and spread to the left and right. The Federals scooped up hundreds of prisoners, numerous colors, and the few remaining cannon. The Confederate artillery batteries ordered to return to the salient arrived just as Hancock attacked and several were captured before firing a shot. A thick mantle of fog settled over the Mule shoe and reduced visibility to about 30 feet. Gordon and other Confederates in the rear were uncertain of what had just happened. Hancock's assault had just captured over 3,000 Confederate soldiers, along with Division Commander, General Edward Johnson and Brigade Commander General George H. Steuart.

Gordon recounted how he first learned of the disaster at the salient. "At about 4:30 or 5 a.m. a soldier, one of the vedettes stationed during the night at different points to listen for any unusual sounds, came hurriedly in from the front and said to me: "General, I think there's something wrong down in the woods near where General Edward Johnson's men are."

"Why do you think so? There's been no unusual amount of firing."

"No, sir; there's been very little firing. But I tell you, sir, there are some mighty strange sounds down there--something like officers giving commands, and a jumble of voices."

In the next few minutes before saddles could be strapped on the officers' horses and cartridge-boxes on the men, report after report in quick succession reached me, each adding its quota of information; and finally there came the positive statement that the enemy had carried the outer angle on General Edward Johnson's front and seemed to be moving in rear of our works. There had been, and still were, so few discharges of small arms (not a heavy gun had been fired) that it was difficult to believe the reports true. But they were accurate." [3]

Gordon submitted a battle report on July 5th, 1864, less than two months after the attack, and described how he deployed his three brigades during the crisis.

"I heard musketry in the direction of the Salient, held by Jones' brigade, of Johnson's division, and at once ordered my other brigade (Johnston's) to move toward the firing. The situation at this time was as follows: Evans' brigade (Evans commanded Gordon's Georgia Brigade) was in position immediately in rear of the left of Johnson's division and Rodes' right. Pegram's brigade was placed by General Johnson in the trenches near his left and to the left of the Salient, and Johnston's brigade was moving from the Harris house toward the Salient. The check given by Jones' brigade to the enemy's assaulting column was so slight that no time was afforded for bringing into position the supporting force. No information was brought to me of the success of the enemy, and in the early dawn and dense fog I was unable to learn anything of the situation until Johnston's brigade met in the woodland between the McCool house and the Salient with the head of the enemy's column.

Brigadier-General Johnston was wounded, and his brigade was soon overpowered and driven back. I at once discovered that the situation was critical, and ordered Colonel Evans to move his brigade (Georgia Brigade) at a double-quick from its position near the trenches to the McCool house, and sent a staff officer to ascertain the position of Pegram's brigade, and, if possible, to withdraw it to the same point. This was promptly done. The fog was so dense that I could not ascertain the progress of the enemy, except by the sound of his musketry and the direction from which his balls came. At this point (the McCool house) I ordered Colonel Evans (Evans commanded the Georgia Brigade) to send in three of his regiments to ascertain the enemy's position and check his advance until the other troops could be gotten into line." [4]

Evidence suggests the three regiments sent to immediately charge against Hancock's massive advance, were the 38th Georgia, 31st Georgia, and 60th Georgia. Captain Henry R. Deadwyler commanded Company H, the Goshen Blues. He led his company forward in the charge against the Federals inside the Confederate works, "after leading his gallant company to the breastworks that had been taken by the enemy. The regiment was ordered to hold the works at all hazards. The enemy reformed and surrounded us, and at the same time, captured Captain Deadwyler and eleven others of the company. The commander and others cut their way out and made their escape to a new line in the rear of Horse Shoe Bend, here General Lee was forming a line to meet the enemy as they advanced." [5]

The captured men of the Goshen Blues probably endured an experience similar to another member of the Georgia Brigade captured that morning, then escorted to the Federal rear. "As I retired through the army of the enemy I found that they had thirteen solid columns of troops massed in our front. We, the prisoners, stood up all the following night in the rain without rations and were closely guarded. The next day we were marched to Aquia Creek, put on a transport for Point Lookout and thrust into prison after being deprived of everything we had."[6]

Sergeant Milton A. Clark, of the Battey Guards, was shot through the right leg. His leg was later amputated and he would never return to active service.

Lieutenant John Goswick, of Company I, the Dawson Farmers, was shot through the neck near the left shoulder, by a rifle ball, the ball ranged downward and came out on the right side of his shoulder. Private David W. Simmemon watched as Isaiah Peter Millsaps, of Company I, the Dawson Farmers, was shot through the body with a Minié ball and then through the shoulder. He died just a few minutes later. Simmemon and Private Zebulon C. Payne buried Millsaps' body after the battle.

Private Jeptha Campbell of Company H, the Goshen Blues, recalled, "Sergeant John H. Bowers was struck in the forehead by a spent ball, while charging the enemy, this being his third wound of the war. The same ball passed through the beard of Private Francis M. Hendricks and then through the neck of Private Allen Y. Gulley, before striking Sergeant Bowers in the forehead. Both Hendricks and Gulley were captured by the Federals." [7] Private Elijah "Bensie" Higginbotham, of Company F, Thornton's Line Volunteers, was shot by a bullet or shell through both legs just below the knees. Both legs were amputated in a Confederate hospital, yet "Bensie" survived. He lived as a cripple for the remaining 48 years of his life, dying at age 89, in Elbert County, Georgia.

These three regiments paid dearly in retarding the advance of the Federal juggernaut. Gordon wrote, "The attacking column, it was ascertained, had advanced considerably to the right of this point, and the temporary check given by these regiments afforded only time enough for moving the remainder of Evans' and Pegram's brigades farther around to the right." Lt John Caldwell, of the 1st South Carolina, described what awaited his regiment as they charged and arrived at the main Confederate line, now occupied by the Federals. The 38th Georgia may have encountered a similar scene,

The sight we encountered was not calculated to encourage us. The trenches, dug on the inner side, were almost filled with water. Dead men lay on the surface of the ground and in the pools of water. The wounded bled and groaned, stretched or huddled in every attitude of pain. The water was crimsoned with blood. Abandoned knapsacks, guns and accoutrements, with ammunition boxes, were scattered all around. In the rear, disabled caissons stood and limbers of guns. The rain poured heavily, and an incessant fire was kept upon us from front and flank. The enemy still held the works on the right of the angle, and fired across the traverses. Nor were these foes easily seen. They barely raised their heads above the logs, at the moment of firing. It was plainly a question of bravery and endurance now.

We entered upon the task with all our might. Some fired at the line lying in front, on the edge of the ridge before described; others kept down the enemy lodged in the traverses on the right. At one or two places, Confederates and Federals were only separated by the works, and the latter not a few times reached their guns over and fired right down upon the heads of the former. No man could raise his shoulders above the works without danger of immediate death. Some of the enemy lay against our works in front. I saw several of them jump over and surrender during relaxations of firing.

This was the place to test individual courage. Some ordinarily good soldiers did next to nothing, others excelled themselves. The question became, pretty plainly, whether one was willing to meet death, not merely to run the chances of it." [8]

The charge of the 38th, 31st, and 60th Georgia, bought just enough time for General Gordon to form Pergram's Virginia Brigade and the other three regiments of the Brigade into a line of battle. Gordon placed the 61st Georgia on the extreme left of the brigade. Evidence suggests the other two regiments in line were the 13th

Sketch of "Lee to the rear!" including Col Evans, Gen. Gordon, and Gen. Lee, all on horseback
Library of Congress

and 26th Georgia.

As these men formed their line of battle, there occurred one of the most famous incidents of the war. Lee rode to the front of the Georgia Brigade and prepared to personally lead them into battle. General Gordon vividly recalled the moment, *"Through that wide breach in the Confederate lines, which was becoming wider with every step, the Union forces were rushing like a swollen torrent through a broken mill-dam. General Lee knew, as did everyone else who realized the momentous import of the situation, that the bulk of the Confederate army was in such imminent peril that nothing could rescue it except a counter-movement, quick, impetuous, and decisive. Lee resolved to save it, and, if need be, to save it at the sacrifice of his own life. With perfect self-poise, he rode to the margin of that breach, and appeared upon the scene just as I had completed the alignment of my troops and was in the act of moving in that crucial countercharge upon which so much depended. As he rode majestically in front of my line of battle, with uncovered head and mounted on Old Traveller, Lee looked a very god of war. Calmly and grandly, he rode to a point near the centre of my line and turned his horse's head to the front, evidently resolved to lead in person the desperate charge and drive Hancock back or perish in the effort. I knew what he meant; and although the passing moments were of priceless value, I resolved to arrest him in his effort, and thus save to the Confederacy the life of its great leader. I was at the centre of that line when General Lee rode to it. With uncovered head, he turned his face toward Hancock's advancing column.*

Instantly I spurred my horse across Old Traveller's front, and grasping his bridle in my hand, I checked him. Then, in a voice which I hoped might reach the ears of my men and command their attention, I called out, 'General Lee, you shall not lead my men in a charge. No man can do that, sir. Another is here for that purpose. These men behind you are Georgians, Virginians, and Carolinians. They have never failed you on any field. They will not fail you here. Will you, boys?' The response came like a mighty anthem that must have stirred his emotions as no other music could have done. Although the answer to those three words, 'Will you, boys?' came in the monosyllables, 'No, no, no; we'll not fail him,' yet they were doubtless to him more eloquent because of their simplicity and momentous meaning. But his great heart was destined to be quickly cheered by a still sublime testimony of their deathless devotion. As this first thrilling response died away, I uttered the words for which they were now fully prepared. I shouted to General Lee, 'You must go to rear.' The echo, 'General Lee to the rear, General Lee to the rear!' rolled back with tremendous emphasis from the throats of my men; and they gathered around him, turned his horse in the opposite direction, some clutching his bridle, some his stirrups, while others pressed close to Old Traveller's hips, ready to shove him by main force to the rear.

I verily believe that, had it been necessary or possible, they would have carried on their shoulders both horse and rider to a place of safety. This entire scene, with all its details of wonderful pathos and deep meaning, had lasted but a few minutes, and yet it was a powerful factor in the rescue of Lee's army. It had lifted these soldiers to the very highest plane of martial enthusiasm. The presence of their idolized commander-in-chief, his purpose to lead them in person, his magnetic and majestic presence, and the spontaneous pledges which they had just made to him, all conspired to fill them with an ardor and intensity of emotion such as have rarely possessed a body of troops in any war. The most commonplace soldier was uplifted and transformed into a veritable Ajax.

To say that every man in those brigades was prepared for the most heroic work or to meet a heroic death would be but a lame description of the impulse which seemed to bear them forward in wildest transport. Fully realizing the value of such inspiration for the accomplishment of the bloody task assigned them, I turned to my men as Lee was forced to the rear, and reminding them of their pledges to him, and of the fact that the eyes of their great leader were still upon them, I ordered, "Forward!" With the fury of a cyclone, and almost with its resistless power, they rushed upon Hancock's advancing column.

With their first terrific onset, the impetuosity of which was indescribable, his leading lines were shivered and hurled back upon their stalwart supports. In the inextricable confusion that followed, and before Hancock's lines could be reformed, every officer on horseback in my division, the brigade and regimental commanders, and my own superb staff, were riding among the troops, shouting in unison: 'Forward, men, forward!' But the brave line officers on foot and the enthused privates needed no additional spur to their already rapt spirits. Onward they swept, pouring their rapid volleys into Hancock's confused ranks, and swelling the deafening din of battle with their piercing shouts. Like the débris in the track of a storm, the dead and dying of both armies were left in the wake of this Confederate charge. In the meantime the magnificent troops of Ramseur and Rodes were rushing upon Hancock's dissolving corps from another point, and Long's artillery and other batteries were pouring a deadly fire into the broken Federal rank, driving the enemy with heavy loss from nearly the whole of the captured works from the left of Wilcox's division to the Salient on General Johnson's line, and fully one-fourth of a mile beyond. Several of the lost guns were recaptured by the Thirteenth Virginia Regiment, of Pegram's brigade, and brought back to the branch near the McCool house."[9]

Private Nichols of the 61st Georgia recalled Gordon's charge after General Lee was escorted to the rear, *"Evans' and Pegram's brigades raised a yell and made a most gallant charge in a dense fog. While our dear boys were being mown down at every step, and after a desperate struggle, we succeeded in recapturing our line on the right of where the reserve line connected with the captured line, and had everything safe so far as a solid line was concerned.*

Our regiment (Sixty-first Georgia) was on the extreme left of Evans' brigade. We charged the enemy where we did not have any support on our left, and our color bearer, Francis Marion McDow, planted our battle-flag on the works and we had a hand to hand battle with club-guns. Our line being very thin and the enemy's fully three times as strong as ours, we were overpowered and a great many of our regiment had to surrender. Here the enemy captured sixty-five of our regiment, McDow and our old tattered and torn battle-flag. The flag was nearly shot to pieces by shot and shell. It was nothing but a tattered and torn rag, for it had been in every battle the regiment had ever been engaged in, and the regiment loved it and its noble bearer. We mourned the loss of McDow and the dear old flag. Sergeant McDow died in the Fort Delaware prison. Evans' brigade was cut to pieces, for we had lost severely. When I saw such officers as Captains J. J. Henderson of Company A, Daniel McDonald of Company C, Adjutant J. J. Mobley, and many others, whose names I do not recollect, begin to surrender, I turned and ran back for dear life. I just abhorred the idea of being a prisoner. It seemed that I almost preferred death, and would take my chances in getting away, with several others. There were not many shots fired at us, for the fog was so dense that the Yankees could not see us more than 30 feet away. We ran back to the reserve line, which was about 75 yards in the rear, and found it full of our men, who were in readiness. This helped my feelings, for I was afraid that the Yankees would get the best of us." (This charge was made just at daylight in a dense fog.)

These three regiments, the 13th, 26th, and 61st, Georgia Regiments were fighting on the Eastern side of the salient. General Gordon reported these regiments pushed the Federals out of the Confederate lines and fully 1/4 a mile beyond the original Confederate earthworks. [10] The 13th and 26th Georgia Regiments were more successful than the 61st Georgia, at least for an hour or so, but these two regiments were also soon cut to pieces and overwhelmed. The attacking Federals were Burnside's 9th Corps, Potter's Division, 17th Vermont Regiment and 48th Pennsylvania Regiments. Interestingly, the 48th Penna. Vols. was the Federal regiment that dug the mine at Petersburg, Virginia, which led to the battle of the Crater, fought July 30th, 1864.

Union Major W. B. Reynolds of the 17th Vermont reportedly engaged the 26th Georgia Regiment early that morning. Major Reynolds wrote, "At 7 a.m. having exhausted 40 rounds per man, as well as all that could be procured from the dead and wounded, we were relieved by the 48th Penna. Vols. and withdrew about twenty paces, where we remained with fixed bayonets, while ammunition was being brought forward. During this time, about fifty of the 26th Georgia, who had been in our front, were sent back as prisoners of war. The few survivors of the regiment made good their escape from the ravine, leaving in our hands a number of dead and wounded." [11]

Union Sergeant Joseph Gould, of the 48th Penna. Vols. wrote of his regiment engaging the 13th Georgia Regiment, *"It was very foggy morning when Captain McKibben of General Potter's staff ordered Colonel Pleasants to follow him with the 48th. Captain McKibben led us up the hill until he had us under fire, when we formed line of battle behind one of the advance regiments. There was a rebel regiment behind the brow of the hill, directly in our front, and our position did not suit our Colonel We moved forward past the right of the advanced regiment until we got about half way between it and the enemy, which proved to be the 13th Georgia. Before we commenced firing about twenty of the rebel troops came in and surrendered. When within about seventy-five yards of the enemy we were ordered to halt, and commence firing, when for a short time the engagement was very lively. The enemy were at a decided disadvantage, they being down the slope of the hill, we at the top. About the time we opened fire another, or part of a rebel regiment, came to their support. We hammered away at them until someone from the centre of our regiment called out that they wanted to surrender, but Colonel Pleasants ordered us to continue firing, which we did until the rebels threw down their arms and came in as a body. We captured fully two hundred prisoners. They left one colonel, three line officers and seventy-five men killed, and a large number of wounded on the field."* [12]

Private Nichols explained the heavy losses and actions after retreating from the main breastworks, *"The reason why our regiment lost so many prisoners was because we ought to have stopped and occupied the reserve line when we drove the Yankees past it. Our orders were to do that, but we did not understand them. Generals Gordon, Evans and Pegram succeeded in establishing Ewell's line on the reserve line and on down the line to the right angle, while General Hancock held the most of the toe of the shoe the rest of the day.*

When we got to the reserve line and knew that our regiment was so badly hurt in killed, wounded and captured, we asked the men where the rest of Evans' brigade was, for the fog was so dense we could not see. They said they did not know. We could not see where to go for the fog. About one hundred of our regiment and brigade turned up on the line towards the left angle. We were badly confused and mortified. About 100 yards from the angle we again found General Lee sitting on his horse trying to look and listen to find out what the enemy was doing.

I went to him and asked him where we could find Evans' brigade. He said it was on the right. He asked us, (for there were about one hundred of us with two officers) if we belonged to that brigade; we told him that we did. He then told us to fall in line and occupy a piece of the line near him that was not occupied. We fell in and just filled up the gap and made the line solid. General Lee told us to hold it at all hazards, and we promised him to do our best, and we did hold it, though we had a hard struggle. After driving the Yankees back our brigade occupied the line that they had recaptured the rest of the day."[13]

Lee's army, along with the Georgia Brigade, fought on all day long in the pouring rain, while the enemy held tight to the outsides of the main breastworks and also held the "toe" of the mule shoe salient. While the men fought fiercely on the front line all day long, Lee's engineers labored a half a mile to the rear constructing a new line of Confederate breastworks. At about 3 a.m. the following morning, word was quietly passed along the Confederate lines to retreat to the new line in the rear. The retreat was accomplished with little loss. When morning dawned, the Federals found the works abandoned and General Grant was uncertain of whereabouts of Lee's army, although they had only retreated a short distance.

William C. Mathews, of Company G, the Battey Guards, was promoted to Captain on May 12th, 1864. He lamented the regiment's losses, "In these battles Col Davant, Lieuts Mattox, Vaughn, and Adjt. Law were captured and the brave Captain Levin W. Farmer of Company G, and Lieut. Sid Farmer of the same company, were both killed. In their death their company and the regiment sustained a serious loss. They were young, gallant, brave and devoted to the cause and fell with their faces to the foe." [14]

Private Nichols wrote of the Georgia Brigade's return to the mule shoe a few days after the battle, "*In this battle at Spotsylvania, Evans' brigade (The Georgia Brigade) had one of the hardest struggles it ever had, and was placed at a sad disadvantage, for it had to fight the enemy after they had been flushed with victory. We had to drive them out of our own fortifications that they had previously captured from General Johnson. They killed and wounded a great many of the brigade, though our losses in killed and wounded were not as great as it was at Manassas, Sharpsburg, Fredericksburg or even Gettysburg; but more were captured in this battle than in all of the other battles.*

About three or four days after the battle the Yankees all seemed to have left our front. Our brigade went to reconnoiter. We marched over the ground on which they had charged and re-charged our batteries so many times on the 12th. In places, the dead Union soldiers were lying almost in piles. They were as thick as corn hills. I saw an officer's horse lying on five men and two or three men were lying against the horse. Here in this death angle there were several acres of ground the worst strewn with dead men I saw during the war. I saw an oak tree, probably eighteen or twenty inches in diameter, that was very nearly cut down with canister shot and minnie balls, and the ground around it was covered with dead men, it being on the lowest ground the enemy had to advance over and where our batteries and small arms had a full sweep on them. Those who were not very badly mutilated were swollen as large as they could swell. Their faces were nearly black and their mouths, nose, eyes, hair and the mutilated parts were full of maggots! This is a horrible picture, but I know it is not over drawn. What an awful scent. It just makes me feel bad to-day as I write about it. I know that hundreds, and probably thousands, died for the want of attention. There were many mothers deprived of kind, loving husbands; their children were made fatherless; their darling sons, whom they had nursed and dandled on their knees while babes, and watched carefully over them while children, and gave them good advice, praying to the great God to spare them to return to them in peace, lying on the cold ground in this horrible condition." [15]

Another Georgia Confederate soldier described what he saw when he returned to mule shoe, possibly on the same day as the Georgia Brigade, May 15th.

I think was Sunday, May 15th, that I went over the field of battle, which the enemy declined to occupy. Dead bodies, most them enemies, thickly strewed the ground, swollen to twice the size of men. The air was horribly foul with the stench of decaying bodies. In some places they lay so thick on the ground I had to pick my way amongst them. And to shun the fearful gaze of their open eyes, I held my head up, looking ahead, afraid to look down. Nearly all of our men had been buried. I saw the graves of Lt Col and Lt Shooter, two brothers, buried together.

The breastwork where our men had fought had been built in crescent shape, with traverses inside. The traverses were invaluable to our line, suffering from flanking fire. Without them, the line would have been untenable. The throwing up so much breastwork had made great pits between the traverses. There having been rain, these pits were now pools of water, so saturated with powder as to be nearly black as ink. In this black water and on the ground outside, lay rifles-broken and unbroken - cartridge boxes, cap boxes, belts, knapsacks, canteens, haversacks, cartridges, ammunition boxes, clothing of all kinds, blankets, oilcloths, tent cloths, bayonets, ramrods,- torn, broken, bent, scarred, black, and bloody-- a very hell. I said to myself if a man wants to see hell upon earth, let him come and look into this black, bloody hole-- upon this horrid confusion, these wet, muddy graves- this reeking mass of corruption of rotting corpses, that fill the air with intolerable stench. How a man can look upon such a scene and still take pleasure in war seems past belief. [16]

On May 18th, the Confederate army was waiting behind powerful earthworks when the Federals attempted an infantry attack through the mule shoe. Their attack was shattered by Confederate artillery before they even came close to main Confederate line. The Confederate infantrymen patted the barrels of their smoking cannons in admiration.

On the afternoon of May 19th, Lee suspected the Federals were moving from his front. He ordered Ewell to conduct a reconnaissance in force to determine the Federal's intentions. Ewell's Corps was reduced to about 6,000 men, including the Georgia Brigade now under General Evans. Ewell moved to the northeast and soon encountered the Federals at the nearby Harris Farm. The Confederates attacked and faced the green troops of the Federal "heavy artillery," troops recently pulled from the forts around Washington, D.C. and now reluctantly assigned to serve as Federal infantry troops. The Confederates and Federals fought for hours before the Confederates retreated back towards Lee's main army. The 38th lost about half a dozen men during this small battle, all being captured.

Private Nichols recalled the march and reconnaissance, "We found the rear of the Yankee column about four miles from the field of battle, and they were moving towards Lee's right. It was nearly dark. A heavy column of them turned and came meeting us. We moved out about 100 yards from the road and lay down in ambush. We kept very quiet and it was too dark for them to see us, but they halted when they got to the part of the road nearest where we were lying. Many of them were cursing, and some shot towards us. We, privates, expected to have to fire a volley into them and charge them, but we did not. They soon all went back and we returned to our command." [17]

The Chestatee Artillery, one of the two artillery companies detached from the 38th when it departed Savannah during June of 1862, rejoined the regiment after the battles at Spotsylvania Court House. They had served for some time near Charleston, S.C., taking a prominent part in the defense of Battery Bee and Morris Island. It was an unhealthy climate and sickness devastated the company. On October 12th, 1863, the company commander, Captain Thomas H. Bomar wrote, "Owing to the large amount of sickness that is now in my command; that was last summer, and that will probably be every summer while I remain on the coast, I herewith make application to be transferred or exchanged with some artillery company…My men being from the mountainous portion of Georgia are unable to stand the emaciating effect of the coast climate and are very anxious to be transferred to colder regions. Should there be no chance to exchange or be transferred as artillery, I would not object to go as cavalry or light infantry. My men being accustomed to handle the rifle from infancy would make good sharp shooters and scouts" [18]

On May 5th, 1864, the Chestatee Artillery was ordered by the Confederate War Department to leave their cannons behind and rejoin with the 38th in Virginia to serve as infantry. At the same time, Company C, the Ben Hill Guards, was ordered on May 6th, 1864, by Special Order #105, to be permanently detached from the 38th, but this order was never carried out and Company C remained with the 38th until the end of the war. The probable intent of these Special Orders was to have the Ben Hill Guards swap assignments with the Chestatee Artillery, as the Ben Hill Guards had trained as an artillery company early in the war.

The Chestatee Artillery rejoined the 38th by May 21st, 1864, as several of its men were captured at Milford Station, Virginia on May 21st. Captain Bomar, being the senior captain in the regiment, was promoted to Major and assumed command of the 38th Georgia Regiment. Gordon provided an analysis of this campaign, counter to General Grant's famous order to General Gordon Meade, of "Where Lee goes, there you will go also."

"General Porter says: "It was the understanding that Lee's army was to be the objective point of the Army of the Potomac, and it was to move against Richmond only in case Lee went there." General Porter further adds that General Grant's own words to Meade were, "Where Lee goes, there you will go also." And yet on the failure of these last desperate assaults upon Lee at Spotsylvania, General Porter represents his chief as writing "an order providing for a general movement by the left, flank toward Richmond, to begin the next night."

I submit that this order of May 18th is hardly consistent with his previously announced plans of looking for Lee's army, and for nothing else, nor with his instructions to Meade: "Where Lee goes, there you will go also." Lee was not going toward Richmond except as Grant went toward Richmond. He was not going in any direction. He was standing still at Spotsylvania and awaiting the pleasure of General Grant. He had been there for about ten days, and was showing no disposition whatever to run away. There was no difficulty in finding him, and it was not necessary for General Meade to go to the North Anna or toward Richmond to find Lee in order to obey intelligently the instructions, "there you will go also."

General Lee first went into the Wilderness because General Grant had gone there, and Lee did not "get out of the Wilderness" until his antagonist had gone out and moved to another place. Lee moved to Spotsylvania because the Union commander was moving there; and any movement of General Meade away from Spotsylvania would be going where Lee was not. He was not on the Rappahannock, where the Union commander proposed to make his base; he was not retreating, he was not hiding. He was close by on the field which had been selected by his able antagonist, and was ready for a renewal of the struggle.

Verily it would seem that Grant's martial shibboleth, "Where Lee goes, there you will go also," had been reversed; for, in literal truth, Meade was not going where Lee went, but Lee was going where Meade went. It was General Grant's intention that General Lee should learn from every Union cannon's brazen throat, from every hot muzzle of every Union rifle, that nothing could prevent the Army of the Potomac from following him until the Confederate hosts were swept from the overland highways to Richmond. The impartial verdict of history, however, and the testimony of every bloody field on which these great American armies met in this overland campaign, from the Wilderness to the water route and to the south side of the James, must necessarily be that the going where the other goes was more literally the work of Lee than of Grant." [19]

Battles of Spotsylvania Court House, 8 - 21 May 1864
Killed, Mortally Wounded, or Captured and Died in Prison

Company A – "The Murphy Guards," DeKalb County, Georgia

1. Private John A. Adams - Captured at Spotsylvania, VA, 5/20/1864. Died of pneumonia at Elmira, NY, Sept. 1 or Oct. 10, 1864.
2. Private Robert W. Miller - Killed at Spotsylvania, VA, 5/18/1864.
3. Sergeant William D. Stewart - Captured at Spotsylvania, VA, 5/20/1864. Died at Elmira, NY, 9/16/1864.

Company B, "The Milton Guards," Milton County, Georgia (Currently Fulton County)

4. Color Corporal Joseph H. "Joe" Hall - Killed at Spotsylvania, VA, 5/12/1864.

Company C, "The Ben Hill Guards," Bulloch & Emanuel Counties, Georgia

5. Private James Daniel Brinson - Killed at Spotsylvania, VA, 5/12/1864.
6. Private Elisha Meeks - Killed at Spotsylvania, VA, 5/10/1864.
7. John R. Woods - Captured at Spotsylvania, VA, 5/18/1864. Died of typhoid fever at Elmira, NY, 2/17/1865.

Company D, DeKalb & Fulton Counties

8. Private Pressley Lanier ("Press") - Killed at Spotsylvania, VA, 5/12/1864.

Company E, "The Tom Cobb Infantry," Oglethorpe County, Georgia

9. Private George Nicholas Coile - Killed at Spotsylvania, VA, 5/12/1864.
10. 4th Corporal William W. Johnson - Killed at Spotsylvania, VA, 5/12/1864.
11. Private Thomas Martin - Killed at Spotsylvania, VA, 5/12/1864.
12. Private John W. Smith - Captured at Spotsylvania, VA, 5/12/1864. Died of typho-malarial fever at Fort Delaware, 6/14/1864.
13. Private John Bennett Young - Captured at Spotsylvania, VA, 5/12/1864. Died of smallpox at Fort Delaware, 7/31/1864.

Company F, "Thornton's Line Volunteers," Hart and Elbert Counties, Georgia

14. Private John H. Fleming - Captured at Spotsylvania, VA, 5/15/1864. Died of disease at Point Lookout, Md., 7/15 or 8/1/1864.

Company G, "The Battey Guards," Jefferson County, Georgia

15. Private Joshua P. Arrington - Killed at Spotsylvania, VA, 5/10/1864.
16. Private David R. Dye - Wounded at Spotsylvania, VA, 5/12/1864. Died of wounds at Lynchburg, VA, 7/12/1864.
17. Captain Levin W. Farmer (or Leven) - Killed at Spotsylvania, VA, 5/12/1864.
18. Private Sidney T. Farmer - Killed at Spotsylvania, VA, 5/10/1864.
19. Private Samuel M. McGahee, (or McGhee) - Captured at Spotsylvania, VA 5/19/1864. Died of general debility at Elmira, NY, 5/4/1865.
20. Private Alexander A. Rodgers - Killed at Spotsylvania, VA, 5/12/1864.
21. Private William A. Thompson - Killed at Spotsylvania, VA, 5/18/1864.

Company H, "The Goshen Blues," Elbert County, Georgia

None.

Company K, "DeKalb & Fulton Bartow Avengers," DeKalb and Fulton County, Georgia

22. Private Amos Wheeler - Killed at Spotsylvania, VA, 5/12/1864.

Company N, "The Dawson Farmers," Dawson County, Georgia

23. Private Isaiah Peter Millsaps - Pvt. 5/6/1862. Killed at Spotsylvania, VA, 5/12/1864.
24. Private John M. Martin - Captured at Spotsylvania, VA, 5/20/1864. Sent to Elmira Prison and died there of chronic diarrhea 10/24/1864.

Chapter 16
From Spotsylvania To Cold Harbor, May - June 1864

Lieutenant Lester wrote of the Army of Northern Virginia's retrograde from Spotsylvania, as they marched south to block General Grant's advance on Richmond. *"Our brigade commander, the noble hero, Gen. John B. Gordon, was promoted to Major-General, Col. C. A. Evans, of the 31st Ga. Regiment, was promoted to Brigadier-General and took command of our brigade. A braver soldier never drew sword in defense of the "lost cause" than Clement A. Evans. About the 19th we began to move from Spotsylvania towards Chickahominy. We were continually skirmishing with the enemy down the North Anna River."* [1]

Private Nichols wrote, "General Grant started on a flank movement, but we cut him off at North Anna River and Hanover Junction and fortified. Grant did not try to move us, but he tested the strength of some parts of Lee's lines. A heavy skirmish battle was in our front. Grant moved on to our right not far from South Anna River and we had to cut him off again at a place called 'Turkey Ridge,' near South Anna River. The Thirty-Eighth was sent out to run off the enemy's sharpshooters and skirmish line. They did so and held their position." [2]

The 38th lost at least eight soldiers, all captured, at the battle of the North Anna River, between May 22nd and May 24th, 1864. The circumstances of their capture are unknown. Two Milton County soldiers from Company B, were captured, Private Green B. Bostick and Private Green B. Bruce. Private David Truitt of Company C, the Ben Hill Guards, was captured May 22nd. Private Jesse P. Owen of Company L, the Chestatee Artillery, was captured May 23rd and died of chronic diarrhea Elmira Prison Camp, August 23rd, 1864. Other captured soldiers included Corporal Thomas H. Stewart of Company A, from DeKalb County. Sergeant Andrew J. Newton and Private Jesse W. Martin of Company B, the Milton Guards, were captured on May 24th. Private William H. Brisendine, of Company K, from DeKalb County, was captured May 24th and was held at Point Lookout Prison until February 14th, 1865, when he was exchanged. [3]

From the North Anna River Grant moved further south, attempting to slide around the right flank of Lee's army at Cold Harbor. Lee realized Grant's continuous march against Richmond would only end if Grant's army was destroyed. General Lee prophesied, *"We must destroy this army of Grant's before he gets to James River. If he gets there, it will become a siege, and then it will be a mere question of time."* Lieutenant Lester wrote, "There came to no general engagement until we reached Cold Harbor, the place where we were first initiated into the inglorious pomp and circumstance of war (on the 27th of June, 1862) where we were engaged more or less for several days. It is remarkable that our forces now held about the same position on the field of battle that the Yankees held at the first battle."

Private Bradwell wrote of a Georgia Brigade attack on Federal earthworks at Cold Harbor and told of reinforcements, the 12th Georgia Battalion, joining Evans'

Georgia Brigade, *"The Federal force holding these forts for some reason kept their heads down, and no sign of life could be seen in them. General Gordon, either through curiosity or to make a diversion to relieve pressure on the right of the army, decided to move his command forward. Accordingly the brigade, which had during my absence been reinforced by the 12th Georgia Battalion, was formed for battle and moved across the open space until it came within thirty feet of the enemy's works. Suddenly a whole line of infantry arose and fired a volley. When their heads popped up our men fell flat on their faces immediately, and not a man was hurt. Our men with their loaded guns jumped up and were on top of their works in a moment before the enemy could reload, shooting and bayoneting them without mercy. This was too much for them, and they broke for their reserve line, followed closely by our men, who entered this line with them. Frightened by this demoralized mass of friends and their foes all mixed up, they made little or no resistance and fled to their third line. Thinking they had gone far enough, our men stopped their pursuit at this line with the loss of but one man from our regiment. But not so with the soldiers of the 12th Battalion.*

These soldiers, sent to us from Charleston, SC, had a wonderful exaggerated idea of the prowess of our brigade from what they had heard of our fighting. I don't know what their orders were, but these fellows, fighting on the right of the brigade, after they had driven the enemy from two lines of their works, assaulted the third line. Taking these also, they continued to press the fleeing enemy, and in ardor of pursuit, lost some men. Their commander, Lt Col Capers, was badly wounded in the early part of the engagement and never again returned to his command. Some said our new soldiers never stopped in their charge until they went on to Washington. When they first arrived and began to detrain, our ragged veterans looked at these splendid fellows in their neat gray uniforms and decided they were 'pets' and had very little confidence in the fighting quality; but after Cold Harbor they never entertained a doubt that they would stand the test. In every engagement which our brigade took part, they showed the same dash and courage as in this their first fight in Virginia. When these brave fellows came to our brigade from Charleston, where they had been holding the pile of bricks in Charleston Harbor which was Fort Sumter, each company contained from one hundred and twenty-five to one hundred and eighty-five men and was as large as one of our decimated regiments. They were the flower of the fighting population of Georgia and showed the same fighting qualities to the end." [4]

The war had entered a new phase on May 5th, 1864. Before that time, the opposing armies had came together to fight a great battle, then separated, often for months before the next battle was fought. Grant introduced a new mode of warfare, continuous warfare. The armies remained in nearly continuous contact since the first day of the Wilderness. The men of the Army of Northern Virginia fought countless, unnamed skirmishes and endured the fire of Federal sharpshooters nearly every day.

Both armies were exhausted from over a month of marching and fighting with little rest. The terms "post traumatic stress" or "shell shock" were unknown, but the soldiers of both armies suffered from these maladies nevertheless. Union Captain Oliver Wendell Holms wrote, *"I tell you, many a man has gone crazy since this campaign began from the terrible pressure on the mind & body."* Many in General Lee's army undoubtedly felt the same way. Grant suffered a stunning defeat on June 3rd, when he launched a massive infantry attack against the entrenched Confederates, losing over 7,000 soldiers with little to show for the effort. Grant sustained 65,000 casualties between May 5th and June 4th, 1864. The main bodies of both armies would remain in almost constant contact until the surrender at Appomattox Court House, on April 9th, 1865.

Captain John G. Rankin, of Stone Mountain, Georgia, the Commander of Company D, and recently commanding the regiment, wrote a letter to Lieutenant George Riley Wells, also of Company D. He told of the terrible strain of the campaign and how the continuous marching and fighting had worn upon him and the men of the 38th Georgia Regiment.

Mechanicesville, Va, Sunday Evening, June 6th, 1864,

Lt G. R. Wells, I have just received yours of the 30th May from Columbia S.C. and was glad to see you were fortunate enough to get a furlough and have no doubt you are before this safely landed at home. I hope you will enjoy the time allowed to you there but you did not say for how long your leave was. The casualties in our company are as near as I can give them as follows. T. P Smith (Tilnon P. "Sally" Smith), Presley Lanier killed, A. Rosenthal (Adolphus Rosenthal), wounded in the foot and died at Richmond May 27th. It is possible, but I do not think probably that Lanier is wounded and in the hands of the enemy. I wish I could think so, but do not. Lt. G. R. Wells, R. M. Simpson, (Robert M. Simpson), J. A. Lafoy (John A. Lafoy), S. G. Harris (Starling G. Harris) C. A. Mason (Camilus A. Mason), J.A. McCanless (Jesse A. McCanless), J. J. Lawhorn (Joel J. Lawhorn) , J. M. and W. L. Singleton (James M. & William L. Singleton) wounded. Lawhorn and McCandless have come in. These comprise all our wounded unless some were wounded yesterday by an accidental shot from the enemy sharp shooters. F. M. Mehaffey (Francis M. Mehaffey), T. J. Williams (T. John Williams), M. H. Nash (Miles H. Nash) and W. L. Thomas (Wyett L. Thomas) are no doubt prisoners of war. We have had some hard times since you left and some hard fighting too, but I am worn out and have done no good for the last week relief and am now staying with the wagon train just outside of the fortifications at Mechanicsville. Hope I will be able to resume my duties by the end of this week if not before. Col. Davant (Phillip E. Davant), Adjutant Law (John Gordon Law), Lieut's Vaughan (Isaac C. Vaughn), Goss are suppose to be prisoners. Capt. Farmer (Levin W. Farmer), Deadwyler (Henry R. Deadwyler), and Lt. Farmer (Sidney T. Farmer) killed, also Lt Wiggins (Jackson C. Wiggins), Co. K, and Lieut, Crane (Solomon F. Crane, killed June 1st) of Capt. Bomar's company and Lt Hendricks (William Hendrix) of same co. badly wounded in leg. Capt. Bomar (Thomas H. Bomar) and seventy men of his company came up with us at Hanover Junction and he has taken command of

the regiment, greatly to my relief. I am not able to give you a list of, nor the remainder of the number killed in the regiment. George Braswell, Robt. Miller, James Gardner in Co. A are killed. Amos Wheeler, Co. K one leg off. Higginbotham (Elijah Benson "Bense" Higginbotham) of Company "F" Ambulance Corp, both legs off below the knees (poor fellow.) When I get back to the Regt. I will try and send you a list of all the killed and wounded from the three Co's from DeKalb. Lt Stubbs (George W. Stubbs, Co. K), Capt. Maxwell (Jackson O. Maxwell, Co. F), and Higdon (John B. Hidgon, Co. C) are with their commands. Capt. Robinson (George F. Robinson, Co. E) in Charlottesville. The 12th Ga. battalion all present except Wash. Johnson's Co. are now in our Brigade have some 460 men and are good fighters. Previous you have heard that Lt. Col. McAuther and Lt Col Jones, 13th Ga., both killed and several other field officers of the Brigade.
Write often won't you.
J. G. Rankin [5]

Little was written of the 38th Georgia's specific roles in the battles at Cold Harbor, between May 31st and June 12th, though the regiment suffered 25 casualties, the majority occurring on June 1st, according to Confederate service records. The Chestatee Artillery Company was particularly hard hit in their first major battle after rejoining the 38th Georgia as infantry, losing 13 soldiers either killed, or wounded, on June 1st. Captain William Mathews of the Battey Guards, wrote, *"Among the killed, the gallant young Lieut. Wiggins (Jackson C. Wiggins) of Company K and Capt. Miller of Company A, both brave, meritorious, and gallant men."* [6]

Casualties of Battles at Cold Harbor, VA, 1- 3 June, 1864[7]

Company A – "The Murphy Guards," DeKalb County, Georgia

1. Private George A. Braswell - Killed 6/3/1864.
2. Private Charles Gardner - Wounded, 6/1/1864. Died of wounds at Rich, VA 6/2/1864.
3. Private James Gardner - Killed, 6/1/1864.

Company F, "Thornton's Line Volunteers," Hart and Elbert Counties, Georgia

4. Private James Stansell - Wounded in right hip, 6/3/1864. Discharged for disability.
5. Private John A. Bailey - Wounded and captured, 6/3/1864.

Company H, "The Goshen Blues," Elbert County, Georgia

6. Sergeant Jeptha E. Campbell - Wounded in left arm and left lung, 6/1/1864.
7. Private W. J. Hendricks - Captured 6/3/1864. Paroled at Elmira, NY, 10/1/1864. Exchanged 10/29/1864. Died at sea while en-route to Venus Point, Savannah River, GA, 11/10/1864.
8. Private William James Hansard - Killed 6/1/1864.

9. Lieutenant Jasper L. Harbin - Captured 6/1/1864.

Company K, "DeKalb & Fulton Bartow Avengers," DeKalb and Fulton County, Georgia

10. Sergeant William Aron Wright - Captured 6/1/1864. Released at Elmira, NY, 6/16/1865.

Company L, Chestatee Artillery

11. Corporal Theodore M. Andoe - Wounded, 6/1/1864, died from wounds 9/4/1864.
12. Lieutenant Solomon Crane - Killed at Cold Harbor, VA 6/1/1864.
13. Private Charles M. Elrod - Wounded 6/1/1864.
14. Private George W. Garrett - Killed 6/1/1864.
15. Private John B. Gibson - Wounded 6/1/1864.
16. Private Ezekiel L. Henderson - Wounded 6/1/1864.
17. Captain William Hendrix - Wounded 6/1/1864.
18. Private Simeon Morgan - Killed 6/1/1864.
19. Private William Jasper Owen - Killed 6/1/1864.
20. Corporal Enoch Patterson - Wounded left leg, 6/1/1864, permanently disabled.
21. Private Samuel Westbrook - Wounded 6/1/1864.
22. Private William R. Westbrook - Wounded, left arm, 6/1/1864, arm amputated above elbow.
23. Private Egbert Baldwin McDaniel - Captured 6/3/1864.

Company N, "The Dawson Farmers," Dawson County, Georgia

24. Private Ervin Alexander Loggins - Wounded and disabled, 6/1/1864.
25. Private John Wesley Jay - Wounded and disabled at Turkey Ridge, Cold Harbor, gunshot wound to bowels, admitted to hospital June 2nd, 1864.

Chapter 17
From Cold Harbor to Washington, D. C., June - July 1864

Private Bradwell wrote of the time after the battles at Cold Harbor, "But now a dark cloud was rising for Lee's army from the west which' threatened to sweep down on Richmond, capture the city, and force the Confederate army to surrender. It was a desperate situation and lacked only bold leadership to succeed; but our old general was equal to the occasion. When Lee's army was grappling with Grant's there was only a small force in the Valley to protect it, and Hunter came out of the mountains of West Virginia and had his own way, destroying and burning that beautiful section of country-After doing this to his satisfaction he crossed the Blue Ridge unopposed and marched to Lynchburg, which place was undefended except by a few old men too weak in number and equipment to offer any resistance. Here he paused and began to fortify his position as if he were face to face with a formidable foe." [1] The Georgia Brigade and 38th Georgia were "withdrawn from the defenses and placed in camps. in the rear of the army, where we were allowed to rest two days."[2]

Major General David Hunter, USA
Library of Congress

Grant realized Lee's earthworks at Cold Harbor were impregnable, so he sought a way around Lee's defenses. Grant began marching his army around Lee's right flank to the James River, on June 12, 1864. He planned to bypass Richmond and seize the railroad junction south of Richmond at Petersburg, Virginia, isolating Richmond. Grant threw a massive pontoon bridge across the James River and marched his army towards Petersburg, setting the stage for a siege that would last from June of 1864 until April 2nd, 1865.

Lee dispatched Early and his 2nd Corp to thwart Hunter's attack on Lynchburg and if possible, to draw troops away from Grant's army. The Georgia Brigade, as part of the Early's 2nd Corp, was sent to confront Hunter's army at Lynchburg, Virginia.

"About the 14th of June we were detached and sent under Gen. Early again to the Valley. Our boys were delighted at this change. The Valley was our favorite campaigning ground," [3] wrote Lieutenant Lester. The 38th had, "no idea that we were about to enter upon a great campaign in an entirely different direction, which would eventually take us to the defenses at Washington and in sight of the Capitol."[4]

"We then set out on the march and soon reached the railroad, which we found to be completely destroyed by Sheridan's army. When we reached the vicinity of Trevilians Station dead horses were everywhere in evidence, and the whole country showed the effects of the battle that had been fought there a few days before between Gen. Wade Hampton and Sheridan." [5]

Lieutenant Lester recounted the 38th Georgia's march from Cold Harbor to Lynchburg, *"We marched all the way to Gordonsville; there we took the train for Lynchburg and arrived on the 19th and in the very nick of time to meet the Federals under Gen. Hunter, who declined to meet us and could not even be detained long enough to give us battle, but ingloriously fled the field in the utmost and wildest confusion, abandoning everything. He continued his flight into Ohio. We captured thus, without a struggle, a large number of prisoners and among other things thirteen pieces of artillery.*

We now took up our line of march and marched continually for days and days, averaging about 20 miles a day. On this march we passed through Lexington, the home of the immortal Stonewall Jackson. We visited his grave. He had fought his last battle, had met the great enemy of our race, and "crossed over the river and was resting under the shade. The green sods of the Valley now pressed on his bosom. With our hearts sad and our eyes filled with tears we marched round his grave at reversed arms, Here were the battle-scarred veterans who had met death in a thousand forms on a score of bloody battlefields unused to weeping, all melted in tears. General Gordon stood over the grave with head uncovered and tears streaming down his face, while the troops marched around the grave of our beloved commander.

Continuing our march we reached our old stamping ground (Winchester) and without any unusual occurrence. We found things in the Valley greatly changed since we left it. The Yankees had control of the country for a long time - ruin and devastation met our eyes on all sides. This fertile and blooming Valley was nothing now but desolation and a mass of shouldering ruins. From Winchester we continued on to Martinsburg, where we arrived on the 3rd of July. Here the Yankees had a large force under Gen. Seigell, but they fled across the Potomac on the approach of old Jubal A, Early.

Here at Martinsburg Gen, Seigell and the Yankees were preparing to celebrate, on a large scale, the glorious old 4th of July. But a barbecue and the 4th of July are institutions peculiar to the old Mother of States and statesmen. They had a table spread about 200 yards long, loaded with all the good things that could be imagined. At this feast the rebels appeared as uninvited guests and proposed to join in the celebration. Dutchy took this in high dudgeon and retired across the Potomac in perfect disgust, leaving their uninvited guests masters of the situation.

Our boys had nothing to do but to fall aboard of those delicious viands, barbecued meats and everything in profusion which they discussed to the fullest extent, while the Yankees on the other side of the rivers watched us with mouths watering and eyes filled with tears. Our boys thought it rather an ill mannerly trick that we had played on the Dutch, but consoled themselves with the thought that the Dutch had no right to eat this fine barbecue on Virginia soil. Here we not only got the 4th of July barbecue, but also a large amount of quartermaster's stores, clothing, hats, shoes and blankets-in fact, everything to delight a soldier. The citizens were greatly rejoiced to see us return to the Valley and many acquaintances were renewed which we had formed before.

At Martinsburg, the Tom Cobb Infantry were detached and left at this place to guard the captured stores, the balance of Gen. Early's command going on into Maryland and marched near Washington." [6]

Sergeant William F. A. Dickerson of Company D, the McCullough Rifles, noted, "On the 4th, five companies of the 38th Regiment were left at Martinsburg, West Virginia, on provost guard, Captain J. G. Rankin (John G. Rankin) in command." [7] In addition to the McCullough Rifles and Tom Cobb Infantry companies, evidence indicates the other three companies remaining at Martinsburg were Company B, the Milton Guards, Company C, the Ben Hill Guards, and Company I, the Dawson Farmers. These five companies suffered no casualties in the battle soon to come at Monocacy, Maryland.

Lieutenant Lester wrote of the fine time the five companies had while the other five companies marched with Early's army towards Washington, D.C., "During this raid of General Early's we were at Martinsburg for a short time, and were then ordered to Winchester, where we did provost duty for some weeks or two. Here we had a splendid time." [8]

Major Gen. Lewis Wallace, USA
Library of Congress

Private Nichols recalled, "On July 5th we crossed the Potomac at Shepherdstown into Maryland. We crossed Antietam creek, and on the 6th Gordon's division drove the enemy into their works on the Maryland Heights. Here we drew plenty of good shoes, of which we were in great need. The Yankees shelled us severely from the Heights, killing and wounding some in the brigade. On the morning of the 8th we marched through South Mountain, at Fox Gap, to Frederick City, Maryland." [9]

Union Major General Lewis "Lew" Wallace had been removed from field command, by Grant, for his poor performance at the battle of Shiloh, in April 1862. His punishment was banishment to command the Middle Department and Eighth Army Corps, in Baltimore, Maryland. When Wallace learned a Confederate army was marching towards Washington, he formed a small army of 5,800 men and posted them in a defensive position along the east side of the Monocacy (pronounced "min-AH-ka-see") River, about three miles from Frederick, Maryland.

Major Thomas Bomar commanded the five companies of the 38th Georgia remaining with the Georgia Brigade on the march to Washington D.C. Casualties incurred at the battle of Monocacy suggests these five companies were Company L, the Chestatee Artillery (now serving as infantry) Company A, the Murphy Guards,

Company F, Thornton's Line Volunteers, Company G, the Battey Guards, and Company H, the Goshen Blues. Just after passing through the city of Frederick, the Confederates found their advance blocked by General Wallace's small army. Private Nichols wrote, "Here we found a large force of Yankees commanded by General Lew Wallace, who retreated across Monocacy River and took a position in a road leading towards Washington. McCausland's brigade of cavalry crossed the river in pursuit and raised a row with them." [10]

Gordon reported, "About 2.30 p. m. July 9th, I was ordered by Major-General Breckinridge, commanding corps, to move my division to the right and cross the Monocacy about one mile below the bridge and ford on the Georgetown pike, which was then held by the enemy." [11] Gordon's division consisted of his old Georgia Brigade, now commanded by General Clement A. Evans, Terry's Virginia Brigade, and York's consolidated Louisiana Brigade.

Private Bradwell recalled the layout of field and the position of the Georgia Brigade, "The road from Frederick City to Washington, D.C., crosses Monocacy two or three miles southeast of that place. A short distance east of the bridge a road leads off in a northwestern direction to Baltimore. Wallace formed his army with these two roads to his back, so that he could retreat on either of them if necessary. His right was protected by the Monocacy, and we could not attack him from that direction, nor get our artillery over to assist us in the fight. Gordon took our brigade by a wide detour to the south, were we managed to ford the river and come to the help of the cavalry, which had already been used up in the initial clash with the Federal cavalry." [12]

General Gordon wrote of deploying his troops for battle, "On reaching the river I directed my brigade commanders to cross as rapidly as possible, and then to file to the left in the direction of the enemy's line, and I rode to the front in order to reconnoiter the enemy's position. I found that Brigadier-General McCausland's cavalry brigade (dismounted) had been driven back by superior numbers, and that the enemy was posted along the line of a fence on the crest of the ridge running obliquely to the river. In his front lay an open field, which was commanded by his artillery and small-arms to the extent of their range, while in his rear ran a valley nearly parallel with the general direction of his line of battle. In this valley I discovered from a wooded eminence in front of his left another line of battle in support of the first. Both these lines were in advance of the Georgetown road. The enemy's line of skirmishers covered the front of this first line and stretched far beyond it to the left. Having been ordered to

Gen. Zebulon York, CSA
Library of Congress

attack this force, I had the division skirmishers, under Captain Keller, of Evans' brigade (the Georgia Brigade), deployed, and directed one brigade (Evans'), under the protection of a dense woodland about 700 yards in front of the enemy's left, to move by the right flank and form so as to overlap the enemy's left. These dispositions having been made, I ordered the command to advance in echelon by brigades from the right." [13]

An attack by echelon brigades was designed to entice the enemy to weaken other parts of their line, to support the troops confronting the first attacking column. The Georgia Brigade, under General Evans, was the first brigade ordered forward to attack. Once the Brigade was fully engaged, the next brigade in line, York's Virginia Brigade, would attack. Once York's Brigade was fully engaged, Terry's Louisiana Brigade would advance. Each brigade's successive, sledgehammer blows, would strike a weakened enemy line, finally crumbling their line of battle. Gordon's plan of attack, after crossing the river, was to cross a small ridge line and strike the flank of the Federal force deployed along the river. However, the Federals discovered Gordon's Division crossing and quickly reformed their lines to face the Confederate attack head on. Private Bradwell wrote, "Our brigade was formed behind a low wooded mountain range out of sight of the enemy. Orders were given to advance quietly over the ridge, where it was thought we would strike their left flank in the open field. We were then to charge them with the usual yell. We advanced in fine style over the mountain, but when we came to the open field we found their lines adjusting to meet us." [14]

Though Wallace commanded the entire Union force, the Federals facing the Georgia Brigade were General James B. Ricketts Division of the VI Corps. These battle-hardened veterans had been pulled from lines around Petersburg and rushed north to defend Washington.

The Georgia Brigade had to advance over 700 yards of open ground. The brigade line of battle was quickly broken, as the fields were interspersed with fences and shocks of wheat. The Federal infantry unleashed a murderous volley on the brigade as they advanced, further adding to the confusion. Private Nichols described the advance, "We formed in line and advanced on their position about three-quarters of a mile through an open field. We charged them and our yell was answered by a well-aimed volley which seriously wounded General Clement Evans and Captain Gordon, his aid, and killed or wounded every regimental commander in the brigade, besides many of our company officers." [15]

Gordon wrote, "The troops emerged from the woods 700 yards in front of the enemy's left under heavy fire from infantry and artillery, and had advanced but a short distance when, on account of the wounding of one brigade commander (Evans), to whom explicit instructions had been given as to the movement of his (the leading) brigade, and the killing of several regimental commanders, and the difficulty of advancing in line through a field covered with wheat-shocks and intersected by fences, the perfect alignment of this brigade was necessarily to some extent broken. However, this temporary confusion did not retard its advance, which, as I had anticipated, forced the enemy to change his front under fire." [16]

Sergeant Josiah C. Langford of Company A, the Murphy Guards, was shot in back of the left elbow, the ball going through the elbow joint, breaking the bone and tearing off the end of the joint. He was left at Monocacy with the other wounded and fell into the hands of the enemy when the brigade moved on towards Washington. [17] General Evans was shot off his horse in the charge, the ball passing through his body. Colonel Atkinson, commanding the 26th Georgia Regiment, took command of the brigade. The 61st was posted to the right of the 38th, and suffered severely from the Federal fire. Private Nichols wrote, "Colonel J. H. Lamar and Lieutenant Colonel J. D. Van Valkinburg (the hero of the wilderness) were both killed on the field...We advanced to within thirty yards of the line of Yankees, but we would have had to fall back, for so many of our men were killed and wounded until we did not have but a mere skirmish line...we could not see a Yankee on our part of the line during the whole advance. All that we could shoot at was the smoke of their guns, they were so well posted." [18]

Gen. Clement A. Evans, CSA
Georgia Brigade Commander
Library of Congress

The Chestatee Artillery, serving as infantry, suffered severely. Sergeant Phillip S. McDaniel was killed in action. Forty-eight year old Private Reuben Partin was shot in the face, the ball breaking his jawbone. Also seriously wounded were Private William J. Bennett, Private Francis M. Borin, and Corporal Andrew Jackson Garrett, all were left behind in a Frederick field hospital and fell into the hands of the enemy the day following. [19]

Private Nichols wrote, "I heard a member of the Twenty-sixth Georgia Regiment say that a Yankee shot at them from behind a shock of wheat, and that several of his regiment shot at the shock of wheat and they hit the Yankee with eighteen balls. Colonel J. H. Baker, of the Thirteenth Georgia Regiment, was wounded. Captain W. H. Harrison, Company E, Thirty-first Georgia Regiment, was severely wounded while leading' his company so bravely. He was left at Frederick City a prisoner. It looked like half of the Twelfth Georgia Battalion were killed or wounded. Company D had the sad misfortune of getting Lieutenant James Mincy severely wounded. He was carrying our battle flag. He had picked it up after the fifth man had been shot down while carrying it in this battle and he was likewise shot down at once. He had already been wounded at Manassas and severely wounded at Gettysburg. Here he was shot through the left lung, the ball just missing his back bone. Bloody froth from his lungs would come out of his mouth and nose, and in the front and back where the ball passed through. He has since told me that the Yankee doctors drew a silk handkerchief through him and treated him very kindly." [20]

The Georgia Brigade stalled and was being shot to pieces, it desperately needed help. The time had come for Gordon to order the York's Louisiana brigade forward. Gordon noted, "At this point the Louisiana brigades, under the command of Brigadier-General York, became engaged, and the two brigades (Evans' and York's) moved forward with much spirit, driving back the enemy's first line in confusion upon his second. After a brief halt at the fence from which this first line had been driven I ordered a charge on the second line, which was equally successful. At this point I discovered a third line, which overlapped both my flanks,

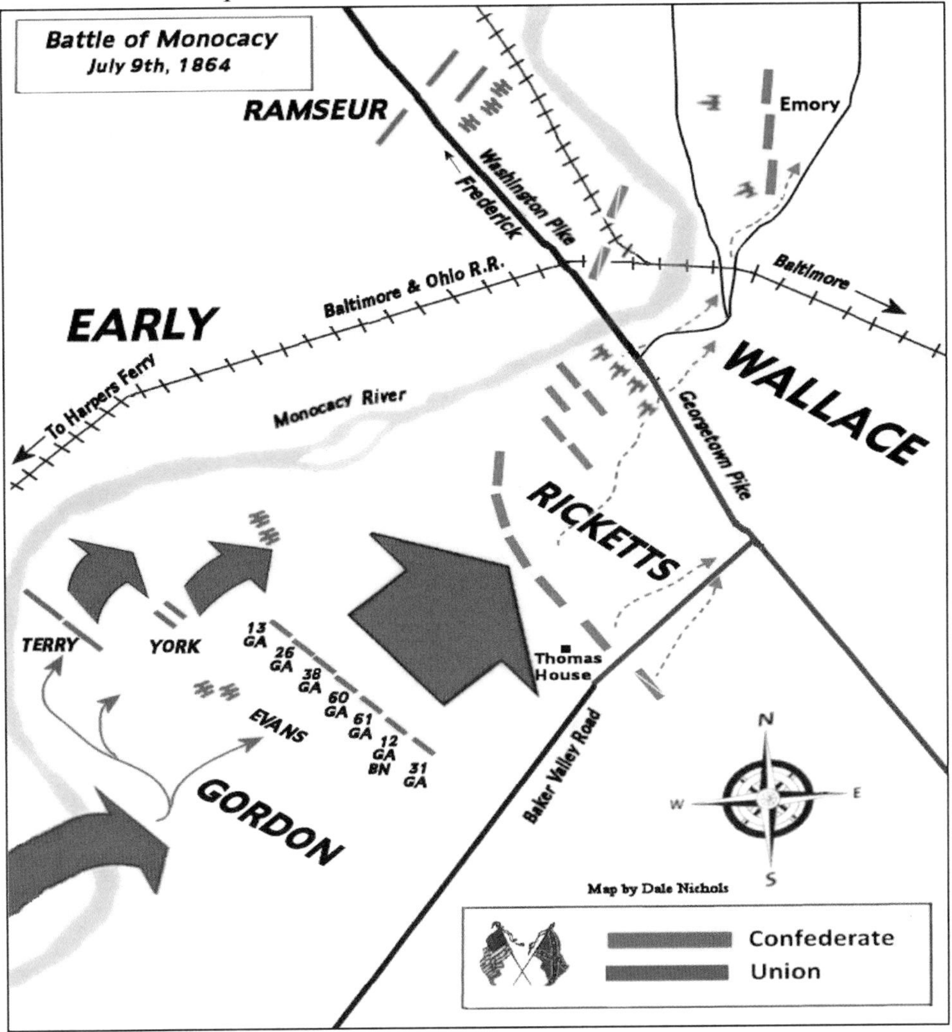

and which was posted still more strongly in the deep cuts along the Georgetown road and behind the crest of the hill near the Monocacy bridge." [21] Private Nichols observed, "Wallace's men were well posted in a road that was washed out and graded till it was as fine breast works as I ever saw." [22] Gordon said, "At once I ordered Brigadier-General Terry, who as yet had not been engaged, to attack vigorously that portion of the enemy's line nearest the river, and from which my troops were receiving a severe flank fire. This brigade advanced with great spirit and in excellent order, driving the enemy from his position on a portion of the line. He still held most stubbornly his strong position in front of the other two brigades and upon my right. He also advanced at the same time two fresh lines of troops to retake the position from which he had been driven by Terry's brigade. These were repulsed with heavy loss and in great confusion." [23]

General Evans and many other key officers had been shot down, but the veteran soldiers of the Georgia Brigade pressed forward. Private Bradwell wrote, "The private soldiers were all veterans and knew what to do. They rushed at the enemy in the field and drove them to a sunken road, which they held against our left so well that our men became discouraged at their heavy loss, and finally succeeded in driving them out of their position only with the timely assistance of the gallant little remnant of the Louisiana ("Tigers") Brigade.

Gen. William Terry, CSA
Library of Congress

On the right the enemy was well posted in the Thomas residence and in the grove around the house, and as the 12th Battalion came up in the open field to attack them, they sustained considerable loss. Finally our artillerymen got one of their guns across the river and placed in position in the back yard of the Worthington residence. With this they opened on the enemy in the Thomas house with such deadly effect that the 12th routed the enemy from that place." [24]

Gordon reported, "Having suffered severe loss in driving back two lines, either of which I believe equal in length to my command, and having discovered the third line longer than either of the others and protected by the cuts in the road, and in order to avoid the great loss it would require to drive the enemy from his position by a direct front attack, I dispatched two staff officers in succession to ask for a brigade to use upon the enemy's flank. Ascertaining, however, that a considerable length of time must elapse before these could reach me, I at once ordered Brigadier-General Terry to change front with his brigade to the right and attack the enemy's right. This movement promptly executed with a simultaneous attack from the front, resulted in the dislodging of their line, and the complete rout of the enemy's forces. This battle, though short, was severe." [25]

Private Isaac Bradwell wrote of this point in the battle, *"The enemy took the road to Baltimore, pursued by our cavalry to that place, and the next day we resumed our march to Washington. Before the fight closed in the center there were only four men still offering resistance to the enemy behind the banks of the road. When the last squad of these ran away, my three companions and I stood in amazement and looked around. The fighting had come to a close so suddenly that we could not first take in the situation. In every direction scattered over the field could be seen guns, army blankets, and other equipment cast off by the enemy, and on a hillside to the left a number of wounded Federals.*

A fire burning the wheat straw was making its way slowly towards these. Seeing their danger and hearing their cries for help, we picked up tent flies and fought the flames, which we finally subdued. The wounded men expressed their gratitude to us and begged us to fill their canteens for them. This we did and left them there for the litter bearers who we knew would be along after awhile. We had won a complete victory, but at a great sacrifice to us. The casualties of the Confederates in this engagement fell almost entirely on Gordon's Brigade (the Georgia Brigade) and amounted to not less than five hundred men killed and wounded, while that of the enemy was much less. All of these veterans of many battle fields, some of whom bore the scars of many wounds. Except the Louisianans, this battle was fought entirely by Gordon's Brigade." [26]

In this short, but bloody battle, the Georgia Brigade fought along a stream near the second line of the enemy and the stream actually ran red with blood. General Gordon witnessed the horrific event, *"I desire in this connection to state a fact of which I was an eye-witness, and which, for its rare occurrence and the evidence it affords of the sanguinary character of this struggle, I consider worthy of official mention. One portion of the enemy's second line extended along a branch, from which he was driven, leaving many dead and wounded in the water and upon its banks. This position was in turn occupied by a portion of Evans' brigade (the Georgia Brigade) in the attack on the enemy's third line. So profuse was the flow of blood from the killed and wounded of both these forces that it reddened the stream for more than 100 yards below. It has not been my fortune to witness on any battle-field a more commendable spirit and courage then was exhibited on this by both officers and men."* [27]

Georgians in Hospital, Frederick, Maryland, July 29, 1863.

We find in the Wilmington Journal a list of sick and wounded Confederates at U. S. General Hospital at Frederick, Md. July 29th. The following are from Georgia. G. H. Miller, Co. I, 61st Ga., T. J. Rutledge, Co. C, 6th Ga., A. L. McNair, Co. G, 31st Ga., J. Knowles, Co. G, 59th Ga., J. S. Leath, Co. G, 13th Ga., S. C. Giddeon, Co. H, 26th Ga., H. L. Paul, Co. F, 12th Ga., J. A. Hicks, Co. G. 1st Ga. Cavalry, D. L. Brewton, Co. E. 36th Ga., John Langford, Co. E, 60th Ga.

The following Georgians have died at said Hospital. Capt. J. T. Lane, 4th Ga., July 25th, 1863, Private A. Boyd, Co. C, 38th Ga., July 25th, 1863. Reported by Col. T. S. Kenan, Col. of 43d North Carolina troops.

Confederate Union, Aug. 18, 1863 -- page 3

Once the battle ended, Private Nichols recalled how Hay's Louisiana's Brigade had turned the tide of battle, "If it had not been that Hay's Louisiana Brigade (who) formed a line on our left and flanked the Yankees out of their position, we would have suffered worse than we did. It made our hearts ache to look over the battle-field and see so many of our dear friends, comrades and beloved officers, killed and wounded. Our loss was terrible, while the Yankees lost but few. I only saw three dead Union soldiers and I did not see one that was wounded, though I did not go over the field." [28] Captain William C. Mathews of the Battey Guards recalled the casualties of the battle at Monocacy, "We had one of the hottest fights for the numbers engaged during the war. Our loss was five killed and thirty wounded." [29] Private Nichols wrote of the Georgia Brigade casualties, "It was called our victory, but it was a costly one, for it cost Evans' Brigade (the Georgia Brigade) over five-hundred men, in wounded and killed. The Sixty-first Georgia Regiment went into the battle with nearly one hundred and fifty men, and after the battle was over we could not stack but fifty-two guns by actual count. We camped on the battle-field, drew and cooked rations and left early the next morning (July 10th). We took the road leading directly to Washington City, with McCausland's cavalry in front." [30]

Many of the wounded were carried to Frederick, Maryland, about three miles from the battlefield. These men were left to the mercy of the enemy and were quickly captured when Early's army departed on the morning of July 10th, 1864. Surgeon G. W. Graves, of the 38th was left behind to care for the wounded, fully knowing he would soon become a prisoner of war. [31]

A memorial marker stands in Mount Olivet Cemetery in Frederick, Maryland, over the remains of 408 unknown Confederate soldiers who died in the battle of Monocacy. Many of these soldiers undoubtedly belong to the Georgia Brigade. The battle of Monocacy was the first and only significant Confederate victory on northern soil, the Richmond Examiner newspaper reported, "We at home will not forget Gordon and Gordon's men, the man and the men have destroyed the prestige of Northern success on Northern soil…it is our first undisputed success on Northern soil, and it is an earnest hope it will not be the last." [32] Though a Confederate victory, the battle of Monocacy had far greater strategic value than many of the much larger battles of the war. Wallace's small army had been defeated, but had delayed the advance of the Confederate Army towards Washington, D.C. for a full day, a day that would prove crucial. In the aftermath, Wallace was relieved of command for his defeat at the battle of Monocacy.

Unfinished dome of U.S. Capital, 1860 - *Photographic History Collection, National Museum of American History*

Early's army departed early the next morning for Washington, D.C., Private Nichols wrote of the extreme, hot weather that slowed the Confederate advance, "It was extremely hot, dry, dusty weather. We camped near Rockville, and were started by daylight next morning. It was one of the hottest days I ever felt. It had not rained for several weeks and we could not get but little water on the road. We were enveloped in clouds of dust and a great many of our dear boys fell by the roadside from exhaustion and we had several sun-strokes, but we pulled on the best we could." [33]

Captain Mathews of the Battey Guards wrote of the 38th, "The 38th had been marched and fought until it did not number 100 men on the march to Washington, near which place, in sight of the dome of the capital, we had a fight in the yard and grounds of Montgomery Blair, losing one man killed and four wounded." [34]

Early's army arrived at the outskirts of Washington, D.C., on July 11th, at about noon. he wrote, *"I rode ahead of the infantry, and arrived in sight of Fort Stevens on the road a short time after noon, when I discovered that the works were but feebly manned. Rodes, whose division was in front, was immediately ordered to bring it into line as rapidly as possible, throw out skirmishers, and move into the works if he could. My whole column was then moving by flank, which was the only practicable mode of marching upon the road we were on, and before Rodes' division could be brought up, we saw a cloud of dust in the rear of the works towards Washington, and soon a column of the enemy filed into them on the right and left and skirmishers were thrown out in front, while an artillery fire was opened on us from a number of batteries. This defeated our hopes of getting possession of the works by surprise, and it became necessary to reconnoitre... I knew that troops had arrived from Grant's army, for prisoners had been captured from Rickett's division of the 6th corps at Monocacy. After dark on the 11th I held a consultation with Major Generals Breckenridge, Rodes, Gordon and Ramseur, in which stated to them the danger of remaining where we were, and the necessity of doing something immediately, as the probability was that the passes of the South Mountain and the fords of the upper Potomac would soon be closed against us. After interchanging views with them, being very reluctant to abandon the project of capturing Washington I determined to make an assault on the enemy's works at daylight next morning, unless some information should be received before that time showing its impracticability, and so informed those officers.*

Fort Stevens, near Washington, D. C., 1865
Library of Congress

During the night a dispatch was received from Gen. Bradley Johnson from near Baltimore informing me that he had received information, from a reliable source, that two corps had arrived from General Grant's army, and that his whole army was probably in motion. This caused me to delay the attack until I could examine the works again, and as soon as it was light enough to see, I rode to the front and found the parapets lined with troops. I had, therefore, reluctantly to give up all hopes of capturing Washington, after I had arrived in sight of the dome of the Capitol, and given the Federal authorities a terrible fright. On the afternoon of the 12th, a heavy reconnoitering force was sent out by the enemy, which, after severe skirmishing, was driven back by Rodes' division with but slight loss to us." [35]

President Abraham Lincoln rode out to Fort Stevens to observe the action. He mounted the parapet to get a closer view of the action and a man standing beside Lincoln was shot down by a Confederate sharpshooter. Legends says a young Captain Oliver Wendell Holms, (future U.S. Supreme Court Justice) shouted to President Lincoln as he stood erect on the parapet, 'Get down you damned fool before you get shot!'

About dark we commenced retiring and did so without molestation." General Early remarked to a staff officer, 'We didn't take Washington, but we scared Abe Lincoln like Hell.'"

Captain Pearce of the Georgia Brigade wrote of the retreat from the gates of Washington and of the re-crossing the Potomac River into Virginia,

"On Thursday night, our forces, which had much scattered driving cattle, having been got together, we began our retrograde movement towards Virginia. This was the most quiet and leisurely march of the expedition. Our troops moved slowly, driving the cattle and horses in front. The enemy's cavalry followed in our track, but at a very safe distance behind. If any of our men were picked up by the enemy it was because, in violation of orders, they had wandered off from the main column in search of plunder.

Our army brought south of the Potomac five thousand horses and twenty-five hundred splendid beef cattle; besides our cavalry and artillery all are supplied with new and valuable horses. Our men are all in great spirits and charmed with the success of their expedition. They represent the time they spent in Maryland as "glorious." The only regret connected with the expedition is the necessity we were under of leaving at Monocacy bridge such of our wounded as could not sit on their horses." [36]

Though Early didn't capture Washington, the campaign was successful in several respects. Early won the only Confederate victory of the war on Northern soil, he forced Grant to dispatch two infantry corp from the trenches at Petersburg, and he also successfully gathered supplies desperately needed by the Confederate army, most importantly, five thousand horses and twenty-five hundred beef cattle.

Grant returned Wallace to command once he realized Wallace had saved the capital from capture, by delaying the Confederate army. Grant noted that Wallace's initiative and daring likely prevented the Confederates from capturing Washington, D. C., "General Wallace contributed on this occasion, by the defeat of the troops under him a greater benefit to the cause than often falls to the lot of a commander of an equal force to render by means of a victory." [37]

Chapter 18
Return to the Shenandoah Valley, Aug - Sept 1864

The five detached companies of the 38th Georgia Regiment selected to guard the captured stores at Martinsburg, rejoined the Georgia Brigade at White Post, Virginia. The 38th was reunited. Lieutenant Lester of the Tom Cobb Infantry wrote of the trying times in the Shenandoah Valley, "We rejoined Gen, Early at White Post, and were now continually after the Yankees, marching and skirmishing alternately. About this time rations gave out entirely and we were forced to forage for a living. It was during this famous tramp around Winchester that we came upon the largest blackberry patch that was ever seen. The army was halted and Generals and all went for the blackberries in a business like way, effectually filling the vacuum that several days of fasting had created." [1]

Gen. George Crook, USA
Library of Congress

"The Confederate army now became a shuttlecock in the game of war, marching and counter marching up and down, in and across the valley of the Shenandoah, in military maneuvers, with scarcely a day of rest. This fruitful valley was to be the granary for its supplies. From it, as a base of operations, Early would make his frequent forays – a constant menace to the peace of the authorities at Washington."

Sergeant William F. A. Dickerson of Company D, the McCullough Rifles, 38th Georgia, left a detailed account of the marching and countermarching of Early's army and the Regiment, "At twelve o'clock on the night of the 20th (July) we marched in the direction of Middletown. Arrived there on the 21st (July) and stayed until the 22d. Marched to Strasburg. Camped until the 24th (July) when we marched back to Middletown, where we had a fight." [2] Gen. Early's army defeated Gen. George Crook's VIII Corps on July 24, 1864, and Crook fled across the Potomac River. Gen. Early's army was now in full possession of the Shenandoah Valley." [3] The 38th Georgia had at least one soldier killed in the battle of July 24th. Captain George W. Stubbs of Company K, Bartow Avengers, was killed. Captain William C. Mathews lamented, "He had been a brave and valiant soldier and his death left his company without an officer." [4] He is buried at Stonewall Confederate Cemetery in Winchester, Virginia, just a few miles from where he was killed.

Gen. John McCausland - CSA
W. Va. State Archives

Sergeant Dickerson recalled marching and tearing up the Baltimore & Ohio Railroad near Martinsburg, "At Winchester we camped until the evening of the 25th, then marched near Bunker Hill, W. Va. On the 26th we passed through Martinsburg, W. Va. Went to work tearing up Baltimore & Ohio R. R." [5]

Early ordered his cavalry, under McCausland, to cross the Potomac and proceed to the town of Chambersburg, Pennsylvania, with instructions to collect a ransom from the town's inhabitants, and if refused, to burn the town to the ground. Early's orders were in response to Federal General David Hunter's burning of Lexington, Virginia a few months earlier. McCausland arrived at Chambersburg on July 30th, 1864 and demanded $100,000 in gold, or $500,000 in greenbacks. His demands were refused and his men reduced the town to ashes. [6]

Ruins of Chambersburg, Pa., after burning by General McCausland's Cavalry
Library of Congress

It was a trying time for Early's army and the Georgia Brigade, a time of constant marching and skirmishing with the enemy. Sergeant Dickerson meticulously documented the 38th's travels, *"On the 31st went to Darksville, where we camped until August 3d, then to Martinsburg, W. Va.*

On the 4th (August 1864) to Shepherdstown, W. Va. Crossed the Potomac; had some skirmishing and camped at Strasburg. On the 6th we recrossed the Potomac at Williamsport. On the 7th to Darksville, where we camped until the 9th when we marched to Bunker Hill, W. Va. On the 10th to Jordan Springs. On the 11th we skirmished all day between there and Newtown, W. Va. On the 12th we fell back three miles to Fisher's Hill, where we remained until the 15th (August). Advanced our line, when we had two killed and five wounded." [7] The two soldiers killed at battle of Deep Bottom, on August 15th, 1864, were Captain Skirving Price, Assistant Surgeon and Private James Washington Coile of Company E, from Oglethorpe County, Georgia, Sergeant Dickerson eulogized the death of Private Coile, *"In one of these skirmishes we lost another good soldier, James W. Coile, killed and fell into the hands of the enemy."*

A private recalled the loss of Captain Skirving Price, assistant Surgeon of the 38th, during the skirmish at Deep Bottom, *"In one of these battles Dr. Price, assistant surgeon of the Thirty-eighth Georgia Regiment, was killed while riding up and down our skirmish line. He had no business there and was doing us no good, and he made an excellent target for the enemy to shoot at. I heard brave officers and men say that he acted more like a fool than a brave man."* [8]

The trials of the Georgia Brigade and 38th Georgia Regiment were far from over, Sergeant Dickerson of the McCullough Rifles wrote, *"On the 16th we laid fortifications. On the 17th (August) we had a fight near Winchester; had one wounded in our company and four wounded in the regiment. On the 18th (August) we stayed in camp. On the 19th we marched to Bunker Hill, W. Va., where we camped until the 21st. Marched to Smithfield and formed battle line. On the 22d marched by Charles Town, where we camped until the 25th."* [9]

Lieutenant Simpson A. Hagood of Company B, the Milton Guards, sat down and wrote a letter to wife Adaline, on Tuesday, August 23rd, while the regiment was camped at Charlestown: :

Camp of 38th Geo Regt
Charlestown, Va.
August 23d 1864

Dear Adaline,

I once have the honor to devote a few moments time to you & you may be assured that it is with great degree of anxiety concerning you & yours that I pen these lines for I have not been able to hear a single word from you either written or verbal since the Yankees first came into that Country. I hope you have fared better than I have any reason to believe they treat people of Georgia. If I may judge from the treatment which the Citizens of Virginia received at their hands: On their last retrograde movement from Strasbourg to Harpers Ferry they (after destroying almost everything in the way of Furniture) even fired the Barns of Hay belonging to the Citizens, not even allowing the owners to take out their Carriages Cattle or Horses. all were consumed with the Barn houses. Such conduct should Kindle the fires of indignation and bitterness in every heart not yet moved by their vandalism. I can't imagine how any one born and raised South of the Potomac can look upon such conduct with anything but loathing and disgust. how they can say to these miserable wretches, "We are with you" is infinitely more than I ever thought a mind reared in the South could consent to. But I am mortified & grieved to say that there are those in this State (and even Georgia too) that can & do express such sentiments.

Now I am in favor of making our Country too warm a place for all such characters after this War is ended. Such people are not worthy a place among those who have fought and bled for the Freedom of their County. Let every Woman & Child keep a watchful eye on their conduct, holding them at all times in utter contempt. be prompt to report all cases and I assure you that they will fare badly at the hands of those noble spirits now around me who have honorable wounds by the score, and whose blood have watered so many fields in defense of the Sunny Lands those Traitors now infest. Lest I weary you I will say no more at present on this subject.

We have not had any regular Engagements with the Enemy since the Battle of Kernstown on the last of July although we have had several Skirmishes with them & I am very much pained to have to state that in one of these at Newtown, Va, John Y. Bagwell was killed. John was a kind, generous, truthful & brave young man. In fact he was too brave to protect or take proper care of himself before the Enemy. he was a member of the Sharpshooting Battalion of our Brigade and it was his privilege as well as his duty to seise on every opportunity of sheltering himself while under fire, yet he wouldn't do so, but at all times rushed forward far in advance of the line. he only lived a few moments after the fatal Ball pierced his Breast. he only spoke one or two words but they could not be caught by his friends. we had him as nicely buried as the circumstances would permit.

Jas. R. Neal arrived safely in Camp a few days ago but was unable to tell me any thing of you. I have not received a letter since the one you sent out to Gainesville by Pensen W. Hazzard. Write as often as possible. I must Close for the Mail is now going out. Myself & Levi are both well and doing well.

My Love to all the Family. I want Mac to write to me. When you see him tell him to write without fail. Send this down to Father after you have read it. Ever remember me in all your petitions.

As ever yours,
S A Hagood [10]

Sergeant Dickerson continued his account of the marching and fighting, *"We had picket fighting from there to Shepherdstown, W. Va. Had a fight on the 26th at Leedstown, Va. On the 27th (August) to Bunker Hill, where we camped until the 29th. Marched in the direction of Smithfield, where we had a little fight with the cavalry. Back to Bunker Hill, where we camped until September 2d. Marched to right of Smithfield, camped near Brucetown. Had picket fighting on the 5th. From the 6th to the 9th (September) we camped three miles below Winchester. On the 13th marched by Brucetown where we formed battle line and had heavy skirmishing. Artillery firing kept up all day. Back to camp at night. On the 17th to Bunker Hill. On the 18th to Martinsville (Martinsburg.)"*[11]

Grant realized Early's small army had wreaked havoc in the Shenandoah Valley, Maryland and Pennsylvania. Early had chased Union General Hunter clear into West Virginia, defeated Wallace at Monocacy, burned Chambersburg, Pennsylvania, and nearly captured Washington, D.C. Early's threat to the capital had caused Grant to detach the VI Corps from around Petersburg and the XIX Corps, just returning from New Orleans to protect the city. Up to this point, Early had been eminently successful and had accomplished all this with the U.S. Presidential election just months away.

Grant left City Point, Virginia, near Petersburg, to meet with Major General Phillip Sheridan on September 15th, 1864, at Charles Town, West Virginia. Grant said of the meeting, "My purpose was to have him (Sheridan) attack Early, or drive him out of the valley and destroy that source of supplies for Lee's army." [12] Grant placed Sheridan in charge of three infantry Corps (Wright's VI, Crook's VIII, Emory's XIX) and two cavalry divisions, along with considerable artillery, totaling over 40,000 men. This newly created army, under Sheridan, was called "The Army of the Shenandoah."

Early had about 14,000 men, organized in four divisions, to face Sheridan. Scouts reported the approach of Sheridan and on September 19th, his army attacked Ramseur's Division near Winchester. Gordon wrote,

"Faithful and enterprising scouts, those keen-eyed, acute-eared, and nimble-footed heralds of an army who, "light-armed, scour each quarter to descry the distant foe," and who had been hovering around the Union army for some days after it crossed the Potomac, reported that General Sheridan was in command and was approaching Winchester with a force greatly superior to that commanded by General Early. The four divisions of Early's little army were commanded at this time respectively by General John C. Breckinridge, the "Kentucky Game-cock," by General Rodes of Alabama, who had few equals in either army, by General Ramseur of North Carolina, who was a most valiant and skillful leader of men, and by myself.

These divisions were widely separated from one another. They had been posted by General Early in position for guarding the different approaches to Winchester, and for easy concentration when the exigencies of the campaign should require it. The reports of the Federal approach, however, did not seem to impress General Early, and he delayed the order for concentration until Sheridan was upon him, ready to devour him piecemeal, a division at a time." [13]

Chapter 19
The Battle of 3rd Winchester (or Opequon Creek)
September 19th, 1864

A Private in the 61st Georgia noted the disparity in numbers Early faced at Winchester, *"It was said that Sheridan had 35,000 infantry and 10,000 cavalry. Early did not have over 8,000 or 10,000 infantry and 2,000 cavalry."* [1]

Private Bradwell, of the 31st, recalled events leading up to the battle, "The day before this engagement (September 18) our brigade (Gordon's) was encamped near where the battle took place, facing Sheridan's army, only two or three miles to the east of us across the Opequon River and behind a low mountain range. The weather was warm…someone abruptly shouted to us to come…the army was moving. This was about noon. When night came we were at Martinsburg, twenty miles away, tired from our rapid march. We had just entered the town and were standing in line awaiting orders when a courier from General Rodes came in great haste to inform General Early of his critical situation. Sheridan had crossed the Opequon in force and was threatening the destruction of his small division and was about to cut off our only way to escape to the south. Nothing was left now but to countermarch that night, tired as we were, and make the best we could of a bad situation. Sheridan's signalmen could see from the mountain tops that Early had marched away far to the north with the bulk of his army, having only a small rear guard to protect his communications. Seeing his opportunity to destroy his enemy, Sheridan at last took courage to fight and put his army in motion; but instead of falling on Rodes that evening with his whole force and taking possession of the Valley pike, our only road to the south, he dallied in a small skirmish with Rodes. Early marched his weary soldiers back halfway that night, and at two o'clock in the morning we were again on the pike. By eight o'clock we could see to the south the white puffs of smoke in the sky made by the exploding shells, indicating that Rodes was engaged with the enemy. The pike was given up to our wagon trains, and the infantry marched alongside in fields and woods, with a line of skirmishers to the left to protect us from a sudden attack from that direction." [2]

Major Gen. Phillip Sheridan - USA
by Mathew Brady, National Archives

Private Nichols was part of the line of skirmishers advancing on the left of the brigade and wagon train. He wrote a detailed account of the battle,

"We marched several miles with quick-step and began to see that something was wrong, for we would march a short piece and stop. The couriers and staff officers were in a stir, riding at rapid gaits carrying orders. Our wagons and

artillery were being moved as fast and cautiously as possible. We would march a short piece and have to wait for the wagons to move on. Evans' brigade (Georgia Brigade) was marched out about a quarter of a mile on the left of the road and we privates knew that trouble was up. We had to march through open fields, woods, etc., along with the wagon trains, which had to keep the road. We marched here in order to protect them. Four other men and I were sent about one hundred yards to the left of Evans' lines and had to march along for some time to look for the Yankees.

We could hear heavy cannonading on our left front. We five walked up on a line of Yankees lying down in thick high clover. We were in ten paces of them before we saw them. They raised up and captured us and made a dash on our brigade. They fired a volley into and charged it, and threw our men into considerable confusion. It was the first time that our brigade had ever retreated without orders. The front Yankee line followed Evans' retreating men.

A little Dutch officer of one of the Maine regiments ran up to me and said: "You dosh throw town your cun." He jerked it away from me. It was a very -fine one and I had had it nearly two years, and had shot about eight hundred rounds at the Yankees with it. The officer saw that it was a fine gun and said: "Here, Yokup - here dosh be one d - d fine cun, an you dosh trow town your old von, and take the good von."He then told us to go to the rear, but sent no guards with us. He asked us what brigade we belonged to. I told him Gordon's old brigade (the Georgia Brigade) He halloed out to his men: Gordon's brigade is retreating; let us press it."

I was in a terrible strait, for I knew it was only Evans' brigade retreating. I was satisfied that our boys would rally when they got in proper shape to do so. I knew that the rest of our division was not more than 500 yards away, marching by the side of our wagon trains. So I decided to risk consequences and take chances, and follow on behind the reserve line of Yankees. I was satisfied that our men would send the Yankees back faster than they were advancing...Some of our artillery had gotten into position and were throwing grape and canister shot into the Yankee line at a rapid rate. By this time the reserve line of the enemy had gotten to near the top of the hill. Their officers commanded them to lie down, which they did, and it paid them to do so, for our grape and canister shot were flying like hail just over their heads.

I saw that none of them were paying any attention to me, so I walked up and lay down, about twenty yards in their rear, like I was dead. By this time the front line of Yankees had a decent whipping put on them and they were retreating in a run, badly confused, and the fighting seemed to cease. I heard one of the Yankee officers say, "the Rebs are flanking- us on the left." I turned my head and looked up the valley about 600 yards away and could see our men swinging around them in excellent order at right shoulder, shift arms; their guns shining in the morning sun like silver. I felt then that if I could act dead I would soon be back with the Confederates, and I suppose I did it very well. I could hear our men coming over the hill, the officers commanding - forward - forward - forward. And along came the front line of Yankees into the reserve line.

The Union officer in command of the reserve line commanded them to fall back. They jumped up and ran right over me...The two lines of Yankees were all mixed up and running for dear life...The Yankees jumped over a ditch of pure spring water and were rising the next hill, I heard our men very close by, and heard a few guns fire. All at once our men got to the top of the hill and in full view of the retreating Yankees. And I have never seen such a deadly volley fired as those noble Alabamians fired at the retreating enemy. It was so terrible that it really looked sickening. It seemed that the first volley cut down half of their line. Our men went on with a yell, and some of the Yankees acted bravely, for they stopped and shot back at the Confederates.

Among the bravest was their color-bearer. He stopped, drew his pistol and emptied it at the advancing Confederates, and he did this walking backwards. Our wagons and artillery got back to a safe position. We soon fell back to a new line nearer Winchester and made connections with the balance of Early's command. I followed about fifty yards in their rear. There were a few of our men killed and wounded all along. There was a good sized hill between the Yankees and the road where our men were traveling. By this time our men had all gone over the hill and the Yankees were following, rising and going over the hill, and the battle soon opened in earnest." [3]

Private Bradwell recounted the advance of the Georgia Brigade, *"Veering to the left from the pike, we came to a body of woodland. The brigade was formed for battle in a deep ravine in the edge of this, with high ground in front of us that obstructed the view more than thirty feet away, and our old reliable sharpshooters were thrown out to develop the enemy's position. Almost immediately Sheridan's whole line, extending far to the right and left, was upon them, and these brave fellows came running back to us in a panic, causing our men to laugh at their disorder. Some said: "What's up?" To this they replied: "You'll soon see."*

Colonel Lowe, of our regiment, in command of the brigade, sitting on his horse, could see the long blue lines of the enemy advancing and ordered the brigade to move forward. This we did, and to our surprise we met face to face long lines of splendid infantry advancing, apparently unaware of our presence. The entire brigade brought down their guns, and a flame of fire flashed along its entire length; at the same time a dreadful yell arose that stampeded the enemy the length of our line. We rushed at them and took advantage of their fright and were driving them in fine style through the woods as we did at the Wilderness when, glancing to the extreme right, to my horror I saw the regiment on that part of the line in great confusion giving ground, and then the next. The center and left were still driving the enemy; but the left, seeing themselves outflanked, gave way also, and there was nothing now for all to do but fall back or be surrounded and cut to pieces.

When our regiments reached the ravine in great disorder, General Rodes's Alabamians (Rodes had just been killed) were forming there, and General Gordon, who had the greatest confidence in his old brigade, seeing our disorder and not knowing the cause, galloped to and fro among them, crying: "What in the world, men, is the matter? Fall in here with General Rodes's men and fight." This last sentence appealed to me as the proper thing to do in the crisis, for" the enemy now had taken courage and was advancing over the same ground from which we had just driven them. I stepped up to the men in the front rank and said: "Will you give me a place in your ranks to fight?" They did so as the word was given to advance, and when we reached the top of the high ground we were face to face again with our enemy. These brave Alabamians rushed at the enemy like tigers, and for a time the two lines were so near each other that the paper of their cartridges flew into our faces. At one point to my left the lines came together, and I saw the ensign of one regiment snatch the colors out of the hands of a Federal soldier and drag them along on the ground, while he held his own standard aloft." [4]

Lieutenant Lester noted the 38th's entry to the battle, after a grueling, 40 mile march, "We went at a double quick into the fight, to relieve Gen. Early, but too late to do much good. The part of Gen. Early's forces already engaged were whipped before we reached the field, and our two divisions were completely worn down, having marched continuously 40 miles." [5]

Forty-eight year old Captain John Rankin, company commander of Company D, the McCullough Rifles, was shot in the arm and captured. Captain William Mathews of the Battey Guards eulogized Captain Rankin, "Capt. John G. Rankin, of Company D, was over military age when the war commenced, and was of northern birth, but the South had no officer or man braver or better. We called him "old reliable," for in nearly every engagement our commander would get killed or wounded and Capt. Rankin would then take command. He refused promotion when he could have been made lieutenant colonel, but preferred to stay with his company. After his capture, having being wounded, he was promoted to position of major of the regiment, and his name so appears in the Confederate archives." [6]

Thirty-seven year old Private Andy J. Gardner, of Company A, The Murphy Guards, was shot in thigh with a Minié ball and captured. He died of his

"Sheridan's final charge at Winchester"
Lithorgraph by Thure De Thulstrup, 1886 Library of Congress

wound three weeks later in a Federal Depot Field Hospital. He left behind a wife and a four year old son. [7] Private Hilliary W. Arrington of Company G, the Battey Guards, was killed in action. He had been seriously wounded, shot in the thigh and captured at Gettysburg, but survived, He was exchanged and returned to the regiment, only to be killed in action in this battle.

Major Gen. Robert Rodes - CSA
Library of Congress

After repulsing several charges by the enemy, Private Bradwell rejoined the Georgia Brigade and wrote of General Gordon ordering the Brigade to retreat, but not run, "A faithful fellow soldier told me that they had driven the enemy back in every assault that evening, and, pointing to three cartridges on the ground, he said: 'That's all I have. General Gordon says that when we drive them back this time we will have to retreat.' I told him to share them with me, and I would help. Very soon their line came out of the woods, and as soon as they were about a hundred yards away our men broke up their formation, and they fell back in confusion to the woods out of which they had just come. General Gordon mounted his horse and said: 'Now, men, is our time to retreat, but don't run. Georgians never run from a battle field. We are only without ammunition.' The fleeing enemy, seeing us leave our place of protection and knowing the reason, now turned and opened on us with a great volley that seemed to cut the dirt from under our feet." [8]

The color bearer of the 38th Georgia, Private Gavin Hamilton Farrer, of Company G, the Battey Guards, was shot down and mortally wounded. He died of his wounds four days later in a Confederate hospital at Woodstock Virginia. He was buried at Massanutten Cemetery in Wookstock, his grave marker incorrectly reads C. S. Farrer, 38th Virginia. [9] Thirty-two year old Sergeant Hardin A. Jordan, of Company N, the Chestatee Artillery Company, was shot through the right thigh, near the groin, by a Minié ball. The ball ranged up through his body, fracturing his hip bones. He was taken prisoner and sent to Point Lookout Prison camp. [10]

EVANS' BRIGADE.

Annexed is a partial list, of the casualties which occurred in Evans' Brigade near Winchester, Sept. 19:

THIRTY-EIGHTH GEORGIA REGIMENT.

Captain Rankin, supposed mortally, prisoner

Co A—Wounded: Bowers, B B Christain, severe.

Co C—Killed: Sergt Mullen. Wounded: Capt Higden, Story, J F Dix, Sergt Dudney, Curly, Thompson, all severe.

Co G—Wounded: G Farrer, color bearer severe.

Co K—Wounded: Autrey, Tweedle, severely.

Augusta Chronicle, Nov 11th, 1864, P.1

Lieutenant Lester wrote, "We held the enemy in check for a short time, but were finally overpowered by them, and forced to fall back-after making the most stubborn resistance it was possible to make under the circumstances - and the fight soon became a perfect stampede, on our part every man taking care of himself as best he could. Our casualties were light, from the fact that as soon as we found out the condition of affairs we acted upon the old maxim that 'tis better to make a good run than a bad standing and consequently did not give the enemy a chance to hurt us. Sanders Jackson was slightly wounded in the foot, but he managed to get out of the way safely. Maj. Gen. Rhodes was killed in this battle." [11]

The other companies the 38th were not as fortunate as the Tom Cobb Infantry. Twenty year old Sergeant Francis M. Mulling, of Company C, the Ben Hill Guards, from Emanuel County, Georgia, was shot and killed. Twenty-seven 38th Georgia soldiers were forced to surrender. As previously stated, Captain Rankin suffered a severe gunshot wound to the arm and was captured. [12] Private Benjamin B. Christian of the Company H, the Goshen Blues, from Elbert County, Georgia, was shot in upper third of his thigh, the bullet breaking the bone and creating a compound fracture. He was captured and carried to a Federal field hospital, where the ball was removed by a Federal surgeon. He spent the next nine months in captivity, with his leg resting on pillows. He died of the wound at Fort McHenry, Maryland, May 15th, 1865, over a month after Lee had surrendered at Appomattox Court House. [13]

The 38th served as rear guard during the retreat and the Yankee cavalry closely pressed the retreating column. Lieutenant Lester wrote, "About sundown we collected our scattered forces beyond Winchester and formed in line behind a stone fence, parallel to the turnpike. While thus formed a regiment of Yankee cavalry came along the pike, flushed with victory, in hot pursuit of what they supposed a fleeing enemy. We reserved our fire until they came opposite our line. The order was given to fire, which we did most effectually, unsaddling nearly every one of them and in a measure redeeming our reputation -- convincing the Yankees that we did not propose to stay whipped." [14] Capt. William C. Mathews of the Company G, the Battey Guards, wrote of the losses to the 38th during the battle of Winchester, "In this fight the 38th Regiment lost 15 killed and 46 wounded. Among the wounded was "old reliable" Capt. J. G. Rankin (John G. Rankin), who in all the fights up to this time, had escaped." [15]

The Georgia Brigade, including the 38th, were roughly handled in the battle. Twenty-seven members of the regiment were captured at Winchester, with only five of the captured being wounded. Eleven of the soldiers captured were from the Chestatee Artillery Company, now serving as infantry.

Gordon told of the retreat from Winchester, *"The pursuit was pressed far into the twilight, and only ended when night came and dropped her protecting curtains around us. Drearily and silently, with burdened brains and aching hearts, leaving our dead and many of the wounded behind us, we rode hour after hour, with our sore-footed, suffering men doing their best to keep up, anxiously inquiring for their commands and eagerly listening for orders to halt and sleep.*

Lucky was the Confederate private who on that mournful retreat knew his own captain, and most lucky was the commander who knew where to find the main body of his own troops. The only lamps to guide us were the benignant stars, dimly lighting the gray surface of the broad limestone turnpike. It was, however, a merciful darkness. It came too slowly for our comfort; but it came at last, and screened our weary and confused infantry from further annoyance by Sheridan's horsemen. Little was said by any officer. Each was left to his own thoughts and the contemplation of the shadows that were thickening around us. What was the morrow to bring, or the next month, or the next year?

There was no limit to lofty courage, to loyal devotion, and the spirit of self-sacrifice; but where were the men to come from to take the places of the maimed and the dead? Where were the arsenals from which to replace the diminishing materials of war so essential to our future defence? It was evident that these thoughts were running through the brains of rank and file; for now and then there came a cheering flash of rustic wit or grim humor from the privates: 'Cheer up, boys; don't be worried. We'll lick them Yankees the first fair chance, and get more grub and guns and things than our poor old quartermaster mules can pull.' Distinct in my memory now (they will be there till I die) are those startling manifestations of a spirit which nothing could break, that strange commingling of deep-drawn sighs and merry songs, the marvelous blending of an hour of despair with an hour of bounding hope, inspired by the most resolute manhood ever exhibited in any age or country.

At a late hour of the night on that doleful retreat, the depressing silence was again broken by a characteristic shot at General Breckinridge from Early's battery of good-natured sarcasm, which was always surcharged and ready to go off at the slightest touch. These two soldiers became very good friends after the war began, but previously they had held antagonistic political views. Early was an uncompromising Unionist until Virginia passed the ordinance of secession. Breckinridge, on the other hand, had long been a distinguished champion of what was called 'the rights of the South in the Territories,' and in 1860 he was nominated for President by the 'Southern Rights' wing of the Democratic party. The prospect of establishing Southern rights by arms was not encouraging on that dismal retreat from Winchester. General Early could not resist the temptation presented by the conditions around us; and, at a time when the oppressive stillness was disturbed only by the dull sound of tramping feet and tinkling canteens, his shrill tones rang out:

Major General John C. Breckenridge - CSA
Library of Congress

'General Breckinridge, what do you think of the 'rights of the South in the

Territories' now?' Breckinridge made no reply. He was in no humor for badinage, or for reminiscences of the period of his political power when he was Kentucky's most eloquent representative in the halls of Congress, or pleaded for Southern rights on the floor of the Senate, or made parliamentary rulings as Vice-President of the United States, or carried the flag of a great party as its selected candidate for the still higher office of President.

When the night was far spent and a sufficient distance between the Confederate rear and Union front had been reached, there came the order to halt - more grateful than sweetest music to the weary soldiers' ears; and down they dropped upon their beds of grass or earth, their heads pillowed on dust-covered knapsacks, their rifles at their sides, and their often shoeless feet bruised and aching. But they slept. Priceless boon - sleep and rest for tired frame and heart and brain! General Sheridan graciously granted us two days and a part of the third to sleep and rest and pull ourselves together for the struggle of September 22."[16]

The 38th suffered 45 casualties at the battle of 3rd Winchester, killed, wounded, or captured. Three soldiers were killed in action. Fifteen soldiers were wounded. Twenty-seven soldiers were captured, with five of the captured being wounded. [17]

Casualties at Battle of 3rd Winchester – September 19th, 1864

Company A – "The Murphy Guards," DeKalb County, Georgia

1. Private Robert Eidson - Captured at Winchester, 9/19/1864.
2. Private Andy J. Gardner - Wounded & captured at Winchester, 9/19/1864. Died of wounds at U. S. A. Depot Field Hospital 10/23/1864.
3. Private James M. Holmes - Captured at Winchester, 9/19/1864.

Company B, "The Milton Guards," Milton County, Georgia (Currently Fulton County)

4. Private John A. Bruce - Captured at Winchester, 9/19/1864.

Company C, "The Ben Hill Guards," Bulloch & Emanuel Counties, Georgia

5. Private James Frederick Dix - Wounded at Winchester, 9/19/1864.
6. Sergeant James Dudley - Captured at Winchester, 9/19/1864.
7. Captain John B. Higdon - Wounded and disabled at Winchester, 9/19/1864.
8. Sergeant Francis M. Mulling - Killed at Winchester, 9/19/1864.
9. Private Josiah Story - Wounded at Winchester, 9/19/1864.

Company D, DeKelab & Fulton Counties

10. Sergeant James C. Harris - Wounded at Winchester, 9/19/1864.
11. Private William T. Nash - Wounded at Winchester, 9/19/1864.
12. Musician Julius Perlinski - (or Perlinsky) Captured at Winchester, 9/19/1864.
13. Captain John G. Rankin - Wounded in arm & captured at Winchester, 9/19/1864.
14. Private S. J. Woods - Killed at Winchester, 9/19/1864.

Company F, "Thornton's Line Volunteers," Hart and Elbert Counties, Georgia

15. Private James Washington Duncan - Wounded at Winchester, 9/19/1864.
16. Private John William Dutton - Captured at Winchester, 9/19/1864.
17. Private Jeptha Mercer Thornton - Captured at Winchester, 9/19/1864.
18. Private James William Thornton - Captured at Winchester, 9/19/1864.

Company G, "The Battey Guards," Jefferson County, Georgia

19. Private Hilliary W. Arrington, - Killed at Winchester, 9/19/1864.
20. Private William Arnold Brown - Wounded at Winchester, 9/19/1864.
21. Private Gavin Hamilton Farrer - Color bearer, wounded at Winchester, 9/19/1864. Died of wounds 9/23/1864 at Woodstock, Virginia.
22. Private William J. Phillips - Captured at Winchester, 9/19/1864.
23. Private Luther C. Smith - Wounded at Winchester, 9/19/1864.
24. Private Jesse W. Weeks - Wounded at Winchester, 9/19/1864.

Company H, "The Goshen Blues," Elbert County, Georgia

25. Private John C. Booth - Wounded at Winchester, 9/19/1864.
26. Private Asa Marion Bowers - Wounded at Winchester, 9/19/1864.
27. Private Benjamin B. Christian - Wounded and captured at Winchester, 9/19/1864. Died of wounds in U. S. A. Post Hospital, Ft McHenry, MD, 5/15/1865.
28. Private William H. Hall - Captured at Winchester, 9/19/1864.
29. Private William Gable Moon - Captured at Winchester, 9/19/1864.

Company K, "DeKalb & Fulton Bartow Avengers," DeKalb and Fulton County, Georgia

30. Private William Allen Autry - Wounded in right leg and permanently disabled at Winchester, 9/19/1864.
31. Private William Thomas Goss - Captured at Winchester, 9/19/1864.
32. Sergeant John W. Jones - Wounded and captured at Winchester, 9/19/1864. Place of capture also shown Harrisonburg, Virginia 9/25/1864.
33. Sergeant Marion J. Tweedell - Wounded at Winchester 9/19/1864.

Company L, "Chestatee Artillery," Forsyth County, Georgia

34. Private James J. Arthur (or James S. Arthur) - Captured at Winchester, 9/19/1864.
35. Corporal John Ratliff Brice - Captured at Winchester, 9/19/1864.
36. Private Paschal Hawkins Dacus - Captured at Winchester, 9/19/1864. Died of measles at Point Lookout, Maryland 3/10/1865.
37. Private John Freeman - Captured at Winchester, 9/19/1864.
38. Private Joseph L. Jones - Wounded in thigh and captured at Winchester, 9/19/1864.
39. Sergeant Harden A. Jordan - Captured at Winchester, 9/19/1864.
40. Private William L. McKinney - Captured at Winchester, 9/19/1864.
41. Private Francis Marion Owen - Captured at Winchester, 9/19/1864.
42. Private Hiram Patterson - Captured at Winchester, 9/19/1864.

43. Private Elias Earl Whitmire - Captured at Winchester, 9/19/1864.
44. Private William Hasley Wood - Captured at Winchester, 9/19/1864.

Company N, "The Dawson Farmers," Dawson County, Georgia

45. Corporal Absalom B. Martin - Wounded at Winchester, 9/19/1864.

Chapter 20
The Battle of Fisher's Hill – September 21-22, 1864

Private Bradwell recalled the night after the defeat at the battle of 3rd Winchester, "As we marched slowly along the pike that night toward Strasburg, our stragglers came up, and by morning all the regiments of Early's army were fully reorganized. We bivouacked at Newtown, about ten miles from the scene of the battle, and the next day marched to Hupp's Hill, south of Cedar Creek and north of Strasburg, where Early began to fortify; but as this position could be flanked easily on the right and left, he withdrew to Fisher's Hill, south of that town, where we found excellent breastworks already constructed extending from the Valley pike at the foot of Massanuttan Mountain on the east to Little North Mountain on the west. Here he deployed his small force, now not more than eight or ten thousand men, in a thin line entirely across the valley." [1]

Lieutenant Lester recalled, "We now retreated leisurely to Fisher's Hill and here fortified ourselves and waited the approach of the enemy. As we were entrenched and in a strong position we felt pretty secure." [2]

Fisher's Hill, just south of Strasburg was a natural fortress, spanning the Shenandoah Valley for four miles from Massanuttan Mountain to the east, to Little North Mountain to west. Breastworks were previously constructed so all the Confederates had to do was fall into the existing fighting positions. While the position was extremely strong, a Confederate officer noted, "The position was a very strong one, but our army was too small to man it." [3]

Early realized he was in trouble by September 22nd, he wrote, "In the afternoon of the 20th, Sheridan's forces appeared on the banks of Cedar Creek, about four miles from Fisher's Hill, and the 21st, and the greater part of the 22nd, were consumed by him in reconnoitering and gradually moving his forces to my front under cover of breastworks. After some sharp skirmishing, he attained a strong position immediately in my front and fortified it, and I began to think he was satisfied with the advantage he had gained and would not probably press it further; but on the afternoon of the 22nd, I discovered that another attack was contemplated, and orders were given for my troops to retire, after dark, as I knew my force was not strong enough to resist a determined assault." [4]

Early had lost about 4,000 at 3rd Winchester and the war department had ordered Breckinridge's command be detached from Early's army. His remaining 10,000 soldiers could only defend about three miles of the four mile chain of hills. Early wrote of his dispositions, "Wharton's division being on the right, then Gordon's, Ramseur's and Rodes', in the order in which they are mentioned. Fitz. Lee's cavalry, now under Brigadier General Wickham, was sent up the Luray Valley to a narrow pass at Millwood, to try to hold that valley against the enemy's cavalry. My infantry was not able to occupy the whole line at Fisher's Hill, notwithstanding it was extended out in an attenuated line, with considerable intervals. The greater part of Lomax's cavalry was therefore dismounted, and placed on Ramseur's left, near Little North Mountain, but the line could not then be fully occupied." [5] The cavalry force on the left flank was a mere skirmish line and this is where the disaster began. The Georgia Brigade was positioned just to the west of the Valley Turnpike.

Sheridan emplaced the bulk of his army directly facing Fisher's Hill and opposing the main Confederate line. With the main line held in place, Sheridan sent Brigadier General George Crook's VIII Corp to march around Early's left flank to Little North Mountain to the southwest. Marching under the cover of hills and forests, to prevent observation from Confederate signal stations, Crook's Corp of 5,500 Federal soldiers were in place by 4 p.m. At about 4:30 p.m., Crook ordered his men forward to attack Early's left flank. The Federals easily rolled over, or pushed aside, the small Confederate cavalry and struck Ramseur's Division on the flank. The entire Confederate line quickly unraveled from left to right. Early wrote of the attack on Fisher's Hill, "Just before sunset, however, Crook's corps, which had moved to our left on the side of Little North Mountain, and under cover of the woods, forced back Lomax's dismounted cavalry and advanced against Ramseur's left. Ramseur made an attempt to meet this movement by throwing his brigades successively into line to the left, and Wharton's division was sent for from the right, but it did not arrive. Pegram's brigades were also thrown into line in the same manner as Ramseur's, but the movement produced some disorder in both divisions, and as soon as it was observed by the enemy, he advanced along his whole line and the mischief could not be remedied. After a very brief contest, my whole force retired in considerable confusion, but the men and officers of the artillery behaved with great coolness, fighting to the very last, and I had to ride to some of the officers and order them to withdraw their guns, before they would move. In some cases, they had held out so long, and the roads leading from their positions into the Pike were so rugged, that eleven guns fell into the hands of the enemy. Vigorous pursuit was not made, and my force fell back through Woodstock to a place called the Narrow Passage, all the trains being carried off safely." [6]

Gordon wrote just a few lines of the disaster at Fisher's, *"The battle, or, to speak more accurately, the bout at Fisher's Hill, was so quickly ended that it may be described in a few words. Indeed, to all experienced soldiers the whole story is told in one word—'flanked.' We had again halted and spread our banners on the ramparts which nature built along the Shenandoah's banks. Our stay was short, however, and our leaving was hurried, without ceremony or concert. It is the old story of failure to protect flanks. Although the Union forces more than doubled Early's army, our position was such that in our stronghold we could have whipped General Sheridan had the weak point on our left been sufficiently protected. Sheridan demonstrated in front while he slipped his infantry around our left and completely enveloped that flank. An effort was made to move Battle and Wharton to the enveloped flank in order to protect it, but the effort was made too late. The Federals saw their advantage, and seized and pressed it. The Confederates saw the hopelessness of their situation, and realized that they had only the option of retreat or capture. They were not long in deciding. The retreat (it is always so) was at first stubborn and slow, then rapid, then--a rout. It is not just to blame the troops. There are conditions in war when courage, firmness, steadiness of nerve, and self-reliance are of small avail. Such were the conditions at Fisher's Hill."* [7]

Lieutenant Lester wrote, "As we were entrenched and in a strong position we felt pretty secure, but to our great surprise we were not attacked in front but in rear. By some mismanagement on the part of our commanders, that a common soldier cannot understand, we were entirely flanked out of our position. We made a stubborn fight for a short time but were compelled to retreat again, and this affair was almost as bad as that at Winchester. In this engagement Willis B. Jackson was wounded in the neck." [8]

Sergeant Dickerson of the McCullough Rifles, noted, the losses in Company D, "Lt. G. R. Wells (George R. Wells) was captured and S. H. Huff and S. J. Woods of Co. D, were killed." [9]

Private Bradwell of the 31st left a detailed account of the retreat of the Georgia Brigade, *"When we left the breastworks, the enemy was only a few yards in front of them. In the scuffle and confusion there a comrade, after firing at the enemy, jerked his gun back and struck me above the right eye with the sharp piece of metal on the butt end of his gun. From the wound the blood gushed out and ran down over my eye that I could see with difficulty. Turning around, I saw my comrades running toward the pike. When I got there, to my surprise just on the other side of the road I saw Col. E. N. Atkinson, of the 26th Georgia Regiment holding on to the wheel of a piece of artillery that had been abandoned there when the men whose duty it was to serve it saw that there was no possible means to save it from falling into the hands of the enemy. The firing was hot and at short range, but there stood this brave, though indiscrete, officer trying to halt every man that passed to get this gun away by hand. In this storm of bullets not a man stopped except the writer. Being young and having been taught to obey orders, and especially by a regimental officer, I seized hold of the opposite wheel and held on to it until the enemy were only a few feet away and commanding us to surrender. I now deemed it about time to consult my own safety and left the colonel, still holding on to the wheel of the gun in his insane effort to get it away. For the next two hundred yards and until I placed a small hill between me and the enemy was only a brief period. Here I sat down to take a breath. Looking back, I saw coming towards me a brave man of the regiment. He had held his ground until the last minute; but once he turned his back on the enemy, he had become completely demoralized. Like a runaway horse, he had lost control of his mind. He had thrown away his gun and accouterments, lost his coat, and was now divesting himself of his shirt. His long beard stood out in the air on each side of his face, and his eyes were wild and staring. As he passed me in his flight, I called to him to stop and rest, as we were now safe. But he only glared at me and passed on.*

From this incident I have decided that the bravest men may sometimes lose their heads in great danger. Colonel Atkinson held on to the piece of artillery in his delirium to save it until he and it both fell into the hands of the enemy, where he remained until the war closed. He was a brave and very popular with the soldiers, but a man of poor judgment." [10]

Private Nichols of the 61st also wrote of the disaster, *"On the evening of the 22nd Sheridan sent out a flanking column around Little North Mountain, and broke our line where we had a few dismounted cavalry, and made a dash on our rear. We did not know what it meant at first, for we saw our men leaving our works on our left and begin to fall back in a run. We were ordered to leave our works and fall back. "Sheridan advanced his whole line and the rout became general. All of our men left the works badly confused, though all of them got their guns and other equipments. To make it worse, the Yankees gave three huzzas and began to shoot at us with a vim. The shooting did but little harm, only to add to the confusion. I was never fleet footed and could hardly keep up. Some tired down and stopped and surrendered. We ran about two miles, and got out past the flanking party. We readily stopped and formed in line; but it looked like we were ruined. Our artillery was nearly all captured and a great many of our wagons and many prisoners had been taken."* [11]

Brigadier General Evans returned to command the Georgia Brigade just a few days after the battle of Fisher's Hill. He said of the retreat of his Georgia Brigade, *"Part of the brigade retreated safely, but about 400 had to flee to the mountains for safety."* [12]

"As soon as it began to grow dark the enemy quit pressing us and camped. If night had not put an end to the contest I suppose we would have been ruined worse than we were. We fell back twelve miles that night and camped near Woodstock. All of the army was nearly exhausted. I was so fatigued till I spit blood. The wound that I received at Winchester on the 19th was paining me and was swollen badly, but I did all I could to keep out of prison." [13]

"It was now a running fight (we the retreating party), until darkness ended the disastrous affair. Not being pressed by the Yankees, we collected our scattered forces next morning, reforming and fell back to Mt. Jackson, where Gen. Early's army all got together again, continuing the retreat to Brown's Gap, where we halted and remained, [14] wrote Lieutenant Lester.

The 38th suffered 15 casualties at Fisher's Hill. Eleven soldiers were captured and only one of the captured was wounded. Three other soldiers were killed in action and one soldier was wounded, but escaped capture. Company D, the McCullough Rifles, suffered the heaviest of all the companies, having two killed in action, and two captured. In just three days (19 - 22 Sept), the 38th Georgia had lost thirty-eight soldiers captured, six killed in action, and sixteen soldiers wounded. Many of the soldiers missing and thought possibly captured, trickled into camp in the days following the battle.

Casualties at Battle of Fisher's Hill, September 22nd, 1864

Company A – "The Murphy Guards," DeKalb County, Georgia

1. Private Asbury Hull Power - Captured at Fisher's Hill, 9/22/1864.

Company B, "The Milton Guards," Milton County, Georgia (Currently Fulton County)

2. Sergeant William J. Cook - Captured at Fisher's Hill, 9/22/1864.

Company C, "The Ben Hill Guards," Bulloch & Emanuel Counties, Georgia

3. Private Reuben T. Westbrook - Captured at Fisher's Hill, 9/22/1864.

Company D, DeKalb & Fulton Counties

4. Sergeant James C. Harris - Wounded 9/19/1864, captured Fisher's Hill, 9/22/1864.
5. Private J. H. Huff - Killed at Fisher's Hill, 9/22/1864.
6. Lieutenant George Riley Wells - Captured at Fisher's Hill, 9/22/1864.
7. Private S. J. Woods - Killed at Fisher's Hill, 9/22/1864.

Company E, "The Tom Cobb Infantry," Oglethorpe County, Georgia

8. Private William Nelson Evans - Captured at Fisher's Hill, 9/22/1864.
9. Private Willis Benjamin Jackson, - Wounded in head at Fisher's Hill, 9/22/1864.

Company F, "Thornton's Line Volunteers," Hart and Elbert Counties, Georgia

10. Sergeant James C. Campbell - Captured at Fisher's Hill, 9/22/1864.
11. Private Elias Pearce Jenkins - Captured at Fisher's Hill, 9/22/1864.

Company G, "The Battey Guards," Jefferson County, Georgia

12. Private William F. Atkinson - Captured at Fisher's Hill 9/22/1864.

Company H, "The Goshen Blues," Elbert County, Georgia

13. Private William Jasper Rice - Captured at Fisher's Hill, 9/22/1864.

Company K, "DeKalb & Fulton Bartow Avengers," DeKalb and Fulton County, Georgia

14. Private John B. Boyd - Captured at Fisher's Hill, 9/22/1864.

Company N, "The Dawson Farmers," Dawson County, Georgia

15. Private Starling Emanuel Hill - Captured at Fisher's Hill, 9/22/1864. Died at Point Lookout Federal POW camp.

Chapter 21
The Valley in Flames, September - October, 1864

Gordon wrote of the pause in combat after Fisher's Hill and the Confederate army's recovery at Brown's Gap, *"Nearly a month--twenty-six days, to be exact--of comparative rest and recuperation ensued after Fisher's Hill. General Sheridan followed our retreat very languidly. The record of one day did not differ widely from the record of every other day of the twenty-six. His cavalry maneuvered before ours, and ours maneuvered before his. His artillery saluted, and ours answered. His infantry made demonstrations, and ours responded by forming lines. This was all very fine for Early's battered little army; and it seemed that Sheridan's victories of the 19th and 22d had been so costly, notwithstanding his great preponderance in numbers, that he sympathized with our desire for a few weeks of dallying. He appeared to be anxious to do just enough to keep us reminded that he was still there. So he decided upon a season of burning, instead of battling; of assaults with matches and torches upon barns and hay stacks, instead of upon armed men who were lined up in front of him."*

On September 27th, the Georgia Brigade was involved in another unnamed skirmish with the enemy near Port Republic, Virginia. This was the first engagement of the brigade since Evans had returned to command the brigade on the night of September 24th, following his recovery from a serious gunshot wound at the battle of Monocacy. General Evans wrote, *"A lot of new clothing & shoes also received and distributed this morning. So we will have very few barefooted or ragged men. I gave the boys several talks as I rode among them, and I think their old spirit is with them again. I have never been in a retreat from the battle field during action and I feel like I never shall be. I am very glad I did not witness the defeat at Winchester & Strasburg. Yesterday when I first formed my line I did it very coolly just like we were drilling, and then rode along the front rank & talked to the troops. You would have been amused to hear their assurances. 'We won't leave you, General.' 'Don't be uneasy with us, General.' 'We'll go with you, General.' 'We can whip them' &c &c - And when I gave the command forward, they moved off in beautiful order, not wavering though it was across an open field and the Yankees firing upon them from a battery - I felt glorious again. My old invincible brigade is itself again."* [1]

With Early's army suppressed, Sheridan set out to wreck the Shenandoah Valley commencing at Staunton and moving north down the Valley to Strasburg, burning a swath 70 miles long and 30 miles wide. From September 26th to October 8th, Sheridan put away his cannon and launched his "assaults with matches and torches upon barns and hay stacks." He burned and plundered the better part of four Valley counties, Shenandoah, Augusta, Page, and Rockingham Counties. Private Nichols wrote of Gen. Sheridan's depredations, "Our army was now reduced to only five or six thousand men and seemed badly discouraged and looked like they thought it was useless to fight any longer. We had an elevated position and could see Yankees out in the valley driving off all the horses, cattle, sheep and killing the hogs and burning all the barns and shocks of corn and wheat in the fields, and destroying everything that could feed or shelter man or beast." [2]

Sheridan reported to Grant, October 7th, from Woodstock, "The whole country, from the Blue Ridge to the North Mountain, had been made untenable for a rebel army. I have destroyed over two thousand barns filled with wheat, hay, and farming implements; over seventy mills filled with wheat and flour; four herds of cattle have been driven before the army, and not less than three thousand sheep have been killed and issued to the troops. A large number of horses have been obtained, a proper estimate of which I cannot now make. Lieut. Jno. R. Meigs, my engineer officer, was murdered beyond Harrisonburg, near Dayton. For this atrocious act, all houses within an area of five miles were burned." A correspondent, who was with the army, thus describes the scenes of their march:

"The atmosphere, from horizon to horizon has been black with the smoke of a hundred conflagrations, and at night a gleam, brighter and more lurid than sunset, has shot from every verge. The orders have been to destroy all forage in stacks and barns, and to drive the stock before them (the Federal army) for the subsistence of the army. Indiscriminating (for with such swift work discrimination is impracticable), relentless, merciless, the torch has done its terrible business in the Valley. Few barns and stables have escaped. The gardens and corn-fields have been desolated. The cattle, hogs, sheep, cows, oxen, nearly five thousand in 'all, have been driven from every farm. The wailing of women and children, mingling with the crackling of flames, has sounded from scores of dwellings. I have seen mothers, weeping over the loss of that which was necessary to their children's lives—setting aside their own—their last cow, their last bit of flour pilfered by stragglers, the last morsel they had in the world to eat or drink. Young girls, with flushed cheeks, and pale, with tearful, or tearless eyes, have pleaded with and cursed the men whom the necessities of war have forced to burn the dwellings reared by their fathers, and turn them into paupers in a day. The completeness of the desolation is awful. Hundreds of nearly starving people are going North. Our trains are crowded with them. They line the wayside. Hundreds more are coming. Absolute want is in mansions used in other days to extravagant luxury." [3]

The Georgia Brigade and 38th witnessed the burning of the Valley. Lieutenant Lester wrote, "After Sheridan had burnt out the upper valley, he fell back to Strasburg and we followed him to Woodstock and then returned to New Market. We found that he had burnt every barn and nearly every dwelling house from Staunton to Strasburg. Most of the dwelling houses in the towns were spared. The distance, I believe, from Staunton to Strasburg is seventy miles, and we heard that Sheridan reported to Lincoln and Grant that 'a crow would perish in the valley.' And it was one of the finest and richest places I ever saw before this merciless 'burn out.'[4] The Confederate army rested and recovered at Brown's Gap. While there, Lee dispatched Major General Joseph Kershaw's Division to re-enforce Early's depleted army. Gordon wrote of Kershaw's arrival, "General Kershaw, who was one of the ablest division commanders in Lee's army, came with his dashing South Carolinians to reenforce and cheer Early's brave and weary men. The arrival of Kershaw's division awakened the latent enthusiasm with which they had pommelled Sheridan at the beginning of the battle of Winchester, but which had been made dormant by the subsequent disastrous defeats on that field and at Fisher's Hill. The news of Kershaw's approach ran along the sleeping ranks, and aroused them as if an electric battery had been sending its stimulating current through their weary bodies. Cheer after cheer came from their husky throats and rolled along the mountain cliffs, the harbinger of a coming victory. "Hurrah for the Palmetto boys!" "Glad to see you, South Ca'liny!"

The addition of Kershaw's Division replaced the Confederate losses suffered at Winchester and Fisher's Hill. Nichols wrote, "General Kershaw's division, from Richmond, consisting of about 3,000 infantry and a brigade of cavalry. This made our command about as strong as it was before the disasters at Winchester and Fisher's Hill." [5] Gordon wrote, "The arrival of reënforcements under Kershaw not only revived the hopes of our high-mettled men, but enabled General Early and his division commanders to await with confidence General Sheridan's advance, which was daily expected. He did not come, however. Our rations were nearly exhausted, and after holding a council of war, General Early decided to advance upon the Union forces strongly entrenched on the left bank of Cedar Creek."[6]

Gen. Joseph B. Kershaw – CSA
Library of Congress

Captain Mathews of the Battey Guards, wrote of the organizational changes to the 38th Georgia from May 12th, 1864, to this point, *"Colonel James D. Mathews, having never recovered from his wounds, resigned, and Colonel Phillip E. Davant, having been exchanged and reported for duty, was made Colonel. Major Bomar (Thomas Bomar) was made Lieutenant Colonel, and Captain J. G. Rankin (John G. Rankin) Major. Lieut. Benjamin Morris promoted to 1st Lieutenant Company C, Lieut. McCurdy (John Wilson McCurdy) Captain of Company D, Lieut. Maxwell, (Jackson O. Maxwell) Captain of Company F, William C. Mathews Captain of Company G, Captain McAfee (William H. McAfee), accepted chaplaincy of 22nd Georgia regiment, and Lieut. Goswick (John W. Goswick), was made Captain of his Company (Dawson Farmers Company.) Sergt. R. H. Fletcher (Richard H. Fletcher), of the 31st Georgia Regiment, was elected Captain of Company K."* [7] Many of these promotions are not reflected in Confederate service records of these soldiers and the promotions may have been made to fill vacancies, but never formally approved. For example Phillip E. Davant was promoted to Lieutenant Colonel, but not Colonel, according to his service record, however he served as the "acting" Colonel of the regiment for the rest of the war.

Chapter 22
Battle of Cedar Creek – October 19th, 1864

Early described his army's organization and strength just before the battle, "The reports of the Ordnance officers showed in the hands of my troops about 8,800 muskets in round numbers, as follows: in Kershaw's division 2,700, Ramseur's 2,100, Gordon's 1,700, Pegram's 1,200 and Wharton's 1,100. Making a moderate allowance for the men left to guard the camps and the signal station on the mountain, as well as for a few sick and wounded, I went into this battle with about 8,500 muskets and a little over forty pieces of artillery." [1] Sheridan fielded about 33,000 soldiers, about three times the size of Early's army.

Maj. Gen. John B. Gordon - CSA
Library of Congress

Early said, *"It may be asked why with so small a force I made the attack. I can only say we had been fighting large odds during the whole war, and I knew there was no chance of lessening them. It was of the utmost consequence that Sheridan should be prevented from sending troops to Grant, and General Lee, in a letter received a day or two before, had expressed an earnest desire that a victory should be gained in the Valley if possible, and it could not be gained without fighting for it. I did hope to gain one by surprising the enemy in his camp, and then thought and still think I would have had it, if my directions had been complied with, and my troops had awaited my orders to retire."* [2]

General Gordon created the battle plan for the battle of Cedar Creek. He recounted the events leading to his development of the plan of battle, *"On the right of the Confederate line, as drawn up at Fisher's Hill, was Massanutten Mountain, rising to a great height, and so rugged and steep as to make our position practically unassailable on that flank.. I decided to go to the top of the mountain...to survey and study Sheridan's position and the topography of the intervening country. I undertook the ascent of the rugged steep, accompanied by that superb officer, General Clement A. Evans of Georgia, in whose conservatism and sound judgment I had the most implicit confidence, and by Captain Hotchkiss of General Early's staff. We finally reached the summit...It was an inspiring panorama. With strong field-glasses, every road and habitation and hill and stream could be seen and noted. The abruptly curved and precipitous highlands bordering Cedar Creek, on which the army of Sheridan was strongly posted; the historic Shenandoah, into which Cedar Creek emptied at the foot of the towering peak on which we stood, and, most important and intensely interesting of all, the entire Union army--all seemed but a stone's throw away from us as we stood contemplating the scene through the magnifying lenses of our field-glasses. Not only the general outlines of Sheridan's breastworks, but every parapet where his heavy guns were mounted, and every piece of artillery, every wagon and tent and supporting line of troops, were in easy range of our vision. I could count, and did count, the number of his guns. I could see distinctly the three colors of trimmings on the jackets respectively of infantry, artillery, and cavalry, and locate each, while the number of flags gave a basis for estimating approximately the forces with which we were to contend in the proposed attack. If, however, the plan of battle which at once suggested itself to my mind should be adopted, it mattered little how large a force General Sheridan had; for the movement which I intended to propose contemplated the turning of Sheridan's flank where he least expected it, a sudden irruption upon his left and rear, and the complete surprise of his entire army.*

It was unmistakably evident that General Sheridan concurred in the universally accepted opinion that it was impracticable for the Confederates to pass or march along the rugged and almost perpendicular face of Massanutten Mountain and assail his left. This fact was made manifest at the first sweep of the eye from that mountain-top. For he had left that end of his line with no protection save the natural barriers, and a very small detachment of cavalry on the left bank of the river, with vedettes on their horses in the middle of the stream. His entire force of superb cavalry was massed on his right, where he supposed, as all others had supposed, that General Early must come, if he came with any hope of success. The disposition of his divisions and available resources were all for defence of his right flank and front, or for aggressive movement from one or both of these points. As to his left flank--well, that needed no defence; the impassable Massanutten, with the Shenandoah River at its base, was the sufficient protecting fortress. Thus reasoned the commanders of each of the opposing armies. Both were of the same mind, and Early prepared to assail, and Sheridan to defend, his right and centre only.

I expressed to those around me the conviction that if General Early would adopt the plan of battle which I would submit, and would press it to its legitimate results, the destruction of Sheridan's army was inevitable. Briefly, the plan was to abandon serious attack of Sheridan's forces where all things were in readiness, making only a demonstration upon that right flank by Rosser's cavalry dismounted, and upon the centre by a movement of infantry and artillery along the pike, while the heavy and decisive blow should be given upon the Union left, where no preparation was made to resist us. This movement on the left I myself proposed to make with the Second Army Corps, led by General Clement A. Evans's division, followed by Ramseur's and Pegram's.

"But how are you going to pass the precipice of Massanutten Mountain?"

A dim and narrow pathway was found, along which but one man could pass at a time; but by beginning the movement at nightfall the entire corps could be passed before daylight." [3]

Gordon's plan of attack was approved by Early. Private Nichols wrote of the Georgia Brigade's movements to attack, "On the 12th of October we advanced to Fisher's Hill, and on the 13th we moved out in line to Hupps' Hill, and had a small engagement with some of the Yankees' cavalry and drove them across Cedar creek, and found that their infantry was well posted in a good fortified position, and we returned to Fisher's Hill and went into camps near our works. We stayed here and rested until the evening of the 18th (October.) About 3 or 4 o'clock all of our regimental and company officers were called for. They assembled at Generals Gordon's and Evans' headquarters. Here the plans of a great battle were made known to them and they made it known to us.

We were commanded to cook two days' rations and be ready to leave at dark, which we did. We were not allowed to carry anything that would make a noise, so we had to leave our canteens and the officers had to leave their swords. We marched and crossed the Shenandoah River and went around the Yankees through a trail in the Three Tops Mountain. We got within 300 yards of their pickets, and got around to another ford in the Shenandoah River, which was almost in the rear of the enemy." [4]

Lieutenant Colonel Davant, commanding the 38th said half the regiment was assigned picket duty the night before the battle of Cedar Creek, he wrote, "Only half my regiment was actively engaged, the other half, being on picket the night before, were deployed as skirmishers on the right to protect the flank of the army." [5] The high number of casualties in Companies A, B, C, E, K, and N suggest these were the companies actively engaged and accompanying the Brigade into battle. The remaining companies, were posted as skirmishers on the extreme right flank of the army.

Lieutenant Lester recalled, "Here General Gordon took his own division and two others and started one night on the march. We were not allowed to carry anything but arms and ammunition, not even a blanket. We knew that Gen. Gordon was on an important errand. We marched all night in a single file, along the mountain side." [6] General Gordon commanded three Divisions during the attack and General Evans commanded Gordon's Division.

Private Nichols wrote of the attack, *"Our division (Gordon's) had gotten there and was massed by 3 o'clock in the morning- and all ready to cross the river at the proper time. We had not made enough noise to cause the enemy's pickets to notice us, and it was reported to us private soldiers that General Gordon managed to capture one of the Yankee pickets, and that he gave General Gordon the countersign correctly, and that Gordon had all of their pickets relieved near the ford and had Confederates put in their places, and had all of the Yankee pickets captured. I do not yet know whether this report was true or not.*

Just at the break of day we marched across the river in four lines. Evans' brigade (the Georgia Brigade) started up the road at double quick step ahead of all the rest, but the rest were following immediately in the rear." [7]

Private Bradwell of the 31st recalled wading the icy Shenandoah River early that morning, "Well do I remember how we plunged into the icy waters of the

Shenandoah as day was beginning to dawn, the struggle to get to the bank on the other side, and the effort to reach the top of the high embankment, now made slick by our wet clothing; how some comrade jostled me just as I reached the top and I slid back into the cold water and had to try it all over again. We were trotted up the road and formed facing the wagon camps at the enemy's left, where they were trying to burn wagonloads of supplies." [8]

Private Nichols continued, *"There was a big frost on the ground, and as we had to wade the river and were wet up to our waists, the run up the hill did us good, for it warmed us up. We advanced up the road at double quick to a certain place and stopped. We faced the Yankees' camp and advanced through the woods in fine order. The enemy was not aware of our being anywhere near them till a few of their camp guards began to shoot at us. It was then light enough for us to see their camps. We advanced in a run and raised the Rebel yell. At this signal Kershaw advanced from a different place and raised a terrible yell. The Yankees fired a few cannon shot at Kershaw's men and then fled. We were soon in their camp. The most of them were still in bed when we raised the yell and began firing at them. They jumped up running, and did not take time to put on their clothing, but fled in their night clothes, without their guns, hats or shoes.*

We were shooting them as fast as we could, and yelling as loud as we could to see them run. It was the worst stampede I ever saw. We captured about twenty stands of colors (battle flags); and some of Sheridan's men told me, in Savannah, Ga., the next June, that some of the Yankees ran to Winchester, which is fifteen miles away, in their night clothes, without shoes or hats. We took many prisoners and captured nearly all of their wagons, artillery, ambulances, horses, mules and a great deal of clothing, shoes, blankets, tents, etc. I saw one Union soldier who had been killed with only one shoe on. We ran in pursuit of them till we had gotten about two miles from their camp, and then everything was halted. A great many of us went back to their camp after blankets, shoes, clothing, etc., and to my surprise, General Early had his own wagons and artillery brought over Cedar creek." [9]

"This was a perfect route to the Yankees. They were thrown into utter confusion and completely demoralized. Many of them were killed in their tents. Here we captured everything - provisions in abundance an immense amount of all kind of army supplies, a large number of prisoners and about 60 pieces of Artillery'-most effectually wiping out the Winchester and Fisher's Hill stain; the credit being due to the consulate skill and Generalship of the gallant Gordon," wrote Lieutenant Lester." [10]

Private Nichols recalled the looting that followed the brilliant victory, "It seemed that there were no Yankees in our front. Everything was as quiet as death. Some of our boys went to sleep while the others were plundering the camps. I got two nice new tent flies, two fine blankets, a fine rubber cloth, two new over shirts and two pair of new shoes. In fact, we could get anything we wanted except Yankee money." [11]

Gen. Gordon wrote of the successful attack and his efforts to complete the victory.

"Two entire corps, the Eighth and Nineteenth, constituting more than two thirds of Sheridan's army, broke and fled, leaving the ground covered with arms, accoutrements, knapsacks, and the dead bodies of their comrades. Across the open fields they swarmed in utter disorganization, heedless of their officers' commands--heedless of all things save getting to the rear. There was nothing else for them to do; for Sheridan's magnificent cavalry was in full retreat before Rosser's bold troopers, who were in position to sweep down upon the other Union flank and rear.

At little after sunrise we had captured nearly all of the Union artillery; we

The Surprise at Cedar Creek – From a War - Time Sketch
Battles of Leaders of the Civil War, Vol IV, 1888

had scattered in veriest rout two thirds of the Union army; while less than one third of the Confederate forces had been under fire, and that third intact and jubilant. Only the Sixth Corps of Sheridan's entire force held its ground. It was on the right rear and had been held in reserve. It stood like a granite breakwater, built to beat back the oncoming flood; but it was also doomed unless some marvelous intervention should check the Confederate concentration which was forming against it. That intervention did occur will be seen; and it was a truly marvelous intervention, because it came from the Confederate commander himself. Sheridan's Sixth Corps was so situated after the other corps were dispersed, that nothing could have saved it if the arrangement for its destruction had been carried out. It was at that hour largely outnumbered, and I had directed every Confederate command then subject to my orders to assail it in front and upon both flanks simultaneously.

At the same time I had directed the brilliant chief of artillery, Colonel Thomas H. Carter of Virginia, who had no superior in ability and fighting qualities in that arm of the service in either army, to gallop along the broad highway with all his batteries and with every piece of captured artillery available, and to pour an incessant stream of shot and shell upon this solitary remaining corps, explaining to him at the same time the movements I had ordered the infantry to make. As Colonel Carter surveyed the position of Sheridan's Sixth Corps (it could not have been better placed for our purposes), he exclaimed: "General, you will need no infantry. With enfilade fire from my batteries I will destroy that corps in twenty minutes." At this critical moment, General Early arrived on the battlefield and assumed command of the entire army, including the three Divisions being pushed forward by General Gordon. Gordon reverted back to commanding his own Division and General Evans relinquished command of Gordon's Division and returned to command of the Georgia Brigade.

Gordon wrote of his meeting with Early on the battlefield and the conversation that ensued, at this moment General Early came upon the field, and said, *"Well, Gordon, this is glory enough for one day. This is the 19th. Precisely one month ago to-day we were going in the opposite direction."*

His allusion was to our flight from Winchester on the 19th of September. I replied,

"It is very well so far, general; but we have one more blow to strike, and then there will not be left an organized company of infantry in Sheridan's army."

I pointed to the Sixth Corps and explained the movements I had ordered, which I felt sure would compass the capture of that corps--certainly its destruction. When I had finished, he said,

"No use in that; they will all go directly. That is the Sixth Corps, general. It will not go, unless we drive it from the field."

"Yes, it will go too, directly."

My heart went into my boots. Visions of the fatal halt on the first day at Gettysburg, and of the whole day's hesitation to permit an assault on Grant's exposed flank on the 6th of May in the Wilderness, rose before me. And so it came to pass that the fatal halting, the hesitation, the spasmodic firing, and the isolated movements in the face of the sullen, slow, and orderly retreat of this superb Federal corps, lost us the great opportunity, and converted the brilliant victory of the morning into disastrous defeat in the evening." [12]

Lester recalled the advance of the Confederates and the looting of the Federal encampment by his comrades, "We now pursued the enemy some two or three miles. By this time Gen. Early had crossed the river at Strausburg and arrived with Kershaw's division. General Early now assumed command, and while he was maneuvering our troops and our men being somewhat demoralized by the immense amount of the spoils." [13]

Sheridan was on his way to Washington, D.C. and twenty miles from Cedar Creek. When he heard the roar of battle to the south he immediately turned around and galloped back towards the battlefield. He soon encountered his broken, fleeing legions and began to rally them. Seldom in the course of military history has the presence of one man so drastically changed the course of a battle. Sheridan exhorted his men to rally and he would lead them to victory. Almost like magic, his army began to reform and prepare for a counterattack against Early's Confederates. At about 4 p.m., Sheridan launched his devastating counterattack.

He placed a division of cavalry on each of his flanks, with his infantry massed in the center, then ordered his army forward. Initially, the Confederate line held solid and exchanged volleys with the Union infantry. Then Federal General George Armstrong Custer, who later died in the battle of Little Big Horn, led the cavalry charge and thundered around the Confederate right flank.

Private Nichols wrote, "Late in the afternoon we heard the pickets or cavalry begin to fight on our right. Evans had his brigade well in hand and all well composed. The firing was about 800 yards from us, and it soon got faster and nearer and we saw the Yankees advancing." [14]

General Evans commanding the Georgia Brigade reported, "While It was a real attack, as all who were present knew, but one which might have been repelled if there had been troops enough in position on the left of the line occupied by Gordon's Division to have covered ground 400 yards. As soon as this attack was made my brigade advanced at the double quick from the field where it was posted; attacked the near right of this line driving them back a short distance; but in the

Philip Henry Sheridan making his famous ride from Winchester.
By Thure de Thulstrup, 1886 - Library of Congress

meantime a small open field on the left of the division had been penetrated by the enemy, who thus filled the gap which had been made between the remainder of the Division. This caused a retreat, at first in confusion, but the brigade soon organized and held itself under good control ready for any orders it might receive. I conjecture now that the enemy did not immediately continue to advance that portion of his left line which had attacked Gordon's Division, but employed his force gradually turning the left of each brigade in succession. My brigade being organized, was directed by Genl. Gordon to advance to the support of Ramseur's Division before it also gave way in retreat. My brigade retreated slowly in line of battle to Belle Grove hill where it halted and supported two pieces of artillery which were here placed in position and firing on the enemy. These pieces being withdrawn and the retreat still continuing, the brigade under my command fell back in orderly line of battle, halting from time to time to confront the enemy."

Private Nichols recalled the moment Early's army began to disintegrate, "We could see the Yankees advancing and our men falling back on our right. At first it was in order, though it seemed to be a poor, thin line. They began to retreat faster and finally began to run. General Evans then ordered us to fall back, which we did in perfect order for nearly a mile, but were finally ordered to double quick, as the other men were so far ahead of us and their officers had lost all control of them. Sheridan's cavalry was making a bold charge on them, shooting with their repeating rifles and making a lot of them prisoners." The retreat soon became a route and became even worst when a small bridge over the valley turnpike collapsed, preventing the removal of wagons, the captured artillery, along with all the Confederate artillery Early had moved forward." [15]

Gen. George A. Custer - USA
Library of Congress

Evans ordered the brigade to fall back, but Cedar Creek loomed ahead, blocking their retreat. He wrote, "In this order the earthworks of the enemy near Cedar Creek on the west side of the Valley Pike which had been captured during the early morning, were reached when another stand was made by the brigade and by rapid firing checked the enemy until it was supposed that all artillery and wagons were safely in the rear. This position was held so long that a further organized retreat across Cedar Creek by using the difficult crossings above the regular ford became impractical. The enemy, although checked in our immediate front, had moved cavalry around our left flank; Cedar Creek was only about 200 yards distant with no crossing except a bridge of poles and fence rails. To attempt to wade the river in order by the flank would have subjected half the command at least to capture at the creek; or else the enemy on reaching his works which perfectly commanded the crossing in easy rifle range, could have killed and wounded large numbers of our men."

Private Nichols continued the account, "We finally got into a perfect run, with the cavalry getting nearer all the time, and some of our men surrendering. To tell the truth, it all turned into a panic. The gallant generals, Gordon and Evans, were showing all personal bravery mortal men could do. Nearly every man would stop, face the enemy and shoot at them, while the rest of the line was running as fast as a herd of wild, stampeded cattle. "We just had to get out or be captured, and we saw it, our officers losing all control over us." [16]

Macon Telegraph, November 1st, 1864

For the Telegraph and Confederate.
CASUALTIES IN 38th GEORGIA REGIMENT, EVANS' BRIGADE.

CAMP 38th GA. REG'T, Oct. 22, 1864.

MR. EDITOR. Below you have a list of the casualties in my regiment. Only half of my regiment was actively engaged, the other half, being on picket the night before, were deployed as skirmishers on the right to protect the flank of the army. I have been too busily engaged to-day to give you any incidents of the engagement. The portion of my regiment engaged moved as the brigade moved, and shared its glory as well as defeat.

I am, very respectfully,
Your ob't serv't,
P. E. DEVANT,
Lt. Col. Commanding 38th Ga. Reg't.

Field and Staff—Wounded: Serg't Major W S Robertson, dangerously. Missing: Major Thomas H Bomar.

Co. A—Wounded: Privates John Eidson, head and arm, prisoner; Newton Wates, thigh broken, prisoner; J F House, leg amputated, prisoner.

Co. B—Wounded: B M Densmore, leg.

Co C—Wounded: Corporal E A Odom, left arm, severe, privates Francis Wilks, shoulder, severe, T B Cocksay, breast, slight.

Co. E—Wounded: Sergeant J M Hawkins, side, slight, private M E Sorrows, head, slight.

Co. K—Killed: Private A W Almond.
Wounded: Privates M O Wiggins, shoulder, W A Childress, ankle.

Co. N—Wounded: Privates W J Childress, back, severe, J M Phillips, head, slight.

Major Thomas Bomar from Atlanta, Georgia and of Company N, the Chestatee Artillery Company, was forced to surrender to the Yankees. Sixteen year old Private William Jackson Childers, also of Company N, was shot in the left side of his back near the spine by a Minié ball, the ball exiting the front of his right hip. He was captured and carried to a Federal hospital and then sent to Point Lookout prison camp. [17] He miraculously survived the ordeal, living another 67 years, dying in 1931 of old age. Private A. W. Almond, of the Goshen Blues, from Elbert County, Georgia, was shot and killed in action. Private Thomas J. House of Company A, from DeKalb County, Georgia, was shot by a Minié ball in the lower left leg, the bullet shattering the bone. His leg was amputated below the knee at a Confederate field hospital. [18] Fifteen 38th Georgia soldiers were captured, with three of the captured dying of their wounds in the hands of the enemy. Nine 38th Georgia soldiers were wounded, total, but avoided being captured during the retreat.

"It was now our time to fall back and we did so, and did not stand on the order of our going - thus losing all the fruits of our brilliant victory of a few hours before. In this battle we lost two of our best and bravest soldiers. Orderly Sergeant F. M. Landrum (Francis Marion Landrum), wounded and captured by the enemy. He recovered from his wounds and bore the harsh treatment in the hands of the enemy, which caused the death of many a Confederate soldier, 2nd Sergeant Walter S. Robertson, mortally wounded, and died shortly after in the hands of the enemy. We lost, in Landrum and Robertson, two of our comrades who had never flinched from any danger or duty; tried men and true as steel," wrote a 38th Georgia soldier."[19]

Union Captain John W. De Forest's regiment fought the Georgia Brigade in the Shenandoah Valley on several occasions. He wrote of fighting the Confederates, "Three points I noted with regards to our opponents. They aimed better than our men; they covered themselves (in case of need) more carefully and effectively; they could move in a swarm, without much care for alignment and touching of elbows. In short, they fought more like (Indians), or like hunters, than we. The result was that they lost fewer men, though they were far inferior in numbers." De Forest also noted of the Confederates, "When they retreated, they went in a swarm and at full speed, thus presenting a poor mark for musketry. We on the contrary, tried to retire in regular order and suffered heavily for it." [20]

Early blamed his troops for the defeat at Cedar Creek. He said their bad conduct in stopping to plunder the Union camps and their panic after being flanked caused the defeat. He never acknowledged his own failure in halting the advance and allowing the Union army to reform. He wrote "But for their bad conduct I should have defeated Sheridan's whole force." [21] Generals Gordon and Evans disputed Early's claims of Confederate soldier's "looting," as being the cause of the loss of the battle. Both fully laid the blame for the disastrous defeat squarely upon General Early for ordering the "fatal halt."

Casualties at Cedar Creek, October 19th, 1864

Company A – "The Murphy Guards," DeKalb County, Georgia

1. Private John Eidson - Wounded and captured. Died of gangrene of brain, 10/24/1864 at U.S.A. Depot Field Hospital, Winchester, Virginia.
2. Private Thomas J. House - Wounded in leg, amputated below knee.
3. Private Newton Waits - Wounded, thigh broken, captured. Died 3/10/1865 of wounds in West's Bldgs. Hospital at Baltimore, Maryland.

Company B, "The Milton Guards," Milton County, Georgia (Currently Fulton County)

4. Private Salathiel Adams - Captured.
5. Private Malachi Polk Dinsmore - Wounded in right leg.
6. Private Reuben P. Phillips - Captured.

Company C, "The Ben Hill Guards," Bulloch & Emanuel Counties, Georgia

7. Private James E. Boyd - Captured.
8. Private Thomas B. Cooksey - Slightly wounded in right breast.
9. Corporal Mathew Curl Jr. - Wounded, died of wounds 9/18/1864 at Mount Jackson, Virginia.
10. Sergeant John Hayes - Captured.
11. Corporal Elijah Anderson Odom - Severely wounded in left arm.
12. Private Francis Wilkes - Wounded.

Company E, "The Tom Cobb Infantry," Oglethorpe County, Georgia

13. Sergeant John Milner Hawkins - Slightly wounded in side.
14. Sergeant Francis Marion Landrum - Captured. Shot just below left shoulder blade by a rifle ball, ball removed from small of back near spine, according to CSA pension application.
15. Sergeant Major Walter S. Robertson - Dangerously wounded, died of hemorrhage and pyaemia, 11/5/1864, at U.S.A. Depot Field Hospital, Winchester, Virginia. 10/20/1864,
16. Private Mathew E. Sorrow- 9/29/1861. Slightly wounded in head. Died 3/14/1865 of pneumonia at C.S.A. General Hospital, Farmville, Virginia.

17. Private Aaron Shirley - Captured.

Company H, "The Goshen Blues," Elbert County, Georgia

18. Private Thomas Newton Butler - Captured.

Company K, "DeKalb & Fulton Bartow Avengers," DeKalb and Fulton County, Georgia

19. Private A. W. Almand, A. W. (or Almond) (or A. W. Allman) - Killed.

20. Private John Sylvester Burdett - Captured.

21. Private William Anderson Childress - Wounded in left shoulder (back) and right leg (ankle) at Cedar Creek.

22. Private Mathew O. Wiggins - Wounded in shoulder and disabled at Cedar Creek; reported in Augusta Chronicle Newspaper, 22. 11/4/1863.

Company N, "The Chestatee Aritllery," Forsyth County, Georgia

23. Major Thomas Hayne Bomar - Captured.

24. Private William Jackson Childers - Wounded through both hips, ball passed through his body, and captured at Cedar Creek.

25. Private William H. Crane - Captured.

26. Private James M. Phillips - Slightly wounded in head. Captured, date and place not given.

Chapter 23
In the Trenches at Petersburg, Dec 1864 - April 1865

Lieutenant Lester of the 38th expressed the common sentiment among the men of the Georgia brigade, concerning the lost battle at Cedar Creek, "Had General Gordon continued in command he would no doubt have pressed his victory and would most probably have destroyed Sheridan's army. We now fell back across the river and continued our retreat to New Market. Not being pressed we remained at New Market some considerable time skirmishing now and then." [1]

A Georgia Brigade soldier wrote of the movements of the Georgia Brigade following the Battle of Cedar Creek, "On the 14th of November we advanced to Cedar Creek and felt for the Yankees. We found them in a good fortified position at Newtown. We returned to New Market without any engagement except some light skirmishing." [2]

Gordon told of the sharp skirmishes and long marches that followed the battle of Cedar Creek, "The Cedar Creek catastrophe did not wholly dispirit Early's army nor greatly increase the aggressive energy of Sheridan's. It was the last of the great conflicts in the historic Valley which for four years had been torn and blood-stained by almost incessant battle. Following on Cedar Creek were frequent skirmishes, some sharp tilts with Sheridan's cavalry, a number of captures and losses of guns and wagons by both sides, and an amount of marching--often twenty to twenty-five miles a day--that sorely taxed the bruised and poorly shod feet of the still cheerful Confederates. On November 16th, Captain Hotchkiss made this memorandum in his Journal: All of the encounters which followed Cedar Creek, however, would not have equaled in casualties a second-rate battle; but they served to emphasize the fact that neither commander was disposed to bring the other to a general engagement." [3]

"We remained in the valley with but little to do till we received orders to pack up and march out of camps on December 11th, 1864. This closed the ever memorable Valley campaign of 1864. Our little command of never over twelve thousand men had marched about two thousand and five hundred miles; had killed, wounded and captured about twenty-two thousand Union soldiers in the valley campaign, besides what we had killed, wounded and captured at the Wilderness, Spotsylvania C. H., North Anna, Turkey Ridge and Cold Harbor before we were sent to the valley. We had invaded Pennsylvania and Maryland; had been to the very walls of Washington; had been in thirty-five hard battles and skirmishes, and had lost about five thousand men, killed, wounded and captured. I do not think there was ever a little army which had done more real hard service, or could have gone through with more hardships, privations, and sometimes almost starvation, than we had ; nor do I believe that we could have done much more than we had done. We were the worst set of broken down men I ever saw. I have never gotten over it. I was a mere boy, and was broken down before I matured into manhood, and I was too old to grow out of it," [4] wrote a Georgia Brigade soldier.

Early was relieved of command by Lee after the disaster at Cedar Creek. The soldiers had lost faith in his leadership abilities. Gordon was appointed commander of the 2nd Corps and served valiantly in this capacity for the remainder of the war. He recalled the 2nd Corp, along with the Georgia Brigade, being ordered to the defense of Petersburg by Lee,

"During this period of Union and Confederate convalescence I was transferred, by General Lee's orders, to the lines of defence around the beleagured Confederate capital and its sympathizing sister city, Petersburg. My command, the Second Corps, consisted of the divisions of Evans, Grimes, and Pegram. Before dawn on December 8th the long trains were bearing two divisions of my command up the western slope of the Blue Ridge range which separated that hitherto enchanting Valley from the undulating Piedmont region, which Thomas Jefferson thought was some day to become the most populous portion of our country because so richly endowed by nature.

As I stood on the back platform of the last car in the train and looked back upon that stricken Valley, I could but contrast the aspect of devastation and woe which it then presented, with the bounty and peace in all its homes at the beginning of the war. Prior to 1862 it was, if possible, more beautiful and prosperous than the famed blue-grass region of Kentucky. Before the blasting breath of war swept over its rich meadows and fields of clover, they had been filled with high mettled horses, herds of fine cattle, and flocks of sheep that rivaled England's best. These were all gone. The great water-wheels which four years before had driven the busy machinery of the mills were motionless--standing and rotting, the silent vouchers of wholesale destruction. Heaps of ashes, of half-melted iron axles and bent tires, were the melancholy remains of burnt barns and farm-wagons and implements of husbandry. Stone and brick chimneys, standing alone in the midst of charred trees which once shaded the porches of luxurious; and happy homes, told of hostile torches which had left these grim sentinels the only guards of those sacred spots. At the close of this campaign of General Sheridan there was in that entire fertile valley--the former American Arcadia--scarcely a family that was not struggling for subsistence." [5]

Lieutenant Lester wrote, "This valley campaign as slightly sketched in this article - its long toilsome marches, arduous duties and fierce and hotly contested battles, can never be transferred to paper. Ours was a small force, put against overwhelming numbers, and commanded by a brave and dashing officer, (General Phil Sheridan,) whose large force of cavalry was handled with commutate skill. Nothing saved the Confederate forces from annihilation but good Generalship and veteran soldiers, long trained and used to toil and hard fighting, who knew no such word as fail as long as there remained the remotest chance of victory.

We continued our march up the valley, crossed the Blue Ridge again at Brown s Gap, and continued on to Charlottesville, where we took the train and were soon landed on Hatcher's Run near Petersburg. Thus after marching hundreds of miles and almost continuous fighting all over the valley of Virginia, we now rejoined General Lee's army to make a last desperate stand for the defense of

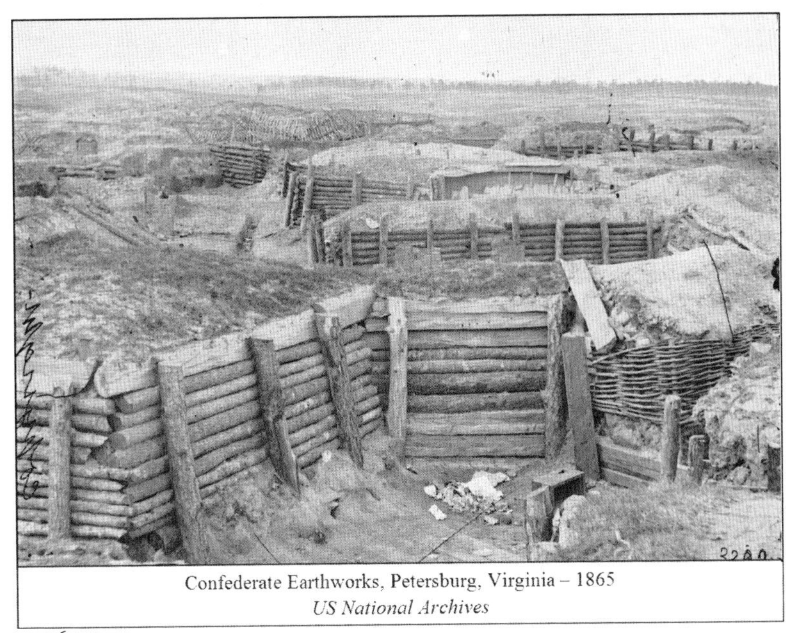

Confederate Earthworks, Petersburg, Virginia – 1865
US National Archives

Richmond." [6]

The siege of Petersburg was the longest siege of the war, lasting nine months. Lee's army was arrayed to protect the city, which was critical to defending Richmond and the railroads supplying the army. The siege lines extended over 35 miles and Grant's strategy was to continuously extend his lines until the Confederate line was but a skirmish line. When the Confederate 2nd Corps and Georgia Brigade arrived at the trenches in December 1864, the rest of Lee's army had already been there for six months.

Captain Mathews of the Battey Guards, recalled of the 38th near Petersburg, *"We were in winter huts, and many members of the regiment were furloughed and our convalescents returned making the fighting force of regiment 250 muskets."* [7]

Lieutenant Lester noted the inflated price of food stuffs and of not being paid, *"Here we had a harder time than we had ever had before rations were scarcer than ever. Sometimes the boys would go out and buy a little meal at $10 a peck, or a few black peas at $4 per quart and everything else that could be bought at all at proportionate prices; and this, too, on a salary of $11 per month, and that not paid."* [8]

Evans' Brigade, of Gordon's division, was commanded by Colonel John H. Baker, of the Thirteenth Georgia Regiment; Evans was in command of Gordon's division, and Gordon in command of the 2nd Corps. [9] Lieutenant Colonel Phillip E. Davant commanded the 38th Georgia Regiment.

The Georgia Brigade and 38th Georgia had suffered severe losses during the valley campaign. The 60th and 61st Georgia Regiments were skeletons of their former selves. Private Nichols recalled, *"On the 18th of January, 1865, the Sixtieth and Sixty first Georgia Regiments were consolidated into one regiment with only one stand of colors. Major W. B. Jones (Old Red) of the Sixtieth Georgia was promoted to the rank of colonel."* [10]

Lieutenant Lester told of the 38th Georgia Regiment being posted in the "Crater" at Petersburg, where the enormous mine was exploded the previous July,

"We were soon moved from here and put in the trenches in front of Petersburg, immediately the celebrated Crater that had been blown up not long before.

We remained in the Crater at Petersburg for some length of time. Here we had the hardest time we had ever yet seen, being confined in the breast-works night and day. Our cooking was done in bomb-proofs, which were nothing but holes dug in the ground and covered over with earth. In these dens we cooked what little rations we could get. We used coal for fuel there being no wood to be had. Our water and rations were brought by means of deep ditches, dug from the breast-works out in the open country, beyond the range of the enemy's sharp-shooters. Our only way of ingress and egress was through these ditches. The position of the Tom Cobb Infantry was peculiarly hard at this place. We occupied the centre of Lee's army.

As the army was very much reduced and a long line of breast-works to be protected, General Lee's small number (comparatively) were so stretched that we stood some five feet apart in the works, and being in the centre of (Gordon's) division were constantly used to reinforce either wing that might be threatened by the enemy. [11]

Grant received intelligence the Confederates were receiving supplies by wagon trains on the Boydton Plank Road in early February of 1865. He sent General David Gregg's cavalry division on a raid to the Boydton Plank Road, by way of Ream's Station and Dinwiddie Court House, to intercept the Confederate supply trains, on February 5th, 1865. He pushed two army corps forward to seize Vaughan Road and block any attempt by the Confederates to interfere with the raid. The Confederate 2nd Corps, commanded by Gordon, stood directly in the path of this Federal advance.

Lieutenant Lester told of the 38th's participation in the two day battle at Hatcher's Run, near Petersburg, "We were not here in long before we were fighting again. We had two considerable engagements on Hatcher's Run, one being a hotly contested night fight, in which we lost John B. Jackson, killed, and Wm. D. Scisson, wounded. [12] "They engaged the Yankees near Burges' mill, on Hatcher's Run, and drove them back to their works after a very stubborn battle of one hour and a half. In the battle Major B. F. Grace, of the Twenty-sixth Georgia Regiment, was killed." [13]

Casualties.

We are indebted to the Georgia Hospital and Relief Association for the following list of casualties in Evans' Brigade, February 5th, 1865:

38th Ga Regiment—J C Wormack co A bowels mortal, George Baxton co D thigh slight, J B Jackson co E killed, W D Sisson co E arm slight, Allen Brown co K leg slight, capt R H Fletcher co K thigh slight, J N Owen co A hip mortal,

Daily Constitutionalist (Augusta, Ga.) March 12th, 1865, P 3.

Private James Cicero Daniel, of the Tom Cobb Infantry, distinguished himself by conspicuous gallantry during the two day battle and was awarded a battlefield promotion to 1st Lieutenant for his courage, the only recorded instance of a battlefield promotion in the 38th Georgia. Lieutenant Colonel Davant, commanding the 38th wrote, *"There being very few commissioned officers present with the regiment, Private James C. Daniel was detailed to act as adjutant of the regiment. During the engagement of Sunday, Feb 5th, 1865, when the regiment was advancing in line of battle under fire, Private Daniel was very conspicuous all along the line encouraging the men and keeping them in their places – at one time when the line was about to give way, Private Daniel, by his coolness and energy, stopping it. The same coolness and energy was displayed by him in the fight on the 6th of February. The gallant manner in which the regiment drove back the enemy in their front twice on that day is attributable in some measure to the coolness and courage exhibited by him on that occasion.* [14]

The following day, February 6th, 1865, Private Nichols recounted the second day of the battle of Hatcher's Run, "On the 6th the Yankees flanked around on the extreme right, and was met by Ewell's corps. Our brigade was, as usual, first to get into the battle. It charged their works, but failed to rout them on the first assault. It had to fall back across a little boggy branch about 100 yards in the rear. Colonel Baker was wounded. Here General Gordon met and rallied it and made the second charge, and routed the Yankees badly in its front. This was a very stubborn battle, and the consolidated regiment had the sad misfortune of getting its gallant Lieutenant-Colonel S. H. Kennedy severely wounded while leading the regiment. He had to be carried from the field. This was on Deep Run."

Twenty year old Private John C. Warnock of Company A was shot through the bowels and mortally wounded, he died on February 7th. He had enlisted at the age of sixteen and fought through the entire war. Private John B. Jackson, of the Tom Cobb Infantry, was killed in action. [15] Private George Baxter of Company D was reportedly slightly wounded. [16]

Lee wrote to the Confederate Secretary of War James Seddon on February 8th, 1865, of the dire condition of his men after Hatcher's Run, *"All the disposable force of the right wing of the army has been operating against the enemy beyond Hatcher's Run since Sunday. Yesterday [the 7th, this date], the most inclement day of the winter, they had to be retained in line of battle, having been in the same condition the two previous days and nights. I regret to be obliged to state that under these circumstances, heightened by assaults and fire of the enemy, some of the men had been without meat for three days, and all were suffering from reduced rations and scant clothing, exposed to battle, cold, hail, and sleet.*

I have directed Colonel Cole, chief commissary, who reports that he has not a pound of meat at his disposal, to visit Richmond and see if nothing can be done. If some change is not made and the commissary department reorganized, I apprehend dire results. The physical strength of the men, if their courage survives, must fail under this treatment. Our cavalry has to be dispersed for want of forage. Fitz Lee's and Lomax's divisions are scattered because supplies cannot be transported where their services are required. I had to bring William H. F. Lee's division forty miles Sunday night to get him in position. 'Taking these facts in connection with the paucity of our numbers, you must not be surprised if calamity befalls us.'" [17] The situation for the Confederate army was becoming more desperate by the day.

Gordon wrote of Lee calling him for a meeting in the middle of the night during February 1865, *"General Lee sent a messenger, about two o'clock in the morning, to summon me to his headquarters. He opened the conference by directing me to read the reports from the different commands as he should hand them to me, and to carefully note every important fact contained in them.*

The revelation was startling. Each report was bad enough, and all the distressing facts combined were sufficient, it seemed to me, to destroy all cohesive power and lead to the inevitable disintegration of any other army that was ever marshaled. Of the great disparity of numbers between the two hostile forces I was already apprised. I had also learned much of the general suffering among the troops; but the condition of my own command, due to the special efforts of which I have spoken, was not a fair measure of the suffering in the army. I was not prepared for the picture presented by these reports of extreme destitution--of the lack of shoes, of hats, of overcoats, and of blankets, as well as of food. Some of the officers had gone outside the formal official statement as to numbers of the sick, to tell in plain, terse, and forceful words of depleted strength, emaciation, and decreased power of endurance among those who appeared on the rolls as fit for duty. Cases were given, and not a few, where good men, faithful, tried, and devoted, gave evidence of temporary insanity and indifference to orders or to the consequences of disobedience--the natural and inevitable effect of their mental and bodily sufferings.

My recollection is that General Lee stated that, since the reports from A. P. Hill's corps had been sent in, he had learned that those men had just been rationed on one sixth of a pound of beef, whereas the army ration was a pound of beef per man per day, with the addition of other supplies; that is to say, 600 of A. P. Hill's men were compelled to subsist on less food than was issued to 100 men in General Grant's army. When I had finished the inspection of this array of serious facts, and contemplated the bewildering woe which they presented, General Lee began his own analysis of the situation. He first considered the relative strength of his army and that of General Grant. The exact number of his own men was given in the reports before him--about 50,000, or 35,000 fit for duty. Against them he estimated that General Grant had in front of Richmond and Petersburg, or within his reach, about 150,000. Coming up from Knoxville was Thomas with an estimated force of 30,000 superb troops, to whose progress General Lee said we could offer practically no resistance--only a very small force of poorly equipped cavalry and detached bodies of infantry being available for that purpose. "From the Valley," he said, "General Grant can and will bring upon us nearly 20,000, against whom I can oppose scarcely a vedette." This made an army of 200,000 well-fed, well-equipped men which General Grant could soon concentrate upon our force of 50,000, whose efficiency was greatly impaired by suffering. Sherman was approaching from North Carolina, and his force, when united with Schofield's, would reach 80,000. What force had we to confront that army? General Beauregard had telegraphed a few days before that, with the aid of Governor Vance's Home Guards, he could muster probably 20,000 to 25,000. But General Joseph E. Johnston had just sent a dispatch saying in substance that General Beauregard had overestimated his strength, and that it would be nearer the truth to place the available Confederate force at from 13,000 to 15,000. So that the final summing up gave Grant the available crushing power of 280,000 men, while to resist this overwhelming force Lee had in round numbers only 65,000."[18]

The Army of Northern Virginia was slowly starving to death and growing weaker by the day, while Grant's Federal army was growing stronger. General Lee was slowly running out of options. He ordered Gordon to conduct a detailed reconnaissance of the enemy lines, seeking a weak point. Gordon discovered the weakest point in the Federal lines was Fort Stedman, also known as Hares Hill. The entire 38th Georgia Regiment numbered only about 250 men. Companies were further reduced by sickness, and furloughs. Company H, the Goshen Blues, from Elbert County, Georgia was probably typical of most companies in the 38th Georgia. Lieutenant Colonel Davant said of Company H, on Monday, March 6th, 1865, *"Company H has but one officer, Captain Deadwyler who is a prisoner of war and has fifty one (51) effective men on the rolls besides prisoners of war, sixteen of this number are present with their company."* [19] On the same day Lieutenant Colonel Davant was summarizing the strength of the Goshen Blues, Lieutenant Simpson A. Hagood, of the Milton Guards, sat down and to write a letter to his wife Adaline. He had not heard from his wife and children for months and was worried they were starving to death. He wrote of the coming spring campaign, little dreaming the war for General Lee's army would suddenly end in just over a month.

Camp of Evans Brigade
Near Burgess Mills Va.
March 6th 1865

Dear Adaline,

I drop you another letter though I have but little hope of your ever receiving it. It seems that we are doomed to never more hear of each other while the War lasts. I have not had a letter from you nor heard a single word concerning you since Bennet Bagwell came here some three months ago. You don't know anything about how uneasy I am about you. I am afraid you are almost if not entirely destitute of something to live upon. You know not how many sleepless hours I spend when all around is deeply wrapped in slumber on this very account. The thought of my poor little babes crying for bread and none to give them seems sometimes almost more than I can bear. There is but one thing which I can do for them and that is to commend them to an all wise and Merciful God. This I can do and I do it daily in my poor weak manner. Until I can get to come or send by some reliable person some thing for your relief, you must do the very best you can. I know you have plenty of friends there who will give you all the assistance in their power-but they have been torn up and so bled by the Yankees (and Men of our own who are meaner than Yankees) until they have nothing at all which they can spare to assist those who may be needy. They can, and will, of course give you their sympathies and advice.

I have no hope now of getting a Furlough this Spring for active operations will soon be again resumed in this Department and then there are no officers scarcely with the Regiment to which I belong. A certain number of Officers are required to be present before one can get a Furlough and there are not more than one fourth of that number now or listed to be present with this Regiment, soon though it will be impossible for me to get a furlough of (-----) this Spring If I had a furlough the Communication with Georgia is Kept Cut so that it would be almost if not entirely impossible for me to get home.

There being so few officers present it makes my duties very heavy for I have the Command of three Companies all the time in my hands and must act as Adjutant of the Regiment besides acting a great part of the time as Assistant Quarter Master of the Regiment. besides all these things nearly every private in the Regt comes to me for all the writing on business of any sort they may have to transact. Therefore my Pen is nearly dry from daylight to dark every day. I am not boasting but telling you of my regular daily duties so if my letters appear brief or hastily written you may have an idea of the manner in which I have to transact the private affairs of my own. I have to snatch a moment now and then. sometimes I can write a few lines then stop for something else and perhaps not have the chance of touching the letter again until the next day.

You can not therefore expect me to write a consistent and well composed letter at all times. The times do indeed look very gloomy at present. It appears that Sherman will be allowed to March his Army anywhere pleases. I cant understand why it is that our Authorities suffer this for I am satisfied that a force could be spared from this and other points sufficient to check him (Old Grant cant more than hold his own with Lee at this place. Now if Lee would so reinforce Johnston as to whip Sherman (even if he lost Petersburg at the time it would be infinitely better than to let Sherman go and destroy when and where he pleases. But I am confident that Lee can reinforce Johnston enough to check Sherman without the loss of Petersburg or Richmond either.

Well, I will try now to finish my letter. I have just had to stop writing and issue a lot of Socks to the Regiment and it has so frustrated me that I scarcely know how to Commence again. I am very happy to be able to state that those four Boys of this Company who ran off a few nights ago have returned voluntarily to the company (I spoke of them having deserted in a letter a few days ago to Father) Their punishment will be a great deal lighter than if they had been brought back by force. If they had been caught and brought they would doubtless have suffered death but as it is their punishment will perhaps be nothing more than a short confinement and the forfeiture of some Pay. The disgrace attached to the Crime of desertion in their case have by this act been very greatly and very justly mitigated although it can never be entirely obliterated nor should it be. They should be made to feel keenly the disgrace of this thing. I think they are heartily sick of the situation and will not be likely to repeat it again. for fear you may not get the letter containing their names I will give them again in order that they may as far as possible be made public as a warning to others. (-----------) are (-------------). All Men who would (----) have seemed to commit dastardly Acts before the War. I am very much ashamed for them for (-----------------) and Absent without leave from the Army and all those nearly who are brought back under Guard are Shot to Death while those who return voluntarily are either pardoned entirely or are punished lightly. This should induce all absentees without leave to return at once to Their proper Commands. I must hurry to a close for it is meal time for Battalions (----).

Do the best you can for our dear little babes and ever remember them and me in your Prayers. Tell Pappa (your Father) that I would be very glad indeed to get a letter from him. Give him my best Love and give to Mamma and Sarah Jane my kindest regards and warmest Love also. Tell me if you know where Mac is I would love to write to him if I know how to direct a letter so that it would reach him. Give my affections to Levi and Mary Edwards and to Sally also.

I remain ever yours

SA H [20]

General Gordon spent a few weeks carefully looking over the Federals lines. Lieutenant Lester described the weak point Gordon discovered, Fort Stedman, "Fort Stedman was a large dwelling house and stood immediately in front of our brigade, and not more than 100 yards from our works, the opposing armies were not more than that distance apart on this. part of the line." [21] The Georgia Brigade was posted, "from the Crater to the left toward the Appomattox River across the railroad, and at the nearest point it was less than one hundred yards from this great fort." [22]

Gordon explained the purpose of this immensely important attack, "The purpose of the movement was not simply the capture of Fort Stedman and the breastworks flanking it. The tremendous possibility was the disintegration of the whole left wing of the Federal army, or at least the dealing of such a staggering blow upon it as would disable it temporarily, enabling us to withdraw from Petersburg in safety and join Johnston in North Carolina." [23]

The attack on Fort Stedman was the last offensive attack of the war by the Army of Northern Virginia.

Gordon planned to assault the fort with three divisions of his Second Corps (under Brigadier General Clement A. Evans, Major General Bryan Grimes, and Brigadier General James A. Walker.) The assault force would also include two brigades from Major General Bushrod R. Johnson's Fourth Corps to provide close support. Two other additional brigades, from, Major General Cadmus M. Wilcox's Third Corps, would serve as a reserve, General George Pickett was ordered to join Gordon's attack with his 6,500 soldiers, but arrived too late to help.

At 4 a.m. on the morning of March 25th, 1865, Gordon's assaulting column rolled forward. Leading the assault were stalwart men with sharp axes, who cleared the abatis and other obstructions between the lines, opening paths for Gordon's columns to pass through the obstructions and capture Fort Stedman. Three columns of 100 handpicked men would then move from Fort Stedman to capture three other forts located behind the fort.

Evans' Georgia Brigade was commanded by Colonel John H. Baker during the attack, as Evans was commanding Gordon's Division. Evans wrote, "Fort Stedman was penetrated by this division, one by the division sharpshooters, one by the 31st and 13th Georgia regiments, Colonel John H. Lowe commanding, and one by the Louisiana troops Col. Eugene Waggaman, 10th Louisiana, commanding." Lieutenant George Lester of the 38th Georgia wrote, "At daybreak on the 25th of March General Gordon made this fierce assault, which was a complete success, driving the enemy from the fort and making a breach in his line, capturing a large number of prisoners, among the rest a Federal General and a large amount of stores and several pieces of artillery." [24]

Fort Stedman was captured, along with a large numbers of prisoners. However, the three handpicked columns of 100 men tasked to capture the three forts in the rear of Fort Stedman, became lost in the darkness and the maze of trenches behind the fort. Dawn revealed the Confederates inside Fort Stedman were nearly surrounded and heavy Federal re-enforcements were rushing from all points to contain the attack. Privates Moxley, of Company C, Smith, Company G, and Wiggins, of Company K, of the Thirty eighth Georgia Regiment, captured and carried out one of Grant's Morton (mortar) guns. [25] Mathew O. Wiggins of Company K, Bartow's Avengers, *"seized and threw a Morton (mortar) shell with a burning fuse out of our line and over our works before it exploded, and probably saved the lives of more than a dozen men."* [26]

The 38th soldiers inside Fort Stedman were forced to either surrender, or run the gauntlet of enemy fire back to Confederate lines. The open space between the lines was swept by Union artillery and rifle fire, *"We had to fall back again to our old position. In this assault we lost Whit Hopper and John G. Tiller (Red Tiller). They were wounded and left in the enemy's hands and no doubt died as they have never been heard of since. These were two good soldiers and had done much hard service for their country,"* [27] wrote Lieutenant Lester.

Private Charles Tabor Berryman was severely wounded, shot in the back, probably during the retreat, and captured. A Minié ball entered 3 inches from the last dorsal vertebrae and exited his shoulder, near the head of the left humerus bone.[28] Captain Mathews noted the other losses in the 38th, "seven killed and seventeen wounded" at the battle of Fort Stedman. He noted, "Robert Wiggins was conspicuous for gallantry in every engagement and was killed at Fort Hare." [29] At least eight men were captured, including Privates Tinsley Rucker Adams, William R. Powell, James Frederick Dix, George W. Pughsley, Isaac H. Smith, Sgt. John W. Green, James Monroe Shiflet, and Ebenezer P. Stewart. [30]

General Evans sat down just a few hours after the end of disastrous attack, and wrote an after action report:

Hdqtrs. Gordon's Division March 25, 1865
Captain V. Dabney, A.A.G.

I have the honor to submit the following general account of the action of this division in the affair of this morning on Hare's Hill. In accordance with instruction, Fort Stedman (the enemy's work opposite Colquites Salient) was penetrated by this division, one by the division sharpshooters, one by the 31st and 13th Georgia regiments, Col. J. H. Lowe commanding, and one by the Louisiana troops Col. [Eugene] Waggaman [Tenth Louisiana] commanding. Immediately afterwards Terry's brigade was thrown across the enemy's works, formed and advanced along the breastworks in rear of them, driving the enemy from them for a space of about six or seven hundred yards, capturing a number of prisoners and several pieces of artillery. Evans' Brigade commanded by Col. [John H.] Baker [Thirteenth Georgia] co-operated with Brig. Gen. [William] Terry as soon as the Brigad was crossed from our breastworks to those of the enemy.

Day was breaking just as Terry's brigade was formed, and the enemy had been aroused. The division advanced as far as the orders which I received permitted and as far as prudence justified under the circumstances existing. Their orders were not to assault a line of breast-works and a fort directly in front, especially if protected by abatis or other obstructions. I communicated my situation to the Major-General commanding [General Gordon] with the statement that to advance further I must charge a strong line of the enemy aided by their artillery and was instructed to await further orders. The orders to retire were afterwards received and the Division withdrawn under a galling fire of artillery and small arms to their original position. The division is entitled to claim the capture of the south half of Fort Stedman and about six or seven hundred yards of the breastworks in which were very many mortars and other pieces of artillery—and at least four hundred of the prisoners captured. Among the prisoners was Brig. Gen. [Napoleon B.] McLaughlin, captured by Lieut. Gwin of Col. Lowe's regiment. The Division lost 43 killed and 122 wounded.

I regret to state that Brig. Gen. Terry is severely wounded. To his skill and cool courage I was very greatly indebted during the morning. I could not but observe his own efficiency and of his admirable staff. Col. Baker commanding Evans brigade was slightly wounded also while in the brave discharge of his duty. We have to mourn over the loss of several gallant officers and good men among those killed and severely wounded. To Brig. Gen. Terry, Col. Baker, Col. Waggaman, Col. Lowe I feel peculiarly indebted for their conspicuous zeal in executing the orders given. Many of the troops behaved well, but in this as in former actions, I could but observe how sadly we need re-organization and discipline. It is almost impossible at times to maneuver the troops at all

C. A. Evans, Brig. Gen.
Commanding Division[31]

HEADQUARTERS NINTH ARMY CORPS,
March 27, 1865.

Bvt. Maj. Gen. A. S. WEBB,
Chief of Staff:

Five deserters have been examined—four from Thirty-eighth Georgia and one from Ninth Louisiana. They have nothing new. Heard that General Terry was slightly wounded and that another general officer was killed. This statement is also made by the deserter mentioned in my first dispatch.

JNO. G. PARKE,
Major-General.

The War of the Rebellion: Series 1, Vol. 46, (Part II) P. 204

The 38th was hemorrhaging soldiers, in addition to the losses incurred at Fort Stedman, three soldiers were reported as deserting on a Union report from March 23rd, 1865. Four more 38th soldiers deserted the day after the battle of Fort Stedman, on March 26th. As Gordon said, "Starvation, literal starvation was doing its' deadly work." [32] Private David W. Simmemon of the Dawson Farmers slipped between the lines, during the night of March 26th and surrendered to the Yankees as a deserter. Private John Childers, of the Chestatee Artillery company also deserted on March 26th. Private William J. Little of Company K, from DeKalb County was reported captured on March 26th, these may have been three of the four deserters mentioned in the March 27th report of Union Major General John Parke. [33] Private Ebenezer P. Stewart of the Murphy Guards was captured during the battle of Fort Stedman and sent to Point Lookout Military Prison, Maryland. He wrote a letter to Colonel Hoffman, the commandant of the prison, shortly after his arrival at Point Lookout, March 29th, 1865. He requested parole and stated he was a deserter.

"Dear Col,
Previous to the 25 day of March, I had made my arrangements to come into your lines. Though an opportunity did not present itself, and on the morning of the 25th '65 my command was order to charge fort Stedman which they did though they was forced back and I remained on purpose and I desire to become a citizen of the United States. You will please grant my request as soon as possible.

Very Respectfully,
E. P. Stewart, Co. A, 38th Ga. Regt" [34]

The letter didn't help Private Stewart get home any sooner than the others, as he was released from prison on May 14th, 1865.

Grant launched a massive attack on the Confederate lines in the early morning hours of April 2nd, 1865. Four Federal corps of infantry attacked and the Union 6th Corps punched a gaping hole in the Confederate line. When the fighting ended that night, Lee ordered the evacuation of Richmond and Petersburg. As he had prophesized nearly a year earlier, "We must destroy this army of Grant's before he gets to James River. If he gets there, it will become a siege, and then it will be a mere question of time." [35] General Lee was correct, the siege of Petersburg ended in disaster for the Confederates, though they had held the line for 292 days. The Army of Northern Virginia began their retreat towards Appomattox Court House that evening. Captain William C. Mathews wrote of the strength of the 38th Georgia at the start of the retreat, *"When the retreat commenced on April 2nd the regiment numbered about 200."*[36]

Chapter 24
The Road to Appomattox Court House, April 2nd - 12th 1865

Sergeant William F. A. Dickerson, of Company D, the McCullough Rifles, wrote, "On April 1st there was heavy cannonading all night. On the 2d our lines were attacked at different places. Fighting continued all day. We kept moving our line from right to left. We were not engaged on that night." [1] "Here we remained repelling assaults almost night and day, until the evacuation of Petersburg, which took place on the 2nd day of April 1865."[2] "Took up line of march, in the direction of Amelia Court House, and continued to march day and night until the night of the 4th when we stopped and drew rations. On the 5th we lay in battle line all day and marched all that night. At daylight we took two hours' rest. On the 6th fought all day and retreated at night. Marched to High Bridge at Farmville." [3]

Lieutenant Lester recalled the trying time after the battle of Fort Steadman and the retreat to Appomattox Court House,

"Here we remained repelling assaults almost night and day, until the evacuation of Petersburg, which took place on the 2nd day of April 1865. Our little band now left the breastworks and began the retreat going in the direction of Amelia Court House. The first day of the retreat we were not pressed, but on the second day the enemy over took us, and it was then a constant fight on to Appomattox Court House. On this retreat we were sorely pressed by the enemy, and it seemed that we suffered more than a human being could stand. We had no rest night or day until the fifth day, when we arrived at Amelia Court-House. Here General Lee expected to get rations for his men but no rations were there and he detailed men to forage the country around for food for his famished soldiers, but with poor success.

After remaining one night at Amelia Court-house we started on the retreat again, more hardly pressed, if possible, than before. Hundreds and hundreds of our men were constantly falling out of line, completely worn out and exhausted from hunger and fatigue.

The trying times of this retreat can only be realized by those who were there. No description can give an adequate conception of the suffering endured. In the rear and on both sides the victorious enemy were pressing us; every little hill and mud-hole blocked up with hurrying wagons; the air filled with bursting shells and exploding ammunition wagons; dense columns of smoke ascending to the heavens and burning wagon trains, exhausted men and worn out mules; dead horses everywhere. Death being welcomed by many a worn out soldier from their sufferings, and it is but reasonable that it should be so, for is it any wonder that many hearts, tried in the fiery furnace of four years, unappalled never before found wanting, should have quailed in the presence of starvation, fatigue, sleeplessness, misery and suffering untold for six, eight days, continuously and knowing that it was to all end in disaster and ruin to the cause for which they had fought. Men were so worn out that they, while engaged in battle, would throw down their guns, and fall down to sleep, regardless of the fighting that was going on all around them." [4]

Disaster struck Lee's retreating army on April 6th, at Sailor's Creek, Virginia. The battle of Sailor's Creek was actually three large separate actions covering several square miles on the same day. The Confederate wagon trains moved westward on two parallel roads. They became bottlenecked at two small bridges crossing Sailor's Creek. The delays in the advance created gaps in the Confederate column. These gaps were quickly exploited by Yankee cavalry under General Custer and then Federal infantry joined the fray. One fourth of Lee's Army was cut off and 7,000 soldiers were forced to surrender after hours of furious combat. Six Confederate generals were captured, including Generals Ewell, Custis Lee, and Kershaw. [5] Among the 7,000 captured were about a dozen soldiers from the 38th Georgia.

Three soldiers were captured from Company A, of DeKalb County, Georgia, Sergeant John J. Carpenter, Private William Harris Mitchell, and Private James A. J. Reeves. Private John W. Logan of Company B, The Milton Guards, was also captured. Company C, the Ben Hill Guards, lost Sergeant Norris N. Durden. Private Bennett Rainy Jefares, an ambulance wagon driver of Company D, from DeKalb County, Georgia. Sergeant Joseph "Joe" Bentley of Company F, Thornton's Line Volunteers, was captured along with Private Joseph Chandler Terry, also of Company F. Three soldiers from Company G, Battey's Guards, from Jefferson County, Georgia were captured, Private James King, Sergeant Thomas J. Peeples, and Private William A. Wright. Private John Poole Ray, of Company D, escaped capture at Sailor's Creek on April 6th, but was captured the following day. [6]

Sergeant Dickerson wrote, *"On the 7th we lay in battle line all day. At night we took line of march and continued until the evening of the 8th. Stopped to within 28 miles of Lynchburg. At Appomattox Court House we drew short rations. While cooking cannonading began and we were called in line. At nine o'clock we marched to Court House and lay in battle line the remainder of the night."* [7]

Sergeant Hudgins remembered the last night of the war, as the 38th marched into the village of Appomattox Court House at midnight, on April 8th, *"The preceding night had been cold and chilly and we reached the village about midnight, when we halted on the court house square, built fires and remained until day."* [8] While the exhausted soldiers of the 38th Georgia slept on the town square at Appomattox Court House, Yankee cavalry worked through the night constructing blockades across the Richmond Stage Road to Lynchburg, blocking the Confederate escape route.

General Gordon, commanding the 2nd Corps, wrote of the precarious position the Confederates, *"The Federals had constructed a line of breastworks across our front during the night. The audacious movement of our troops was begun at dawn. The dashing cavalry leader, Fitzhugh Lee, swept around the Union left flank, while the infantry and artillery attacked the front. I take especial pride in recording the fact that this last charge of the war was made by the footsore and starving men of my command with a spirit worthy the best days of Lee's army."* [9]

Lieutenant Lester wrote of this last charge by the Georgia Brigade and 38th Georgia, *"General Gordon formed his command and ordered us to charge the enemy's cavalry, who were immediately in our front. We believed that it was a useless waste of life, but being always ready to obey orders we made the charge and repulsed the cavalry and drove them back some distance, but soon came upon a whole Yankee corps and were compelled to fall back."* [10]

Gordon recounted of how his troops celebrated their small success, as legions of other Federals closed in from all sides. A dangerous gap was opened between Gordon's Corps and Longstreet's Corps and Federal cavalry attempted to drive a wedge between the two wings of the Confederate army, *"The Union breastworks were carried. Two pieces of artillery were captured. The Federals were driven from all that portion of the field, and the brave boys in tattered gray cheered as their battle-flags waved in triumph on that last morning. The Confederate battle-lines were still advancing when I discovered a heavy column of Union infantry coming from the right and upon my rear. I gathered around me my sharpshooters, who were now held for such emergencies, and directed Colonel Thomas H. Carter of the artillery to turn all his guns upon the advancing column. It was held at bay by his shrapnel, grape, and canister. While the Confederate infantry and cavalry were thus fighting at the front, and the artillery was checking the development of Federal forces around my right and rear, Longstreet was assailed by other portions of the Federal army. He was so hardly pressed that he could not join, as contemplated, in the effort to break the cordon of men and metal around us. At this critical juncture a column of Union cavalry appeared on the hills to my left, headed for the broad space between Longstreet's command and mine. In a few minutes that body of Federal cavalry would not only have seized the trains but cut off all communication between the two wings of Lee's army and rendered its capture inevitable. I therefore detached a brigade to double-quick and intercept this Federal force."* [11]

Sergeant Hudgins vividly recalled the moment General Evans ordered the sharpshooters battalion of his division to make the final charge of the war for the Army of Northern Virginia, *"Suddenly there appeared a long line of blue, infantry moving down on the right, with fixed bayonets and flags gaily flying in the morning breeze. General Evans came riding down his thin line watching the Federal infantry as they advanced until he came opposite Kaigler's battalion sharpshooters, when he ordered them to deploy and charge. The idea of this battalion, 26 men and two officers, charging an army corps and them flushed with victory, none but Gens. Gordon or Evans would have thought of, and I am sure no men but those who had followed Lee, Johnston and Jackson would have complied.*

Soon the roar of musketry told us that Kaigler had attacked as that good soldier knew how. The Federal line now seemed to halt and Kaigler was now returning with 71 prisoners who had surrendered. Lieut. Logan of that gallant band, was cool in the charge, having been on all the battlefields of the Civil War that his command was engaged." [12]

Gordon wrote of the famous dispatch he sent Lee that sealed the fate of the

Army of Northern Virginia, *"Such was the situation, its phases rapidly shifting and growing more intensely thrilling at each moment, when I received a significant inquiry from General Lee. It was borne by Colonel Charles S. Venable of his staff. The commander wished me to report at once as to the conditions on my portion of the field, what progress I was making, and what encouragement I could give. I said: "Tell General Lee I have fought my corps to a frazzle, and I fear I can do nothing unless I am heavily supported by Longstreet's corps." Colonel Venable has left on record this statement: General Lee received my message, he said: 'There is nothing left me but to go and see General Grant, and I had rather die a thousand deaths.'"*[13]

George Custer rode into the Confederate lines near the 38th Georgia's line of battle. Sergeant Hudgins wrote, "We then re-formed, marched out of Appomattox in a south westerly direction, passing Carter's, Starke's and Poagues' battalions of Confederate We formed another line along the road east of the village and General Geo. A. Custer, commanding United States cavalry, rode into our line in a few feet of the writer and asked to see some general officer in the Confederate army." [14]

Custer probably never realized he was nearly numbered among the very last men killed in the war. A soldier of the Georgia Brigade wrote of Custer's approach to the Confederate line of battle and told how Custer was nearly shot dead by a member of the 31st, *"We saw a man in blue uniform riding toward us and waving a red handkerchief before him, while his long curly, flaxen hair stood out behind him in the morning air. As he passed us he inquired who was in command. Some one replied: 'General Gordon.' A young soldier with tears streaming down his cheeks brought up his gun to shoot was in the act of doing so when someone knocked it up and said, 'Don't! Perhaps General Lee has surrendered and it might cause trouble.' He was a splendid marksman, and if he had not been hindered, the bloody minded tyrant Custer, the incendiary, who helped burn out the valley of Virginia, the murdered of our military scouts and inoffensive citizens, would never have lived to fight the Sioux Indians. He was a brave man, but cruel; his cruelty overshadowed his heroic conduct in battle."* [15]

The men of the 38th Georgia Regiment witnessed the very last shot fired by the Army of Northern Virginia, fired by an artillery piece. Sergeant Hudgins wrote, *"The last gun fired was from 1st Virginia artillery just east of the court house, at the head of Evans' Brigade (Georgia Brigade) just after General George A. Custer rode into the Confederate line with the flag of truce. While Custer was talking to General C. A. Evans, I was within a few feet of the gun as she fired her last shot, and at the same time I cut the center start of the 38th Georgia Regiment flag with a case knife. I still keep the star as a relic. After General Custer spoke with General Clement A. Evans, he "was directed to General Gordon. We soon had orders to uncap our guns, march out into the field and stack arms."* [16]

Private William H. Mitchell of Company A, was captured at the battle of Sailor's Creek along with two other members of his company, Sergeant John J. Carpenter, and Private James A. J. Reeves. They witnessed from afar the surrender of the Army of Northern Virginia, as prisoners of war, *"On the sixth day of April, we were captured at Sailor's Creek. We were kept right up at the head of the enemy's line; were in plain view when the surrender was made at Appomattox Court House, Va. We were sent back about one mile into a grove and a guard placed around about 400 of us as prisoners. This was on a Sunday morning. We had nothing to eat for eight days, only a pint of corn divided among four. He [Carpenter] got so feeble he could hardly get up when he was down and could not talk above a whisper."* [17]

Lieutenant Lester recounted Lee's return from the meeting with General Grant, "General Lee was soon seen riding to the rear, dressed better than usual with his sword buckled on, which was unusual for him. Later on in the day it was said that we had surrendered. The feeling of the men cannot be described. There was a feeling of relief that the long and unequal struggle was over, mingled with regret for the failure of the cause we had so long upheld. The men were silent in their grief. They moved and spoke with a subdued air, as though in the presence of death. Soon General Lee was seen returning from his interview with General Grant. The men could not be controlled. Whole lines would break ranks and rush up and crowd around their beloved commander, speaking words of comfort and affection, striving to lighten the burden they knew must be upon him. We knew him to be great in council, in the field and in any position in which he was placed." [18]

General Robert E. Lee surrendered to Federal General Ulysses S. Grant on April 9th, 1865, at Appomattox Court House. The following day, General Lee gave his farewell address to the Army of Northern Virginia. He wrote:

Headquarters, Army of Northern Virginia, 10th April 1865.
General Order No. 9

After four years of arduous service marked by unsurpassed courage and fortitude, the Army of Northern Virginia has been compelled to yield to overwhelming numbers and resources. I need not tell the survivors of so many hard fought battles, who have remained steadfast to the last, that I have consented to the result from no distrust of them. But feeling that valour and devotion could accomplish nothing that could compensate for the loss that must have attended the continuance of the contest, I have determined to avoid the useless sacrifice of those whose past services have endeared them to their countrymen. By the terms of the agreement, officers and men can return to their homes and remain until exchanged. You will take with you the satisfaction that proceeds from the consciousness of duty faithfully performed, and I earnestly pray that a merciful God will extend to you his blessing and protection. With an unceasing admiration of your constancy and devotion to your Country, and a grateful remembrance of your kind and generous consideration for myself, I bid you an affectionate farewell. [19]

— R. E. Lee, General, General Order No. 9

Chapter 25
The Surrender and Prisoners of War

While Generals Lee and Grant had already departed Appomattox Court House, the Federal surrender terms demanded a formal surrender of arms ceremony. General John B. Gordon, once the commander of the Georgia Brigade, now leading the 2nd Corps, was selected to lead the Confederate surrender ceremony and parade. At about sunrise on April 12th, 1865, General Gordon formed the remnants of the Confederate brigades in column and marched up the Richmond – Lynchburg Stage Road towards the village of Appomattox Course House. Gordon rode in front of the Stonewall Brigade leading the way, to surrender their arms and colors. The Georgia Brigade followed, contained the pitiful handful of survivors of the 38th Georgia Regiment. The 38th numbered over 1,300 men when first formed, but less than 10% remained at the surrender. *"When the retreat commenced from Petersburg on April 1st, the regiment numbered about 200, but in the numerous skirmishes from then until the surrender at Appomattox on 9th April, our number were reduced to 105 men,"* wrote Captain William C. Mathews, of the Battey Guards. [1] Of the 105 soldiers present, sixty-seven enlisted men and six officers were armed for battle.

Sergeant Hudgins was commanding his company and recalled of the 38th, *"General John Gibbons' division of 3rd corps, U. S. A., was drawn up on the north side of the public road east of the village to receive our surrender. General Chamberlain commanding the 1st brigade, as we came opposite, ordered his men to present arms, the highest compliment from the victor to his defeated foe."*[2]

Another soldier in the Georgia Brigade remembered the surrender ceremony quite different, *"We were ordered to take our arms and march and were marched to the road...there on the east side and about 100 feet from it was a long line of Yankee soldiers facing towards us. We were formed in front of them and stood quietly for some time. We were now so weak from our long fast that we could hardly stand up in the ranks and were dressed in rags, so our appearance made a poor contrast with that of our well-feed and well clothed enemies. The silence was finally broken by someone in their ranks, and the whole line then began to curse and use the most opprobrious language. This continued for some time, when an officer, riding to and fro in the rear of their line, spoke to them and said: "These Confederate soldiers are brave men. If you were half as brave as they are, you would have conquered them long ago. If I hear another cowardly scoundrel curse these men again, I will break my sword over his head." We all gave a shout for the major, and silence prevailed."* [3]

The 38th Georgia Regiment halted and faced their enemies; aligned just a few paces away. Sergeant Hudgins wrote, *"We halted, stacked arms, furled our banner that we carried from the beginning to the end.* [4] *Colonel John Lowe, commanding the Georgia Brigade, watched as the men stacked their rifles in front of the Yankees. Once the rifles were stacked, he said, "If you have anything on your person that belongs to the Confederacy, put it on the stack."* [5] The men of the 38th Georgia were issued their paroles and began the long trip home. Sergeant Hudgins said, *"We received our paroles on the 13th, and then commenced the long tedious march homeward to find this once happy land a wilderness, made so by Sherman in his march to the sea. The writer with four of his company reached Atlanta, Ga. on April 24th, the first of Lee's army to arrive."* [6] Sergeant Dickerson wrote of arriving at his home in DeKalb County on April 30th, after a journey of eighteen days. [7]

"Parole of Sergeant F. L. Hudgins. Appomattox Court House, Virginia, April 10th, 1865. The bearer, F. L. Hudgins, Sergeant of Co. K, 38th regiment. Ga. volunteers, a paroled prisoner of the army of Northern Virginia, has permission to go to his home and there remain undisturbed.
P. E. Davant, Lieutenant Col. Com'dg."

Paroles were issued to 26,765 Confederate soldiers of the Army of Northern Virginia. Captain William C. Mathews wrote, "Thus the curtain fell and the 38th Georgia Regiment, numbering at one time over 1,300 men, retired from the stage of action covered with glory, and started on the homeward march beaten, but not vanquished, conquered, but not subdued. As a regiment they had done their full duty and in their disbandment it was without a blemish on their record." [8]

Appomattox Roster of 38th Georgia Regiment

The 38th Georgia Regiment surrendered 105 soldiers at Appomattox Court House, six officers and sixty-seven enlisted men were armed for battle. The names on this roster have been extracted from the regimental roster of the Confederate Soldiers of Georgia, complied by Lillian Henderson. [9] Full names have been added by the author and cross referenced with other sources, such as pension records, burial records and family records of descendants.

38th Georgia Regiment Officers

Lieutenant Colonel Phillip E. Davant, Commander

1st Lieutenant Jacob "Jake" Perry Pughsley, Co. C

Captain Robert Toombs Dorough, Co. E

1st Lieutenant Samuel A.Thornton, Co. E

1st Lieutenant Thomas D. Thornton, Co. F

1st Lieutenant James C. Daniel, Co. H

Captain William Hendrix, Co. N

Company A, Murphy Guards, DeKalb County Georgia

Sergeant Robert M. Robinson,

Private Joseph E. Elliott

Sergeant Azariah Milton Holcombe
Corporal John W. Ball
Private John Thomas Austin
Private Lewellin J. Cash
Private David A. Chesnut
Private William Larkin Chamblee
Private William J. Johnston, Teamster
William H. Jenkins, Drummer
Private Josephus B. Harmon
Private Russell D. Lord
Private Newton R. Miller

Company B, Milton Guards, Milton County Georgia, (Now Fulton County)

Sergeant James Richard Neal
Corporal Harvey Jasper Gibbs
Private Henry Columbus Emerson
Private Allen O. Landrum
Private Joseph W. McDaniel,
Private John Carr McMakin
Musician William Jefferson McMakin
Private Reuben P. Reese
Private James A. Sims
Private James M. Turner
Private John J. Tillotson
Private James W. Tillotson (Tillison)
Private John Orr, Teamster

Company C, Ben Hill Guards, Bulloch & Emanuel Counties, Georgia

Corporal Elijah Anderson Odom
Private Peter Curl
Private Benjamin Faircloth
Private Ebenezer G. Higdon,
Private Henry M. Moxley
Private Benjamin Murray
Private John H. Steele
Private Archibald R. Woods

Company D, McCullough Rifles, DeKalb County, Georgia

Sergeant James F. Coggin
Sergeant William F. A. Dickerson
Private Camilus A. Mason
Private Samuel “Sam” Wylie Cochran
Private Josiah T. Grogan
Private William D. Grogan
Private Abner Harrison Hix
Private Moses H. Newsom
Private Ledford N. Verdin
Private William Verdin, Teamster

Company E, Tom Cobb Infantry Oglethorpe County, Georgia

Commissary Sergeant Amaziah C. Daniel
Sergeant William T. Burt
Sergeant John M. Hawkins
Sergeant Willis Benjamin Jackson
Corporal William Sanders Jackson
Private William H. Burt
Private John C. Chafin
Private William Johnson Fleeman
Private Henry G. Farmer, Provost Guard
Private Booker W. Hopper, Provost Guard
Private Charles T. Hardeman
Private John T. Pittard

Company F, Thornton Line Volunteers, Hart & Elbert Counties, Georgia

Sergeant William M. Thornton
Private John D. Bentley
Private Judge Middleton Barrett Hunt
Private Thomas Jefferson Howland
Musician William Nelms, Drummer
Private James C. Thornton

Company G, Battey Guards, Jefferson County, Georgia

Sergeant Benjamin Beasley
Private Seaborn F. Baggett
Private Robert A. Martin
Private James B. Stewart
Private Armory Solon Smith, Acting Ord Sgt.
Private John William Penrow

Company H, Goshen Blues, Elbert County, Georgia

Sergeant John H. Bowers
Sergeant Jeptha E. Campbell
Corporal Thomas A. Moore
Private John D. Brown
Private David S. Brown
Private Joseph A. Carrouth, Wagoner
Private Gideon A. Carruth
Private James E. Carrouth, Wagoner
Private William G. Campbell
Private Singleton S. Ginn
Private Edwin Jackson "Jack" Mann
Private J. G. Leemore (not found in CSA Rcrds)

Company I, Dawson Farmers, Dawson County, Georgia

Corporal Absalom Martin

Company K, Bartow Avengers, DeKalb & Fulton Counties, Georgia

Sergeant Francis L. Hudgins
Corporal James M. Wiggins
Corporal James M. Swiney
Private William Brooks
Private Lewis Brown
Private Allen Brown
Private Zachariah J. Cowan
Private Robert Daniel Francis Jones
Private Benjamin Seals
Private William "Bill" T. Vaughn
Private Enoch D. Wade
Musician George W. Wade, Drummer

Company N, Chestatee Artillery, Forsyth County, Georgia

Sergeant Horatio R.Tatum
Corporal William Henry Taylor
Private Elijah B. Bennett
Private Obediah P. Booker
Private James S. Bolton
Private David O'Shields,
Private D. Perryman
Private Richard Watson
Private George W. Watson
Private Moses Mulchy Taylor

Lieutenant Colonel Phillip E. Davant, commanding the regiment at the surrender, signed an affidavit at the bottom of the roster that read, *"I certify, on honor, that the above number of men, they were present, actually armed and in the line of battle, six officers and sixty-seven enlisted men on the morning of the 9th inst., the day of the surrender of this army."*

P. E, Davant
Lt Col Com'dg

Private William H. Mitchell, of Company A, was captured at Sailor's Creek, along with Sergeant John J. Carpenter and Private James A. J. Reeves, of the same company. They were held prisoners near Appomattox Court House until Friday, April 14th, then they were sent by train back to Farmville. Private Mitchell wrote, *"They kept us there until the next Friday and sent us by rail back 31 miles to Farmville, Virginia. Late the next evening, (Saturday, April 15th) we got our parole. Then they gave [us] a spoonful of coffee and one of sugar and three hard – tacks apiece. We started for home. Our progress was slow for a few days. We had no money and the army scattered over the country ahead of us. We had a hard time to get something to eat. We hardly ever got three meals a day. We lived mostly on milk and bread the trip home and we were thankful to get it."* [10]

The war was over for the men surrendered at Appomattox Court House, but 110 other soldiers of the 38th Georgia still languished in Northern prisons. [11] These soldiers were imprisoned at Point Lookout Prison, Maryland, Fort Delaware, Delaware, Camp Chase, Ohio, Newport News, Virginia, Camp Douglas, Illinois, and Elmira Prison camp, New York. Eight of these 110 soldiers died before being released.

Confederate prisoners were required to take an oath of allegiance to the United States in order to be released from prison. Major Thomas Bomar, of the Chestatee Artillery Company, captured at the battle of Cedar Creek, October 19th, 1864, was the last soldier of the 38th Georgia to "swallow the oath" and be released from a Yankee prison camp. He was released from Fort Delaware prison on July 24th, 1865. One soldier stayed in prison even longer than Major Bomar, Private Garnett White, of the Goshen Blues, from Elbert County, Georgia. He was very sick and died at Elmira Prison, New York, on July 27th, 1865. Private White was the very last soldier of the 38th Georgia to die on duty in the war.

Private Calvin Pruitt, of Company B, the Milton Guards, from Milton County Georgia, was captured at Spotsylvania on May 19th, 1864. He was released from Elmira Prison in New York on May 24th, 1865, but never returned home to his family in Milton County. Perhaps he was so sick he died on the way home, or was killed. Or, perhaps for reasons known only to him, he started a new life somewhere far from Georgia, His fate remains unknown to his descendants to this very day.

Captain William C. Mathews of the Battey Guards wrote a sketch of the 38th Georgia Regiment and it was published in the Sunny South Newspaper on January 10th, 1891. Captain Mathews noted many exceptional soldiers of the 38th Georgia Regiment, he wrote, *"While I dislike to discriminate where all generally acted so well, yet I must make mention of the following who were particularly noted and distinguished for their brave acts, their soldierly qualities, and their manly bearing through all the trying scenes in which they acted their part.*

Sergeant Azariah M. Holcombe of Company A, was one of the finest soldiers in Lee's Army. All the officers of his company having been killed or disabled, he commanded the company for the last year of the war and should have been captain of it. Private John W. Ball of the same company was in every battle the regiment was engaged and was truly a brave and noble soldier, was never wounded. Peter M. Ball, David A. Chesnut, John T. Austin, Robert M. Robinson, William L. Chamblee,, James P. Harman and Russell D. Lord of Company A, and all now honored citizens of DeKalb County were at the surrender and were conspicuous for their gallantry. Also may be mentioned as splendid solders, William H. Mitchell and John J. Carpenter of Company A.

Private Harvey J. Gibbs of Company B was another hustler in a fight and in the charge was ever among the foremost. Lieut. Joe Mattox of Company B, now one of Atlanta's wholesale merchants on Alabama Street, was noted for his splendid soldierly qualities and coolness in action.

Lieutenants Benjamin "Ben" T. Morris and Jacob "Jake" Pughesly, of Company C as well as Captain Miller Wright and Sergt. Andrew "Drew" Thompson, of Company C, were all in the way of soldiers and officers that the army could wish, and nobly performed their duties. Privates Henry M. Moxley, of Company C, and Amory S. Smith, of Company G, carried out a mortar from Fort Hare (or Fort Stedman.) Captain John G. Rankin, of Company D, was over military age when the war commenced, and was of northern birth, but the South never had officer or man braver or better. We called him the "old reliable," for in nearly every engagement our commander would get killed or wounded, and Captain Rankin would then take command. He refused promotion when he could been made lieutenant colonel, but preferred to stay with his company. After his capture, having been wounded, he was promoted to the position of major of the regiment, and his name so appears in the Confederate archives.

Chaplain Jabez M. Brittain both as chaplain and lieutenant was ever at his post, and I am glad to note that he was lately accepted a pastorship in Atlanta. As a devout and eloquent divine he deservedly stands among the foremost as he did in war times.

Cicero Daniel, Private, Company E, was promoted on the field to a lieutenancy for his bravery. Private Burnside, the Campbell brothers and color bearer Powell, were whole teams in a fight and noted for their staying qualities.

Sergeant Patrick H. "Pat" Smith, Milton A. Clark and Swan, of Company G, as well as Privates Armory S. Smith, Hampton J. Hudson, Chelsea S. Wise, William A. Wright, the Stewart brothers, William A. Brown, James "Jim" King, Benjamin "Ben" Beasley, Augustus A. Murphy, and Robert A. Martin, of the same company, were the equal of any one in Lee's Army, fully possessed for all the qualities that go to make good soldiers and perfect chivalric men. Robert A. Martin was in every fight that the Thirty-eighth was in, and was never wounded.

Captain Robert P. Eberhart, of Company H, now of your city, was an officer admired and beloved by his whole command and was behind none in bravery and gallantry. The two Harbin boys of Company H, Jasper and Marion -(the latter as salesman with J. A. Anderson, Whitehall street), were noted as any of the regiment for their bravery and dash. The latter (Marion) was the youngest and smallest in the regiment, enlisting when he was only thirteen years old.

Company K, was one of the best fighting companies in the regiment. We sometimes called them the "orphans," because in every fight they would lose their commander and a sergeant would command them. This duty always devolved upon Sergeant Francis L. Hudgins, and was always equal to the occasion, no matter what it was. Wounded several times severely, he was always back in time to be in the next fight. He commanded the company for the last six months of the war, and it has no better commander than he. He was a truly brave and gallant Soldier and now lives near Clarkston, in DeKalb County, and is one of her most prosperous farmers, and with a most excellent wife has reared a family of boys and girls who are as handsome, intelligent and enlightened as their father was gallant and brave. He with the following named surrendered at Appomattox: William Brooks, Zachariah J. Cowan, Benjamin A. Seals, James M. Wiggins, Lewis Brown and Allen Brown. Mathew O. Wiggins of Company K seized and threw a shell with a burning fuse from the lines before its' explosion. James M. Wiggins of the same company shot one of the enemy and relieved Captain Rankin who had been captured. He also carried from Fort Hare (Fort Stedman) a mortar. Isaac H. Smith of Company D also shot one at the same time. Robert Wiggins was conspicuous for gallantry in every engagement and was killed at Fort Hare (Fort Steadman.) Sergeant James H. Gazaway, Privates Hendley V. Bayne, William A. Brooks, and Brown were know and recognized throughout the regiment as brave and daring soldiers. Corporal James E. Chandler commanded the company at Sharpsburg (Antietam) and died bravely in the thickest of the fight. John Baxter Company D, and Private Isaac N. Nash were also noted for their splendid soldierly qualities. The latter lost an arm at Gettysburg, and is now DeKalb's Tax Collector.

For the numerous heroes dead of the regiment who fell fighting on every field in Virginia, and those who died of disease yet none the less heroes, the writer cherishes the hope that the survivors will ever cherish and keep green their memory. Many of them sleep their last sleep in unmarked graves, but invisible friends will hover around them until the dawn of that great day when all the armies will be marshaled into line again, and the great eternity of Peace commenced.

Let me urge upon the survivors to attend our annual reunions and talk over again those days of trials and hardships endured at the war's horrid front. Wishing you all long prosperous lives, I remain your friend and comrade.

W. C. Mathews
Captain Co G, 38th Ga. [12]

THE CONQUERED BANNER

by Abram Joseph Ryan

Furl that Banner, for 'tis weary;
Round its staff 'tis drooping dreary;
Furl it, fold it, it is best;
For there's not a man to wave it,
And there's not a sword to save it,
And there's no one left to lave it
In the blood that heroes gave it;
And its foes now scorn and brave it;
Furl it, hide it--let it rest!

Take that banner down! 'tis tattered;
Broken is its shaft and shattered;
And the valiant hosts are scattered
Over whom it floated high.
Oh! 'tis hard for us to fold it;
Hard to think there's none to hold it;
Hard that those who once unrolled it
Now must furl it with a sigh.

Furl that banner! furl it sadly!
Once ten thousands hailed it gladly.
And ten thousands wildly, madly,
Swore it should forever wave;
Swore that foeman's sword should never
Hearts like theirs entwined dissever,
Till that flag should float forever
O'er their freedom or their grave!

Furl it! for the hands that grasped it,
And the hearts that fondly clasped it,

Cold and dead are lying low;
And that Banner--it is trailing!
While around it sounds the wailing
Of its people in their woe.

For, though conquered, they adore it!
Love the cold, dead hands that bore it!
Weep for those who fell before it!
Pardon those who trailed and tore it!
But, oh! wildly they deplored it!
Now who furl and fold it so.

Furl that Banner! True, 'tis gory,
Yet 'tis wreathed around with glory,
And 'twill live in song and story,
Though its folds are in the dust;
For its fame on brightest pages,
Penned by poets and by sages,
Shall go sounding down the ages--
Furl its folds though now we must.

Furl that banner, softly, slowly!
Treat it gently--it is holy--
For it droops above the dead.
Touch it not--unfold it never,
Let it droop there, furled forever,
For its people's hopes are dead![13]

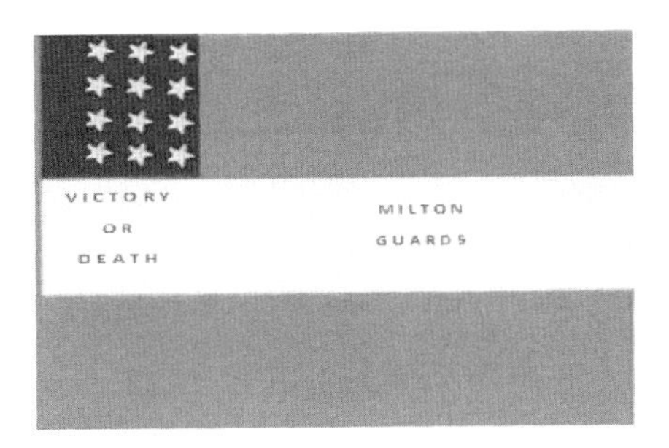

Chapter 26
Deaths, Deserters & "Galvanized Yankees"

Over the course of war, 1630 soldiers were assigned to the 38th Georgia Regiment, this number includes companies that were later detached, new recruits and conscripts.

38th Ga. Regt Deaths During War	Assigned	KIA or Mortally wounded	Died Disease	Total died in War	Percentage Died
Staff	16	1	0	1	6%
Company A. Dekalb Co.	122	22	40	62	51%
Company B, Milton Co.	172	24	19	43	25%
Company C, Bulloch & Emanuel Co.	113	23	22	45	40%
Company D, Dekalb Co.	86	13	19	32	37%
Company E, Oglethorpe Co.	106	37	28	65	61%
Company F, Hart & Elbert Co.	141	18	31	49	35%
Company G, Jefferson Co.	130	32	39	71	55%
Company H, Elbert Co.	144	20	49	69	48%
Company I (Dawson Co)	85	10	24	34	40%
Company I Henry Co., AL [1]	104	13	10	23	22%
Company K, Dekalb Co.	127	32	16	48	38%
Company L, Forsyth Co. [2]	187	7	19	26	14%
Company M, Fulton Co. [3]	97	0	4	4	4%
Total	**1630**	**252**	**320**	**572**	**35%**

At least 252 soldiers were killed in battle, or died of wounds during the war. Another 320 soldiers died of disease, the total of all deaths equals 572 soldiers, or

35% of all who served. An additional 172 soldiers were disabled by wounds, or discharged due to disease. There were 361 soldiers captured during war, but many were exchanged, or paroled from Federal prisons, and returned to duty with the regiment. Four companies suffered extraordinary death rates of nearly 50% or higher, from both battle and disease. Company A, McCullough Rifles, lost 62 men of the 122 assigned, a casualty rate of 51%. Company E, The Tom Cobb Infantry, was the company that carried the regimental flag and they suffered 65 deaths of the 106 assigned, a horrific death rate of 61%, the highest in the entire regiment. Company H, The Goshen Blues, lost 69 soldiers out of the 144 assigned, nearly 50%. Overall, one out of every three soldiers who served in the 38th Georgia Regiment died in the war.

A total of 35 soldiers from the 38th Georgia deserted during the war, of those 19 soldiers took the oath of allegiance to the U.S. government and agreed to not fight again, under penalty of death, if recaptured. Of the 361 soldiers captured from the regiment during the war, 16 took the oath of allegiance to the United States and enlisted in the U.S. Army. The following proclamation was ordered read to the Confederate prisoners at Point Lookout POW camp in Maryland during January of 1864:

HDQRS. DEPT. OF VIRGINIA AND NORTH CAROLINA,
Fort Monroe, January 9, 1864.

Brigadier General G. MARSTON:

You will cause every prisoner at Point Lookout to answer one of the following questions, taking his answer, after he has heard them all:

First. Do you desire to be sent South as a prisoner of war for exchange!

Second. Do you desire to take the oath of allegiance and parole, and enlist in the Army or Navy of the United States, and if so in which!

Third. Do you desire to take the oath and parole and be sent North to work on public works, under penalty of death if found in the South before the end of the war!

Fourth. Do you desire to take the oath of allegiance and go to your home within the lines of the U. S. Army, under like penalty of death if found South beyond those lines during the war!

You will adopt the form set forth in this book, and let each signature be witnessed, causing the oath and parole to be read to each man, the questions to be propounded to these men alone and apart from any other rebel prisoner.

Truly, yours, BENJ. F. BUTLER, Major-General, Commanding.[4]

Sixteen soldiers 38th Georgia captured took the oath of allegiance to U.S. Govt. at Camp Chase, Illinois, or Point Lookout, Maryland and enlisted in the U.S. Army:

Company	**Name**
B	Grogan, Richard W. - 2d Sgt. Enlisted at Point Lookout, Md. in 1st U.S. Volunteers 1/24/1864.
B	Herring, Wiley B. - Pvt. Enlisted in U.S. service at Point Lookout, Md. 5/19/1864.
B	Maddox, Henry G. - Pvt. Enlisted in 1st U.S. Volunteers at Point Lookout, Md.
C	Dix, John H. - (*Or John H. Dicks*) 3d Corp. Enlisted in Co. D, 2nd U.S. Volunteers, at Point Lookout, Md. 10/17/1864.
C	Thompson, Berry A. - Pvt. enlisted in the 1st Regiment Connecticut Cavalry prior to 10/4/1863.
D	Goss, John W. - Pvt. 9/26/1861. Enlisted in 1st U.S. Volunteer Regiment, 2/23/1864.
D	Lang, William B. (or Long, Wiley B.) - Pvt. Enlisted in U.S. Army at Point Lookout, Md. 2/1/1864
D	Luckey, J. R. - (*John R. Luckey*) Pvt. Enlisted in Co. G, 1st Regt., U.S. Volunteer Infantry, at Point Lookout, Md. 4/5/1864. Died of scurvy, 5/28/1865 at Fort Rice, Dakota Territory
E	Bryant, Robert M. - Pvt. Enlisted in Co. B, 2nd U.S. Volunteer Infantry 10/17/1864.
F	Almond, Isaac B. - Pvt. Enlisted in 1st U.S. Volunteer Infantry 1/24/1864.
F	Pearson, George W. - Pvt Enlisted in 6th Regiment U.S. Vols. 4/1/1865.
F	Stamps, James J. *(James Johnson Stamps)* - Pvt. Enlisted in Co. G, 1st Regiment of US Volunteer Infantry at Point Lookout, Md. 1/24/1864.
G	Hudson, Hezekiah H. - Pvt. Enlisted in Co. B, 1st Regiment, U.S. Volunteer Infantry.
K	Gazaway, William *(William Marion Gasaway)* Pvt. Enlisted in 6th U.S. Volunteers at Camp Chase, OH, 3/16/1865.
L	Arthur, James J - (or James S. Arthur) Pvt. Enlisted in Co. D, 2nd US Volunteers 10/17/1864.
L	McKinney, William L. - Pvt. Enlisted in Co. D, 2nd U.S. Volunteer Infantry.

Confederates soldiers who enlisted in the U.S. Army were known as "galvanized Yankees." The reasons these men enlisted in the Federal army have been lost to history, but the death rate at northern POW camps was horrific and the choice may have seemed "life or death" to the men faced with the decision. Camp Chase was called by one northerner an "extermination camp" and Point Lookout, Maryland was little better. The exchange of prisoners had been halted by General Grant in April 1864, so there was little hope of being released by exchange. These men were considered traitors to the Confederate cause. They knew when they enlisted in the U.S. Army they might face their old comrades in the line of battle, but Grant wisely decreed Confederate prisoners enlisting in the U.S. Army would be sent west to suppress the American Indian uprising.

Private John R. Luckey, of Co. D, proved to be the most "unlucky" of all. He enlisted in Company G, 1st Regiment of U.S. Volunteer Infantry and was sent west to fight Indians at Fort Rice, in the Dakota Territory. He died there of scurvy May 28th, 1865, just two months after Lee surrendered. His remains are buried at Custer National Cemetery, Big Horn County, Montana.

Whatever the reasons, these men knew that by enlisting in the Federal army, they could not easily return to their homes in Georgia, regardless of the outcome of the war. Should they return to Georgia, they would have been ostracized as "traitors" by the local community and possibly even marked for death.

It's little wonder that at least four of these sixteen men, probably more, moved west after the war to start anew. Private James J. Stamps of Company F, left a wife and baby in Elbert County, Georgia and settled at Meeker Co., Minnesota, to start a new life with a new wife. Private William M. Gasaway, of Co. K, settled in Bates, Missouri. Private Robert M. Bryant of Co. E, settled in the small town of Dunbar, Otoe Co., Nebraska. Private John W. Goss of Co. D, moved to Cleburne Co., Arkansas, and even applied for a Confederate pension in 1906.

Some soldiers refused to fight against their former comrades, yet also refused to continue to serve the Confederacy. Perhaps they clearly saw the cause was doomed and further resistance was hopeless. The four men listed below either deserted, or were captured at home by Sherman's army, and took the oath of allegiance to the U.S. Government. They requested to remain in the north for the duration of the war.

Company	Name
A	Edwards, William A. - Pvt. Deserted to enemy at Strasburg, Va. 8/14/1864. Took oath of allegiance to U.S. Govt. and furnished transportation to Philadelphia, Pa., date not stated.
A	Harmon, Francis M. - Pvt. Captured at DeKalb Co., Ga. Released north of Ohio River, date not given.

A	Wallace, George W. (George Washington Wallace) - Pvt. Captured in Gwinnett Co., in 1864, released to remain north of Ohio River during war, 8/6/1864.
L	Prater, Benjamin G. - Pvt. Deserted at Charleston, S. C. 5/15/1864. Captured in Hall Co., Ga. in 1864. Released to remain north of Ohio River during war, 9/26/1864

Three of these four soldiers probably returned to Georgia, since Prater, Wallace, and Edwards are all buried in Georgia. What became of Private Francis M. Harmon is unknown. These cases show the various stages of loyalty to the Confederacy and also demonstrate things weren't always as plain as "black and white." While deserting was looked down upon by many in local communities, it's wasn't a death sentence. However, to have enlisted in the Federal army was treason in the minds of many former soldiers and probably made life much more difficult for those returning to Georgia after the war.

Chapter 27
Reunions And Life After The War

The war was essentially over when General Robert E. Lee surrendered the Army of Northern Virginia on April 9th, 1865. Other Confederate armies were still in the field, but they soon realized the hopelessness of further resistance. Among the last to surrender was General Stand Waitie, a Cherokee Indian chief who formally signed a cease fire agreement on June 23rd, 1865, in Oklahoma.

To the Officers and Soldiers
—OF—
EVAN'S GEORGIA BRIGADE,
—FORMERLY—
Gordon and Lawton's Brigade, A. N. V

YOU are requested to meet at the Court-house in the city of Macon, on THURSDAY EVENING, the 26 inst., at 8 o'clock for the purpose of organizing an association for the preservation of the records of your several commands, and that such of your number as may be in a helpless or destitute condition may be relieved, and to provide for the destitute families of those who were killed or who died during the late war.

Officers and soldiers will please bring such muster rolls of their regiments, companies or mess as they can prepare from memory, or from any records in their possession.

The following regiments were in the brigade at the surrender of the army:

Thirteenth Georgia Regiment—Cols J M Smith and Baker.
Twenty sixth Georgia Regiment—Col Atkinson,
Thirty first Georgia Regiment—Col J T Lowe,
Thirty-eighth Georgia Regiment—Col P E Davant,
Sixtieth Georgia Regiment—Cols J D Mathews and W B Jones,
Sixty-first Georgia Regiment—Col J Hill Lamar,
Twelfth Georgia Battalion—Col H D Capers.

COMMANDING OFFICERS:
Gen A R Lawton, Gen J B Gordon, Gen C W Evans.
oct25 td

Macon Telegraph & Messenger, Oct. 25th, 1871

The war had ended, but many Confederate and Union soldiers continued to die from the effects of wounds and disease. Private Bennett R. Jeffers of Company D, from DeKalb County, died of tuberculosis just a few days before Christmas, on Decemeber 20th, 1865. He likely contracted the disease while serving as a nurse in a Confederate hospital.

Other former 38th Georgia soldiers met a voilent end after the war. Captain Richard H. Fletcher, of Company K, was murdered by a gunman on a Georgia highway on May 11th, 1869, the result of personal dispute. "It seems the weapon used was a shotgun, loaded with large shot, eight of which entered his neck, killing him instantly."[1] His killer was captured and inprisioned, but was dragged from the jail and lynched when the wheels of justice turned too slowly for Fletcher's friends.

Corporal Private Elijah A. Odom, of Company C, The Ben Hill Guards served through the entire war, from November of 1861 and was present at the surrender at Appomattox Court House on April 9th, 1865. He was shot and killed at a social gathering in Emanuel County, Georgia, on October 10th, 1884. "It was said to be a clear case of murder; originating from an old feud and too much whiskey."[2]

Many of the families of 38th Georgia soldiers were forever changed by the war. At least 572 soldiers died during the war, either from disease or battle. Each of these 572 soldiers represented an irreplaceable loss to their familes. The Wiley Powell family from Elbert County, Georgia, was devastated by the war. Forty-seven year old Wiley Powell and his two teenage sons, William Joseph Powell and James Lewis Powell, all joined Company F, Thornton's Line Volunteers. All three perished in the war. Corporal William Joseph Powell was killed in action at

Gettysburg on July 1st, 1863. His brother, Private James Powell died at home of wounds three months later, on October 8th, 1863. The father, Private Wiley Powell, died of tuberculosis four months after William, on January 24th, 1864, in a Lynchburg, Virginia Confederate hospital. In the short span of just seven months, Mrs. Mariah Powell lost her husband and two beloved sons to the cruel hand of war. She was left alone to raise eight minor children in rural Elbert County, Georgia. She never remarried and lived for another 47 years, dying February 5th, 1912, in Elbert County.

At least 172 soldiers were disabled by wounds, or discharged due to disease. Many of these soldiers lost arms or legs due to battle wounds and amputations, or had their health premanently ruined, leaving them unable to provide for their familes.

Most 38th Georgia soldiers who survived the war with their health intact, returned to their former professions, the majority being farmers. However, several former soldiers were elected to high politcal offices after the war. Captain William H. McAfee of the Dawon Farmers company served two terms in the Georgia State Senate. Sergeant Major John Cottle Chew, of the Ben Hill Guards, was elected to serve in the Georgia House of Represenatives. Private Henry C. Emerson of the Milton Guards, moved to Van Burn County, Arkansas after the war. He was elected to the Arkansas State Senate in 1890.

Lieutenant John Baxter of Company D, was elected Sheriff of DeKalb County, Georgia. He married the widow of Private Bennett Jeffares, another 38th Georgia soldier, who died in December of 1865. General John B. Gordon, the most famous Georgia Brigade commander, was elected as a Georgia Senator, serving from 1873 through 1880. He was elected Governor of the State of Georgia in 1886 and returned to the state senate 1891, serving through 1896.

For many of the surviving soldiers, the war was the watershed event in their lives. The bonds of brotherhood, forged on bloody fields of battle, could never be recaptured in their lives as civilians. Many sought to recapture the comradarie of the war experience by attending reunions or joining Confederate veteran groups.

The first documented instance of the Georgia Brigade forming a veteran's association was the annoucement of a meeting to be held at Macon, Georgia, on October 26th, 1871. The annoucement noted the purpose of the meeting was to form an association "for the preservation of the records of your several commands, and that such of your number as may in a helpless or destitute condition may be relieved, and to provide for the desititute families of those who were killed or who died in the late war."[3]

This was a very noble gesture, as there were no Federal pensions for Confederate soldiers or widows, since the Federal government considered them traitors to the Union. The State of Georgia was too shattered and poor to provide pensions for the crippled soldiers, widows, and poor families of ex-Confederates. In 1877 the State of Georgia began issuing pensions to Confederate soldiers who lost a limb in the war. The law was slowly broadened to include other soldiers disabled by military service and then to indigent soldiers. Indigent widows finally became eligible to receive a Confederate pension in 1890, based on their husbands' honorable service. Ex-Confederate soldiers fortunate enough to own land, or other substantial property, were not eligible to receive pensions.

Another notice of a Georgia Brigade reunion was published in August of 1874. It called for the survivors of the Georgia Brigade to convene in the City of Atlanta, on Wednesday, October 21st, 1874, during the State Fair. This reunion was attended by many prominent officers of the Georgia Brigade, including all three former brigade commanders, General Alexander R. Lawton, General John B. Gordon, and General Clement A. Evans. Colonel James D. Mathews, former commander of the 38th Georgia Regiment also attended and the other regimental Colonels of the brigade were invited to attend. The brigade held a meeting in the State capital senate chamber on October 29th. The purpose of this meeting was to formally create a Georgia Brigade association. The organization was called "The Survivor's Assocation of Lawton's-Gordon-Evans' Brigade, Army of Northern Virginia. This organization was probably an attempt to formalize and expand on the brigade organization proposed in 1871. General Clement A. Evans was elected President of the organization.[4] The Georgia Brigade had several reunions over the next 30 years, often in conjunction with the annual Georgia State Fair.

The men of the 38th Georgia Regiment formed their own veterans organization in 1883, it was called the 38th Georgia Survivors Association. This organization held annual reunions for at least 45 years, the last documented reunion was held in 1928 at Sandy Springs, Fulton County, Georgia. Three companies of the 38th Georgia were from DeKalb County and one company hailed from Milton County, now Fulton County. Most of the 38th Georgia reunions were held near Atlanta.

These reunions were large festive community events. The reunion held at Lithonia, Georgia in August of 1906 was attended by over 2,000 people.[5] Before each reunion, a notice was printed in local newspapers in the counties the soldiers enlisted, announcing the date and place of the reunion. The three famous leaders of the brigade, Generals Lawton, Gordon, and Evans, were often invited to attend and did attend many of the reunions. All Confederate and Union veterans were invited to several reunions. Preachers were invited to give invocations. Music was provided by local bands and families attending would bring a basket dinner to eat on the ground.

The 1910 Thirty-Eight Georgia Regiment reunion was held at Sandy Springs Campground, Fulton County, Georgia. "The members and their families were present and brought ample dinner baskets with them and there was plenty to go around. A number of speeches were made by some of the members present, and after dinner the old boys warmed up and told again stories of life around the camp fire."

While the 38th Georgia reunions continued for 45 years, each year a few more 38th Georgia soldiers passed into the great beyond, until there was only one surviving soldier. Corporal William “Billy” Thomas Broadwell, of Abbeville, South Carolina, died in 1943 at age 99. Broadwell witnessed amazing advances in techology that were never dreamed of when the war began. He witnessed the invention of the telephone, automobile, airplane, radio, movie theaters, and television. He witnessed world changing events such as World War I, the crash of the stock market, Prohibition, the great depression and the opening of World War II.

The detailed records of all these reunions, if any were ever created, have been lost to history. Perhaps there is a treasure trove of material stored in an old trunk or attic somewhere in Georgia, detailing the war stories and anecdotes the old 38th Georgia soldiers told at these reunions. Perhaps there are transcripts of the many speeches given at the reunions, by the soldiers of the regiment, or Generals of the Brigade. Timeless historical materials documenting the history of the regiment may still exist, just waiting to be discovered.

Officers of the 38th Georgia Regiment, Circa 1894

Front Row, L - R, Unknown, Capt. John Goswick, Capt. John G. Rankin, Lt. Phillip W. McCurdy, Lt. George R. Wells, Lt. John Baxter
Back Row, John A. Booth, Lt. Miller Wright, Lt. Simpson Hagood, 1st Sgt. Francis L. Hudgins, Capt. John W. Brinson

Photo courtesy of anonymous descendant.

Appendix
Roster of the 38th Georgia Regiment

This roster contains important information to GUIDE the reader in finding additional information concerning these soldiers. The first step in your further research should be to obtain the full Confederate Service Record (CSR) for the soldier. The second step should be to determine if your ancestor, or his widow, filed a Confederate Pension Application (CPA). These two records are invaluable to researchers and are available online at Ancestry.com and Fold3.

The foundation of this roster is Lillian Henderson's Roster of Georgia Confederate Soldiers, created for the State of Georgia in 1950's. While Henderson's roster is the starting point, it has been analyzed and compared to actual Confederate service records, Confederate pension records, Confederate casualty lists, and newspaper articles of the day. Unfortunately, Henderson's roster is rife with errors. It often lacks critical information and analysis, due to the monumental scale of the work. In many cases, soldiers are missing, or simply identified by their first and middle initials, along with their last name, such as: Kidd, W. B. H., which research revealed to be Willis Benjamin Henderson Kidd. The full names of the soldiers are provided when found. Again, numerous sources were examined and in many cases descendants provided the full names of soldiers, along with other details. In this roster you will find the full text from the Henderson roster for each soldier, along with additional information added by the author. Notations for each soldier describe important documents contained in the Confederate Service Record, such as a discharge record, or a death claim filed by surviving family members.

Confederate Service Records often contain death claims of soldiers that died during the war. These claims were filed by the surviving widow, parent, sibling, or guardian, to collect outstanding pay due to the soldier. Death claims often reveal if the soldier was married or single, if he had children, his age, where he died and the circumstances of his death. Confederate Service records also frequently contain discharge records for soldiers discharged during the war. These records provide a treasure trove of information concerning the soldier. Discharge records contain the soldier's name, place of birth, age, a physical description of the soldier, where he enlisted, where he lived before enlistment, previous occupation, and why he was being discharged.

All Confederate Pension Applications identified in this roster were filed in the state of Georgia, unless otherwise noted. Confederate pension records contain a wealth of information that may not be found elsewhere. Details of how a soldier was killed, or wounded, along with the extent of his injuries are often included. Pension records often contain the nature of sickness and wounds, how and where acquired, the effects on the soldier since the war, witnesses to the soldier's service, dates of marriage in some cases, dates of death, along with details of their financial situation.

Widows of CSA veteran's were also allowed to file pension claims in Georgia beginning in 1891, if the soldier had died in the service, or before 1891. Widow's pension applications often contain information documenting when and where they were married, and in some cases the files contain a copy of their marriage license. Widow's applications also include the date a soldier died, a fact often hard to find from other sources. When searching for a Georgia Confederate pension application, it's important to include the name of the veteran and/or widow, the county the application was filed and the year the application was filed.

Confederate pensions were only paid to soldiers who could prove honorable service during the war. If your ancestor deserted, took the oath of allegiance to the U.S. government, or could not explain his whereabouts at the end of the war, they may have never applied for a Confederate pension. Not all soldiers have a Confederate pension application record, as the State of Georgia only allowed pensions for indigent soldiers, or those who lost limbs in the war, or their surviving widows. If your ancestor had significant financial resources, such as cash, farm acreage, or other valuable property, they were probably ineligible to receive a Confederate pension.

Obituaries are annotated by newspaper and date, when found, Obituaries, as we know them today, only came about in the late 1800's and early 1900's. Before obituaries, death tributes, or death notices, were rarely provided to newspapers. Death tributes during the war are rarely seen for Private soldiers and were more commonly published for prominent officers and men of wealth. These tributes often extol the virtues of the deceased, talk of their good character, patriotism, courage, and their sacrifices made during the war. If you ancestor died during the late 1800's or early 1900's he may have an obituary. Obituaries often provide in-depth information on a soldier and provide family lineage not found elsewhere.

Lastly, I've attempted to locate the burial site of all the soldiers. Many Confederate burial records contain numerous errors. Surnames, plus first and middle initials, dates of death, and regiments assigned, often contain errors. When possible, I've attempted to gather available evidence from service records, pension records, cemetery burial records, and descendants of soldiers, to determine their burial locations. In some cases, evidence was analyzed, with the likely burial location determined by the author. For example, the notation from Henderson's Roster of Georgia Confederate Soldiers, for Private Joel Whiddon, of Company I, The Irwin Invincibles, reads:

Whiddon, Joel - Pvt 5/10/1862. Died of chronic diarrhea at Camp Lee, near Richmond, VA 10/16/1862.

Research revealed a soldier named Whitden, no first name noted, Co. I, 27th Georgia Regiment, died 10/20/1862, and was buried at Hollywood Cemetery, Richmond, Virginia. This is likely Joel Whiddon, as no other soldier named "Whitden" or a soldier with a similar name, could be found on the roster of the 27th Georgia Regiment.

Unfortunately, many soldiers, particularly those killed in battle, lie in unmarked graves scattered across Virginia, Maryland, and Pennsylvania. Even for the men who returned home to Georgia and died soon after the war, many families were left so destitute by the war they couldn't afford stone grave markers. In many cases wooden grave markers were all families could afford and the ravages of time have destroyed these markers, leaving their graves unmarked today. Many of the soldiers and their families moved west after the war and their graves are scattered across the Southern states from Georgia to Texas. If you know the burial location of a soldier and I don't have it listed here, please contact me (38thga@gmail.com) and I'll add the information in my next updated edition of the book. There are sure to be errors in this roster, due to the monumental scale of this work (over 1,600 solders). All errors are solely the responsibly of the author.

Things to understand about this roster:

- All counties listed in this roster are in Georgia, unless otherwise noted.
- Widow's of Confederate veterans were allowed to file pension application in Georgia in 1891.
- Common abbreviations used in his roster are explained below:

Appt -	Appointed	Inf. -	Infantry
AWOL	Absent Without Leave	K -	Killed
AWL -	Absent With Leave	LOF -	Last On File
B -	Born	Lt. -	Lieutenant
BSU -	Burial Site Unknown	NLR -	No Later Record
Bur -	Buried	OOA -	Oath of Allegiance
C -	Captured	P -	Paroled
Cem -	Cemetery	PA -	Pension Application
Chur -	Church	PLO -	Point Lookout Prison Camp
Co. -	County	Pvt -	Private
Corp -	Corporal	POW -	Prisoner of War
CPA -	Confederate Pension Application	Rcv -	Received
Cpt -	Captain	Regt -	Regiment

CSR -	Confederate Service Record	Rich.	Richmond, Virginia
D -	Died	S -	Surrendered
Date	Month/Day/Year	Sav.	Savannah, Georgia
Des -	Deserted	Sgt -	Sergeant
E -	Exchange	Trn -	Transferred
Fred. -	Fredericksburg, Virginia	UVA -	University of Virginia
Hos -	Hospital	W -	Wounded

ROSTER OF FIELD, STAFF AND BAND,

Alexander, Philemon W. (or Pleasant W.) - Quartermaster Sgt. See Pvt Co. F.
Arrington, W. J. (William J. Arrington) - Surgeon 10/12/1861. P Augusta, 5/20/1865. Death reported in Augusta Chronicle, 3/28/1878. BSU.
Bailey, Joseph E. - Quartermaster Sgt. See Pvt Co. M.
Beals, W. H. - Musician. 4/29/1862. Discharged prior to 10/31/1862. Full name unknown. BSU.
Bomar, Thomas H. - Major. See Company L
Booth, John W. G. - Drum Major. See Co. H.
Booth, W. A. - (William Absolam Booth.) See Co. H.
Brittain, Jabez Mercer - Chaplain. See 1st Sgt, Co. E.
Brown, Beniah - Hos Steward. See Pvt Co. G.
Chew, John C. - Quartermaster Sgt. See Pvt Co. C.
Daniel, Amaziah C. - Commissary Sgt. See Pvt Co. F.
Davant, Phillip E. - (Phillip Edwin Davant) Elected Cpt of Co. B, 24th Regt GA Inf. 6/9/1861; Maj 3d Battn. GA Sharpshooters 6/8/1863; Elected Lt Col of 38th Regt Inf. 9/18/1863. C Spotsylvania, VA, 5/12/1864. Trn from Ft. Delaware, DE to Hilton Head, SC, E 6/25/1864. S Appomattox, VA, 4/9/1865. D 10/9/1906. One of the immortal 600. Bur Butler Memorial Cem, Butler, Taylor Co.
Dean, Edward C. - Hos Steward. See Pvt Co. L.
Devore, Andrew W. - Quartermaster Sgt. See Pvt Co. B.
Endies, T. M. (or Endier) - Musician 12/29/1861. First name unknown. BSU.
Farmer, Sidney T. - Quartermaster Sgt. See Pvt Co. G.
Fisher, Frederick - Musician. See musician Co. H. Bugler.
Flowers, Asbury P. (Anderson "Preston" Flowers) - See Co. A.
Flowers, John Y. - (John Yancey Flowers) Major. - See Co. A.
Graves, G. W. - Surgeon. Left at Frederick City, MD with W, 7/10/1864, taken prisoner. BSU.
Hanleiter, Cornelius R. - 1st Lt 9/26/1861. Elected Cpt 10/25/1861. See Co. M.
Harbin, Jasper L. - Ensign. See Pvt Co. H.
Holloway, Robert G. (Robert Green Holloway) - Appt Assistant Surgeon, C.S.A., 6/10/1863. Assigned this Regt 10/5/1864. Appt Surgeon and reassigned this Regt 4/5/1865. P Bowling Green, VA, 5/12/1865. D Oct/1919, bur Saint Peters Church, Port Royal, Caroline Co., VA.
Houston, James H. - Assistant Surgeon. BSU.
Humphrey, J. R. - Assistant Surgeon. 10/12/1861. Resigned 5/12/1862. Replaced by Skirving Price. BSU.
Hutcherson, James M. - Musician. See Musician Co. F.
Jernigan, William H. - Cpt & A.Q.M. See Pvt Co. E (Assistant Quartermaster.)
Jones, George A. - Quartermaster Sgt. See Pvt Co. E.
Law, John - (John Gordon Law) Adjutant 11/21/1863. C Spotsylvania, VA, 5/10/1864. Released Ft. Delaware, 6/15/1865. 1st cousin of General John B. Gordon. D 3/29/1916, bur West View Cem, Oconee Co., SC. Obit in SC State Newspaper, 3/30/1916, page 3.
Lee, Barney Drewry - Sgt Major. See Pvt Co. M.
Lee, George Washington - Colonel. Enlisted as a Pvt in Co M, 9/26/1861. See Co. M.
Lewis, L. (or Louis) - Chief Musician, 4/29/1862. Discharged prior to 10/31/1862. First name unknown. BSU.

Mashburn, John H. (John Harvey Mashburn) - Chaplain 12/24/1861. Resigned, 7/23/1862. Died 6/13/1876, Hall Co. Buried Ebenezer Cemetery, Forsyth Co.
Mathews, James Davant - Colonel. Cpt in Co. E, 9/29/1861. See Co. E.
Mathews, William C. - Ordnance Sgt. See Pvt Co. G.
McDaniel, R. S. - Ordnance Sgt. See Pvt Co. L.
McMakin, William - Musician. 10/6/1861. Transferred to Co. B, as a Pvt. See Co. B.
Mendez, F. Augustus - Musician. See Musician, Co. L. Listed as Bugler in CSR. BSU.
Mitchell, Horace M. - Commissary. See Pvt Co. M.
Parr, Lewis J. - Lt Colonel. Major. See Co. M.
Perlinski, Julius - (or Perlinsky) Musician 4/29/1862. Transferred to Co. D, 1864. C Winchester, VA, 9/19/1864. Released PLO, MD, 5/14/1865. BSU.
Prescott, John R. - Commissary Sgt. See Pvt Co. C.
Price, Skirving - Cpt & Assistant Surgeon. 12/4/1862 (to rank from 8/11/1862). K Deep Bottom, VA, 8/15/1864. Bur Massanutten Cem, Woodstock, VA.
Pughsley, George W. - Ordnance Sgt. See 3d Sgt Co. C.
Quinn, John M. - A.C.S. 11/16/1861. Appt Commissary. Relieved from duty by Act of Congress Jul/1863. D 10/14/1897, B in MS. Bur Myrtle Hill Cem, Floyd Co. No grave marker.
Robertson, Walter S. - Sgt Major. See 5th Sgt, Co. E.
Robinson, George F. - Enlisted as a Pvt in Co. E, 4th Regt GA Inf. 4/28/1861. Appt Adjutant the 38th GA. May/1863. Elected Cpt of Co. K, 8/13/1863. Retired to Invalid Corps 12/12/1864. BSU.
Rosenthal, Adolphus - Ordnance Sgt. See Pvt Co. D.
Shaw, Augustus - Acting Adjutant. See Pvt Co. M.
Sherrod, John H. - Adjutant. See 1st Lt Co. C.
Smith, Amory S. - Acting Ordnance Sgt. See Pvt Co. G.
Toney, Charles W. - Musician. 9/26/1861. Transferred to Co. K, see Co. K.
Vaughan, Alexander - Hos Steward. See Pvt Co. H.
Williams, Thomas H. - A.Q.M. (Assistant Quartermaster.) See Pvt Co. M.
Wright, A. R. (Augustus Romaldus Wright) - Elected as Democrat to 35th U.S. Congress, served from 3/4/1857 - 3/3/1859, served as delegate to GA Secession Convention, opposed secession, and to Confed. Secession Convention. B in Wrightsboro, GA, 6/16/1813. A demand by civil government for his services induced him to resign his commission 2/14/1862. D 3/31/1891, home "Glenwood," near Rome, bur Myrtle Hill Cem.

*MUSTER ROLL OF COMPANY A, DeKALB COUNTY, GA, DeKALB MURPHY GUARDS, OR MURPHY GUARDS

*This company was originally known as Co. C. It was divided 4/1/1862, and part of it became Co. D. The Murphy Guards was raised and organized by Captain John Y. Flowers.

Adams, John A. - Pvt 9/26/1861. Captured Spotsylvania, VA, 5/20/1864. Died of pneumonia Elmira, NY, Sept. 1 or 10/10/1864. Buried Woodlawn National Cemetery, Grave #687, Elmira, NY.
Akins, Charles Thomas (or Aiken) - Pvt 9/26/1861. Transferred 4/1/1862 to Co. D, see Co. D.
Anderson, George J. - Pvt 9/26/1861. Transferred 4/1/1862 to Co. D, see Co. D.
Arnold, B. W. - Pvt 9/26/1861. D Camp Bartow, 12/16/1861. Full name unknown. BSU.
Austin, A. C. - (Alexander C. Austin) Pvt 9/26/1861. K Gaines Mill, VA, 6/27/1862. CSR shows father, Alexander C. Austin, filed death claim, Fulton Co. Died of wounds, 9/26/1862. BSU.
Austin, John Thomas - Pvt 5/15/1862. S Appomattox, VA, 4/9/1865. (B GA in 1844) Filed CPA Cobb Co., 1913. D 9/27/1922, Cobb Co. Bur Sandy Springs United Meth. Cem, Fulton Co.
Austin, W. F. - (William F. Austin) Pvt 3/1/1862. K Antietam 9/17/1862. First name found in CSR. CSR states widow, Elizabeth F. Austin, filed death claim in Fulton Co. BSU.
Baker, W. M. - (William M. Baker) Pvt 5/15/1862. D disease Rich, VA, hos 2/14/1863. Bur UVA CS Cem. Charlottesville, VA. The bronze plaque on the Confed. monument lists William M. Baker, 36th GA CSR shows father, Alexander Baker, filed a death claim Gwinnett Co.

Ball, John W. - Pvt 9/26/1861. Appt 4th Cpl 4/1862; 1st Cpl 8/1/1863. Wounded Antietam 9/17/1862, according to CSR. Captured Fred 12/13/1862, paroled for exchange 12/17/1862. Surrendered Appomattox, VA, 4/9/1865. (Born in GA) Filed CPA Fulton Co., 1919, Died 10/5/1920, buried there Ball Family Cemetery. Obituary in Confederate Veteran Magazine, Dec/1920, pg.468.
Ball, Peter M. - (Peter Marion Ball) Pvt 9/26/1861. Wounded 2d Manassas, VA, 8/28/1862. Det hos in 1863. Roll dated 11/9/1864, last on file, shows "Absent, wounded, in Franklin Co., furlough." Filed CPA Grayson, TX, 1913. Died 10/13/1914. Buried Tioga Cemetery, Grayson Co., TX.
Ball, William M. - Pvt 3/1/1862. Wounded in left leg and captured Gettysburg, PA, 7/1/1863, and P in 1863. Retired to Invalid Corps Athens, 9/10/1864. (B GA 2/15/1837) Filed CPA, Milton Co., 1888, died 1/6/1919, buried Ebenezer Church Cem, DeKalb Co. Obit in Atlanta Constitution, 1/7/1919.
Baxter, Francis M. - Pvt 3/1/1862. Transferred 4/1/1862 to Co. D, see Co. D.
Baxter, John (John M. Baxter) - Pvt 3/1/1862. Transferred 4/1/1862 to Co. D, see Co. D.
Binion, Joseph D. - Pvt Wounded in side Wilderness, VA, 5/6/1864. Sent to hospital and frl home. Could not reach command when frl expired. Enlisted as Pvt, Co. A, Carroll's Battalion GA Militia, 1864. S Kingston, GA 3/12/1865. (B Prince Edward Co., VA, 11/9/1842. Killed by fall from train 6/8/1917) Filed CPA 1902, Fulton Co. BSU.
Braswell, B. S. (Brasil S. Braswell) - Pvt 5/15/1862. D Rich, VA, July 16 or 18, 1862. CSR has death claim filed by widow, Martha S. Braswell. DeKalb Co. BSU.
Braswell, G. A. (George A. Braswell) - Pvt 5/15/1862. K Cold Harbor, VA, 6/3/1864. BSU.
Braswell, Samuel H. (Samuel House Braswell) - Pvt 9/26/1861. Arm disabled 2d Manassas, VA, 8/28/1862. W Fred., VA, 5/4/1863. Roll dated 11/9/1864, shows Absent, W; now in SC on frl. CPA shows home, W, close of war. Filed CPA, Fulton Co., 1895. D. 2/19/1914. Obit in Atlanta Constitution, pg 20, 2/20/1914. Bur Prospect Meth. Church, DeKalb Co.
Campbell, Benjamin M. (or Benjamin W.) - Pvt 9/26/1861. Wounded and captured 2d Manassas, VA, 8/28/1862. Exchanged, wounded May/1864. Roll dated 11/9/1864, last on file, shows Absent, wounded, supposed home, DeKalb Co." Paroled Augusta, GA 5/29/1865. BSU.
Campbell, George M. (George W. Campbell) - Pvt Jun/1864. K Monocacy, MD 7/9/1864. Widow, Lucy G. F. Campbell, filed CPA, 1891, DeKalb Co. BSU.
Carpenter, John J. - Pvt 9/26/1861. Appt 3d Sgt, captured Farmville, VA, 4/8/1865 and paroled there 4/17/1865. (Born in GA, 1838) Died 3/1/1918. Buried Old Roswell Cem, Roswell, Fulton Co. Widow, Sylvania Carpenter, filed a CPA 1918, in Cobb Co.
Carroll, J. M. (James J. Carroll, or Carrell) Pvt 9/26/1861. Appears last on roll for Feb/1863. Detached as Regtal sutler Nov - Dec 1862, according to CSA records. BSU.
Carter, John D. - Pvt 9/26/1861. Transferred 4/1/1862 to Co. D, see Co. D.
Carter, William J. - Pvt 9/26/1861. Transferred 4/1/1862 to Co. D, see Co. D.
Cash, L. J. (Lewellin J. Cash) - Pvt 9/26/1861. Wounded Chancellorsville, VA, 5/3/1863. Surrendered Appomattox, VA, 4/9/1865. (B Orange Co., VA, 1835) Filed CPA Cobb Co., Fulton Co., DeKalb Co., Douglas Co., & Cobb Co. Died after Jan/1907. BSU.
Chamblee, W. L. (William Larkin Chamblee) - Pvt 9/26/1861. Surrendered at Appomattox, VA, 4/9/1865. Buried Friendship Baptist Church Cemetery, Forsyth Co. Born 2/19/1832, died 3/27/1923.
Chesnut, D. A. (David Alexander Chesnut) - Pvt 3/20/1862. S Appomattox, VA, 4/9/1865. Filed CPA, DeKalb Co., 1914. B 8/17/1840, D 12/7/1914. Bur there Prosperity Cem. Marker inscribed 38 GA INF.
Chesnut, G. R. M. (George Robert McDill Chesnut) - Pvt 5/15/1862. D Rich, VA, 7/6/1862. Bur there Hollywood Cem, section M, grave #159. CSR contains full name and contains a death claim filed by father, Alexander Chesnut, of DeKalb Co. Born 4/8/1843.
Cochran, S. W. (Samuel "Sam" Wylie Cochran) - 3d Cpl 9/26/1861. Trn 4/1/1862 to Co. D, see Co. D.
Coggin, J. F. (James F. Coggin) - Pvt 3/20/1862. Transferred 4/1/1862 to Co. D, see Co. D.
Conn, William A. - Pvt 9/26/1861. Transferred 4/1/1862 to Co. D, see Co. D.

Copeland, Obediah B. - Pvt 9/26/1861. W knee and back Antietam, MD 9/17/1862. Absent, sick, May 1863 - 7/1864. C Rossville, GA 7/9/1864. Released Camp Douglas, IL, 6/16/1865. (B in GA 1/9/1834) Filed CPA, Milton Co., 1893. D. 9/29/1895. Buried Ebenezer Primitive Baptist Church, Cemetery, near DeKalb/Fulton Co. line.
Corley, R. J. (Robert J. Corley) - Pvt 3/1/1862. D Emory, VA, 1/12/1863. Widow, Ann Corely, filed CPA, DeKalb, 1891. She stated he D typhoid fever at Emory & Henry College hos, VA. BSU.
Davis, William Caleb - Pvt 9/26/1861. Transferred 4/1/1862 to Co. D, see Co. D.
Dees, David - Pvt 3/1/1862. Roll for 8/31/1864, LOF, shows present. Filed CPA, Attala Co., MS, 1921, buried there Shady Grove Cemetery. Died 8/21/1928.
Dickerson, W. F. A. (William F. A. Dickerson) - Pvt 3/1/1862. Trn 4/1/1862 to Co. D, see Co. D.
Dodgen, J. N. (John Newton Dodgen) - Pvt 9/26/1861. Sick frl 60 days from 9/20/1864. D 3/3/1894, DeKalb Co., bur there Prosperity Cem. Margaret A. Sellers Dodgen filed CPA, Fulton Co., 1901.
Dodgen, William O. - Musician 9/26/1861. Transferred 4/1/1862 to Co. D, see Co. D.
Ealum, Joel R. (or Elum) - Pvt 9/26/1861. Transferred 4/1/1862 to Co. D, see Co. D.
Edwards, William A. - Pvt 5/15/1862. Des to enemy Strasburg, VA, 8/14/1864. Took OOA to U.S. Govt. and furnished trans to Philadelphia, PA, date not stated. D 1910. Bur Mt. Salem Bap., Hall Co.
Eidson, A. J. (Andrew J. Eison) - Pvt 9/26/1861. Appt 5th Sgt 3/1/1863. C Spotsylvania, VA, 5/20/1864. P Elmira, NY, 3/10/1865. Rcv Boulware & Cox's Wharves, James River, VA, for E, 3/15/1865. Born 1832, died 1886. Bur Shiloh Cem, Polk Co. CSA marker reads: CO A, 38 GA INF, Name spelled "Eison" in CPA and on grave marker. Widow, Jennett Eison filed CPA, Polk Co., 1902.
Eidson, John - Pvt 9/26/1861. W Antietam, MD, 9/17/1862; Chancellorsville, VA, 5/3/1863. W and C Cedar Creek, VA, 10/19/1864. Admitted USA. Depot Field Hos Winchester, VA, 10/20/1864, D there, brain gangrene, 10/24/1864. Bur Stonewall Confed. Cem, Winchester, Marker reads J. J. Eison.
Eidson, Robert - Pvt 9/26/1861. Appt 4th Cpl 1862; 2d Cpl 1863. Captured Winchester, VA, 9/19/1864. Released Point Lookout, MD, 6/26/1865. Filed CPA, DeKalb Co., 1897. Born 5/20/1830, died 1/19/1906. Buried Prosperity Cemetery, DeKalb Co.
Eidson, R. S. (Robert S. Eidson) - Pvt 9/26/1861. W right arm, arm amputated below elbow, Wilderness, VA, 5/6/1864. At home, W, close of war. (D Kennesaw, Cobb Co. 4/7/1907) Appears W, 2d Manassas, 8/28/1862. Filed CPA, Cobb Co. 1879. D 3/5/1907, bur Kennesaw City Cem.
Eidson, William M. - Pvt 9/26/1861. Captured Antietam, MD, 9/17/1862. Died of typhoid fever at Frederick City, MD, 10/6/1862. Buried there Mt. Olivet Cemetery. CSA marker reads: CO A, 38 GA INF. CSR has death claim by father, Boyce Eidson, DeKalb Co.
Eidson, Z. (Zachariah "Zack" Eidson) - Pvt 9/26/1861. Wounded Antietam, 9/17/1862. Wounded Wilderness, VA, 5/6/1864. Captured Waynesboro, VA, 3/2/1865. Released Ft. Delaware, 6/16/1865. Born 1823, died 4/17/1873, buried Eidson Family Cemetery, DeKalb Co.
Elliott, Joseph E. - Pvt 5/15/1862. Surrendered at Appomattox C. H., VA, 4/9/1865. Born 4/21/1833, died 12/8/1892. Buried Chamblee Cemetery, DeKalb Co.
Estes, Gainham T. - Pvt 9/26/1861. Trn to Co. D, 4/1/1862. D of typhoid fever Charlottesville, VA, 7/7/1862. Bur there UVA Confed. Cem. CSR has death claim widow, Delila Estes, of Jackson Co.
Evans, John L. (or Evins) - Pvt 9/26/1861. D Camp Bartow (near Sav) 12/17/1861. CSR contains death claim by mother, Isabelle A. Evins, in Cobb Co. BSU.
Evins, Nelson H. (Nelson Houston Evins) - Pvt 9/26/1861. Discharged for defective eye, Fred., VA, 12/14/1862. Reenlisted in 1863. Captured Richmond, VA, 4/3/1865. Admitted Hammond USA General Hospital, pneumonia, 5/12/1865. Transferred to General Hos 7/27/1865. (B near Atlanta, GA, 10/22/1839) Filed CPA, Fulton Co., 1895. BSU.
Fields, William A. - Pvt 5/15/1862. C Spotsylvania, VA, 5/23/1864. P at PLO, MD Mar/1865. Rcv Boulware & Cox's Wharves, James River, VA, E, 3/16/1865. (B SC 9/4/1844) Filed CPA, Milton Co., 1907. BSU.
Flowers, Asbury P. (Anderson "Preston" Flowers) - Musician. 2/6/1862. Trn to Co. A, as Pvt 3/1/1862. Appt fifer 1862. Died Staunton, VA, hospital of smallpox, 1/20/1863. CSR contains death

claim filed by father, John Y. Flowers, DeKalb Co. Buried Thornrose Cem, Staunton, VA. Marker reads Anderson P. Flowers, Mus. 38th GA INF.
Flowers, James A. (James Asbury Flowers) - Pvt 9/26/1861. D in camp 2/11/1863. CSR contains death claim filed by widow, Caroline G. Flowers, DeKalb Co. D of typhoid fever. BSU.
Flowers, John Y. (John Yancey Flowers) - Major. - Cpt, Co. A, 9/26/1861. Elected Maj 2/14/1862. Resigned, hernia, June/1862. D 8/6/1887, DeKalb Co., bur there Nancy Creek Primitive Bap. Chur.
Gaddy, G. W. (George Washington Gaddy) - Pvt 9/26/1861. Appt 5th Sgt Apr/1862. K Antietam, MD, 9/17/1862. Full name in CSR. Widow, Emily Gaddy, filed CPA, Fulton Co., 1899. BSU.
Gardner, Andy J. - Pvt 7/31/1863. W and C Winchester, VA, 9/19/1864. D of wounds USA Depot Field Hos 10/23/1864. Widow, Martha Gardner, filed CPA DeKalb Co., 1891. BSU.
Gardner, Charles - Pvt 5/15/1862. W Cold Harbor, 6/1/1864. D wounds Rich, VA, 6/2/1864. BSU.
Gardner, James R. (or James K.) - Pvt 3/1/1862. K Cold Harbor, VA, 6/1/1864. BSU.
Gardner, William - Pvt 3/1/1862. D in Sav, GA, hos 4/30/1862. Cause of death not in CSR. BSU.
Goss, John W. - Pvt 9/26/1861. Transferred 4/1/1862 to Co. D, see Co. D.
Green, David A. - Pvt 5/15/1862. Appt 2d Cpl Dec/1862. K Gettysburg, PA, 7/1/1863. CSR contains death claim filed by father, Washington C. Green, in Cobb Co. BSU.
Gresham, J. W. (Josiah Wheeler Gresham) - Pvt 9/26/1861. Appt 3d Sgt of Co. D, 4/1/1862. K Antietam, MD 9/17/1862. CSR states mortally W 9/17/1862 and D same day. First name in CSR. CSR contains death claim filed by widow, Parthenia Henderson Gresham, DeKalb Co. BSU.
Hall, John W. - Pvt 9/26/1861. Transferred 4/1/1862 to Co. D, see Co. D.
Hambrick, James F. - Pvt 9/26/1861. Transferred 4/1/1862 to Co. D, see Co. D.
Hambrick, J. T. - (James T. Hambrick) Pvt 9/26/1861. Transferred 4/1/1862 to Co. D, see Co. D.
Hambrick, Robert C. - Pvt 9/26/1861. Transferred 4/1/1862 to Co. D, see Co. D.
Hardman, E. C. - (Ewell Chandler Hardman) Pvt 9/26/1861. Appears last on roll for Apr/1862. D 9/16/1883. DeKalb Co. Bur Prospect Meth. Cem, DeKalb Co.
Hardman, J. S. W. (John Silas Uriah Hardman) - Pvt 5/15/1862. D Chancellorsville, VA, July 6 or 7, 1862. CSR contains a death claim filed by father, E.C. Hardman, DeKalb Co., CSR states D Charlottesville, VA, 7/7/1862. Believed to be bur there UVA Confed. Cem, Confed. memorial at cem lists J.J. Hardman, 31st NC; no such soldier assigned to 31st NC.
Harmon, Francis M. - Pvt 5/15/1862. W in 1863. Absent on Surgeon's certificate Jul/1864. Took OOA to U.S. Govt. at Louisville, Ky. 8/10/1864. Released north of Ohio River, date not given. Captured in DeKalb Co. BSU.
Harmon, James P. - Pvt 3/1/1862. W, left eye destroyed & right leg disabled, Kernstown, VA, 7/23/1864. W Ft. Stedman, VA, 3/25/1865. C Rich, VA, 4/3/1865. Delivered to Provost Marshal 4/28/1865. (B GA 11/18/1842) Filed CPA, Fulton Co, 1894. D 1925, bur Alta Vista Cem, Gainesville.
Harmon, John T. - Pvt 9/26/1861. W in left foot Fred., VA, 12/13/1862. Appt 3d Cpl 1863. W, leg Gettysburg, PA, 7/3/1863. Absent on Surgeon's certificate of disability 10/1863 - 9/1/1864. CPA shows guard duty Atlanta, GA, disabled by wounds, end of war. (B Spartanburg, SC) Filed CPA, Fulton Co., 1889. D Fulton Co. 10/22/1919. BSU.
Harmon, J. B. (Josephus B. Harmon) - Pvt 2/1/1864. S Appomattox, VA 4/9/1865. D aft. 1901. BSU.
Harmon, R. A. (Robert A. Harmon) - Pvt 9/26/1861. D Rich, VA, 7/18/1862. CSR contains death claim by father, Samuel Harmon, DeKalb Co. Bur Oakwood Cem, Rich, VA.
Harris, Edmond T. - 4th Cpl 9/26/1861. Transferred 4/1/1862 to Co. D, see Co. D.
Harris, J. C. (James C. Harris) - Pvt 3/1/1862. Transferred 4/1/1862 to Co. D, see Co. D.
Harris, Sterling G. - 1st Cpl 9/26/1861. Transferred 4/1/1862 to Co. D, see Co. D.
Hix, A. H. (Abner Harrison Hix) - Pvt 9/26/1861. Transferred 4/1/1862 to Co. D, see Co. D.
Holcombe, A. M. - (Azariah Milton Holcombe) 5th Sgt 9/26/1861. Appt 2d Sgt Dec/1862; 1st Sgt 4/1/1863. Elected Cpt Apr/1865. S Appomattox, VA, 4/9/1865. B Nov/1836, D 8/31/1909, bur Indian Creek Cem, DeKalb Co. Widow, Talitha F. Ayres Holcomb, filed CPA, DeKalb Co., 1911.
Holly, H. J. (Hiram J. Holly) - Pvt 3/1/1862. Transferred 4/1/1862 to Co. D, see Co. D.

Holmes, James M. - Pvt 5/15/1862. Captured Winchester, VA, 9/19/1864. Transferred to Point Lookout, MD, for exchange 2/20/1865. No later record. CRS states age 53 in 1865. BSU.
House, David C. - Pvt 9/26/1861. Died near Culpeper, VA, 9/2/1862. CSR contains death claim filed by father, Samuel House. BSU.
House, George W. - Pvt 5/15/1862. Absent, sick, Feb/1863. NLR Bur Chamblee Cem, DeKalb Co.
House, John T. - Pvt 8/1/1862. W 2d Manassas, VA, 8/28/1862. Died at home of wounds, Cobb Co. 11/28/1862. CSR contains death claim filed by widow, first name illegible, A. House. BSU.
House, Thomas J. - Pvt 9/26/1861. Appt color bearer 1862; 5th Cpl 4/1/1862. Wounded in leg, amputated below knee, Cedar Creek, VA, 10/19/1864. Mount Jackson, VA, hospital, 11/9/1864. Filed CPA, DeKalb Co., 1886, Died 5/12/1911. Buried Prospect Meth. Cem, DeKalb Co. Widow, M. A. House, filed CPA, DeKalb Co., 1920.
Ivey, Aaron H. - Pvt 9/26/1861. Transferred 4/1/1862 to Co. D, see Co. D.
Ivey, Joseph J. - Pvt 9/26/1861.Transferred 4/1/1862 to Co. D, see Co. D.
Jenkins, J. D. - (Jeremiah D. Jenkins) Pvt 9/26/1861. Detached as drummer for 6 months 4/1/1862. AWOL Sept 1864. Took OOA to U.S. Govt. Atlanta, GA, 9/27/1864. Reported W in battle of Gettysburg, according to Savannah Republican newspaper, dated 7/30/1863. BSU.
Jenkins, William H. (William "Bill" H. Jenkins) - Musician 9/26/1861. Surrendered Appomattox, VA, 4/9/1865. Kettle drummer. Died 10/2/1899. Widow, Nancy JINKINS, filed CPA, Butts Co., 1911. Buried Jenkinsburg Methodist Church, Butts Co.
Jenkins, W. C. - (William Cary Jenkins) Pvt 9/26/1861. W & C Gettysburg, PA, 7/2/1863. E PLO, MD 2/13/1865. At Camp Lee near Richmond, VA, 2/17/1865. On wounded list Fredericksburg, 12/13/1862. First name in CSR. D. 9/18/1902, buried Winters Chapel Meth. Church Cemetery, DeKalb Co. Marker reads 38 GA INF. Widow, Sarah Jane Jenkins. filed CPA, Fulton Co., 1919.
Jett, James S. - 1st Sgt 9/26/1861. Elected Jr. 2d Lt 4/15/1862; 2d Lt 10/7/1863. Wounded 2d Manassas, 8/28/1862, Died 2/7/1864. Widow, Elizabeth A. Jett, filed CPA, Gwinnett Co, 1891. She stated he was sent home on sick furlough and died of chronic diarrhea. BSU.
Johnson, G. F. M. - Pvt 9/26/1861. Died 8/1/1862. Full name unknown and BSU.
Johnson, J. T. M. - (James T. M. Johnson) Pvt 9/26/1861. Sick in hospital Feb/1863. NLR. BSU.
Johnston, W. J. (William J. Johnston) - Pvt 9/26/1861. Wounded Gaines Mill, VA, 6/27/1862. Surrendered at Appomattox, VA, 4/9/1865. Listed as "teamster" at Appomattox. (Born DeKalb Co., GA, 1845) Filed CPA, Fulton Co., 1913. BSU.
Jones, C. M. - (Charles M. Jones) Pvt 9/26/1861. Roll dated 11/9/1864, shows "absent without leave since 9/18/1864; where unknown." NLR. BSU.
Kite, Claiborn - Pvt 9/26/1861. Discharged, disability, 6/6/1862. CSR has discharge record. BSU.
Lafoy, John A. - Pvt 9/26/1861. Transferred 4/1/1862 to Co. D, see Co. D.
Lang, John W. (or Long) - Pvt 9/26/1861. Transferred 4/1/1862 to Co. D, see Co. D.
Lang, William B. (or Long, Wiley B.) - Pvt 3/1/1862. Transferred 4/1/1862 to Co. D, see Co. D.
Langford, Josiah C. (Josiah Clint Langford) - Pvt 9/26/1861. Trn to Co. B, 5/15/1862. See Co. B.
Lanier, (or Lanear) Pressley - Pvt 3/1/1862. Transferred 4/1/1862 to Co. D, see Co. D.
Lanier, Joseph W. - Pvt Enlisted 9/20/1861, Died 8/23/1862 Winder Hospital, Richmond, VA. CSR contains death claim filed by mother, Nancy Lanier. BSU.
Lanier, William L. - Pvt 9/26/1861. Transferred 4/1/1862 to Co. D, see Co. D.
Lawhorn, Joel J. - (or Lawhon) Pvt 9/26/1861. Transferred 4/1/1862 to Co. D, see Co. D.
Leavell, E. F. - (Edward "Ed" Franklin Leavell) Pvt 9/26/1861. Trn 4/1/1862 to Co. D, see Co. D.
Leavell, F. M. (Francis Marion Leavell) - Pvt 9/26/1861. Trn 4/1/1862 to Co. D, see Co. D.
Lord, Russell D. (Russell Daniel Lord) - Pvt 9/26/1861. Wounded in head Fredericksburg, VA, 12/13/1862. Surrendered Appomattox, VA, 4/9/1865. (B in Jackson Co., GA, 2/14/1838) Filed CPA Fulton Co., 1888, died 6/9/1914. Buried Mt. Paran Cem, Fulton Co. Marker reads: 38 GA INF.
Lord, S. M. (Stephen M. Lord) - Pvt 9/26/1861. Discharged, disability, 6/6/1862. CSR contains discharge record. Filed CPA, DeKalb Co., 1899. BSU.
Lord, William J. - Pvt 9/26/1861. Died in Empire Hospital, Vineville, Macon, GA, 9/21/1864. BSU.

Luckey, J. R. - (John R. Luckey) Pvt 9/26/1861. Appt Musician in 1861. Trn 4/1/1862 to Co. D, see Co. D.
Lynch, Thomas - Pvt 9/26/1861. Transferred 4/1/1862 to Co. D, see Co. D.
Maddox, Thomas L. - Pvt 9/26/1861. Transferred 4/1/1862 to Co. D, see Co. D.
Marable, John J. - Pvt 9/26/1861. Elected Jr. 2d Lt 3/8/1862; 2d Lt 4/15/1862. Wounded at Gaines Mill, VA, 6/27/1862. Resigned 10/8/1863. BSU.
Marbut, John S. (or John T. Marbut) - Pvt 9/26/1861. Trn 4/1/1862 to Co. D, see Co. D.
Marlow, William T. (or William L.) (or William F. Marlow) - Pvt 9/26/1861. Captured at battle of Fredericksburg, 12/13/1862 and exchanged 12/17/1862. Died 2d Div, General Hos Camp Winder, Richmond, VA, chronic diarrhea and typhoid fever, 10/22/1863. Buried there Hollywood Cemetery.
Martin, Benjamin S. - Pvt 9/26/1861. Transferred to Co. K, see Co. K.
Martin, James - Pvt 9/26/1861. Captured at Gettysburg, PA, July 1st or 3rd, 1863. Died of scrofula at Ft. Delaware, 5/7/1864. Buried Finn's Point National Cemetery, Salem, NJ.
Mason, Cornelius A. - (Camilus Anderson Mason) Pvt 9/26/1861. Trn 4/1/1862 to Co. D, see Co. D.
McCandless, Jesse A. - (Jesse Andrew McCanless) Pvt 9/26/1861. Trn 4/1/1862 to Co. D, see Co. D.
McClendon, William - Enlisted as a Pvt in Co. K, 60th Regt GA Inf. 5/10/1862. Appt Musician in 1863. Trn to Co. A, 38th Regt as Musician Mar/1864. Roll for 9/1/1864, shows present. NLR. BSU.
McCurdy, John Wilson - Pvt 3/1/1862. Transferred 4/1/1862 to Co. D, see Co. D.
McCurdy, Phillip Burford - Pvt 3/1/1862. Transferred 4/1/1862 to Co. D, see Co. D.
McElroy, Samuel B. (Samuel Bryson McElroy) - Pvt 5/15/1862. Died Richmond, VA, 8/30/1862, buried there Hollywood Cemetery. Cemetery records show: McMoy, S. B., 38 GA INF. Widow, Jane G. McElroy, filed CPA DeKalb Co., 1891.
Medlock, Eli W. (Eli Wren Medlock) - Pvt 9/26/1861. Trn 4/1/1862 to Co. D, see Co. D.
Miller, Andrew J. - Pvt 3/1/1862. Wounded Antietam, MD, 9/17/1862. Captured Fredericksburg 12/13/1863 and exchanged 12/17/1862. Age 24 at capture. Appt 5th Sgt Dec/1862; 4th Sgt Feb/1863. Wounded 1864, gunshot to arm. Captured Harrisonburg, VA, 9/25/1864. Released Point Lookout, MD, 6/29/1865. BSU.
Miller, Charles - Pvt 9/26/1861. Died Staunton, VA, 8/12/1862. Bur there in Thornrose Cem.
Miller, David Y. (David Young Miller) - Pvt 9/26/1861. Killed Antietam, MD, 9/17/1862. CSR contains death claim filed by widow, Mahulda Miller, DeKalb Co. M. H. Ross, widow, filed CPA, DeKalb Co., 1911. BSU.
Miller, H. B. (Henry Bryce Miller) - Pvt 5/15/1862. Died Staunton, VA, 8/24/1862. Buried there Thornrose Cemetery. Widow, Mary E. Flowers Miller, filed CPA, DeKalb Co, 1891.
Miller, N. R. (Newton R. Miller) - Pvt 9/26/1861. S Appomattox, VA, 4/9/1865. Died 5/17/1889. Widow, Amanda E. Miller filed CPA, DeKalb Co., 1911. Bur Prospect Meth. Chur Cem, DeKalb Co.
Miller, Robert W. - Pvt 7/31/1863. Killed at Spotsylvania, VA, 5/18/1864. Widow, Jane L. Miller, filed CPA, DeKalb Co., 1891. BSU.
Miller, W. A. C. Jr. - (William Alexander Caldwell Miller) 2d Lt 9/26/1861. Elected 2d Lt.; 1st Lt 1/1863; Cpt 1/28/1863. Home, Newton Co., GA, disabled, 6/6/1864 - 2/27/1865. D disease 3/20/1865. Bur Prosperity Cem, DeKalb Co. Widow, Elizabeth McDonald Miller filed CPA, DeKalb Co., 1891.
Mitchell, William H. (William Harris Mitchell) - Pvt 3/1/1862. Wounded 2d Manassas, VA, 8/28/1862; wounded Winchester, VA, 6/13/1863; wounded Spotsylvania, VA, 5/18/1864. Captured at High Bridge, VA, 4/6/1865. Paroled at Farmville, VA, Apr. 11-21, 1865. Filed CPA Fulton Co., 1889. Died 4/08/1932, buried there at Sandy Springs United Methodist Church.
Morris, D. C. (David C. Morris) - Pvt 9/26/1861. Died Richmond, VA, 7/23/1862. First name found in CSR. CSR contains death claim filed by father, Gideon Morris, in DeKalb Co. BSU
Nash, John Miles - Pvt 9/26/1861. Transferred 4/1/1862 to Co. D, see Co. D.
Nash, J. N. (Isaac Newton Nash) - Pvt 3/1/1862. Transferred 4/1/1862 to Co. D, see Co. D.
Nash, Miles H. - Pvt 9/26/1861. Transferred 4/1/1862 to Co. D, see Co. D.

Newsom, Moses H. - Pvt 9/26/1861. Transferred 4/1/1862 to Co. D, see Co. D.
Parrish, Isaac N. - Pvt 3/1/1862. D Rich, VA, 7/19/1862. Bur there in Hollywood Cem. Cem records read Parish, J. W., Co A, 38 GA, D 7/19/1862. Widow, Martha Parish filed CPA, DeKalb Co., 1891.
Pool, A. J. H. - (Alfred J. H. Pool) 1st Lt 9/26/1861. Elected Cpt 2/14/1862. Accidentally K Atlanta, GA Jan. 15 or 27, 1863. First name in CSR. Bur Lively Family Cem, Brownings Dist, DeKalb Co.
Pool, W. H. - (William H. Pool) 3d Sgt 9/26/1861. Appt 1st Sgt 4/1/1862. Wounded Fredericksburg, 12/13/1862.Wounded Feb/1863. Elected Jr. 2d Lt 1/10/1863; 1st Lt 10/7/1863. Roll dated 11/7/1864, shows Absent since 5/4/1864, home in DeKalb Co. Dropped from roll 1/16/1865. Wounded May/1864, Died 1/5/1919, buried Stone Mountain Cemetery, DeKalb Co.
Poss, James - Pvt 9/26/1861. Killed at Fredericksburg, VA, 12/13/1862. CSR contains death claim filed by widow, Elendor Poss, of Milton Co. BSU.
Powers, Asbury H. - (Asbury Hull Power) Pvt 9/26/1861. Captured Fisher's Hill, VA, 9/22/1864. P at Point Lookout, MD in 1864. Rcv Venus Point, Savannah River, GA for exchange, 11/15/1864. Filed CPA, Edwards Co., TX, 1907. D 4/5/1915, Bur TX State Cemetery, Confederate Field, Austin, TX.
Powers, J. W. (Joseph W. Power) - Pvt 9/26/1861. D of typhoid fever, Chimborazo Hos #2, Rich, VA, 7/13/1862. CSR contains death claim filed by father, Samuel Power, in DeKalb Co. BSU.
Powers, Patrick H. (Patrick H. Power) - Pvt 9/26/1861. Appears on list of wounded at Antietam, 9/17/1862. Died at Bunker Hill, VA, 10/9/1862. CSR contains death claim filed by father, Samuel Power, in DeKalb Co. BSU.
Powers, William H. (or Power) - Pvt 9/26/1861. Died at Staunton, VA, 8/3/1862. Buried there in Thornrose Cemetery.
Rainey, John G. - 4th Sgt 9/26/1861. Appt 2d Sgt 4/15/1862. Pvt Dec/1862. C Gettysburg, PA, July 1, or 4, 1863. P at PLO, MD and Trn to City Point, VA, for E 3/16/1864, Rcv there 3/20/1864. Widow, Sallie Rainey filed CPA, DeKalb Co., 1891. She said he D of disease PLO Prison, MD. BSU.
Rainey, John M. - Pvt 9/26/1861. Captured at South Mountain, MD, 7/3/1863. Paroled Ft. McHenry, MD, and Transferred to Ft. Delaware, 7/9/1863. Paroled at Point Lookout, MD, 2/18/1865. At Camp Lee near Richmond, VA, 2/27/1865. Died 1880 in GA. BSU.
Rankin, John G. - 2d Lt 9/26/1861. Transferred 4/1/1862 to Co. D, see Co. D.
Ray, John P. - Pvt 5/15/1862. Transferred 4/1/1862 to Co. D, see Co. D.
Ray, Robert A. - Pvt 9/26/1861. Transferred 4/1/1862 to Co. D, see Co. D.
Reeves, James Andrew Jackson - Pvt 9/26/1861. Captured near Farmville, VA, 4/7/1865 paroled there Apr. 11 - 21, 1865. Died 2/8/1902, buried Winters Chapel Methodist Church Cemetery, DeKalb Co. Widow, Martha E. "Reeves" filed CPA, DeKalb Co., 1903.
Reeves, William P. (William Posey Reeves) - Pvt 3/1/1862. Roll dated 9/1/1864, shows him "Absent on Surgeon's certificate since May 1863." Born 6/11/1822, died 10/13/1894. Buried Lebanon Baptist Church Cemetery, Roswell, Fulton Co. CSA marker reads: 38 GA INF.
Ridling, M. V. - (Martin Van Buren Ridling) Pvt 9/26/1861. Appt 3d Cpl 4/1/1862. Wounded Chancellorsville, VA, 5/3/1863. Roll dated 11/9/1864, shows him "AWOL, last heard from with an artillery company Sav GA" Filed CPA, Fulton Co., 1898. D 2/16/1913. Bur Hollywood Cem, Atlanta.
Ridling, Thomas C. - Pvt 9/26/1861. Appt 5th Cpl Aug/1863. Home, Jackson Co., on sick furlough 9/1/1864. (Born GA, 12/24/1840) Filed CPA, Fulton Co., 1900. Died 6/2/1900. Widow, Mary Jane Pirnnell Ridling filed CPA, Fulton Co., 1901. BSU.
Robinson, Robert M. - Pvt 9/26/1861. W 2d Manassas, 8/28/1862. Appt 2d Sgt 5/15/1863. Wounded & captured 1864. Exchanged, S Appomattox, VA, 4/9/1865. (Resident GA since 1850) Filed CPA, Fulton Co, 1911. Died 9/1/1932, buried West View Cem, Atlanta. Obit in Atlanta Journal, 9/1/1932.
Sellers, John A. - Pvt 9/26/1861. Appt 3d Cpl in 1862. Killed at Antietam, MD 9/17/1862. CRS contains a death claim by his mother, Jennett Sellers, of DeKalb Co. BSU.
Simpson, Robert M. - Pvt 9/26/62. Transferred 4/1/1862 to Co. D, see Co. D.

Simpson, William M. - Pvt 9/26/1861. Transferred 4/1/1862 to Co. D, see Co. D.
Singleton, William Leonard - Pvt 3/1/1862. Transferred 4/1/1862 to Co. D, see Co. D.
Smith, Tilnon P. - Pvt 9/26/1861. Transferred 4/1/1862 to Co. D, see Co. D.
Stewart, E. P. (Ebenezer P. Stewart) - Pvt 3/1/1862. Sick in Staunton, VA, hos 11/9/1864. C near Petersburg, VA, 3/25/1865. Released PLO, MD, 5/14/1865. CSR contains a letter to Officer In Charge of PLO Prison, stated he deserted at battle of Ft. Stedman and desired to take OOA. BSU.
Stewart, Thomas H. - Pvt 9/26/1861. Appt 1st Cpl in 1862; 2nd Cpl in 1862. Captured at Jericho Ford, VA, 5/23/1864. Paroled at Elmira, NY, and sent to James River, VA, for exchange, 2/20/1865. Filed CPA, Woodruff Co., AR, 1904, veteran #5766. Died 3/6/1910. BSU.
Stewart, William D. - Pvt 9/26/1861. Appt 2d Sgt in 1862; 3d Sgt 4/1/1862. C Spotsylvania, VA, 5/20/1864. D Elmira, NY 9/16/1864. Buried Woodlawn National Cem, Grave #173, Elmira, N.Y.
Tankersley, J. M. (John M. Tankersley) - Pvt 5/15/1862. Died at Lynchburg, VA, 7/18/1862. Buried there in Old City Cem. No. 2, 2d Line, Lot 161. Died of Rubella, (German measles) CSR contains death claim by widow, C. L. Tankersley, DeKalb Co.
Thomas, Thomas Jefferson - Pvt 3/20/1862. Transferred 4/1/1862 to Co. D, see Co. D.
Thomas, Wyett L. - Pvt 9/26/1861. Trn to Co. D, 4/1/1862. C Spotsylvania, VA, 5/12/1864. P Ft. Delaware, Feb/1865. E at Boulware & Cox's Wharves, James River, VA, Mar. 10-12, 1865. BSU.
Thomas, Zephaniah - Pvt 3/1/1862. Transferred 4/1/1862 to Co. D, see Co. D.
Thompson, G. F. M. - Pvt Died of typhoid fever in General Hos #l, Lynchburg, VA, 8/1/1862. Buried there in Old City Cem, No. 10, 1st Line, Lot 162, D Warwick House Hos. Full name unknown.
Thompson, Joseph Maxine - (or James M. Thompson) Pvt 9/26/1861. Wounded Gaines Mill, VA, 6/27/1862. D, pneumonia, Winder hosp, Rich, VA, 12/30/1862. Bur there Hollywood Cem, Rich VA.
Thompson, Richard Govan - Pvt 3/20/1862. Trn from Co. A, 4/1/1862. Died Winchester, VA, hospital, 10/31/1862. Buried at Mount Hebron Cemetery, Stonewall Confederate Cemetery section, Winchester, VA. Georgia Section, Plot #192. CSR contains death claim filed by widow, Lodusky C. Thompson. Widow L. C. Nash filed a CPA, DeKalb Co., 1914.
Tucker, John T. - Pvt 9/26/1861. Transferred 4/1/1862 to Co. D, see Co. D.
Verdin, William J. (or William V.) - Pvt 9/26/1861. Transferred 4/1/1862 to Co. D, see Co. D.
Waits, Newton - (or Wates) Pvt 5/15/1862. Wounded & captured Cedar Creek, VA, 10/19/1864, thigh broken, Died of wounds West's Bldgs. Hos Baltimore, MD 3/10/1865. Widow, Salina Waits, filed CPA, Gwinnett, Co., 1891. Bur Confed. Hill, Loudon Park National Cem, Baltimore City, MD.
Wallace, George W. (George Washington Wallace) - Pvt 9/26/1861. Appt 5th Sgt 1862; 4th Sgt 4/1/1862. W Antietam, MD, 9/17/1862. C Gwinnett Co., 1864, sent to Louisville, KY, took OOA to U.S. Govt., and released to remain north of Ohio River during war. D 3/18/1884 in Gwinnett Co. Widow, Mary Rebecca Wallace filed CPA, Fulton Co., 1901. Full name in pension record. BSU.
Warnock, John C. - Pvt 9/26/1861. Wounded Federicksburg, 12/13/1862. Wounded, Lynchburg, VA, 6/17/1864. Killed at Hatcher's Run, VA, 2/6/1865. Buried Hollywood Cemetery Richmond, VA. Cemetery records read: Warmack, CO. A, 38th GA, died 2/7/1865.
Warnock, Patrick - Pvt 7/31/1863. Died 8/5/1864. Widow, Elizabeth filed CPA, DeKalb Co., 1891. She stated he died on the way home on a furlough from VA, at Pike Co., GA. D of diarrhea. BSU.
Warnock, William R. - Pvt 9/26/1861. Appt 3d Sgt 4/1/1863. Wounded, disabled Gettysburg, PA, 7/2/1863. Captured, date and place not given. Exchanged Jan/1864. Retired to Invalid Corps 1/25/1864. Filed CPA, DeKalb Co. 1889. Died 12/16/1908. Widow, Martha Warnock, filed CPA, DeKalb Co., 1911. Buried New Hope Cemetery, DeKalb Co.
Wells, George R. (George Riley Wells) - Transferred 4/1/1862 to Co. D, see Co. D.
Wells, Willis Virgil - Pvt 3/1/1862. Transferred 4/1/1862 to Co. D, see Co. D.
West, John P. - Pvt 3/1/1862. Died in Staunton, VA, hospital, VA, 7/24/1862. Buried there in Thornrose Cem. CSR contains a death claim filed by widow, Elizabeth West, in Fulton Co.
Williams, William Jasper - Pvt 9/26/1861. Transferred 4/1/1862 to Co. D, see Co. D.
Williams, William W. - Pvt 5/15/1862. D in Hos 2d Corps, Orange C. H., VA, 2/8/1864. BSU.

Wilson, John Hamilton - Pvt 9/26/1861. Discharged 3/10/1862. Filed CPA, Floyd Co., 1893. Stated wounded by gunshot through left hand, June or July, 1862. Resident Georgia since 11/30/1831. BSU.
Wilson, Thomas - Pvt 9/26/1861. Absent without leave 9/22 - 11/9/1864. No later record. BSU.
Wilson, Thomas M. - (Thomas McDill Wilson) Pvt 9/26/1861. D Staunton, VA, 8/5/1862. Bur there Thornrose Cem. CSR contains a death claim filed by father, Robert M. Wilson, in DeKalb Co., GA
Wilson, R. M. - (Richard M. Wilson) 2d Cpl 9/26/1861. Appt 1st Cpl 4/1/1862. C Mine Run, VA, 5/12/1864. P Ft. Delaware, Feb/1865. E 3/7/1865. (B Jackson Co., GA, 1830) Filed CPA, Fulton Co., 1895. D 12/15/1897, buried Sandy Springs United Meth. Cemetery, Sandy Springs, Fulton Co.
Wilson, W. A. (William Alexander Wilson) - Pvt 5/15/1862. D of pneumonia Hamilton's Crossing, VA, 4/5/1863. CSR contains death claim filed by father, Robert M. Wilson, DeKalb Co. CSR contains discharge record, 3/1/1864. Buried Spotsylvania Confed. Cem. Marker reads: CO A 38 GA INF.

*MUSTER ROLL OF COMPANY B, MILTON GUARDS, MILTON COUNTY, GA.

This company was first organized by Cpt George W. McCluskey. Note: Milton Co., GA was absorbed by Fulton Co., in 1932. *This company became Co. H, reorganization.

Adams, Richard - Pvt 10/6/1861. Died of disease 9/25/1862. BSU.
Adams, Robert W. - Pvt 5/1/1862. Died College Hos Lynchburg, VA, Jan/1863, bur there in Old City Cem 1/5/1863. No. 8, 3d Line, Lot 79. Widow, Nancy Adams, filed a CPA, Fulton Co., 1891.
Adams, Salthiel - Pvt 5/1/1862. Wounded Gaines Mill, VA, 6/27/1862; Fredericksburg, VA, 12/13/1862. Captured Cedar Creek, VA, 10/19/1864. Paroled at Point Lookout, MD, transferred to Aiken's Landing, VA, for exchange 3/17/1865. Received at Boulware's Wharf, James River, VA, 3/19/1865. (In GA since 5/14/1831) Filed CPA Cherokee Co., 1890. BSU.
Adams, Thomas - Pvt 10/6/1861. Appears last on roll for 10/31/1861. BSU.
Bagwell, John Y. - Pvt 10/6/1861. W 2d Manassas, 8/28/1862. K Newtown, VA, 8/11/1864. BSU.
Bagwell, Nathan B. (Nathan Bennett Bagwell) - Pvt 10/6/1861. Appt 4th Sgt in 1862. W in left leg and C Antietam, MD 9/17/1862. P Ft. McHenry, MD, Nov/1862. Rcv City Point, VA, for E 1862. Leg amputated below knee, 1862. Absent, wounded, close of war. CSR contains discharge record. Filed CPA, Milton Co., 1879. Died11/8/1908, buried Union Hill Cemetery, Forsyth Co.
Bagwell, William G. - Pvt 10/6/1861. Killed at Winchester, VA, 6/13/1863. CSR contains death claim filed by father, Nathaniel Bagwell, Fulton Co. BSU.
Baker, William A. - Pvt 10/6/1861. Wounded at battle of Gaines Mill, VA, 6/27/1862. Mortally wounded at the battle of Fredericksburg, VA, 12/13/1862, left leg amputated close to body, died of wounds. CSR contains death claim filed by wife, Elizabeth Baker in Cherokee Co. BSU.
Barrett, John A. - Pvt 5/1/1862. Wounded and disabled in 1863. On detail duty as Sub Enrolling Officer June/1863 - 8/31/1864. BSU.
Bates, Joseph C. (Joseph Christian Bates) - Pvt 10/6/1861. Wounded Gaines Mill, VA, 6/27/1862. Wounded 1863. Captured Waynesboro, VA, 3/2/1865. Released Ft. Delaware, 6/16/1865. (Born in GA, died Alpharetta, 5/7/1920) Filed CPA, Milton Co., 1900. Buried Big Springs United Methodist Church Cemetery, Cherokee Co. CSA marker reads, CO B, 38 REGT GA INF, born 1/23/1841.
Bates, George W. - Pvt, enlisted 5/1/1862 Sav, GA, sick in Chimborazo Hos #2, 7/23/1862, Richmond, VA. Returned to duty 8/6/1862. Died 9/5/1862. BSU.
Bates, William A. - Pvt 10/6/1861. Wounded Fred., VA, 12/13/1862. AWOL Sept 1 - 11/4/1864. Filed CPA, Bartow Co., 1906. B 10/19/1843, D 8/15/1925. Bur Emerson Cem, Bartow Co.
Bell, Augustus C. - Pvt 5/15/1862. Elected Jr. 2d Lt of Co. I, 10/24/1862. Elected 1st Lt 2/25/1864. Roster dated 3/13/1865, shows 1st Lt, Present incumbent. Died 4/2/1913. Buried Ebenezer Methodist Church Cemetery, Forsyth Co. Obit in Atlanta Constitution, 4/4/1913.
Bennett, Francis M. - Pvt 3/22/1862. Died 7/12/1862. BSU.

Beshears, Alexander (or Bechears) - Pvt 5/15/1862. Died 2d GA Hos (Augusta GA) 3/6/1864. CSR contains death claim filed by father, Isaiah Beshears, in Milton Co. Bur Magnolia Cem, Augusta.
Bishop, Abner (Abner H. Bishop) - Pvt 10/6/1861. Discharged on account of chronic dyspepsia 7/28/1862. CSR contains discharge record, age 26 at discharge. Born abt. 1836. Filed a CPA, Etowah County, AL. 1899. BSU.
Bostwick, Green B. (Green Berry Bostick) - Pvt 5/1/1862. Captured Hanover, VA, 5/24/1864. Paroled at Point Lookout, MD Feb/1865. Rcv Cox's Landing, James River, VA, Feb. 14-15, 1865. Filed CPA, Tishomingo Co., MS in 1916. Died 1927 and buried there at Oak Grove Cemetery, CSA marker reads, CO B 38 GA INF.
Bruce, Baylis - Pvt 5/16/1862 Atlanta, GA, Wounded Wilderness 5/5/1864. Born 12/10/1828, died 4/9/1896, buried South Union Cemetery, Fentress, Choctaw Co., MS.
Bruce, Benjamin - Pvt 5/16/1862. Received pay Richmond, VA, 10/10/1862. BSU.
Bruce, Green B. - Pvt 5/1/1862. Captured Jericho Ford, VA, 5/22/1864. E at PLO, MD, 3/22/1865. Filed CPA, Choctaw Co., MS 1903. Bur Mount Moriah Cem, Choctaw Co., MS. CSA marker reads B 1831, D 3/22/1910, CO B 38 GA INF.
Bruce, John (John A. Bruce) - Pvt 2/27/1862. Captured Winchester, VA, 9/19/1864. Paroled at Point Lookout, MD, in 1864. Exchanged 10/29/1864. Died 10/8/1899 Ackerman, MS. Buried there Enon Cumberland Presbyterian Church. Marker reads: 38 GA INF.
Bruce, Mac V. (Mack V. Bruce) - Pvt 10/6/1861. K at battle of Antietam, MD, 9/17/1862. BSU.
Burgess, James H. - Pvt 10/6/1861. Sick in hospital Oct/1863. NLR. BSU.
Burgess, Joel (Joel Jordan Burgess) - 4th Sgt 10/6/1861, 1861. Appears last on roll for 10/31/1861. Pension records show discharged, disability, Apr/1862. (B in GA. 1837) Filed CPA, Milton Co. 1910. D 8/7/1913, bur Midway Meth. Church, Forsyth Co. Grave marker reads: Born 8/28/1836.
Burgess, Lewis - Pvt 2/27/1862. Killed at the battle of Antietam, MD, 9/17/1862. CSR contains death claim filed by widow, Catharine Burgess, in Milton Co. BSU.
Burgess, Pleasant F. (Pleasant Franklin Burgess) - Pvt 10/6/1861. Died of disease Danville, VA, 9/22/1864. Middle name found in CSR. BSU.
Burgess, William - Pvt 10/6/1861. Died Staunton, VA, 11/9/1862. Bur there in Thornrose Cem.
Butler, John K. (or J. R.) (John Kirkland Butler) - Pvt 10/6/1861. Appt 4th Cpl in 1864. C Spotsylvania, VA, 5/20/1864. Released Elmira, NY, 6/19/1865. Filed CPA, Panola Co., TX, 1909. Born 1/27/1843, died Carthage, TX, 7/16/1921, buried Six Mile Cemetery, Panola Co., TX.
Cape, William M. (William Marion Cape) - Pvt 5/16/1862. Captured Gettysburg, PA, July 3 or 4, 1863. Released Ft. Delaware, 5/10/1865. CSR states he was admitted to Farmville hos for gunshot wound to left hand, Jan/1863. B 3/29/1836, D 3/9/1903. Bur Chattahoochee Bapt. Cem., White Co.
Carter, Martin - Private 2/27/1862. K Fredericksburg, VA, 12/13/1862. CSR contains death claim filed by widow, Darcus C. Wayne Carter Holbrooks. She filed a CPA, Murray Co., 1911. BSU.
Cook, William J. - Pvt 10/6/1861. Appt 5th Sgt 2/1/1864. C Fisher's Hill, VA, 9/22/1864. Trn from PLO, MD to US General Hos 1/27/1865, released, June 26 or 28, 1865. D 9/23/1886 Milton Co. Bur Boiling Springs Primitive Bap Chur, Fulton Co. Widow, Ann Johnson Cook, filed CPA, Tift Co.
Cross, Anderson C. (Anderson Caleb Cross) - 4th Cpl 10/6/1861. Wounded Fredericksburg, VA, 12/13/1862. Pvt 1864. Wounded Deep Bottom, VA, 8/16/1864. Absent without leave Sept - Nov 4th, 1864. B 12/19/1840, D 11/22/1902, buried Macedonia Baptist Church Cemetery, Miller Co., AR.
Cross, Kinchen R. - 1st Sgt 10/6/1861. Elected Jr. 2d Lt 7/22/1862; 2d Lt 8/18/1862. C Spotsylvania, VA, 5/12/1864. Released Fort Delaware, 6/16/1865. Born 1836, died 1888. Buried Lone Star Cemetery, Sebastian Co., AR.
Cross, Levi J. - Pvt 10/6/1861. D 12/22/1862. CSR has death claim by father, Garrison Cross. BSU.
Day, Allen Bayless - Pvt 10/6/1861. K Gettysburg, PA, 7/1/1863. CSR contains death claim filed by widow, Tabitha J. Day, in Cherokee Co. Cenotaph Big Springs UMC Cem, Cherokee Co. BSU.
Day, Edgar M. - Pvt 10/6/1861. AWOL Aug 1 - Aug 31, 1864. Pension records show home on sick frl close of war. Born 1/12/1837. (Died in Cherokee Co., 3/7/1910) Buried Big Springs United Meth. Church, Cherokee Co. Widow, C. F. Day, filed CPA, Cherokee Co., 1912.

DeLaney, William Green - Pvt 10/6/1861. Trn to Co. C, Phillips' Legion GA Cavalry 4/11/1862. Roll for 10/31/1864, LOF, shows him on detail for horse by order of General Lee since 9/19/1864. Filed CPA, Cobb Co., 1898. D Fulton Co., 3/25/1911, bur there Boiling Springs Primitive Bap. Cem.
Devore, Andrew W. (Andrew Warren Devore) - Pvt 10/6/1861. Appt Regtal Quartermaster Sgt, 10/14/1861. W Winchester, VA, 8/17/1864. Frl 40 days from General Hos #9, Rich, VA, 9/17/1864. B 10/4/1839, D 8/6/1897. Bur Providence Bap. Chur, Fulton Co., Obit, Georgia Bap. newspaper, 8/17/1897. Widow, Cornelia Devore filed CPA, Milton Co., 1911.
Devore, Eldridge G. (Eldridge Gerry Devore) - Pvt 10/6/1861. Died of disease Marietta, GA, 8/7/1864. BSU.
Devore, James Edward - Pvt 5/1/1862. Wounded at the Wilderness, VA, 5/6/1864. Captured in Cherokee Co., GA, 9/12/1864. Released Camp Douglas, IL, 6/17/1865. (B South Carolina 9/12/1837) Filed CPA, Milton Co., 1901. D 12/14/1914. Buried at Cross Plains Baptist Church Cemetery. Widow, B. A. Devore, filed CPA in Milton Co. CSA grave marker reads: CO B, 38 GA INF.
Dill, James Jasper - Pvt 5/1/1862. Wounded in head Gaines Mill, VA, 6/27/1862. Wounded, hand, two fingers amputated, Gettysburg, PA, 7/2/1863. In Raleigh, NC hospital 4/13/1865. Filed CPA, Habersham Co., 1889. BSU.
Dill, John A. - Pvt 2/27/1862. AWOL in 1864. BSU.
Dinsmore, James F. (or Densmore) (James Foy Dinsmore) - Pvt 10/6/1861.W 2d Manassas, VA, 8/28/1862. Accidentally killed 12/21/1862. CSR states he was "Killed by cars" (railroad cars) CSR contains a death claim filed in Milton Co., by guardian, John G. Cantrell. BSU.
Dinsmore, Malchi (Malchi Polk Dinsmore) - Pvt 10/6/1861. Wounded Cedar Creek, VA, 10/19/1864. In CSA General Hospital Charlottesville, VA, W, 10/24/1864. Trans to Lynchburg, VA, hos 4/5/1865, there until 6/6/1865. (B in GA 1840.) Filed CPA, Tift Co. 1916. Died at Tifton GA, 6/15/1919. Buried Boiling Springs Primitive Bapt Church, Fulton Co. Marker reads: B 5/14/1843.
Dinsmore, William W. (or Densmore) (William Washington Dinsmore) - Pvt 5/1/1862. Wounded 2d Manassas, VA, 8/28/1862. Died of wound. 8/30/1862. CSR contains death claim filed by widow, Mary J. Dinsmore, in Cherokee Co. BSU.
Donald, Asa F. - Pvt 10/6/1861. Trn to Co. C, Phillips' Legion GA Cavalry 4/11/1862. S Greensboro, NC, 4/26/1865. B 5/23/1843 in SC. D 3/9/1874, bur Pine Log Cemetery, Bartow Co.
Dudley, James B. - Pvt 3/7/1863. Captured 1864. Died of disease Point Lookout, MD, 9/15/1864. Buried Point Lookout Confederate Cem, St. Mary's Co., MD, name on CSA memorial at cemetery.
Edwards, Peter I. (Peter Isham Edwards) - Pvt 10/6/1861. Discharged, tuberculosis Richmond, VA, 10/1/1862. CSR contains discharge record. Filed CPA, Forsyth Co., 1900. Born 1/27/1840, died 9/6/1910, buried Westview Cem, Atlanta.
Edwards, Thomas J. - Pvt 10/6/1861. Discharged, disability, 11/12/1861. BSU.
Emerson, Henry C. (Henry Columbus Emerson) - Enlisted as Pvt, Co. C, Phillips' Legion GA Cavalry 7/1/1861. Trn to Co. B, 38th Regt GA Inf. 5/1/1862. S Appomattox, VA, 4/9/1865. Elected as Democrat to the State Legislature in 1890, Van Burn Co., AR. Filed CPA, Van Buren Co., AR, 1902. D 6/2/1909, Van Buren Co, Bur there Emerson Cem. Obit in Van Buren Co. Democrat, 6/11/1909.
Erwin, James A. - 3d Sgt 10/6/1861. W 2d Manassas, VA, 8/28/1862. Reduced to ranks Feb/1864. W Wilderness, VA, 5/5/1864. AWOL Sept. 1 - Nov 4, 1864. NLR. CSR contains application to join "disabled" command performing light duty, July/1863, disabled, gunshot wound through feet. BSU.
Gaines, James H. F. (James Humphrey Francis Gaines) - Pvt 5/1/1862. Discharged Savannah, GA 5/27/1862. Filed CPA, Hamilton Co., TN. Pension # S6686. BSU.
Gaines, John H. - (John Henry Gaines) Pvt 2/27/1862. Admitted to Chimborazo Hos #2, Rich, VA, with hand lacerated by cars, 1/20/1863. Returned to duty 4/9/1863. Discharged 5/17/1863. D 2/15/1907. Buried Corinth Bapt. Chur. Cem, Heard Co. Obit in 2/16/1907, Chattanooga Star. (TN).
Gaines, Reuben W. (Reuben Humphrey Gaines) - Pvt 5/1/1862. W Wilderness, VA, 5/6/1864. AWOL 9/1/1864. NLR. Filed CPA, Winston & Franklin Co. AL. Filed CPA, St. Francis County, AR, 1903. D 11/14/1910, bur Fairview Cem, Winston Co., AL. Obit in Winston New Era, 11/18/1910.

Gibbs, Harvey J. - (Harvey Jasper Gibbs) Pvt 10/6/1861. Appt 1st Cpl 7/28/1862. S Appomattox, VA, 4/9/1865. Filed CPA, Etowah Co., AL, 1914. D 12/16/1927. Bur Gibbs Chapel, Blount Co., AL.
Gibbs, Robert D. - Pvt 10/6/1861. Killed at Gaines Mill, VA, 6/27/1862. CSR contains death claim filed by father, John Gibbs, Cherokee Co. BSU. Cenoptaph at Old Orange Cemetery, Cherokee Co.
Gober, John P. - Pvt 5/1/1862. AWOL Sept. 1 - Nov 4, 1864. NLR. CSR contains request for duty disabled command due to tuberculosis, July/1863. Absent previous 9 month due to the disease. BSU.
Graham, Joseph W. - Pvt 5/1/1862. Died of typhoid fever, Chimborazo Hos #2, Richmond, VA, 4/27/1863. BSU.
Griffin, William F. - 2d Cpl 10/6/1861, 1861. Killed Gaines Mill, VA, 6/27/1862. CSR contains death claim filed by father, Thomas Griffin, Milton Co. BSU.
Grogan, John W. - (John Washington Grogan) Pvt 10/6/1861. Captured High Bridge, VA, 4/6/1865. Released Newport News, VA, 6/25/1865. BSU.
Grogan, Richard W. (Richard Wilborn Grogran) - 2d Sgt 10/6/1861. AWOL and reduced to ranks 7/20/1862. C Gettysburg, PA, 7/3/1863. Took OOA to U.S. Govt. PLO, MD and enlisted in U.S. service 1/24/1864. Served in 1st U.S. Volunteers. D 4/23/1915. Bur Pine Crest Cem, Cass Co., TX.
Grogan, Thomas J. - (Thomas Jefferson Grogan) Pvt 5/1/1862. AWOL 8/1 - 11/4/1864. NLR. Filed CPA, Cass Co., TX, 1913. D 5/4/1927. Bur Salem Bap. Cem, west of Bloomburg, Cass Co., TX.
Hagood, Levi Garrison - Pvt 10/6/1861. W 2d Manassas, VA, 8/28/1862. C Chester Station, VA, 4/3/1865. Released Hart's Island, NY, 6/15/1865. (Appears also as Haygood.) Filed CPA, Cobb Co., 1888. D 7/19/1924, bur Wesley Chapel Cem, Marietta, Cobb Co. Obit in Marietta Journal 7/24/1924.
Hagood, Simpson A. (Simpson Asbury Hagood) Pvt 10/6/1861. Elected Jr. 2d Lt 12/4/1862. C Petersburg, VA, 3/27/1865. Released Ft. Delaware, 6/17/1865. (Appears also as Haygood.) (B SC 1/29/1836) Filed CPA, Gwinnett Co., 1899. D 2/3/1912, Lawrenceville, bur there Shadowlawn Cem.
Hall, Joseph H. (Joseph H. "Joe" Hall) - Pvt 10/6/1861. Appt Color Cpl 1862. Killed Spotsylvania, Courthouse, VA, 5/12/1864. BSU.
Hall, Martin - Pvt 10/6/1861. Wounded Antietam, MD, 9/17/1862. On detail duty on account of disability, in Richmond, VA, hospital, Jan. 1 - 11/4/1864. Captured and paroled Greenville, SC, 5/24/1865. Died 10/11/1880. Widow, Sarah J. Hall, filed CPA, Cobb Co., 1898. BSU.
Hardman, Bennett Mitchell (or Hardeman) - Pvt enlisted 2/27/1862 Skidaway Island, GA Roll for 11/4/1864, last on file, shows present. No later record. (Born in GA in 1824) Filed CPA, Milton Co., 1895. Died 3/4/1904, buried Providence Baptist Church Cemetery, Crabapple, Fulton Co.
Hardman, William R. (or Hardeman) - Pvt 2/27/1862. Died Lynchburg, VA, 8/1/1862. Died of pneumonia. Buried there in Old City Cem. No. 10, 2d Line, Lot 162; Died Warwick House Hospital. CSR contains a death claim filed by widow, Matilda Hardeman, in Milton Co.
Head, John J. - Pvt 5/15/1862. W Gettysburg, PA, 7/2/1863. AWOL Jan. 1 - Aug. 31, 1864 BSU.
Henson, Alfred - Pvt 2/27/1862. BSU.
Henson, William A. - Pvt 10/6/1861. Discharged 6/5/1862. Died 1915, buried Oakwood Cemetery, Dyer, Gibson Co. TN.
Herring, Elijah - Pvt 5/15/1862. Wounded Antietam, 9/17/1862. Wounded Gettysburg, PA, 7/2/1863. Absent without leave 8/31/1864. No later record. BSU.
Herring, Wiley B. - Pvt 5/17/1862. Captured Wilderness, VA, 5/6/1864. Took OOA to U.S. Govt. and enlisted in U.S. service Point Lookout, MD, 5/19/1864. Filed CPA, Paulding Co., 1899. BSU.
Holcombe, Garrison L. - Pvt 10/6/1861. Admitted to CSA General Hos Charlottesville, VA, chronic diarrhea 9/1/1864. Returned to duty 9/16/1864. CPA shows home, disabled, Nov/1864, to end of war. (B Cherokee Co., 10/7/1842) Filed CPA, Gordon Co., 1902. 11/13/1922. Bur Marietta City Cem.
Holcombe, Henry S. (Henry Snow Holcombe) - Pvt 5/15/1862. AWOL 8/31/1864. No later record. Filed CPA, Murray Co., 1908. Died 2/16/1935, buried Hopewell Bap. Church Cemetery, Gordon Co.
Holcombe, Little B. - Pvt 10/6/1861. Discharged from General Hos #1, Lynchburg, VA, on account of anchylosis of left elbow, shortness of thigh, loss left ear, 2/18/1863. CSR contains discharge record. Age 21 at discharge, Born about 1842, at Cassville, Bartow Co., GA. BSU.
Holcombe, Tillman D. (Tillman Dixon Holcombe) - Pvt 5/15/1862. AWOL 8/31/1864. No later record. Filed CPA. Cullman Co., AL, Died 11/13/1913, buried Hopewell Cemetery, Cullman Co.

Hughes, Phillip L. - Pvt 2/22/1862. Died 5/30/1862. BSU.
Hughes, Richard P. - Pvt 10/6/1861. D of pneumonia, 4/26/1862, at Augusta, GA General Hos, buried there in City Cem, resident of Walton Co. before enlistment. Remains removed to home. BSU.
Hunter, Thomas H. - Pvt 10/6/1861. Discharged on account of dropsy, from C.S.A. General Hos Farmville, VA, 10/10/1862. CSR contains discharge record. D 3/2/1902, bur Midway Meth. Church Cem, Forsyth Co. Widow, Margaret P. Hunter filed CPA, Fulton Co., 1920.
Hunter, William M. - Pvt 10/6/1861. W and disabled Wilderness, VA, 5/5/1864. Absent, W, 11/4/1864. NLR. Filed CPA 1900, Forsyth Co. D 3/16/1928, bur Midway Meth. Church, Forsyth Co.
Ingram, Henry C. (Henry Clay Ingram) - Pvt 3/1/1864. C Spotsylvania, VA, 5/12/1864. P Ft. Delaware, Feb/1865. E 3/7/1865. Rcv Boulware & Cox's Wharves, James River, VA, Mar. 10 - 12, 1865. Filed CPA Wood Co., TX, 1920. Died 5/3/1921. Buried Quitman City Cem., Wood Co., TX.
Ingram, J. A. - Pvt Appears on Register of Receiving & Wayside Hos, Rich, VA, Admitted 3/11/1865. Dispatch to Camp Lee. Full name unknown. BSU.
Ingram, Robert N. - Pvt 10/6/1861. Captured Spotsylvania, VA, 5/12/1864. Paroled Ft. Delaware, 9/28/1864, exchanged 9/30/1864. Rcv Varina, VA, 10/5/1864. Admitted Jackson Hos Richmond, VA, typhoid fever, 10/7/1864. Frl 30 days 10/13/1864. (B Cherokee Co., 6/1/1837) Filed CPA, Fulton Co., 1904. D 8/24/1910, buried Hollywood Cem, Atlanta. Obit in Atlanta Constitution, page 7, 8/26/1910.
Jameson, William T. (William Thomas Jameson) - Pvt 10/6/1861. W 2d Manassas, VA, 8/28/1862. D of wounds 9/15/1862. Bur Manassas Cem, Warrenton, Fauquier Co., VA. Cem records state he died 9/21/1862. No grave markers in cem, Federals used wooden head boards as firewood in 1863.
Justice, Alexander - Pvt 5/1/1862. Absent without leave 8/1-11/4/1864. No later record. BSU.
Justice, Allen - Pvt 10/6/1861. Wounded Gettysburg, PA, 7/1/1863. Captured Waterloo, PA, 7/5/1863. Paroled at Point Lookout, MD, 2/18/1865. Rcv Boulware & Cox's Wharves, James River, VA, Feb. 20-21, 1865. CSR states "Wounded, left at Gettysburg." BSU.
Justice, Francis M. - Pvt 5/1/1862. Roll dated 9/14/1864, shows him "Absent, special service in disabled cavalry." Filed CPA, Yell Co., AR, 8/31/1901. Died 4/20/1907. BSU.
King, James Madison - Pvt 5/1/1862. Wounded at Gettysburg, PA, 7/2/1863. Roll for 11/4/1864, shows "Absent, sick; supposed at Mount Jackson Hospital since 8/12/1864." Captured near Petersburg, VA, 3/25/1865. Released at Point Lookout, MD, 6/28/1865. BSU.
King, John - Pvt Enlisted 4/27/1862, in Cherokee Co., for 3 years or duration of war. W 2d Manassas 8/28/1862. D 9/6/1862 of W Gordonsville, VA. CSR has death claim by widow, Avarilla King. BSU.
King, Thomas (Thomas A. King) - Pvt 5/1/1862. Wounded Antietam, 9/17/1862. Wounded Gettysburg, PA, 7/2/1863. Wounded, both arms, permanently disabled Wilderness, VA, 5/6/1864. Roll for 11/4/1864, shows absent sick, Milton Co., since May 1864. (Died Tilton, 6/10/1910) Filed CPA, Whitfield Co., 1893. Buried there Tilton Community Cemetery.
Kinson, George - Appears on Register of Returned Sick in Hos (Sav, GA) since 5/4/1862. BSU.
Landrum, Abner - Pvt 10/6/1861. Wounded Gaines Mill, VA, 6/27/1862. Roll for 11/4/1864, Last on file, shows present. BSU.
Landrum, Alfred W. - Pvt 5/1/1862. Wounded, left hand and arm totally disabled, Fredericksburg, VA, 12/13/1862. Wounded Gettysburg, PA, 7/2/1863. Absent, sick, Oct/1863. CPA shows "served in Cavalry, Wounded in left eye, resulting in loss of sight." (Born GA 7/1/1843) Filed CPA, Forsyth Co., 1891 Died 4/15/1920. Buried Southview Cemetery, Cherokee Co.
Landrum, Allen - Pvt 10/6/1861. Captured Gettysburg, PA, 7/3/1863. Paroled at Point Lookout, MD, and trn to Aiken's Landing, VA, for exchange 9/18/1864. Received Varina, VA, 9/22/1864. Surrendered Appomattox, VA, 4/9/1865. Age 24, at surrender, 4/9/1865. Born abt. 1841. BSU.
Landrum, John T. - Pvt 3/1/1864. Absent without leave, Sept. - Nov. 4, 1864. NLR. BSU.
Langford, Josiah C. (Josiah Clint Langford) - Pvt 9/26/1861. Trn from Co. A, 5/15/1862. Appt 2d Sgt 7/20/1862. W Fred., VA, 12/13/1862. W & C Monocacy, MD, 7/9/1864. P at PLO, MD, In 1864. Rcv Venus Point, Sav River, GA, for E, 11/15/1864. Filed CPA, Dawson Co., 1888. D

10/4/1898. Bur Concord Bap. Church Cem, Forsyth Co. Widow, Sarah A. Langford, filed CPA, Dawson Co., 1903.

Maddox, Henry G. - Enlisted as a Pvt in "The 20th Confed. Vols." 10/6/1861. Trn to Co. B, 38th Regt GA Inf. 10/25/1861. C Gettysburg, PA, 7/1/1863. Took OOA to U.S. Govt., and enlisted in U.S. service PLO, MD, 1/24/1864. (Name not on rolls of 20th Regt GA Inf. or 20th Battalion GA Cavalry.) Enlisted in 1st U.S. Vol., deserted near Troy, NY, 8/23/1864, en-route to St. Louis, MO. BSU.

Maddox, Joseph J. - Jr. 2d Lt (Joseph Jefferson Maddox) 10/6/1861. Elected 2d Lt 7/17/1862; 1st Lt 8/18/1862. C Wilderness, VA, 5/6/1864. Released Ft. Delaware, 6/16/1865. D 1/9/1912, bur Westview Cem, Atlanta, Fulton Co., Obit in Columbus Ledger Enquirer newspaper 1/10/1912.

Martin, James W. (John M. Martin) - Pvt 5/1/1862. Wounded Gaines Mill, VA, 6/27/1862. Killed Fredericksburg, VA, 12/13/1862. CSR contains death claim filed by mother, Sarah Martin, Cherokee Co. CSR records mixed with "Jackson Martin." of same regiment. BSU.

Martin James - CSR shows severely Wounded in the thigh battle of Gaines Mill, 6/27/1862, Died at Confederate hospital, Richmond, VA, 7/5/1862. CSR contains death claim filed by widow, Susannah D. Martin, Jackson Co., Widow, Susan Martin, filed CPA, Jackson Co., 1911. BSU.

Martin, Jesse W. - Pvt 5/1/1862. W in hand, necessitating amputation of two fingers, Malvern Hill, VA, 7/1/1862. C Battle of North Anna, VA, 5/24/1864. P at PLO, MD, sent to Aiken's Landing, VA, for E 3/14/1865. Filed CPA, Cobb Co., 1889. D 5/24/1920, bur Maloney Springs Cem, Cobb Co.

Martin, John - Pvt 10/6/1861. Wounded Gettysburg, PA, 7/2/1863, Captured there 7/4/1863. Released Ft. Delaware, 5/3/1865. Filed CPA, Morgan Co., AL, 1898. Died 1914, buried Grange Hill Cem., Trinity, Morgan Co., AL. CSA grave marker. Widow, Mary Martin, filed CPA, Morgan Co., AL, 1917; pension file #19211.

Martin, Simeon - Pvt 5/15/1862. D 9/10/1862. CSR contains death claim filed by widow, Emma Martin, Gwinnett Co., She stated he died in or near the City of Frederick, MD, 9/10/1862. BSU.

McCleskey, Francis C. - Pvt 2/27/1862. Appt 1st Sgt 7/22/1862. W 2d Manassas, VA, 8/28/1862. Killed at Gettysburg, PA, 7/1/1863. Georgia State Archives contains several letters home. BSU.

McCleskey, George W. (or McLusky) (George Washington McCleskey) - Cpt 10/6/1861. W Gaines Mill, VA, 6/27/1862. D wounds 7/17/1862. Bur Oakwood Cem, Rich, VA.

McDaniel, Joseph W. R. - Pvt 5/1/1862. Wounded 2d Manassas, 8/28/1862. Wounded Gettysburg, PA, 7/2/1863. Captured 1864, exchanged, surrendered at Appomattox, VA, 4/9/1865. BSU.

McMakin, Andrew J. (Andrew Jackson McMakin) - 1st Lt 10/6/1861. Elected Cpt 7/17/1862. Resigned 8/18/1862. Found incompetent by examining board, according to officer report in National Archives. Died 3/21/1900, Parker Co., TX. Buried there Spring Creek Cemetery. Obit printed in Dallas Morning News, 3/22/1900, page 5.

McMakin, John C (John Carr McMakin) - Pvt 5/15/1862. S Appomattox, VA, 4/9/1865. Filed CPA Cobb Co., 1889. D 8/4/1889, Cobb Co. Widow, Sarah McMakin filed CPA, Fulton Co., 1896. BSU.

McMakin, William (William Jefferson McMakin) - Musician. 10/6/1861. Trn to Co. B, as a Pvt Appt Musician. S Appomattox, VA, 4/9/1865. Filed CPA, OK, 1915. D 8/28/1899 Muskogee, OK. BSU.

McNair, Larkin H. (or McNairn) - Pvt 10/6/1861. Transferred to Co. C, Phillips' Legion GA Cavalry 10/9/1862. Surrendered Greensboro, NC, 4/26/1865. BSU.

McVay, Amos H. (or McVeigh) - Pvt 5/15/1862. Appt Chaplain of the 28th GA Inf. 6/9/1863. On frl of indulgence 12/31/1863. Resigned 3/16/1865. Filed CPA Gregg Co., TX, 1916, pension #32688. Died 11/23/1921, buried Pirtle Baptist Cemetery, Rusk Co., TX.

Meaders, William M. (William McKendree Meaders) - 5th Sgt 10/6/1861. Wounded Fred., VA, 12/13/1862. Pvt in 1863. AWOL 9/1/1864. NLR, D 8/21/1876. Bur Old Orange Cem, Cherokee Co.

Moore, Elias B. (or Eliab B.) - Pvt 5/15/1862. Wounded 1863. AWOL Aug 1 - Nov 4, 1864. No later record. Died 2/13/1899, buried Providence Baptist Church Cemetery, Alpharetta, Fulton Co., GA.

Moore, James A. - Pvt 5/15/1862, enlisted Camp Kirkpatrick. Died of typhoid fever, 6/28/1862, at Charlottesville, VA. BSU.

Moore, John M. - Pvt 5/15/1862. Wounded 2nd Manassas 8/28/1862. Listed as "MIA, "Supposed K" Fredericksburg, VA, 12/13/1862." Widow, Mary J. Moore, filed CPA, Jackson Co., AL, 1893. BSU.
Neal, James R. (or Noel) (James Richard Neal) - Pvt 2/27/1862. Wounded Gettysburg, PA, 7/2/1863. Appt Sgt S at Appomattox, VA, 4/9/1865. Died 7/11/1914, buried Dyal Cem, Bradford Co., FL. Widow, Mary Maxville Neal, filed CPA in Bradford Co., FL, 1919, pension # A04516.
Neal, John W. (or Noel) - Pvt 10/6/1861. Appears last on roll for 10/31/1861. Died 11/19/1890. Widow, Sarah C. Neal, filed CPA, Forsyth Co., 1911. Bur Holbrook Campground Cem, Cherokee Co.
Newton, Andrew J. - Pvt 5/15/1862. Appt 4th Sgt Feb/1864. Wounded 2d Manassas, 8/28/1862. C Battle of North Anna, VA, 5/24/1864. Paroled Point Lookout, MD, for E 2/18/1865. Rcv Cox's Landing, James River, VA, Feb. 14 - 15, 1865. Died 11/21/1881. Buried Beech Creek Cemetery, Cass Co., TX. Widow, A. J. Newton, filed CPA, Ouachita Co, AR, 8/20/1915.
Newton, Thomas B. - Pvt 2/27/1862. Appt 1st Sgt 7/1/1863. CPA shows sick frl for 30 days Nov/1864 and couldn't reach command when frl expired. Elected Cpt, GA Militia, 1864. (B Hall Co., 11/15/1836 Filed CPA DeKalb Co., 1908. Died 8/10/1925. Buried Boiling Springs Primitive Bap. Church Cemetery, Fulton Co.
Nix, John W. - Pvt 10/6/1861. Killed Antietam, MD, 9/17/1862. CSR contains death claim filed by widow, Rebbeca E. A. Nix, in Milton Co. BSU.
Noel - See Neal.
Orr, John - Pvt 10/6/1861. Wounded Gettysburg, PA, 7/2/1863. Surrendered Appomattox, VA, 4/9/1865. Listed on Appomattox roster as teamster. Filed CPA, Washington Co., AR, 1909. D 12/28/1923, bur Little Rock National Cem, Pulaski Co., AR. Marker reads CO B, 38 GA INF.
Orr, Thomas - Pvt 10/6/1861. Died Henningsen Hospital, Richmond, VA, Nov. 1 or Dec. 5, 1862. Bur there Hollywood Cemetery. Lot unknown. Widow, H. A. Orr, filed a CPA, Walker Co., 1891.
Peek, James - Pvt CSA Hos Danville VA, 7/23/1862, "rubella." Returned to Duty 8/8/1862. BSU.
Peak, Willis L. - Pvt 10/6/1861. Roll for Feb/1863, shows "Absent, home sick." Roll dated 11/4/1864, shows AWOL since Dec 1862. NLR. Willis L. PEEK, filed CPA, Cobb Co., 1907. BSU.
Peak, Zachariah - Pvt 5/15/1862. AWOL 12/1862 -11/4/1864. NLR. D 12/23/1928. Bur Henagar Bap. Church Cem, DeKalb Co., AL. Marker reads: CO B, 38 GA INF.
Pettit, Daniel - 1st Cpl 10/6/1861. Discharged 7/28/1862. CSR contains discharge record. Filed CPA, Blount Co., AL, 1893. BSU.
Pettit, George Erwin - Pvt 10/6/1861. Discharged, disability, 11/7/1861. Enlisted as Pvt in Co. C, 1st Regt GA State Troops (Galt's) 11/16/1863. Roll dated 3/31/1864, LOF, shows present. Widow filed CPA, Blount Co., AL., D 2/11/1915, buried there at Salem Primitive Bapt. Church Cemetery.
Pettit, Levi H. - (Levy Harvey Pettit) Pvt 5/16/1862. W, arm amputated, C Gettysburg, PA, 7/2/1863. P Baltimore, MD 8/23/1863. Rcv City Point, VA, E 8/24/1863. Filed CPA, Polk Co., GA, 1879. D 8/6/1909, Polk Co., bur there Shiloh Bap. Chur. Widow, Ann Petit filed CPA, Polk Co., 1910.
Pharr, Armenius E. - Pvt 5/15/1862. AWOL Aug. 1 - 31, 1864. NLR. Filed CPA, Carroll Co., 1911. D 1920, bur Mount Zion United Meth. Church, Carroll Co., Marker incorrectly lists 30 GA INF.
Phillips, Andrew J. (Andrew Jackson Phillips) - 2d Lt 10/6/1861. W Gaines Mill, VA, 6/27/1862. Elected 1st Lt 7/17/1862; Cpt 8/18/1862. Resigned, disability, 5/27/1863. Retired, disability, 1/21/1865. Resigned due to gunshot wound in foot received at Gaines Mill, 6/27/1862. BSU.
Phillips, David A. (David Andrew Phillips) - Pvt 10/6/1861. Appt 2d Cpl 7/22/1862. C near Petersburg, VA, 3/25/1865. Released PLO, MD, 6/16/1865. D 3/20/1887. Bur Byrd Cem., Blount Co., AL, CSA grave marker reads, CO B, 38 GA INF.
Phillips, Reuben P. - Pvt 10/6/1861. C Cedar Creek, VA, 10/19/1864. E. C Petersburg, VA, 4/3/1865. Released at Newport News, VA, 6/14/1865. BSU.
Pierce, Caleb W. - Pvt 5/15/1862. On hos duty Lynchburg, VA, on account of disability, Aug 1862 - 1/19/1865. Frl from Lynchburg, VA, 1/19/1865. Bur Mt. Zion Cem, Oscarville, Forsyth Co. Marker reads: 38 GA INF.

Powell, James C. - Pvt 2/27/1862. W left arm, right leg, hip, skull fractured, and C Gettysburg, PA. 7/1/1863. Sent from DeCamp General Hos, NY Harbor, to City Point, VA, for E, 8/28/1863. Detailed as Sub Enrolling Officer in Fayette Co., GA, on account of disability, 1863 - 11/4/1864. Filed CPA, Bartow Co., 1895. D 3/15/1919, bur Myrtle Hill Cem, Rome.
Powell, James P. - Pvt 10/6/1861. W 2d Manassas, VA, 8/28/1862. Absent, W, 8/31/1863. BSU.
Powell, Peter R. - Pvt 10/6/1861. W Gettysburg, PA, 7/2/1863. AWOL Sept. - Nov. 4, 1864. Filed CPA, Lincoln Co. TN. D 2/14/1923. Bur Campbell Cem, Lincoln Co., TN.
Powell, William R. - Pvt 2/27/1862. Captured at Ft. Stedman, VA, 3/25/1865. Released Point Lookout, MD, 6/16/1865. BSU.
Pruitt, Calvin - Pvt 2/27/1862. Captured Spotsylvania, VA, 5/19/1864. Released Elmira, NY, 5/29/1865. Never returned home to Milton Co., Georgia, fate unknown. BSU.
Pruitt, Jonathan L. - Pvt 10/6/1861. AWOL 8/31/1864. Filed CPA, Cherokee Co. D 1891, buried Pleasant Hill Bap. Union Hill, Cherokee Co.
Pruitt, Zilman - Pvt 3/22/1862. Discharged 11/7/1862. D of tuberculosis in Henningsen Hos Rich, VA, 7/25/1863. CSR contains discharge record. CSR contains a death claim filed by widow, Mary D. Pruitt, Milton Co. Bur Hollywood Cem records show Pruitt, L., Company B, 38 GA, D 4/26/1863.
Pugh, Francis C. (Francis Ellison Pugh) - Pvt 5/15/1862. Died 2/15/1863. Buried Confed. Cemetery, Fredericksburg, VA, Section 11, Row 3, Grave 5. Marker cracked in half, repaired.
Reaves, John A. (or John A. Reeves) - Pvt 10/6/1861. Received pay 8/29/1864. NLR. BSU.
Reaves, Joseph W. - Pvt 5/16/1862. C Waterloo, PA 7/5/1863. P Ft. McHenry, MD, sent to Ft. Delaware, 7/9/1863. Roll for Aug/1864, shows "Absent, POW." W 2d Manassas, 8/28/1862. BSU.
Reese, Alfred C. (or Reece) - Pvt 10/6/1861. C Spotsylvania, VA, 5/20/1864. P Elmira, NY, 3/2/1865. CSR shows he appeared on register, Wayside Receiving Hos Or General Hos #9, Rich, VA, 3/10/1865, D there the next day, chronic diarrhea. Bur Oakwood Cem, Rich, VA, cem records read: Pvt Alfred C. Reese, Co. B, 38th GA.
Reese, Jeremiah M. (Jeremiah M. Reece) - Pvt 10/6/1861. AWOL July - Nov. 4, 1864. Pension records show he was at home, sick with fever, close of war. (D Cherokee Co., GA, 1920) Filed CPA Cherokee Co., 1903. D 9/1/1920, bur Hickory Flat Church Cem, Cherokee Co.
Reese, Reuben P. (or Recce) (Reuben P. Reece) - Pvt 5/25/1863. S Appomattox, VA, 4/9/1865. D 6/10/1927. bur at Lebanon Cem, Prentiss Co., MS. CSA grave marker incorrectly lists unit as CO F, AL. CAV. Obit printed 6/17/1927, Booneville Independent, Prentiss Co., MS.
Reynolds, Eldridge J. (Eldridge James Reynolds) - Pvt 10/6/1861. Det on gunboat Sav, GA in 1862. K Marye's Heights, VA, 5/4/1863. CSR contains a death claim filed by widow, Martha E. Reynolds, in Milton Co. D in 1863, memorial marker at Hood Cem, Blount Co., AL. BSU.
Rogers, Thomas N. - Pvt 2/27/1862. C in 1864. D of disease PLO, MD, 9/1/1864. Bur PLO Confed. Cem, St. Mary's Co., MD. Widow, R. C. Rogers, filed CPA, Milton Co., 1891.
Sims, James A. (or Simms) - Pvt 10/6/1861. Wounded Antietam, MD 9/17/1862. Surrendered Appomattox, VA. Letter by Lt Simpson Hagood, 10/12/1864, states he was home on frl, captured by the Federals, Spent two months as POW, exchanged, arrived back camp on 10/11/1864. On roll of CSA veterans, The Daily Ardmoreite, Ardmoreite, OK, 7/25/1902. BSU.
Sims, William M. (or Simms, or Syms) - Pvt 5/15/1862. Wounded Fredericksburg, VA, 12/13/1862. AWOL 8/31/1864. No later record. CSR states "gunshot wound to right hand on 5/6/1863." Died 11/15/1882. Buried Pugh Family Cemetery, Cherokee Co.
Strange, Richard H. - Pvt 5/1/1862. Sick or W in Camp Winder General Hosp, 2nd Div, Rich, VA, 5/4/1863. AWOL Jan. 1 - Nov. 4, 1864. NLR. Filed CPA, Floyd Co., 1895. BSU.
Thaxton, W. H. (or Thackston) - Pvt. On register of Effects of Deceased Soldiers, turned over to Quartermasters, C.S.A., reportedly D in Chimborazo Hos, Rich, VA, 12/10/1862. BSU.
Tillotson, James W. (James W. Tillison) - Pvt 10/6/1861. W Antietam, MD, 9/17/1862. S Appomattox, VA, 4/9/1865. Filed CPA, Pickens Co., 1901, D 1/4/1911, bur Liberty Bap. Church Cem, Gordon Co. Name listed as TILLISON in pension record and on gravestone.
Tillotson, John J. - Pvt 10/6/1861. Appt 3d Cpl in 1864. S Appomattox, VA, 4/9/1865. BSU.

Tucker, Noah - Pvt 10/6/1861. W in 1862. Discharged on account of gunshot wound from General Hos Camp Winder, Rich, VA, 10/16/1862. CSR contains discharge record. BSU.
Tucker, Richard M. (Richard Madison Tucker) - Pvt 5/15/1862. AWOL Jan. 1 - Nov. 4, 1864. NLR. Filed CPA Milton Co., 1910. D 2/18/1918. Bur Hopewell Bap. Cem., Fulton Co.
Tucker, Samuel L. (Samuel Lewis Tucker) - Pvt 3/22/1862. Admitted to 4th Div General Hos Camp Winder, Rich, VA, 10/1/1862, discharged there 10/14/1862. CSR contains discharge record. D 1879. Bur Brooksville, Blount Co., AL. CSA marker reads: 38 REGT GA VOLS.
Turner, James M. - Pvt 10/6/1861. Admitted to General Hos #9, Rich, VA, 3/10/1865, and Transferred to Camp Lee near Rich, VA, 3/11/1865. S Appomattox, VA, 4/9/1865. BSU.
Turner, John H. - Pvt 10/6/1861. Appt 3d Sgt Feb/1864. C Mine Run, VA, 5/12/1864. P Ft. Delaware, Feb/1865. E 3/7/1865. W 2d Manassas, 8/28/1862. BSU.
Turner, William J. - Pvt Discharged, disability, 7/28/1862. (B in NC) CSR contains discharge record. Filed CPA, Milton Co., 1898. D 9/4/1905. Bur Mount Pisgah Meth. Church, Fulton Co. Widow, Mary Turner, filed CPA Milton Co., 1919.
Vinson, George W. - Pvt 5/15/1862. K in action the battle of Gaines Mill, 6/27/1862. CSR contains death claim filed by mother, Elizabeth Martin, of Jackson Co. BSU.
Walker, John A. - Pvt 10/6/1861. Wounded, left hand disabled Gaines Mill, VA, 6/27/1862. Roll for 8/31/1864, shows Absent, Sub Enrolling Officer for Cherokee Co., GA, by order Medical Examining Board, since Jan/1863. Filed CPA, Milton Co., 1890. Died 3/26/1911, Fulton Co., buried there Rest Haven Cemetery, Alpharetta, Fulton Co.
Walker, Joseph W. (Joseph Wesley Walker) - Pvt 10/6/1861. W in left hand, necessitating amputation of third finger, Fred., VA, 12/13/1862. W Gettysburg, PA, 7/2/1863. At home disabled by wounds close of war. Descendant states that he was wounded at Fredericksburg on 12/13/1862 and was at home for the rest of the war, not wounded at Gettysburg. Filed CPA, Milton Co., 1889. D 3/20/1915. Bur Providence Bapt. Chu Cem, Birmingham, GA. Marker reads CO B, 38 GA INF.
Wayne, Henry A. - Pvt 10/6/1861. Wounded Antietam, 9/17/1862. Wounded Fredericksburg, 12/13/1862. AWOL Aug. - Nov. 4, 1864. No later record. BSU.
Westbrook, Charles H. - Pvt 5/1/1862. Wounded 1862. Died of wounds in General Hos, Camp Winder, Rich, VA, 12/22/1862. Buried there in Hollywood Cemetery, buried in Section S, grave #344. CSR contains death claim filed by widow, Rebecca Westbrook, Cherokee Co. Cenotaph at Subligna Methodist. Church Cemetery, Chattooga Co., GA.
Westbrook, Reuben T. - Pvt 2/27/1862. Captured Fisher's Hill, VA, 9/22/1864. Paroled at Point Lookout, MD and Transferred to Aiken's Landing, VA, 3/17/1865. Rcv Boulware's Wharf, James River, VA, for E, 3/19/1865. (Born GA 2/22/1841. Died in Cherokee Co., GA, 1919) Filed CPA, Cherokee Co., 1911, buried there Union Hill Methodist Church Cemetery.
Westbrook, Thompson - Pvt 10/6/1861. No later record. BSU.
Westbrook, Talmon H. (Tillman Howard Westbrook) - Pvt 10/6/1861. C Mine Run, VA, 5/12/1864. Released Ft. Delaware, 5/10/1865. Filed CPA, Levy Co. FL, 1921, pension #A04693. D 1/24/1920, in Levy Co, FL, bur there Orange Hill Cem. Widow, Mary Westbrook filed CPA, Levy Co., FL, 1920.
Westbrook, Wesley E. - Pvt 10/6/1861. AWOL Aug. - Nov. 4, 1864. Filed CPA, Fulton Co., 1893. CPA show he "Contracted rheumatism when forced to wade Rapidan River in Virginia. Sent to hospital, furloughed home, unable to return." Died 9/16/1910, buried Bethlehem Cem., Forsyth Co.
Williams, Alfred - Pvt 6/1/1862. Absent without leave, Dec/1862. Deserted, took Oath of Allegiance to U.S. Govt. at Chattanooga, TN, 4/16/1864. BSU.
Wilson, Hilliard J. - 3d Cpl 10/6/1861. Reduced to ranks and left as nurse Gettysburg, PA, 7/3/1863. Captured in 1863. Paroled Sep/1863. AWOL May 1864 - Aug. 31, 1864. No later record. BSU.
Wood, Thomas J. - Pvt 10/6/1861. Appt Musician. Captured Gettysburg, PA, 7/5/1863. CSR states he was left at Gettysburg as a nurse. Paroled DeCamp General Hos, David's Island, NY Harbor, 8/24/1863. Received at City Point, VA, exchanged 8/28/1863. AWOL May - 11/4/1864. BSU.

Woodliff, Thomas J. (Thomas Jefferson "Jeff" Woodliff) - Pvt 5/17/1862. Wounded Antietam, 9/17/1862. Killed at Fredericksburg, VA, 12/13/1862. Buried Confederate Cemetery, Spotsylvania Courthouse, VA. Marker reads: CO B, 38 GA INF.

*MUSTER ROLL OF COMPANY C, BEN HILL GUARDS, BULLOCH & EMANUEL COUNTIES, GA

*This company was also known as Old Co. A, New Co. D, and Co. M. This company was raised and organized by Captain William L. McLeod. It was originally called "McLeod Artillery," later "Ben Hill Guards." It was ordered to be permanently detached from the 38th Georgia 5/6/1864, by Special Order #105, but this order was never executed and the company remained with the 38th Georgia until the end of the war.

Allen, W. Y. (William Y. Allen) - Pvt 5/8/1862. W Gaines Mill, VA, 6/27/1862. C Gettysburg, PA, 7/3/1863. P at PLO, MD 4/27/1864. Rcv City Point, VA, for E 4/30/1864. D disease Richmond, VA, May/1864. Bur there Oakwood Cem, records reads W.T. Allen, 38th GA, Div F, Row L, Grave 10.
Anderson, Calvin - Pvt 10/1/1861. Discharged in 1862. CSR states "Sick with measles" during Oct 1861. Filed CPA, Emanuel Co., 1910. Born April/1843. BSU.
Atkinson, Alexander (or Adkinson) - Pvt 10/1/1861. Wounded Shepherdstown, WV, 10/1/1862. Died Staunton, VA, 12/15/1862. CSR states Wounded Boteler's Ford, VA, 9/20/1862. BSU.
Baum, H. J. - Buried Hollywood Cem, Richmond, VA, records read Company C, 38th, Ga., D 5/4/1864, bur in section I, grave#136. First name unknown, CSR not found.
Bennett, J. P. (Jesse P. Bennett) - Pvt 10/1/1861. Wounded Antietam, MD, 9/17/1862. On detail duty, hos nurse Feb. - Dec. 1863. NLR. CSR contains discharge record with first name. BSU.
Boyd, Abraham - Pvt 10/1/1861. Appt 4th Cpl in 1862. K 2d Manassas, VA, 8/28/1862. CSR states 22 years old, born in Burke Co., GA. Appears on CSA causality list for 2nd Manassas, 8/28/1862, as wounded. Captured, date/place unknown, Died 7/28/1863 in Frederick, MD, U.S.A. Hospital, from flesh wound to right arm, probably wounded at Gettysburg. Resident Emanuel Co., GA, before enlistment. Buried Mt. Olivet Cem, Frederick, MD, marker reads CO. C, 38 GA INF.
Boyd, Benjamin H. - Pvt 10/1/1861. W Gettysburg, PA, 7/3/1863. C Boonesboro, MD, 7/5/1863. D of wounds in General Hos Frederick, MD, 7/28/1863. Bur there Mount Olivet Cem, Marker reads CO. C, 38 GA INF, CSR states W Gettysburg, left there. Cenotaph Waynesboro Confed. Cem, Burke Co.
Boyd, James - Pvt 10/1/1861. Wounded Gaines Mill, VA, 6/27/1862. Captured Cedar Creek, VA, 10/19/1864. Paroled at PLO, MD, trn to Aiken's Landing, VA, for E 3/28/1865. Bur Cool Springs Church Cem, Candler Co. Cenotaph at Waynesboro Confed. Cem, Burke Co.
Boyd, John - Pvt 1/1/1862. K Gaines Mill, VA, 6/27/1862. CSR contains a death claim filed by mother, Patience Boyd, in Emanuel Co. Cenotaph in Waynesboro Confed. Cem, Burke Co. BSU.
Braddy, James (or Joseph) (James Robert Brady) - Pvt 5/1/1862. Died 11/15/1862. Bur Moxley Cem, Emanuel Co.
Brinson, James D. (James Daniel Brinson) - Pvt 10/1/1861. K Spotsylvania, VA, 5/12/1864. BSU.
Campbell, Isla - Pvt 10/1/1861. Died of typhoid fever Camp Winder, Richmond, VA, 7/22/1862. Buried Hollywood Cemetery, Richmond, VA, buried in Soldiers Section Q, Lot 146.
Chew, John C. (John Cottle Chew) - Pvt 10/1/1861. W 1862. Appt Regtal Sgt Major 7/20/1862; Quartermaster Sgt, trn to Co. G, as Pvt 12/18/1863. Rcv pay Richmond, VA, 4/26/1864. W Antietam, according to CSR. Bur Green Fork Cem, Rich Co., GA. CSA marker reads: D 3/16/1890, 38 GA INF.
Coleman, Isaac W. (Isaac Washington Coleman) - 2d Cpl 10/1/1861. Appt 3d Sgt in 1862. Captured Fredericksburg, VA, 12/13/1862. Paroled for exchange 12/17/1862. Sick frl Jan - Apr/1863. Reduced to ranks 1863. AWOL Apr 1863 - Nov 7, 1864. BSU.
Coleman, Mathew (Mathew M. Coleman) - Pvt 10/1/1861. Appt 1st Cpl in 1862. Killed Antietam, MD, 9/17/1862. CSR contains a death claim filed by father, Elisha Coleman, Emanuel Co. Wounded

& captured Sharpsburg 9/17/1862. Federal records state he died White House Hospital near Sharpsburg & also Line's Farm Hospital. Bur at Washington Confed. Cem, Hagerstown, MD.
Cooksey, Thomas B. (or Cocksay) - Pvt 10/1/1861. Slightly Wounded right breast Cedar Creek, VA, 10/19/1864. Absent, W, 11/7/1864. P Augusta, GA, 5/18/1865. Died Coffee Co., 4/23/1899, buried there Meeks Cemetery. Widow, Eliza Ann. Anderson, filed CPA, Coffee Co., 1902.
Corbin, Nelson - Pvt 10/1/1861. Discharged, disability, Apr/1862. Enlisted as a Pvt in Claghorn's Company, Chatham Siege Artillery, 2/2/1863. Roll for Feb/1865, LOF, shows present. B 7/10/1846, D 5/19/1896. Bur Corbin Cem, Emanuel Co. Widow, Nancy Corbin, filed CPA, Emanuel Co., 1912.
Corbin, Silas S. - Pvt 10/1/1861. Killed at Gaines Mill, VA, 6/27/1862. BSU.
Curl, John - Pvt 10/1/1861. Died in Richmond, VA, hospital, 7/11/1862. BSU.
Curl, Mathew (Mathew Curl Jr.) - Pvt 10/1/1861. Appt 4th Cpl 1/12/1864. W Cedar Creek, VA, 10/19/1864. D of wounds Mount Jackson, Shenandoah Co VA, 9/18/1864. Bur there "Our Soldiers Cem," Date of wounding is after death date. Widow, Missouri Curl, filed CPA, Emanuel Co., GA, in 1891. She stated in CPA he was shot in the left shoulder the battle of the Wilderness (5 - 7 May 1864)
Curl, Peter - Pvt 10/1/1861. Wounded Gaines Mill, VA, 6/27/1862. Surrendered Appomattox, VA, 4/9/1865. Appears on list of Wounded for battle of Winchester, VA, 6/13/1863. Filed CPA, Appling Co., 1910. Died 7/18/1921 in Bacon Co., buried Rose Hill Cem, Bacon Co.
Curl, Reuben - Pvt 10/1/1861. Sick in hos Apr/1862. Filed CPA, Emanuel Co., 1897. Died there 12/27/1911. BSU.
Daniels, Benjamin E. - Pvt 10/1/1861. W Gaines Mill, VA, 6/27/1862. Roll for 9/1/1864, shows Absent, Det for hos duty Augusta, GA, 8/1864." Discharged for flesh wound to hip/leg. Filed CPA, Richmond Co., 1889. D 11/24/1892. Bur at Magnolia Cemetery, Augusta, Richmond Co. GA.
Daniels, Freeman H. - Pvt 10/1/1861. Transferred to Co. D, 48th Regt GA Inf. 4/6/1863. Wounded Petersburg, VA, 6/22/1864. Roll 10/31/1864, last on file, shows present. (Born in Burke Co., GA, 7/18/1832) Filed CPA, 1895, Richmond Co. BSU.
Davis, William Jackson - Pvt 10/1/1861. C Wilderness, VA, 5/6/1864. P Elmira, NY and sent to James River, VA, for E 3/14/1865. Rcv Boulware's Wharf, James River, VA, Mar. 18-21, 1865. D 3/22/1890, bur Davis Family Cem, Emanuel Co. Widow, Philena Davis, filed CPA, Emanuel Co.
Dix, James F. (James Frederick Dix) - Pvt 10/1/1861. W Winchester, VA, 9/19/1864. C Ft. Stedman, VA, 3/25/1865. Released PLO, MD 6/26/1865. (B Houston Co., GA, 11/6/1844. D Fitzgerald, Ben Hill Co., 10/28/1910) Filed CPA, 1909, Ben Hill Co. Bur Old Vidalia (McMillan) City Cem, Toombs Co. Marker reads, CO C 38 GA INF. Obit in Fitzgerald Enterprise newspaper, 10/29/1910.
Dix, John H. - (Or John H. Dicks) Pvt 10/1/1861. Appt 2d Cpl in 1862. Pvt May/1863. Appt 3d Cpl Aug/1863; 2d Cpl Jan/1864. C Port Republic, VA, 9/27/1864. Took OOA to U.S. Govt. and enlisted in U.S. service PLO, MD, 10/17/1864. Enlisted in Co. D, 2nd U.S. Volunteers. BSU.
Dix, W. R. (Or W. R. Dicks) - Pvt 10/1/1861. Captured at Spotsylvania, VA, 5/20/1864. Released Elmira, NY, 6/14/1865 BSU.
Douglass, D. D. (David Daniel Douglas) - Pvt 5/8/1862. Wounded & captured Gettysburg, PA, 7/2/1863. Transferred to Provost Marshal 7/14/1863. Admitted Hammond USA General Hospital, Point Lookout, MD, 1/12/1864. Died of gunshot fracture to right leg, 3/10/1864. Buried Point Lookout Confederate Cemetery, MD. Cenotaph at Moxley Family Cemetery, Emanuel Co.
Dudley, James - Appt 2d Cpl of Co. H, 48th Regt GA Inf. 3/4/1862; 5th Sgt 6/26/1862. W Gaines Mill, VA, 6/27/1862. Appt 4th Sgt Feb/1863; 3d Sgt May/1863. Trn to Co. C, 38th Regt GA Inf. in 1863. Captured Winchester, VA, 9/19/1864. Exchanged Point Lookout, MD, 1/17/1865. At Camp Lee near Richmond, VA, 1/25/1865. No later record. Bur Strange Cem #2, Emanuel Co. GA
Dudley, James W. - Pvt Killed at 2d Manassas, VA, 8/28/1862. CSR contains death claim filed by father, James Dudley, Emanuel Co. BSU.
Durden, Norris N. (Norris Nathan Durden) - Pvt 10/1/1861. Appt 3d Sgt 1/12/1864; 2d Sgt May/1864. Captured Farmville, VA, 4/6/1865. Released Newport News, VA, 6/25/1865. Filed CPA, Emanuel Co., 1910. Born 10/20/1839, died 3/4/1933. Buried Durden Cemetery, Emanuel Co.

Faircloth, Benjamin - Pvt 10/1/1861. Surrendered Appomattox, VA, 4/9/1865. Filed CPA, Emanuel Co. 1902. Died 11/17/1915, Emanuel Co. Buried there Ebenezer Methodist Church Cemetery.
Faircloth, George W. - Pvt 10/1/1861. Died near Port Royal, VA, 1/8/1863. BSU.
Faircloth, James - Pvt 10/1/1861. Wounded at Gaines Mill, VA, 6/27/1862. Wounded at 2d Manassas, 8/28/1862. AWOL Aug 8 - Sept 1, 1864. BSU.
Flanders, John R. (John Rushing Flanders) - Pvt 10/1/1861. W Gaines Mill, VA, 6/27/1862; Fred., VA, 12/13/1862. Appt 5th Sgt Feb/1863. Trn to Co. H, 48th Regt GA Inf. Sept/1863. Appt 3d Sgt Sept/1863. W Riddle Shop, VA, 6/13/1864. C Swainsboro, 11/29/1864. Released PLO, MD, 6/27/1865. (B Emanuel Co., GA 5/6/1843) Filed CPA Emanuel Co., 1909, D there 6/2/1915, bur Swainsboro City Cem. Obit in Augusta Chronicle 6/4/1915, page 2. Marker reads: SGT 48 GA INF.
Grant, William J. - Pvt 10/1/1861. K 2d Manassas, VA, 8/28/1862. D 9/13/1862 Middleburg, VA. B in 1839. Widow, Catharine Grant Hall, filed CPA, Montgomery Co., 1918. BSU.
Gregor, Sherrod - Pvt Appears in CSR on a register of deceased soldier's claims filed for settlement. D 8/18/1862. CSR contains death claim filed by mother, Mirtha Gregor, Emanuel Co. BSU.
Gregory, Ephraim - Pvt 5/8/1862. D 8/1/1862. Bur in Confed. Cem Charlottesville, D of typhoid fever. Bur UVA Confed. Cem, Charlottesville, VA. Soldier's names listed on memorial at cem.
Griffith, William S. - Pvt 10/1/1861. D 8/8/1862. Bur in Confed. Cem Charlottesville, VA. D of typhoid fever. Bur UVA Confed. Cem, Charlottesville, VA. Soldier's names on cem. memorial.
Hamilton, J. H. - Pvt 10/1/1861. Absent, sick, Aug/1863 -11/7/1864. Full name unknown. BSU.
Hampton, Robert C. - Pvt 10/1/1861. Appt pioneer in 1863. Roll for 11/7/1864, LOF, shows him present. NLR. Reported mortally W Fred., 12/13/1862, in newspaper list of casualties, 1/7/1863. BSU.
Harrell, John W. - Pvt 10/1/1861. Absent, sick, Jun/1864. C near Atlanta, GA, 8/28/1864. D Camp Douglas, IL, 12/16/1864. Widow, Elizabeth Harrell, filed CPA, Emanuel Co., 1910. She stated he D 2/27/1879, Emanuel Co. No mention in pension application of being captured. BSU.
Harrell, Robert - Pvt 10/1/1861. Appears last on roll for 12/31/1861. B 1835 in Emanuel Co. BSU.
Harrington, (or Herrington) Drury S. (Drury Stealy Herrington) - Pvt 10/1/1861. Wounded at Winchester, VA, 6/13/1863. Absent without leave 10/28 - 11/7/1864. BSU.
Harrington, James W. (James W. Herrington) - Pvt 10/1/1861. Killed Winchester, VA, 6/13/1863. Cenotaph at Herrington Family Cem, Nunez, Emanuel Co. Bur in Winchester, VA, specific BSU.
Hays, James (or Hayes) - Pvt 10/1/1861. W in 1863. Det for hos duty Lynchburg, VA, Jan/1864, and P there 4/13/1865. Appears on list of casualties as W, battle of Winchester, VA, 6/13/1863. BSU.
Hays, John (John Hayes) - 4th Sgt 10/1/1861. Appt 3d Sgt 10/12/1861. Pvt in 1862. Wounded 2d Manassas, VA, 8/28/1862. Captured at Cedar Creek, VA, 10/19/1864. Prisoner of war, 11/7/1864. Filed CPA, Montgomery Co., 1902. Spelled name "Hayes" in CPA. BSU.
Hays, Leroy (or Hayes) - Pvt 10/1/1861. Appt 2d Cpl in 1862. Died Lynchburg, VA, Buried there in Old City Cemetery, No. 2, 3d Line, Lot 162. Died at Cooledge Hospital.
Higdon, Ebenezer G. (or Rigdon) - Enlisted as a Pvt, Co. F, 4th Regt GA Inf. 4/29/1861. Appt 2d Sgt of Co. C, 38th Regt GA Inf., May 1863; 1st Sgt 8/1863. Surrendered at Appomattox, VA, 4/9/1865, 21 years old at surrender. Died 1865. Buried at New Hope Cemetery, Terrrell Co., GA.
Higdon, John B. - Enlisted as a Pvt, Co. F, 4th Regt GA Inf. 4/29/1861. Elected Jr. 2d Lt of Co. G, 28th Regt GA Inf. 8/2/1861. Resigned 2/12/1862. Enlisted as a Pvt in Co. C, 38th Regt GA Inf. 4/16/1862. Elected 2d Lt 11/1/1862; Cpt Jul/1863. W and disabled Winchester, VA, 9/19/1864. P Albany, GA, 5/15/1865. Died 9/1/1885, Lampasas, TX, buried there Oak Hill Cemetery. Widow, Martha R. Higdon filed CPA, Lampasas, TX, 1909.
Holton, Daniel (or Holden) - Pvt 10/1/1861. D from poison Orange Court House, VA, 5/4/1864. CSR states he D of disease. Widow, Martha Holton, filed CPA, Emanuel Co., 1891. She said he became ill after eating wild poisonous vegetables, carried to Rich, VA; Died in a hos a few days later. BSU.
Hooks, Joseph W. - Pvt 10/1/1861. Killed at Winchester, VA, 6/14/1863. BSU.

Johns, William - Pvt 10/1/1861. Appears on register of General Hos Camp Winder, Rich, VA, for 4/28/1864, with remarks "To report to Surgeon Chambliss for duty." On 7/22/1864, Rich, VA, applied for transfer from Winder Hos to Augusta, GA, for hos duty. NLR. CSR has discharge record. BSU.
Johnson, Henry J. - Pvt 10/1/1861. Died at Camp Mercer, Sav, GA, Apr. 1 or 3, 1862. BSU.
Johnson, Jasper - Pvt 10/1/1861. Died Richmond, VA, Jul 6 or 7, 1862. Buried there in Hollywood Cemetery. Buried in Soldiers Section M Lot: 373.
Jordan, Thomas F. - Enlisted as Pvt, Co. D, 48th Regt GA Inf., 3/4/1862. Trn to Co. C, 38th GA 4/6/1863. Wounded Monocacy, MD 7/9/1864, captured there 7/10/1864. Exchanged Ft. McHenry, MD, 10/30/1864. Received Venus Point, Savannah River, GA, 11/15/1864. (Born in 1844) Filed CPA, Marion Co., MS, 1922. Buried Bartow City Cemetery, Jefferson Co., GA.
Kersey, J. A. (John A. Kersey) - Pvt 10/1/1861. K Gaines Mill, VA, 6/27/1862. BSU.
Kersey, S. J. (Solomon J. Kersey) - Pvt 10/1/1861. Wounded Gaines Mill, VA, 6/27/1862. Roll for 11/7/1864, LOF, shows present. D 9/12/1910. Bur Lamb's Cem, Emanuel Co., GA CSA marker.
Lewis, Joshua K. - Pvt 10/1/1861. Appt 4th Sgt in 1862. Pension records show elected Sheriff of Emanuel Co., GA, and discharged, 10/21/1863. (B in GA 1839) W Ox Ford, VA, 8/31/1862. CSR has discharge. Filed CPA, Emanuel Co., 1916. Bur Lewis-Canady Cem, Emanuel Co. CSA marker.
Lumpkin, Joseph (Joseph S. Lumpkin) - Pvt 10/1/1861. Absent without leave, 10/1/1863. CSR states discharged, 12/24/1863 due to disability, discharge record in CSR. BSU.
McLean, John - Pvt 10/1/1861. Killed at 2d Manassas, VA, 8/28/1862. BSU.
McLean, P. H. (P. Hugh McLean) - Pvt 10/1/1861. Killed at Gaines Mill, VA, 6/27/1862. CSR contains a death claim filed by his father, Hugh M. McLean, in Emanuel Co. BSU.
McLean, William - Pvt 10/1/1861. W in hand, necessitating amputation of two fingers, Seven Pines, VA, 5/31/1862. CPA shows discharged in 1862. (B in GA Feb/1841) Filed CPA Tattnall Co., 1899. Accidently W by gunshot traveling from the Army to a hos. Bur Edward Findley Cem, Emanuel Co.
McLeod, William L. - Cpt 10/1/1861. K Gettysburg, PA, 7/1/1863, shot through the head, during charge against Barlow's Knoll. Remains returned to GA after the war and bur in McLeod family cem, Emanuel Co. Marker reads Lt. Col. Wm. L. McLeod, K Gettysburg, PA, 7/1/1863. Obit in Augusta Constitutionalist 9/5/1863. CSR contains a death claim filed by father, Neill McLeod in Emanuel Co.
Meeks, Elisha - Pvt 10/1/1861. Killed Spotsylvania, VA, 5/10/1864. Bur Confederate Cemetery, Spotsylvania Courthouse, VA. CSA marker reads: CO. C 38 GA INF.
Meeks, John - Pvt 10/1/1861. D Rich, VA 8/15/1862. Bur there Hollywood Cem., Sect. A, grave #74.
Morris, B. T. (Benjamin T. Morris) - Pvt 5/8/1862. Appt Cpl; 3d Sgt May/1863. Elected Jr. 2d Lt 8/14/1863; 2d Lt 1/12/1864. Roster of officers dated 3/13/1865, shows him present. Filed CPA, Toombs Co., 1920. D 8/31/1927. Bur Morris Cem, Emanuel Co., Marker reads, 38 GA Inf.
Moxley, Henry M. - CSR shows: Pvt 10/1/1861. Enlisted Augusta, GA states he was W Fred., 12/13/1862. S Appomattox C.H., 4/9/1865. D 8/2/1900. Bur Moxley Cem Swainsboro, Emanuel Co., marker reads; 38 GA INF. Widow, Mary L. Moxley, filed CPA, Emanuel Co., 1926.
Moxley, Joseph (Joseph Thomas Moxley) - Pvt 10/1/1861. W Feb/1863. Absent, W, Apr/1863. NLR CSR states discharged for a gunshot wound to right arm 8/25/1863. Filed CPA, Emanuel Co., 1900. D 10/27/1916. Bur Moxley Cem, Swainsboro. Marker reads CO C, 38 GA INF.
Mullen, Francis M. (Francis M. Mulling) - Pvt 10/1/1861. Appt 4th Cpl Aug/1863; 3d Cpl Jan/1864; Sgt. Killed at Winchester, VA, 9/19/1864. Born 1844. BSU.
Murray, Benjamin - Pvt 10/1/1861. Absent, sick, June - 10/1864. Admitted to Jackson Hos Rich, VA, 10/14/1864. Returned to duty 12/4/1864. S Appomattox, VA, 4/9/1865. BSU.
Murray, William - Pvt 10/1/1861. AWOL 7/1863 - 11/7/1864. NLR. BSU.
Odom, Elijah A. (Elijah Anderson Odom) - Pvt 10/1/1861. Appt 3d Cpl in 1864. Severely Wounded in left arm Cedar Creek, VA, 10/19/1864. Surrendered Appomattox, VA, 4/9/1865. (B Emanuel Co., 1839) Widow, S. A. Marintha Burden Odom, filed CPA, Emanuel Co., 1911. Shot and killed Oct./1884 at social event, buried at Odom Family Cemetery, Jefferson Co.

Odom, William G. (or Odion) (William Goff Odom) - 3d Cpl 10/1/1861. Appt 5th Sgt in 1862. Pvt in 1863. C Gettysburg, PA, 7/3/1863. D PLO, MD, 1/11/1864. Bur PLO Confed. Cem Scotland, MD.
Phillips, James W. - Pvt 10/1/1861. Appt Cpl of the Color Guard in 1861. Died of chronic bronchitis in General Hos Camp Winder, Richmond, VA, 1/7/1864. Buried Hollywood Cemetery Rich, VA, Soldier's Section, precise burial lot unknown, according to cemetery records.
Phillips, John D. - Pvt 7/21/1862. Discharged on account of defective eyesight 12/7/1862. (Died in Montgomery Co., GA 3/13/1912) Bur Phillips Cemetery, Tarrytown, Montgomery Co.
Phillips, J. R. (James Randall Phillips) - Pvt 7/21/1862. Discharged 12/7/1862. Discharged due to "inability to distinguish objects except at close distances." Bur James R. Phillips Cem, Treutlen Co, CSA grave marker.
Phillips, Thomas L. - Pvt 5/8/1862. W 2d Manassas, 8/28/1862. K Fred., VA, 12/13/1862. BSU.
Prescott, John R. - Pvt 5/12/1862. Appt Commissary Sgt, 1862. Roll for Jun/1863, shows him present. CSR contains discharged record. Later served as Cpt, 2nd GA Regt. Died 1872. Bur McLeod family cemetery, in Emanuel Co., Marker reads: CAPT, CO H, 2 GA INF.
Pughsley, George W. - 3d Sgt 10/1/1861. Appt 2d Sgt 10/12/1861; Ordnance Sgt Feb/1863. W, disabled, C Ft. Stedman, VA, 3/25/1865. In Fair Grounds Post Hos Petersburg, VA, 5/25/1865. BSU.
Pughsley, Jacob Perry - ("Jake") 1st Sgt 10/1/1861. Elected Jr. 2d Lt 10/12/1861; 2d Lt 7/14/1862; 1st Lt 1/12/1864. Captured Fredericksburg, 12/13/1862, Paroled & exchanged, 12/17/1862. Wounded & disabled Wilderness, VA, 5/5/1864. Admitted Stuart Hos Rich, VA, debility & old wounds, 3/28/1865. Returned to duty 3/31/1865. Surrendered Appomattox, VA, 4/9/1865. Died 6/22/1911. Bur Lyons City Cem, Toombs, Co. Widow, Mary E. Pughsley filed CPA Toombs Co., 1911.
Rich, W. J. (William J. "Bill" Rich) - Pvt 5/8/1862. Discharged, disability, 12/20/1862. CSR has discharge. Filed CPA, Emanuel Co., 1906. D 12/30/1916. Bur Hall Cem, Emanuel Co. CSA marker.
Scott, William Thomas - Pvt 10/1/1861. Appears last on roll for 10/31/1861. D in 1892. Bur in Scott Cem #2 Swainsboro, GA) Widow, Nancy Flanders Scott, filed CPA, Emanuel Co., 1911.
Sherrod, John H. - 1st Lt 10/1/1861. Appt Adjutant 10/12/1861. Resigned 4/30/1862. Capt Hanleiter stated Sherrod was offered the choice of resigning, or being court-martialed, for intoxication, 1/22/1862. D 9/11/1903, bur Midway Cem, Treutlen Co. CSA grave marker.
Smith, David M. (David Madison Smith) - Pvt 10/1/1861. AWOL Feb. - Sep. 1, 1864. NLR. Bur Swainsboro Primitive Bap. Church, Jefferson Co. CSA grave marker reads: PVT CO C 38 GA INF.
Smith, Ebenezer P. - Pvt 10/1/1861. C Gettysburg, PA, 7/3/1863. E PLO, MD, 2/18/1865. CSR states deserted Richmond, VA, 7/15/1862, absent for six months. BSU.
Smith, George W. - Pvt 10/1/1861. Killed at Gaines Mill, VA, 6/27/1862. BSU.
Steele, John H., Jr. - See Pvt Co. M.
Story, Josiah - Pvt 8/1/1862. W Winchester, VA, 9/19/1864. Absent, W, 11/7/1864. NLR. BSU.
Thompson, Berry A. - Pvt 10/1/1861. Captured Gettysburg, PA, 7/2/1863, and sent to Ft. Delaware. Took OOA to U.S. Govt. and enlisted in the 1st Regt Connecticut Cavalry prior to 10/4/1863. CSR states, deserted Gettysburg, PA, 7/1/1863. BSU.
Thompson, Andrew J. - 2d Sgt 10/1/1861. Appt 1st Sgt 10/12/1861. C Gettysburg, PA, 7/3/1863. P Ft. Delaware, Feb/1865. Rcv Boulware & Cox's Wharves, James River, VA, for E, Mar. 10-12, 1865. Filed CPA, Tattnall Co., 1901. Descendant say bur Waters Cem, Emanuel Co., but no grave marker.
Thompson, Solomon (Solomon Williamson Thompson) - Pvt 10/1/1861. Appt 5th Sgt 10/10/1861. W in 1862. Captured Richmond, VA, 4/3/1865. Released Newport News, VA, 6/25/1865. Died 1876, Montgomery Co. Buried Water's Cemetery, no grave maker. Emanuel Co. Widow, Elizabeth C. Griffith Thompson, filed a CPA, Emanuel Co., 1911. BSU.
Truitt, David - Pvt 10/1/1861. Wounded Gettysburg, PA, 7/2/1863. Captured Jericho Ford, VA, 5/22/1864. P at PLO, MD, 3/14/1865. Rcv Boulware & Cox's Wharves, James River, VA, for E, 3/16/1865. D 12/23/1885. Widow, Susan J. Collins Truitt filed CPA, Montgomery Co., 1903. BSU.

Truitt, James Valentine - 4th Cpl 10/1/1861. Appt 3d Cpl in 1862; 1st Cpl May/1863. Captured in Petersburg, VA, hospital 4/8/1865. In Fair Grounds Post Hos Petersburg, VA, 5/25/1865. Died 8/13/1908 and buried Coleman-Edenfield Cemetery, Emanuel Co, CSA grave marker.

Truitt, William P. - Pvt 10/1/1861. Discharged, disability, 7/28/1862. CSR has discharge, discharged at age 17, born abt. 1845, Born at Barnwell District, SC. BSU.

Watts, John R. - Pvt 10/1/1861. Discharged in 1862. D 12/15/1862, or 1/1/1863. D 12/28/1863. Bur Huguenot Springs Confed. Cem, Powhatan Co. VA. Marker reads:, PVT CO C, 38 INF, CSA.

Webb, Levi E. - 1st Cpl 10/1/1861. D Sav, GA, 2/5/1862. Cause of death not listed in CSR. BSU.

Wiggins, Ashley E. - Pvt 10/1/1861. W Feb/1863. Absent without leave Dec. 1863 - Sept. 1, 1864. No later record. Bur Mulling-Hayes Cem, Jefferson Co. Marker reads: PVT CO C, 38 GA INF.

Wilkes, Francis - Pvt 5/8/1862. Wounded Antietam, 9/17/1862. Wounded Cedar Creek, VA, 10/19/1864. Absent, wounded, 11/7/1864. No later record. BSU.

Wilkes, F. M. - Pvt Died at Lynchburg, VA, 11/1864. Bur there in Old City Cem 11/4/1864. No. 5, 3d Line, Lot 198; died at Reid's Hospital 4.

Wilkes, Peter J. - Pvt 10/1/1861. Appears last on roll for Feb/1863. Died after 1900. Bur Mosley-Wilkes Cem, Emanuel Co., grave unmarked

Williamson, Andrew J. (Andrew Jackson "Jack" Williamson) - Pvt 10/1/1861. Wounded 1862. Appt 4th Cpl 12/1862; 2d Cpl May 1863; 3d Sgt Jan/1864. Killed at Snicker's Gap, VA, 7/18/1864. BSU.

Williamson, George Washington - Pvt 10/1/1861. D in Lynchburg, VA, hos 8/12/1862. Bur there in Old City Cem. No. 9, 5th Line, Lot 177; D Burton's Factory Hos. Widow, Millie Ann Banks Williamson, filed CPA, 1891 Emanuel Co., states he D of typhoid fever.

Williamson, John A. - Jr. 2d Lt.10/1/1861. Elected 2d Lt 10/12/1861. Resigned 7/14/1862. D. 3/17/1916, bur in Emanuel Co., according to obit . Bur Williamson Family Cem, Toombs Co. Obit printed in Montgomery Monitor, Mount Vernon, GA, 2/23/1916.

Williamson, John G. - Pvt 10/1/1861. Wounded Mar/1863. Appt 4th Sgt in 1863; 3d Sgt Sept/1863. Wounded Wilderness, VA, 5/6/1864. Frl from hos for 60 days Feb/1865. Pension records show him home on wounded frl close of war. (B in GA) Filed CPA, Vidalia, Toombs Co., 1919. Died 12/9/1923, buried Williamson Family Cemetery, Emanuel Co.

Williamson, Riley - Pvt 10/1/1861. AWOL Jan. - 9/1/1864. NLR. Died 1870, buried at Williamson Cemetery, Toombs Co.

Williamson, V. L. (Ulysses L. Williamson) - 5th Sgt 10/1/1861. Appt 4th Sgt 10/12/1861. Appears last on roll for Apr/1862. CSR contains discharge record. D 1873. Bur Corbin Cem, Emanuel Co.

Woods, A. R. (Archibald R. Woods) - Pvt 6/1/1864. S Appomattox, VA, 4/9/1865. B 8/20/1845, d. 2/15/1916. Bur Moxley Cem, Swainsboro, Emanuel Co. Grave has CSA marker.

Woods, John D. - Pvt 10/1/1861. Died of tuberculosis in General Hos #25, Richmond, VA, 12/20/1862. Bur Oakwood Cem, Richmond, VA, cem records show Pvt J. D. Woods, Co. C, 38th GA Regt., buried in Div A, Row M, #158.

Woods, John R. - Pvt 10/1/1861. Appears on a list of casualties for the battle of 2d Manassas, 8/28/1862, W. Captured Spotsylvania, VA, 5/18/1864. Died of typhoid fever Elmira, NY, 2/17/1865. Bur Woodlawn National Cemetery, Elmira, NY, Grave #235, CSA marker

Woods, William J. - Pvt 10/1/1861. C Gettysburg, 7/2/1863. Released Ft. Delaware, 5/3/1865. BSU.

Wright, Alexander H. (Alexander Hamilton Wright) - Pvt 10/1/1861. Det hos duty Greensboro, NC, Sept./1862. In Bell Hos Eufaula, AL. Feb. 29-11/1, 1864. NLR. Filed CPA, Floyd Co., 1901. D 2/16/1919, Bur New Armuchee Cem., Floyd Co. Marker reads M.D. (Medical Doctor.)

Wright, Miller A. (Miller Armstead Wright) - 2d Lt 10/1/1861. Elected 1st Lt 10/12/1861. W Antietam, 9/17/1862. Resigned 11/30/1863. Son of Col. Augustus R. Wright, Pardoned by President Lincoln, 11/5/1864. BSU.

MUSTER ROLL OF COMPANY D, 38th REGT
DEKALB COUNTY, GA

McCULLOUGH RIFLES

Most of the members of this company were enlisted by Captain John Y. Flowers in the DeKalb Murphy Guards and were mustered into the service of the Confed. States 9/26/1861, Camp Kirkpatrick, DeKalb Co. Company A became too large and was divided 4/1/1862, forming Company D. The new company took the name of McCullough Rifles and Captain John G. Rankin was appointed company commander.

Akins, Charles Thomas (or Aiken) - Pvt 9/26/1861. Trn from Co. A, 4/1/1862. K Antietam, MD, 9/17/1862. CSR contains a death claim filed by father, James F. Akins, filed in DeKalb Co. BSU.
Anderson, George J. - Pvt 9/26/1861. Trn from Co. A, 4/1/1862. Wounded Malvern Hill, VA, 7/1/1862. Discharged, disability, Atlanta, GA, 10/29/1864. CSR contains discharge record, 19 yrs old at discharge, B DeKalb Co., GA. Died 3/25/1897, Halstead Cem., Pulaski Co., Little Rock, AK.
Arnold, Oliver H. P. - See Pvt Co. E.
Baxter, Francis M. - (Francis Marion Baxter) Pvt 3/1/1862. Transferred from Co. A, 4/1/1862. K Antietam, MD. 9/17/1862. Widow, Sarah J. Baxter Turner, filed CPA, Milton Co., 1912. BSU.
Baxter, George - Pvt 7/31/1863. W Ft. Stedman, VA, 3/25/1865. On detail as nurse, Farmville, VA, hos, 4/11/1865, and P there Apr. 11-12, 1865. Filed CPA, Yell Co., AR, 1897, bur there Hale Cem.
Baxter, John (John M. Baxter) - Pvt 3/1/1862. Trn from Co. A and elected Sr. 2d Lt, 4/1/1862. C Gettysburg, PA, 7/4/1863. Released Ft. Delaware, 6/12/1865. Captured by Kilpartick's cavalry at Smithburg, MD, according to New York Herald newspaper, 7/12/1863. Filed CPA, DeKalb Co, 1900. D 4/9/1915. Bur Gresham Weed Cem, DeKalb Co. Obit in Atlanta Constitution, page 21, 4/11/1915.
Carter, John D. - Pvt 9/26/1861. Trn from Co. A, and Appt 5th Sgt 4/1/1862. Died near Gordonsville, VA, 8/12/1862. CSR contains death claim filed by father, Tarlton Carter. BSU.
Carter, William J. - Pvt 9/26/1861. Trn from Co. A, 4/1/1862. Died in Macon, hospital 4/25/1862, buried there at Rose Hill Cemetery. CSR record of Pvt John D. Carter contains death claim for John and William Carter, filed by their father, Tarlton Carter, of DeKalb Co., GA.
Cochran, S. W. (Samuel "Sam" Wylie Cochran) - 3d Cpl 9/26/1861. Transferred from Co. A, 4/1/1862. W 2d Manassas, VA, 8/28/1862. Surrendered Appomattox, VA, 4/9/1865. Filed CPA, DeKalb Co., 1893. Died 1922, buried at Cochran Family Cemetery, Tucker, DeKalb Co.
Coggin, J. F. (James F. Coggin) - Pvt 3/20/1862. Trn from Co. A, 4/1/1862. Appt 2d Cpl 5/1/1864. S Appomattox, VA, 4/9/1865. (B DeKalb Co., GA) Sgt. at Appomattox surrender. Filed CPA, Gwinnett Co., 1902. Bur Bethesda United Meth. Cem, Gwinnett Co. Marker reads: B 5/8/1845, D 6/3/1917.
Coggin, William M. (William Marion Coggin) - Pvt 5/1/1862. W Fred., VA, 12/13/1862. Roll for 11/1/1864, LOF, shows present. Filed CPA, DeKalb Co., 1899. Buried Rock Chapel Historic Cemetery, DeKalb Co. Grave marker reads: B 11/21/1834, D 5/3/1903.
Collier, John E. J. - See Pvt Co. K.
Conn, William A. - Pvt 9/26/1861. Transferred from Co. A, 4/1/1862. Discharged, disability, 6/4/1862. CSR contains discharge record. Died in Fulton Co. 8/2/1870. BSU.
Davis, William Caleb - Pvt 9/26/1861. Trn from Co. A, 4/1/1862. D Sav, GA 5/13/1862. BSU.
Dickerson, W. F. A. (William Franklin Anderson Dickerson) - Pvt 3/1/1862. Trn from Co. A, 4/1/1862. Appt Sgt, wounded Fredericksburg, VA, 12/13/1862. Surrendered Appomattox, VA, 4/9/1865. Filed CPA, DeKalb Co., 1906. D there 6/30/1926, buried Stone Mountain City Cemetery. CSA marker reads: CO D 38 GA INF. B 7/26/1840. Obituary in Atlanta Constitution, 7/1/1926.
Dodgen, William O. - Musician 9/26/1861. Trn from Co. A, 4/1/1862. W Fred., VA, 12/13/1862 Roll for 8/31/1864, LOF, shows present. NLR. Listed a "fifer" in CSR. BSU.
Ealum, Joel R. (or Elum) - Pvt 9/26/1861. Transferred from Co. A, 4/1/1862. Wounded at Antietam, 9/17/1862. Died of disease Hamilton's Crossing, VA, 4/2/1863. CSR contains death claim filed by father, J. E. Ealum, DeKalb Co. BSU.

Estes, Gainham T. - Pvt 9/26/1861. Trn to Co. D, 4/1/1862. D of typhoid fever Charlottesville, VA, 7/7/1862. Bur there UVA Confed. Cem. CSR has death claim by widow, Delila Estes, in Jackson Co.

Estes, Hilliard (Hillard H. Estes) - Pvt 4/15/1862. Wounded Jan/1863. Wounded near Wilderness, VA, 5/4/1864. Absent, wounded, 9/1/1864. NLR. Filed a CPA, Sevier Co., AR, 8/14/1911. BSU.

Goss, John W. - Pvt 9/26/1861. Trn from Co. A, 4/1/1862. Captured Mine Run, VA, 11/28/1863. Took OOA to U.S. Govt., enlisted in U.S. Army and released PLO, MD, 2/23/1864. Enlisted in the 1st U.S. Vol. Regt. Filed CPA, Cleburne Co., AR, 1906, veteran #15704, Died 3/20/1922. BSU.

Goza, Robert D. - Pvt 5/1/1862. Captured Berryville, VA, 8/10/1864. Released Elmira, NY, 7/7/1865. Filed a CPA, DeKalb Co., 1910 Bur Pleasant Hill Bap. Church Cem, Tucker, DeKalb Co.

Gresham, J. W. (Josiah Wheeler Gresham) - Pvt 9/26/1861. Trn from Co A and Appt 3d Sgt 4/1/1862. K Antietam, MD 9/17/1862. CSR states mortally W 9/17/1862, died same day; first name found in CSR. CSR has death claim by widow, Parthenia Henderson Gresham, of DeKalb Co. BSU.

Grogan, Joseph T. (or Josiah T.) - See Pvt Co. K.

Grogan, William D. (William Daniel Grogan) - Pvt S Appomattox, VA, 4/9/1865. D 10/13/1903, bur New Georgia Cem, Paulding Co., CSA marker. Widow, Sarah O. Grogan, filed CPA there in 1905.

Hall, John W. - Pvt 9/26/1861. Trn from Co. A, and Appt 1st Cpl 4/1/1862. W in right arm, necessitating amputation, Fred., VA, 12/13/1862. At home, W, close of war. BSU.

Hambrick, James F. - Pvt 9/26/1861. Trn from Co. A, 4/1/1862. D Rich, VA, 7/6/1862. CSR contains a death claim filed by father, James Hambrick, DeKalb Co. BSU.

Hambrick, J. T. - (James T. Hambrick) Pvt 9/26/1861. Transferred from Co. A, 4/1/1862. Died Charlottesville, VA, 11/11/1862, buried there University of Virginia Confederate Cemetery. CSR contains death claim by father, J. M. Hambrick, DeKalb Co.

Hambrick, Robert C. - Pvt 9/26/1861. Trn from Co. A, and Appt 1st Sgt 4/1/1862. W Gaines Mill, VA, 6/27/1862. Died of wounds Rich, VA, 7/14/1862. CSR contains a death claim filed by father, James Hambrick in DeKalb Co. Bur Oakwood Cem, Div., C, Row L, Grave #120, Richmond, VA.

Harris, Edmond T. - (Edmond Thomas Harris) 4th Cpl 9/26/1861. Transferred from Co. A, 4/1/1862. Appt 1st Sgt 7/15/1862. Wounded Antietam, MD, 9/17/1862. Roll for 8/31/1864, last on file, shows present. No later record. Died 11/18/1900, buried Adams Cemetery, Johnson Co., AR.

Harris, J. C. (James C. Harris) - Pvt 3/1/1862. Trn from Co. A, 4/1/1862. Appt 5th Sgt Dec/1863. W Winchester, VA, 9/19/1864. C Fisher's Hill, VA, 9/22/1864. Released PLO, MD, 6/28/1865. Appears on casualty list 2d Manassas, 8/28/1862, W. D 11/19/1899. Bur Stone Mountain Cem, DeKalb Co,

Harris, Sterling G. - (Starling Gilmore Harris) 1st Cpl 9/26/1861. Trn from Co. A, 4/1/1862. Appt 3d Sgt 7/15/1862. Wounded in the hip Wilderness, VA, 5/5/1864. Home on wounded furlough, 8/31/1864. No later record. Filed CPA, Pope Co. AR, 1921. Died 12/29/1930, buried Adams Cemetery, Johnson Co., AR. Cemetery was originally called the Jordan Cemetery.

Hix, A. H. (Abner Harrison Hix) - Pvt 9/26/1861. Trn from Co. A, 4/1/1862. Wounded Fredericksburg, VA, 12/13/1862. S Appomattox, VA, 4/9/1865. (B in GA 7/12/1832) Filed CPA, Madison Co., 1900, died there 2/15/1907, bur Old Hix Family Cem, Commerce, Jackson Co.

Holly, H. J. (Hiram J. Holly) - Pvt 3/1/1862. Transferred from Co. A, 4/1/1862. Sent to hos 6/1/1864. Captured Harrisonburg, VA, 10/1/1864. Released Point Lookout, MD, 5/13/1865. Died 5/23/1883. Bur Stone Mountain Cem, DeKalb Co.

Huff, J. H. (or S. H. Huff) Pvt Killed at Fisher's Hill, VA, 9/22/1864. BSU.

Huff, Thomas P. (Thomas Poole Huff) - Pvt 10/11, 1863. W in left leg near Strasburg, VA, in 1864. Sick in 1st Div Jackson Hos Rich, VA, 2/21/1865. NLR. Filed CPA, Greene Co., 1890. BSU.

Ivey, Aaron H. - Pvt 9/26/1861. Transferred from Co. A, 4/1/1862. Died at Winchester, VA, 10/25/1862. (Place of death also shown as Staunton, VA, 10/28/1862) CSR shows he died of typhoid fever. CSR contains death claim filed by father, Martin B. Ivey, in DeKalb Co. BSU.

Ivey, Joseph J. - Pvt 9/26/1861. Trn from Co. A, 4/1/1862. D in camp Winchester, VA, 10/8/1862, or Staunton, VA, 10/28/1862. CSR has death claim filed by father, Martin B. Ivey, DeKalb Co. BSU.

Jackson, Henry B. - Pvt 11/26/1862. On frl Jan 1 - Feb 29, 1864. Roll for 8/31/1864, shows Absent, sent to Macon, GA, hos 5/4/1864. CPA shows in Atlanta, GA, hos end of war. (B Warren Co., 1825) Filed CPA, Walton Co., 1902. D Nov/1902, buried Snellville Historic Cemetery, Gwinnett Co.
Jeffers, Bennett Rainey (or Jeffares) - Pvt 5/12/1862. Wounded and disabled Gaines Mill, VA, 6/27/1862. Det with ambulance corps. Captured Farmville, VA, 4/6/1865. Released Newport News, VA, 6/16/1865. Died 12/20/1866 of tuberculosis. Bur Cochran Cemetery on Tucker-Chamblee Road, DeKalb Co. Name is spelled "Jeffares" on marker, marker inscribed 38 GA INF.
Lafoy, John A. - Pvt 9/26/1861. Transferred from Co. A, and Appt 4th Sgt 4/1/1862. W Cold Harbor, VA, 6/1/1862; Locust Grove, VA, 5/5/1864. Deserted in 1864. Appears on report of deserters at Clarksburg, WV, 1/4/1865, as having "reported to Colonel Youart at Beverly, WV, 12/19/1864; took OOA to U.S. Govt., released 12/24/1864." (B in Anderson, SC in 1841) BSU.
Lang, John W. (or Long) - Pvt 9/26/1861. Transferred from Co. A, 4/1/1862. Died at Petersburg, VA, 8/23/1862. BSU.
Lang, William B. (or Long, Wiley B.) - Pvt 3/1/1862. Transferred from Co. A, 4/1/1862. Wounded Antietam, MD, 9/17/1862. Captured Waterloo, PA, 7/5/1863. Took oath of allegiance to U.S. Govt., enlisted in the U.S. Army, Point Lookout, MD, 2/1/1864. BSU.
Lanier, (or Lanear) Pressley (Pressley "Press" Lanier) - Pvt 3/1/1862. Transferred from Co. A, 4/1/1862. CSR stated captured Fredericksburg, 12/13/1862, exchanged 12/17/1862. Killed Spotsylvania, VA, 5/12/1864. His name and "Killed at Spotsylvania C.H." inscribed on the back of wife's (Nancy Lainer) tombstone at Stone Mountain Cemetery DeKalb Co. CSR contains death claim filed by father, William L. Lanier. Widow, Nancy Lanier, filed CPA, DeKalb Co., 1891. BSU.
Lanier, William L. - Pvt 9/26/1861. Trn from Co. A, 4/1/1862. Died of typhoid fever in General Hos Lynchburg, VA, 1/26/1863. Bur there in Old City Cem. No. 4, 4th Line, Lot 187. Died Clayton's Hosp. Correct grave position is Lot 187 is 3rd Row, Grave # 3. CSR shows he was 15 years old at enlistment and contains death claim filed in Cobb Co., by mother, Nancy Lanier.
Lawhorn, Joel J. – (or Lawhon) Pvt 9/26/1861. Transferred from Co. A, 4/1/1862. Wounded Fredericksburg, VA, 12/18/1862; Wilderness, VA, 5/6/1864; Newtown, VA, 7/17/1864. At home, wounded, 8/31/1864. Filed a CPA, DeKalb Co., 1891. Died 1/5/1934. Buried Greenville, Henderson Cemetery, DeKalb Co. CSA marker, Name on marker as J. J. Lawhon, CSA grave marker.
Leavell, E. F. - (Edward "Ed" Franklin Leavell) Pvt 9/26/1861. Trn from Co. A, 4/1/1862. Wounded in left hand, necessitating amputation, Jack's Shop, VA, 12/23/1864. At home, wounded, close of war. Accidentally shot by a fellow Confederate soldier. Filed CPA, Paulding Co., 1877. Died 3/30/1900 in Draketown, Harralson Co., buried Gresham Weed Cem, DeKalb Co.
Leavell, F. M. (Francis Marion Leavell) - Pvt 9/26/1861. Trn from Co. A, 4/1/1862. W in arm, necessitating amputation, Antietam, MD 9/17/1862. Retired on account of wounds 11/3/1864. Filed a CPA, Pope Co., AR, 1901, veteran #7057. Died 6/17/1912, Broomfield Community, Pope Co., AR. Bur Adams Cemetery, Johnson Co., AR. Marker reads; FRANCIS M. LEVELL, CO D, 38 GA INF.
Long - See Lang.
Luckey, J. R. - (John R. Luckey) Pvt 9/26/1861. Appt Musician in 1861. Transferred from Co. A, 4/1/1862. C Gettysburg, PA, 7/4/1863. Took OOA to U.S. Govt., enlisted in U.S. service Point Lookout, MD, 4/5/1864. Assigned 1st Regt U.S. Volunteer Infantry, Died of scurvy, 5/28/1865, at Ft. Rice, Dakota Territory. Buried at Custer National Cemetery, Big Horn Co., MT.
Lynch, Thomas - Pvt 9/26/1861. Transferred from Co. A, 4/1/1862. Roll dated 11/1/1864, shows him "Present, Detailed Richmond, VA, hospital Feb/1864." No later record. BSU.
Maddox, Thomas L. - Pvt 9/26/1861. Transferred from Co. A, 4/1/1862. Appt 5th Sgt 7/15/1862. Detailed with Invalid Corps in GA, 1862. Captured in Gwinnett Co., GA, 10/26/1864. Died of pneumonia Camp Douglas, IL, 1/22/1865. Buried Block# 2, Chicago City Cemetery, Chicago, IL.
Marbut, John S. (or John T. Marbut) - Pvt 9/26/1861. Trn from Co. A, 4/1/1862. D Richmond, VA, Hos 2/25/1865. CSR states D of pneumonia. Bur there Hollywood Cem. Cem records show: Marbott, J. T., Company D, 38 GA, D 2/25/1865, bur Section W, grave #11. Marker reads: CO D, 38 GA INF.

Mason, Cornelius A. - (Camilus Anderson Mason) Pvt 9/26/1861. Trn from Co. A, 4/1/1862. Appt Cpl S Appomattox, VA, 4/9/1865. Filed CPA, Citrus Co., FL, 1909, PA #A04096. D 7/28/1914. Bur Lake Lindsey Cem, Brooksville, Hernando Co., FL. Marker reads: Camilus Anderson Mason, 38 GA.
Mason, William S. - Pvt 8/6/1862. Killed at Fredericksburg, VA, 12/13/1862. BSU.
Mehaffey, Francis M. - Pvt 5/15/1862. Captured Spotsylvania, VA, 5/12/1864. Exchanged Ft. Delaware, 3/7/1865. Rcv Boulware & Cox's Wharves, James River, VA, Mar. 10-12, 1865. BSU.
McCandless, Jesse A. - (Jesse Andrew McCanless) Pvt 9/26/1861. Trn from Co. A, 4/1/1862. Wounded Gaines Mill, VA, 6/27/1862. Roll for 9/1/1864, shows him present. Pension records show he was home on sick furlough at close of war. Filed a CPA Cherokee Co., 1907. Died 7/26/1911. Buried in Riverview Cemetery, Canton, Cherokee Co.
McCurdy, John Wilson - Pvt 3/1/1862. Elected 1st Lt of Co. D, 4/1/1862. Wounded Antietam, 9/17/1862, shot in right leg. Roll for 8/31/1864 shows "Absent, Invalid Corps in GA" Dropped from rolls by order Secy. of War. 1/21/1865. Filed CPA, DeKalb Co., 1907. Died 1922, buried at Stone Mountain Cemetery, DeKalb Co.
McCurdy, Phillip Burford - Pvt 3/1/1862. Transferred from Co. A, and Appt 2d Sgt 4/1/1862. Wounded Fredericksburg, VA, 12/13/1862. Shepherdstown, WV, 10/1/1862. With Invalid Corps in 1864. Paroled Augusta, GA Apr/1865. (Born GA 7/1/1837. Died 2/29/1914) Filed CPA, DeKalb Co, 1889. Buried Stone Mountain Cemetery, DeKalb Co.
McCurdy, Stephen C. - (Stephen Cicero McCurdy) Pvt 5/1/1862. Died of chronic diarrhea dysentery Div #2, Jackson Hospital, Richmond, VA, 10/14/1863. Buried there Hollywood Cemetery. Cemetery records show: McCurdy, SC, Company G, 38 GA, D 10/14/1863, bur Section T, grave # 248.
Medlock, Eli W. (Eli Wren Medlock) - Pvt 9/26/1861. Trn from Co. A, 4/1/1862. W Antietam, MD 9/17/1862. W 7/15/1864. On W frl 9/1/1864. NLR. D 8/31/1904, Cobb Co, bur there Rose Hill Cem.
Nash, John Miles - Pvt 9/26/1861. Trn from Co. A, and Appt 4th Cpl 4/1/1862. Hand amputated in 1862. D Petersburg, VA, 6/15/1862. Brother of Isaac Nash of same company. No evidence of hand being amputated in CSR, may be confused with record of Isaac Newton Nash, brother, shown below. CSR contains death claim filed by father, Miles P. Nash. Bur Sweetwater Bap. Church, Gwinnett Co.
Nash, J. N. (Isaac Newton Nash) - Pvt 3/1/1862. Trn from Co. A, 4/1/1862. Appt 2d Cpl Jul/1862; 4th Cpl Jan/1863. W in left arm Gettysburg, PA, 7/1/1863, necessitating amputation above wrist, Winchester, VA. At home, W, close of war. Filed CPA, DeKalb Co., 1879, listed as J. N. Nash. D 11/29/1913. Bur Stone Mountain Cem, DeKalb Co. Obit in Atlanta Constitution 11/30/1913.
Nash, Miles H. (Miles Henry Nash) - Pvt 9/26/1861. Trn from Co. A, 4/1/1862. W Gaines Mill, VA, 6/27/1862. C at Mine Run, VA, 5/6/1864. Sent from Elmira, NY to James River, VA, for E, 2/20/1865. (B GA, 1841. D Gwinnett Co., GA, 1913) Filed CPA, Gwinnett Co. 1912. Bur Sweetwater Bap. Church and Cem, Duluth, Gwinnett Co., GA, CSA marker reads Co D, 38th GA REGT, CSA.
Nash, William T. - Pvt 5/1/1862. W Antietam, MD 9/17/1862; Winchester, VA, 9/19/1864. At home, W, 11/1/1864. (B in DeKalb Co., GA 3/1/1834) Filed CPA, Fulton Co., 1897. BSU.
Newsom, Moses H. - Pvt 9/26/1861. Trn from Co. A, 4/1/1862. Wounded in 1863. Surrendered Appomattox, VA, 4/9/1865. (Born in Morgan Co., GA, 1841) Captured battle of Fredericksburg, 12/13/1862, paroled for exchange 12/17/1862, according to CSR. Died 1/29/1907, buried Sylvester Cemetery, DeKalb Co., widow, Millie Newsom, filed CPA, Fulton Co., 1907.
Perlinski, Julius - See Regtal Musician. BSU.
Rankin, John G. - (John "Red" "Old Reliable" Gray Rankin) 2d Lt 9/26/1861. Elected Cpt of Co. D, 4/1/1862. W in the arm and C Winchester, VA, 9/19/1864. Released Ft. Delaware, DE 6/17/1865. Filed CPA. DeKalb Co., 1900. D 10/6/1902, bur Stone Mountain Cem, DeKalb Co.
Ray, John P. (John Poole Ray) - Pvt 5/15/1862. Transferred from Co. A, 11/1/1862. Captured High Bridge, VA, 4/7/1865. Released Point Lookout, MD, 6/17/1865. Filed CPA, DeKalb Co., 1902. Died 1/18/1915. Buried Indian Creek Cemetery, DeKalb Co.

Ray, Robert A. (Robert Anderson Ray) - Pvt 9/26/1861. Transferred from Co. A, and Appt 3d Cpl 4/1/1862; 1st Cpl Jul/1862. Reduced to 3d Cpl Jan/1863. Appt 2d Sgt. Deserted 12/19/1864. Took OOA to U.S. Govt. and released 12/24/1864. Descendants say he died 8/28/1906, DeKalb Co. BSU.
Rosenthal, Adolphus - Pvt 5/1/1862. Appt Ordnance Sgt 6/1/1862. Wounded at the Wilderness, VA, 5/6/1864. Letter from Capt. Rankin, to Lt G. R. Wells, states he was wounded in the foot at the Wilderness and died at Richmond, VA, 5/27/1864. Born in Germany, about 1835. Remains moved to Savannah, GA, for burial by his Jewish friends, buried at Laurel Grove Cemetery, Savannah, GA.
Simpson, Robert M. - Pvt 9/26/62. Trn from Co. A, 4/1/62. W Gaines Mill, VA 6/27/62; near Winchester, VA 8/11/64. CSR contains certificate of disability /retirement, dated 2/16/1865. Filed CPA, Cleburne Co., AL. 1907. D 4/9/1909, bur Old Davistown Cem, Davistown, Calhoun Co., AL.
Simpson, William M. - Pvt 9/26/1861. Transferred from Co. A, 4/1/1862. Died 4/16/1862. BSU.
Singleton, James Madison - Pvt 5/1/1862. Wounded Gaines Mill, VA 6/27/1862; Wounded left hand Wilderness, VA 5/5/1864. In Macon, GA hos, W, 9/1/1864. CPA shows in Atlanta, GA hos close of war. (B DeKalb Co., 8/23/1843. Filed CPA, DeKalb Co, 1893. D Feb/1905, Clarkston, Bur Zion Cem.
Singleton, William Leonard - (William "Bill" Leonard Singleton) Pvt 3/1/1862. Trn from Co. A, 4/1/1862. Wounded Gaines Mill, VA 6/27/1862; Locust Grove, VA 5/5/1864, Wounded in left hand. CPA shows home on W frl close of war. (B Gwinnett Co., GA 6/29/1841, Died there 3/19/1918) Filed CPA, Gwinnett Co, 1915. Buried in Pleasant Hill Baptist Church Cemetery, DeKalb Co., near Tucker.
Smith, Isaac H. - Pvt 3/20/1862. Appt 3d Cpl 12/1/1863. Captured Ft. Stedman, VA, 3/25/1865. Released Point Lookout, MD, 6/19/1865. BSU.
Smith, Tilnon P. (Tilnon P. "Sally" Smith) - Pvt 9/26/1861. Transferred from Co. A and Appointed 2d Cpl 4/1/1862. Killed at the Wilderness, VA, 5/6/1864. BSU.
Smith, Thomas - Pvt 5/1/1862. D, typhoid fever, Chimborazo Hos #3, Rich, VA, 7/18/1862. BSU.
Thomas, Thomas Jefferson - Pvt 3/20/1862. Trn from Co. A, 4/1/1862. Wounded at battle of Gettysburg and captured at South Mountain, MD, 7/5/1863. Paroled at Point Lookout, MD, 2/18/1865. Received Boulware & Cox's Wharves, James River, VA, Feb. 20-21, 1865. At Camp Lee near Richmond, VA, 2/27/1865. Died 6/19/1891. Buried Stone Mountain Cemetery, DeKalb Co.
Thomas, Wyett L. - Pvt 9/26/1861. Trn from Co. A, 4/1/1862. C Spotsylvania, VA 5/12/1864. P Ft. Delaware, DE Feb/1865. E Boulware & Cox's Wharves, James River, VA, Mar. 10-12, 1865. BSU.
Thomas, Zephaniah - Pvt 3/1/1862. Transferred from Co. A, 4/1/1862. Wounded in left hand, necessitating amputation, Gaines Mill, VA 6/27/1862. Discharged, disability, Nov. 7 or 12/7/1862. CSR contains discharge record. Died 10/15/1904, Bosque, TX. Buried Cranfills Gap Cem, Hamilton Co., TX. Widow, M. A. Thomas, filed a CPA, in Hamilton Co., TX, 1914, pension # 28220.
Thompson, Richard Govan - Pvt 3/20/1862. Transferred from Co. A, 4/1/1862. Died in Winchester, VA, hospital 10/31/1862. Buried there at Mount Hebron Cemetery, Stonewall Confederate Section, Georgia Section, Plot #192. CSR contains death claim filed by widow, Lodusky C. Thompson. Widow filed a CPA, DeKalb Co., 1914.
Tucker, John T. - Pvt 9/26/1861. Transferred from Co. A, 4/1/1862. Killed at Antietam, MD, 9/17/1862. CSR contains death claim filed by mother, Jane Tucker, DeKalb Co. BSU.
Verdin, L. N. (Ledford N. Verdin or Verdon) - Pvt 10/11/1862. S Appomattox, VA 4/9/1865. BSU.
Verdin, William J. (or William V.) - Pvt 9/26/1861. Transferred from Co. A, 4/1/1862. Wounded Gaines Mill, VA 6/27/1862. Surrendered at Appomattox, VA, 4/9/1865. BSU.
Wells, George R. (George Riley Wells) - 2d Sgt 9/26/1861. Elected 2d Lt., Co. D, 4/1/1862. C Fisher's Hill, VA 9/22/1864. Released Ft. Delaware, 6/17/1865. W 2d Manassas, 8/28/1862. Died 1/5/1919, buried Stone Mountain Cemetery, DeKalb Co. Obit in Atlanta Constitution 1/7/1919.
Wells, W. S. - (Willard Simpson Wells) Pvt 5/1/1862. Sick in Macon, GA hospital Dec/1862 - 10/31/1863. On detail duty, nurse, Rome, GA, hospital, Nov - Dec 1863. Absent, sick, 11/1/1864. Descendants say he died 4/16/1865. BSU.
Wells, Willis Virgil - Pvt 3/1/1862. Transferred from Co. A, 4/1/1862. On detail duty in Lynchburg, VA, hospital 11/1/1864. Killed at Ft. Stedman, VA, 3/25/1865. BSU.

Williams, T. J. - (T. John Williams) Pvt 2/28/1864. Captured Spotsylvania, VA, 5/12/1864. Paroled Ft. Delaware, Feb/1865. Exchanged Boulware & Cox's Wharves, James River, VA, 3/10/1865. BSU.
Williams, William Jasper - Pvt 9/26/1861. Transferred from Co. A, 4/1/1862. Wounded in right arm, necessitating amputation above elbow, 2d Manassas, VA, 8/28/1862. Discharged, disability, 10/14/1862. (Born 6/15/1840. Died at Confederate Soldiers' Home Atlanta, GA, 3/3/1931. Buried Lithonia, GA) Filed CPA, Fulton Co., 1879. Buried Lithonia City Cemtery, DeKalb Co., GA.
Woods, S. J. - Pvt 10/11/1863. Killed in action at Fisher's Hill, VA, 9/22/1864. BSU.

*MUSTER ROLL OF COMPANY E, 38th REGT
COMPANY E, TOM COBB INFANTRY OGLETHORPE COUNTY, GA

*This company was also known as Old Co. K, New Co. C, and Co. A. Company originally organized and led by Captain James Davant Mathews.

Adams, Thomas J. (Thomas Jefferson Adams) - Pvt 9/29/1861. C Gettysburg, PA, 7/2/1863. P at PLO, MD in 1864. Rcv Venus Point, Sav River, GA for E, 11/15/1864. C and P Hartwell, GA 5/17/1865. D 11/5/1868, murdered by Robert Ben Cade, Elbert Co., buried Concord United Meth. Church. Marker reads: "A Faithful Confederate Soldier." Widow filed CPA, Elbert Co., 1901.
Allgood, John H. - Pvt 5/3/1862. Wounded and disabled Fredericksburg, VA, 12/13/1862. Wounded 1863. Discharged, disability, 4/3/1864. Discharge record in CSR. Filed CPA, Clarke Co., 1888, Died 10/23/1891, buried there Oconee Hill Cemetery. Marker reads: born 5/24/1834.
Arnold, James Wylie - Pvt 5/1/1862. Wounded Gaines Mill, VA 6/27/1862. Det Feb/1863. Roll dated 11/4/1864, LOF, shows him "Absent, assessor of tax in Oglethorpe Co." Died 12/30/1906, buried Resthaven Cemetery, Wilkes Co., Obit in Augusta Chronicle newspaper 1/1/1907, page 10.
Arnold, Martin L. (Martin Luther Arnold) - Pvt 5/5/1862. Died at home 6/20/1862. Buried Arnold-Pope Family Cem, Wilkes, Co.
Arnold, Moses H. - ("Moch" or "Moke") Pvt 5/3/1862. Discharged 5/16/1862. Died 12/30/1905. Obit in Oglethorpe Echo, 1/5/1906. Buried Crawford City Cemetery, Crawford, Oglethorpe Co.
Arnold, Moses P. - Pvt 5/1/1862. Died of disease Lynchburg, VA Jul/1864. Buried there in Confed. Cem 7/30/1864. No. 1, 4th Line, Lot 194. Died at Camp Davis Hos. Substitute for Jas. W. Adkins.
Arnold, Oliver H. P. - Pvt 5/1/1862. Transferred to Co. D, 5/1/1862. Transferred to Co. E, 6/1/1862. Discharged, furnished substitute, prior to 10/31/1862. Died 9/20/1907. Buried Clark Cemetery, Oglethorpe Co. Obit in Oglethorpe Echo, 9/27/1907.
Arnold, Randolph J. (Randolph Jones Arnold) - Pvt 5/5/1862. Discharged, furnished J. H. Hendricks as substitute, 1862. Died 3/30/1921. Buried Sardis Church Cemetery, Oglethorpe Co. Grave marker reads: born 1/9/1832. Obit in Oglethorpe Echo newspaper, 4/1/1921.
Bacon, Joel Josiah - Pvt 4/29/1862. Appt 1st Sgt Jan/1863. Discharged, furnished George Cusick as substitute, 4/30/1863. Died 12/2/1903. Buried Presbyterian Cemetery, Oglethorpe Co. Obit in Oglethorpe Echo newspaper, 12/3/1903, page 2, also in Atlanta Constitution, 12/3/1903, page 5.
Baughn, Peter Butler - Pvt Reported as a recruit who enlisted Sav, GA. Discharged 6/20/1862, CSR contains discharge. Buried Baughn family cem, Oglethorpe Co. Obit in Oglethorpe Echo 2/12/1904.
Boggs, James S. - Pvt 9/29/1861. Killed at battle of the Wilderness, VA, 5/6/1864. BSU.
Boggs, Richard H. H. - Pvt 9/29/1861. Wounded in lungs Gettysburg, PA, 7/1/1863, Died there of wounds, 7/20/1863. BSU.
Boggs, Samuel H. - Pvt 9/29/1861. Missing Winchester, VA, 9/17/1864. NLR. BSU.
Brittain, Jabez Mercer - 1st Sgt, 9/29/1861. Elected Jr. 2d Lt 4/22/1862; 2d Lt 7/23/1862; 1st Lt 6/14/1863. Appt Chaplain 6/16/1863. Resigned, Appt missionary to Army of Northern VA from Southern Bap. Board of Missions, 7/12/1864. D 2/11/1912, Atlanta, bur there West View Cem. Obit in Macon Telegraph 2/12/1912, page 6. Widow, Ida Callaway Brittain, filed CPA, Fulton Co., 1919.
Brittain, Jack Lumpkin - Pvt 4/30/1862. D of gastritis, General Hos Danville, VA, 7/10/1862. BSU.

Bryant, Robert M. - Pvt 9/29/1861. Captured Gettysburg, PA, 7/3/1863. Took oath of allegiance to U.S. Govt. at Point Lookout, MD and enlisted in U.S. service 10/17/1864. Enlisted 2nd U.S. Volunteer Infantry. Descendants say he died 1/22/1913, Otoe Co., NE . BSU.
Burt, William H. - Pvt 4/27/1862. W in eye resulting in loss of sight; W in finger, Spotsylvania, VA, 5/12/1864. S at Appomattox, VA 4/9/1865. Filed CPA, Oglethorpe Co., 1890. Joel J. Bacon stated in his CPA that William H. Burt S the colors of the 38th GA at Appomattox Court House, at Gen. Lee's surrender. D Sept/1909. Bur Burt family cem, Oglethorpe Co. Obit in Oglethorpe Echo 10/8/1909.
Burt, William T. (William Thomas Burt) - Pvt 4/27/1862. Appt. 1st Sgt 4/6/1864. S Appomattox, VA, 4/9/1865. Filed CPA, Oglethorpe Co., 1920. Born 1844, died 9/1/1934. Buried Burt's Chapel United Meth. Church Cem, Oglethorpe Co. Obit printed in Augusta Chronicle 9/12/1934, page 12.
Busbin, John D. - Pvt 4/27/1862. Died of pneumonia in Hos 2d Army Corps, Orange Court House, VA, 4/7/1864. CSR contains death claim filed by widow, Hennretia Busbinn, Oglethorpe Co. BSU.
Busbin, Peter P. - Pvt 4/27/1862. Discharged in 1862. Confederate Service Record not found. BSU.
Butler, Joe W. (Job W. Butler) - Pvt 9/29/1861. Admitted to CSA. General Hospital Farmville, VA 2/25/1865. Returned to duty 4/2/1865. BSU.
Callahan, James S. - Pvt 4/30/1862. Mortally Wounded in battle of Gaines Mill, Died 6/29/1862. CSR contains death claim filed by widow, W. A. E. Callahan. BSU.
Callaway, Jonathan I. (Jonathan Isaac Callaway) - 4th Sgt 9/29/1861. Discharged, furnished substitute, 3/1/1863. Enlisted in Co. B, 2d Kentucky Special Cavalry Battalion (Duke's Brigade.) Elected Cpt. Requested leave of absence for 7 days, 10/28/1864. No later record. Filed CPA in Morgan Co., 1923. Died 7/9/1929 in Morgan Co. BSU.
Chafin, John C. - Pvt 9/29/1861. Captured Falling Waters, MD, 7/14/1863. Released 12/13/1863. Surrendered Appomattox, VA, 4/9/1865. (Born in Wilkes Co., GA, in 1843) Filed CPA Madison Co., 1900. Died there 8/16/1914; buried at Hitchcock Family Cemetery, Madison Co.
Chandler, Reuben - Pvt 5/3/1862. Absent without leave Apr/1863. No later record. BSU.
Childress, Lafayette B. - Pvt 9/29/1861. Appt. 2d Cpl 4/22/1862. Wounded Fredericksburg, VA, 12/13/1862. Died of wound, gunshot to the arm, at Chimborazo Hos #2, Richmond, VA, 1/7/1863. CRS contains a death claim filed by father, Mastron Childress, of Oglethorpe Co. BSU.
Coile, B. J. (Benjamin J. Coile) - Pvt, 1/25/1864. Absent, on extension of furlough, Aug/1864. Retired 12/3/1864. BSU.
Coile, George N. (George Nicholas Coile) - Pvt 9/29/1861. Killed Spotsylvania, VA 5/12/1864. CSR shows mortally W Spotsylvania, admitted to Jackson Hos in Richmond with gunshot wound to left side 5/19/1864, D 6/14/1864. Bur Hollywood Cem, Rich, VA, Section: Soldiers Section, lot unknown.
Coile, James W. (James Washington Coile) - Pvt 4/28/1862. Wounded Gaines Mill, 6/27/1862. Killed at Deep Bottom, VA, 8/15/1864. Killed during skirmish 8/15/1864; body fell into the hands of the enemy. CSR contains death claim filed by widow, Lucy D. Coile, of Oglethorpe Co. BSU.
Coile, Joseph W. (Joseph Warren Coile) - Pvt 9/29/1861. Killed 2d Manassas, VA, 8/28/1862. CSR contains a death claim filed by father, James Coile, in Oglethorpe Co. BSU.
Cunningham, Frederick C. - 4th Cpl 9/29/1861. Appt 1st Cpl 12/7/1862; 5th Sgt May/1863. W in 1863. Captured Harrisonburg, VA 9/25/1864. Released Point Lookout, MD, 5/13/1865. (B Oglethorpe Co., GA May/1839) C Fred. 12/13/1862, P for E 12/17/1862. Filed CPA, Fulton Co., 1909. BSU.
Cunningham, George T. - 1st Cpl 9/29/1861. Discharged for disability, 12/7/1862. Discharge record in CSR. Born in Madison Co., GA, about 1820. Forty-two years old at discharge. BSU.
Cunningham, John W - (John William Cunningham) Pvt 1/6/1862. Roll for 11/4/1864, LOF, shows him present. NLR. D 1/6/1904. Denton Co., TX. Bur Oakwood Cem, Soldier's Row, Row 10, Space 1, Ft. Worth, TX.
Cusick, George - Pvt 4/30/1863. Substitute for Joel J. Bacon. Deserted Rich, VA, 5/14/1863 BSU.
Daniel, Amaziah C. - Pvt 9/29/1861. Appt. Commissary Sgt. Surrendered Appomattox, VA 4/9/1865. Buried Dillard Cem., Oglethorpe Co., near his home. Obit in Oglethorpe Echo, 6/2/1916.

Daniel, James Cicero - Pvt 9/29/1861. Elected 1st Lt of Co. H, 3/4/1865. Surrendered Appomattox, VA, 4/9/1865. (B in GA 3/28/1843) CSR contains a letter noting his courage in battle and recommending his promotion to Lt. Filed CPA, Henry Co., 1919, died 8/7/1927. Buried there McDonough Cemetery. Obituary in the McDonough Advertiser, 8/11/1927.
Daniel, John J. - Jr. 2d Lt 9/29/1861. Resigned 4/22/1862, due to ill health. Died 10/27/1904, buried Peek Family Cem, behind 2346 Crawfordville Rd., Wilkes Co. Obit in Oglethorpe Echo, 10/28/1904.
Daniel, William G. - Pvt 9/29/1861. Discharged 3/15/1862. Reenlisted 4/6/1862. Discharged 7/11/1862. CSR contains discharge record, 18 yrs old at discharge. BSU.
Davis, John W. - Pvt 4/28/1862. W Antietam 9/17/1862. D in Mount Jackson, VA hos 11/25/1862. D of wound, according to CSR. Bur "Our Soldiers Cem," Mt. Jackson, Shenandoah Co., VA. CSR has death claim filed by widow, Eliza M. Davis, Oglethorpe Co. Widow filed CPA, Madison Co., 1891.
Day, John H. (John Henry Day) - Pvt 9/29/1861. Mortally wounded in the battle of Gaines Mill, VA, 6/27/1862. Died from wounds, 8/1/1862. BSU.
Dorough, Robert T. - (Robert Toombs Dorough) 3d Sgt 9/29/1861. Appt 1st Sgt 4/22/1862. Elected 1st Lt 7/23/1862; wounded "Rappahannock," 8/24/1862. Cpt 6/14/1863. Surrendered Appomattox, VA, 4/9/1865. Died 7/14/1898 Tyler, Smith, TX, buried there at Oakwood Cemetery. Widow, Mary E. Dorough, filed a CPA in Dallas, TX, 1918.
Dorough, Thomas L. - Pvt 3/10/1862. Died in Augusta, GA, hospital of rubella (German measles), 6/9/1862. Buried at Magnolia Cemetery, Augusta, GA.
Doss, William L. - Pvt 5/13/1862. BSU.
Dowdy, William F. (William "Frank" Dowdy) - Pvt 5/3/1862. Discharged May/1862. BSU.
Drake, Eugene M. - Pvt 9/29/1861. W in 1862. Discharged, disability, 11/5/1863. W in the battle of Gaines Mill, 6/27/1862, gunshot wound, knee, discharged for wound. CSR has discharge record. Filed CPA, Oglethorpe Co., 1892. D 1/16/1901, bur there Clark Cem. Obit in Oglethorpe Echo, 1/18/1901.
Eberhart, Eli - Pvt 9/29/1861. Captured 7/6/1863. Paroled Salisbury, NC, 5/3/1865. Filed CPA, Oglethorpe Co., 1900, Died 1916, buried there Burt's United Methodist Church Cemetery.
Eckles, William E. - (William Edward Eckles) Pvt 9/29/1861. Died of disease in General Hos Camp Winder, Rich, VA, 1/29/1863. Treated for typhoid in Dec. 1862, Camp Winder, Richmond, VA. Bur Hollywood Cem, Richmond, VA, cem records show: Died 1/29/1863, bur in Section D, grave #203.
Evans, William N. (William Nelson Evans) - Pvt 11/18/1863. Captured Fisher's Hill, VA, 9/22/1864. Paroled at Point Lookout, MD 1/17/1865. Rcv Boulware & Cox's Wharves, James River, VA, for exchange, 1/21/1865. (Born in Wilkes Co., GA, 9/10/1822) Filed CPA, Jackson Co., 1898. BSU.
Farmer, Ethelred (Ethelred "Dread" Farmer) - Pvt 5/5/1862. Died in Staunton, VA, 8/5/1862. Buried there in Thornrose Cemetery. CSR contains death claim by widow, Amanda Farmer, Oglethorpe Co.
Farmer, Henry G. (Henry George Farmer) - Pvt 5/2/1862. Wounded at Fredericksburg, 12/13/1862. W in 1863. Appt Sgt 8/31/1864. S Appomattox, VA, 4/9/1865. Listed as Provost Guard, Appomattox roster, (Born in GA. 11/13/1840) Filed CPA, 1919, Jackson Co. D 7/29/1923, bur there Antioch Cem.
Farmer, Jesse T. (Jesse Theoderick Farmer) - Pvt 3/10/1862. K Spotsylvania, VA, 5/12/1864. CSR says W at Spotsylvania 5/12/1864, D of wound 7/23/1864. Bur Spotsylvania C. H. Confed. Cem, VA.
Faulkes - See Folks.
Faust, George Washington - Pvt 5/2/1862. Wounded Gaines Mill, VA, 6/27/1862. Discharged, furnished Jesse Jenkins as substitute, 1/4/1863. Enlisted as a Pvt in Co. K, 6th Regt GA Inf. 2/2/1864. Killed near Petersburg, VA, 7/12/1864. Other records show he was Killed 9/30/1864. Widow's obit (Martha Louise Settle) in Oglethorpe Echo, 9/18/1910. BSU.

Fleman, James R. - (James R. Fleeman) Pvt 4/28/1862. Mortally Wounded battle of Gaines Mill, VA, 6/27/1862, Died 7/21/1862 in Richmond, VA hospital. CSR contains death claim by widow, Louisa J. Fleeman, of Oglethorpe Co. BSU.
Fleman, John W. (John Wesley Fleeman) - Pvt 4/28/1862. Killed Gaines Mill, VA, 6/27/1862. Listed among seven members of Co. E. as "dead on the field" Gaines Mill. CSR contains death claim filed by widow, Louisa E. Fleeman, Oglethorpe Co. Widow filed CPA, Oglethorpe Co., 1891. BSU.
Fleman, William J. (William Johnson Fleeman) - Pvt 4/28/1862. Surrendered Appomattox, VA, 4/9/1865. Died 5/12/1905. Buried at Beaverdam Church Cemetery, renamed Carter-Fleeman Cemetery, Oglethorpe Co. Obituary in Oglethorpe Echo 5/19/1905.
Gillespie, J. M. (Miller W. Gillespie) - Pvt 9/29/1863. Discharged 10/12/1864. First name found in CSR. CRS contains discharge. Born Franklin Co., GA, discharged for old age, age 48. BSU.
Glenn, Dred W. (or Dread W. Glenn) - Pvt 9/29/1861. W Gaines Mill, 6/27/1862. Captured Mine Run, VA 5/12/1864. Sent to Ft. Delaware, 5/20/1864, P there for E Feb/1865. E Boulware & Cox's Wharves, James River, VA, 3/10/1865. (B Oglethorpe Co., GA, 1843) Filed CPA, Madison Co., 1895. bur there Mount Hermon Presbyterian Chur Cem. marker reads, B 4/29/1843, died 7/12/1918.
Glenn, James M. - Pvt 9/29/1861. Killed battle of Gaines Mill, VA 6/27/1862. Listed among seven members of Co. E. as "dead on the field" the battle of Gaines Mill. BSU.
Glines, Thomas A. (or Glenn or Grimes) - Pvt 9/29/1861. Discharged 12/7/1862. CSR discharge certificate states he was B Martinboro, New Hampshire, age 46, discharged for hernia. BSU.
Hardeman, Charles T. (Charles Troy Hardman) - Pvt 5/6/1862. Wounded in right leg in VA, 1864. Sent to Staunton, VA hos 10/13/1864. S Appomattox, VA, 4/9/1865. (B Oglethorpe, GA 5/9/1835) Filed CPA, Oglethorpe Co., 1903, died there 6/11/1924. Buried Hardeman Family Cem near his home in Grove Creek District, Crawford, Oglethorpe Co. Obit in Oglethorpe Echo newspaper 6/13/1924.
Harris, Nathan - Pvt 9/29/1861. Wounded Gaines Mill, 6/27/1862, Wounded Fredericksburg, 12/13/1862. Roll dated 11/4/1864, last on file shows him on furlough since Mar/1864. NLR. BSU.
Harris, Woodson B. - Pvt 9/29/1861. Mortally wounded Gaines Mill, 6/27/1862, died at Richmond, VA, 8/4/1862. BSU.
Hawkins, Charles Alexander - (Charlie") 2d Lt 9/29/1861, Elected Cpt 7/23/1862. Killed Winchester, VA, 6/13/1863. Wounded 2d Manassas, 8/28/1862; wounded Antietam 9/17/1862; wounded Fredericksburg, 12/1/3/1862, Buried Stonewall Confed. Cemetery Winchester, VA. Buried in Georgia Section, Plot# 58. Widow, H. H. Hawkins, filed a CPA, Chattahoochee Co., 1891. Death notice in Augusta Chronicle, 7/30/1863, page 2.
Hawkins, John M. - (John Milner Hawkins) 2d Sgt 9/29/1861. Wounded Gaines Mill, 6/27/1862. Det Sub Enrolling Officer in GA in 1863. W Cedar Creek. S Appomattox, VA, 4/9/1865. 6/27/1862. Filed CPA, Jackson Co., 1911, died there 11/23/1929, buried there Apple Valley Bap. Church.
Hendricks, J. H. (John H. Hendricks) - Pvt In 1862. Substitute for Randolph J. Arnold. Died Richmond, VA, of typhoid fever, 7/20/1862 or 8/6/1862. CSR also states he died 8/1/1862. BSU.
Heyward, William G. (or Haygood) - Pvt 5/23/1862. BSU.
Holcombe, William P. (William Pindar Holcombe) - Pvt 9/29/1861. Discharged 8/8/1862. CSR has discharge record. Discharged for feeble health, 59 years old. Descendants say D in 1869. BSU.
Hopper, Booker W. (Booker Woodson Hopper) - Pvt 9/29/1861. W battle of Fred., 12/13/1862. S Appomattox, VA 4/9/1865. Listed as Provost Guard on Appomattox roster. Filed CPA, Autauga Co., AL, 1907. D 5/19/1931, bur there Liberty Baptist Church Cem, Obit in Prattville Progress, 5/21/1931.
Hopper, Daniel H. - Pvt 9/29/1861. Killed 2d Manassas, VA, 8/28/1862. CSR contains a death claim filed by father, Jonathan Hopper, in Oglethorpe Co. BSU.
Hopper, W. W. (William Whit Hopper) - Enlisted as a Pvt in Co. B, 8th Regt AL Inf. 5/13/1861. Transferred to Co. E, 38th Regt GA Inf. as a Cpl (being a citizen of GA,) 4/1/1864. Wounded in right thigh, gunshot, date and place not given. Captured Jackson Hospital, Richmond, VA 4/3/1865, died there of wounds 4/16/1865. Buried at Hollywood Cemetery, Richmond, VA.

Howard, William G. (William Glenn Howard) - Pvt Enlisted 5/2/1862 Lexington, GA. Listed among seven members of Co. E. "dead on the field" battle of Gaines Mill. CSR contains death claim filed by widow, Elizabeth M. Howard, in Oglethorpe Co. BSU.
Huff, John Peter - Killed in the battle of Gaines Mill, 6/27/1862. Listed among 7 members of Co. E. as "dead on the field" battle Gaines Mill, CSR contains a death claim filed by widow, Nancy Huff, Oglethorpe Co. BSU.
Huff, John W. - Pvt 9/29/1861. Wounded Gaines Mill, VA 6/27/1862. Wounded battle of Antietam, 9/17/1862. On frl Sept. - Nov. 4, 1864. CPA shows at home on sick frl close of war. (Born in GA, Died Oconee Co., GA, 1925) Filed CPA Oconee Co., 1917. Buried Hardigree Cemetery, Oconee Co.
Jackson, John B. (John Baughn Jackson) - Pvt 9/29/1861. Sent to Staunton, VA hospital 9/25/1864. NLR. Reported as killed Hatcher's Run, Feb 5-6, 1865, in account written by Lt George Lester. BSU.
Jackson, John M. - Pvt 4/23/1862. Wounded and left Gettysburg, PA, 7/2/1863. Died of pneumonia in Chambersburg, PA, hospital, 7/8/1863. CSR contains a death claim filed by widow, Eleanor J. Jackson, in Oglethorpe Co. BSU.
Jackson, John W. - Pvt 9/29/1861. Killed Gaines Mill, VA, 6/27/1862. Listed among seven member of Co. E. as "dead on the field" the battle of Gaines Mill. CSR contains a death claim filed by father, Edward Jackson, in Oglethorpe Co. Possibly man identified as William Jackson, buried at Oakwood Cemetery, Richmond, VA, Div A, Row J, Grave 11.
Jackson, William S. (William "Sanders" Jackson) - Pvt 4/23/1862. Appt Cpl Surrendered Appomattox, VA 4/9/1865. Filed a CPA, Oglethorpe Co., 1900. Died Aug/1922. State death certificate states buried Stephens Cemetery, Oglethorpe Co., GA Obit in Oglethorpe Echo, 8/25/1922.
Jackson, Willis Benjamin - Pvt 4/23/1862. Wounded in foot Antietam, MD, 9/17/1862. Appt 4th Cpl in 1863; Sgt; wounded in neck Wilderness, VA, 5/6/1864; wounded in head Fisher's Hill VA 9/22/1864. Surrendered Appomattox, VA, 4/9/1865. (B in GA 3/17/1837) Filed CPA, Oglethorpe Co., 1905, died, buried there Salem Baptist Church Cemetery. Obit in Oglethorpe Echo, 11/18/1921.
Jarvis, John W. - Pvt 4/22/1862. Admitted to General Hos Charlottesville, VA. 9/1/1864. Trn to Lynchburg, VA 9/26/1864. Roll dated 11/4/1864, shows him "On frl given Sep/1864." NLR. BSU.
Jenkins, Jesse - Pvt 1/4/1863. Substitute for George W. Faust. C Spotsylvania, VA, 5/12/1864. P Ft. Delaware, 9/14/1864. Sent to Aiken's Landing, VA, for exchange 9/18/1864. Rcv General Hos #9, Rich, VA 9/21/1864. Sent to Jackson Hos Rich, VA 9/22/1864. Frl for 40 days 9/27/1864. NLR. BSU.
Jernigan, William H. - Pvt 9/29/1861. Appt Cpt and A. Q. M. 10/8/1861. Roster dated 3/18/1865, shows he was assigned to duty in Florida by Quartermaster General in 1864, by act of Congress. BSU.
Johnson, Nicholas B. (Nicholas "Nick" Bradley Johnson) - Pvt 9/29/1861. Appt 4th Sgt 2/18/1863. C Spotsylvania, VA, 5/12/1864. Paroled Ft. Delaware, 10/30/1864. Exchanged Point Lookout, MD 10/31/1864. Paroled Salisbury, NC. 5/3/1865. Filed CPA, Oglethorpe Co., 1893. Died 1/4/1916. Buried Salem Baptist Church Cemetery, Candler Co. Obit in Oglethorpe Echo, 1/14/1916.
Johnson, William W. - Pvt 9/29/1861. Wounded Fredericksburg, VA, 12/13/1862, when the Regt, charged a Federal battery. Appt 4th Cpl in 1863. Killed at Spotsylvania, VA, 5/12/1864. BSU.
Johnson, Woodson H. - Pvt 9/29/1861. Died in Augusta, GA hos 5/31/1862, of typhoid fever. From Oglethorpe County, GA. Buried Magnolia Cemetery, Confederate Square Section, Augusta, GA.
Jones, George A. - Pvt 3/15/1862. Appt Quartermaster Sgt, 12/18/1863. Roll dated 11/4/1864, last on file, shows him present. No later record. BSU.
Kidd, John - Listed as a recruit of Co. E, according to a post war roster. No CSR found. BSU.
Kidd, Richard H. - Pvt 4/28/1862. Roll dated 11/4/1864, LOF, shows present. NLR. Filed a CPA Elbert Co., 1900. Bur Rehoboth Bap. Church, Elbert Co., marker reads:, CO E 38 GA INF.
Kidd, Robert M. - Pvt 4/25/1862. Died in Belmont & Grove Hos, Lovingston, VA, 8/29/1862. BSU.

Kidd, W. B. H. (Willis Benjamin Henderson Kidd) - Pvt 1/10/1864. Roll dated 11/4/1864, last on file, shows present. Died 11/10/1900. Buried West Hill Cemetery, Grayson Co., TX. Sarah J. Kidd, filed a CPA in Childress Co., TX, 1914.

Landrum, F. M. (Francis Marion "Frank" Landrum) - Pvt 4/25/1862. Appt 1st Sgt May/1863. Captured Cedar Creek, VA, 10/19/1864. P at Point Lookout, MD, and E 2/10/1865. Admitted Jackson Hos Rich, VA, 2/15/1865. Trn to Camp Lee Rich, VA, 3/1/1865. Filed CPA, Oglethorpe Co., 1899. D 9/1/1921 Oglethorpe Co., buried there Antioch Baptist Church. Obit in Oglethorpe Echo, 10/7/1921.

Landrum, Jabez R. (Jabez "Jabe" R. Landrum) - Pvt 5/2/1862. Discharged 5/10/1862. Wife's obit in Oglethorpe Echo, 3/20/1891; reads in part, "Mary A. Landrum, consort of Jabez R. Landrum, she D 3/19/1891. Married 2/10/1853 and left widow by her husband's death on battlefield, 7/30/1864. BSU.

Lester, George H. (George Harper Lester) - 1st Lt 9/29/1861. Resigned 7/15/1862. Died 11/7/1893. Buried Clark Cemetery, Lexington, Oglethorpe Co. Obituary in Oglethorpe Echo, 11/10/1893.

Lumpkin, Joseph Jackson - Pvt 9/29/1861. Discharged, disability, Liberty Mills, VA, 7/27/1862. CSR contains discharge record. Died 1/12/1892. Buried Salem Bap. Church Cem, Oglethorpe Co.

Martin, Thomas - Pvt 9/29/1861. Killed Spotsylvania, VA, 5/12/1864. Widow, Elizabeth Martin, filed CPA application in Oglethorpe Co. in 1891. BSU.

Martin, William T. J. (William Thomas J. Martin) - Pvt 3/10/1862. Roll dated 11/4/1864, shows him "Absent on extension of frl given 6/6/1864." Wounded Antietam, 9/17/1862. Died 3/14/1885, Oglethorpe Co. Buried there at Burt's Chapel United Meth. Church. Obit in Oglethorpe Echo, 3/27/1885. Widow, L. F. Martin, filed a CPA, in Oglethorpe Co., 1902.

Mathews, Adonivan J. (Adoniram Judson Mathews) - Pvt 5/3/1862. Discharged from General Hospital #13, Rich, VA, 9/23/1862. Died 4/7/1871, bur Butler - Mathews Family Cem, Spalding Co.

Mathews, James Davant - Cpt in Co. E, 9/29/1861. Severely wounded Gaines Mill, VA, 6/27/1862. Elected Major June 1862; Lt Col 7/15/1862; Col 12/13/1862. Retired to Invalid Corps, 10/31/1864. Died 8/2/1878 Hall Co., GA, buried Lexington Presbyterian Church Cemetery, Oglethorpe Co. Obituary in Augusta Chronicle, 8/27/1878, Page 1.

Mathis, George L. (or George L. Mathews) - Pvt 9/29/1861. K Wilderness, VA, 5/6/1864. Listed as a member of Co. E. C the battle of 2d Manassas, 8/28/1862. CSR also list name as Mathews. BSU.

Mathis, Robert H. (or Robert A.) - Pvt 9/29/1861. Killed Antietam, MD, 9/17/1862. BSU.

Mealor, James Asberry - Pvt 9/29/1861. Wounded Gaines Mill, 6/27/1862. Appt 3d Cpl Jan/1863. Wounded in 1864. Sent to Staunton, VA, hos 10/10/1864. CPA shows on wounded frl close of war. Filed CPA, Clarke Co., 1906. Died and bur there 8/3/1928. Marker reads, CO E, 38 GA INF.

Milner, John E. - Pvt 9/29/1861. Discharged, furnished Joseph Walter as substitute, 10/10/1862. CSR contains discharge record. Died Jonesboro, 1/27/1901. Buried McDonough Cem, Henry Co. Widow, Cornelia A. Milner, filed CPA, Clayton Co., 1919.

Milner, Pleasant J. - Pvt 9/29/1861. Captured Spotsylvania, VA, 5/12/1864. Paroled Ft. Delaware, Feb/1865, exchanged 3/7/1865. Admitted General Hos #9, Rich, VA 3/10/1865. Transferred to Camp Lee Rich, VA, 3/11/1865. (B in Oglethorpe Co., GA 1840. D of typhoid fever Palestine, Anderson Co., TX, 1878. Bur there.) Widow, Mary McWhorter Milner, filed a CPA, Fulton Co., 1926. BSU.

Milner, William P. (William Pitman Milner) - Pvt 5/15/1862. Died near Charlottesville, VA, 6/23/1862. BSU.

Murray, Edward H. (Edward Hartsfield Murray) - Pvt 9/29/1861. Died Petersburg, VA, 6/18/1862. CSR contains court document stating his will was probated in Oglethorpe Co., Oct/1862. BSU.

Nash, George W. - Pvt 9/29/1861. Mortally Wounded battle of Gaines Mill, 6/27/1862. Died 7/11/1862. CSR contains death claim filed by father, Thomas Nash, of Taliaferro Co. BSU.

Nash, Joseph S. - Pvt 9/29/1861. Appears last on roll for 10/31/1862, presence or absence not shown, "Remark on roll is illegible on account of faded ink." BSU.

Nash, Thomas J. - Pvt 9/29/1861. Died in Augusta, GA hospital. 4/29/1862. Article in Augusta Chronicle & Sentinel, dated 5/13/1862, shows he died at the Military Hospital at Augusta, but his

remains removed home for burial. Died of Pneumonia. CSR contains a death claim filed by his father, Thomas Nash, Taliaferro Co. Buried Thomas Nash Cemetery, Oglethorpe Co.
Norton, William Jefferson - 2d Cpl 9/29/1861. Appt 3d Sgt 4/22/1862. Wounded 2d Manassas, 8/28/1862. Wounded 1863. Roll dated 11/4/1864, last on file, shows him "Absent, Det Enrolling Officer in Oglethorpe Co., GA Feb/1864." NLR. Bur Crawfordville City Cem, Taliaferro Co.
Park, Jerry M. (Jeremiah "Jerry" M. Parks) - Pvt 9/29/1861. Wounded in the head and permanently disabled, at Gaines Mill, VA, 6/29/1862. Detailed as guard in Atlanta, Nov/1863. Discharged, disability, Feb/1864. (Born in GA abt. 1828) Filed a CPA in Gordon Co., 1889. BSU.
Patridge, Francis B. (Or Francis "Frank" B. Partridge) - Pvt 9/29/1861. Died in Augusta, GA hos 4/25/1862, of pneumonia; 27 years old. Buried Confed. Square, Magnolia Cem, Augusta, Rich Co.
Patridge, Nicholas G. J. - Pvt 3/10/1862. D of typhoid fever in General Hos Staunton, VA Aug. 24, or 10/10/1862. Bur there in Thornrose Cem. CRS has death claim filed by widow, Emma Partridge.
Peek, Archibald P. (or Peeke) - Pvt 9/29/1861. Discharged prior to 10/31/1862. CSR contains discharge certificate dated 7/22/1862. Discharged due to rheumatism and ill health. BSU.
Peek, Thomas (or Peeke) - Pvt 9/29/1861. Roll dated 11/4/1864, LOF, shows present. NLR. BSU.
Peterman, George L. - Pvt 4/28/1862. Died in Chimborazo Hos #2, Rich, VA 3/19/1863, of typhoid fever. CSR has death claim filed in Oglethorpe Co. by widow, Rebecca J. Peterman. Bur Oakwood Cem, Rich, VA. Modern brass grave marker, reads GEORGE L. PETERMAN, PVT, CO. E, 38 GA.
Phararah, John - Pvt 9/29/1861. Discharged prior to 10/31/1862. CSR contains discharge certificate for disability dated 10/14/1862. Discharge states Born in France, 55 years old at discharge. BSU.
Pittard, John T. - Pvt 4/30/1862. Surrendered at Appomattox, VA, 4/9/1865. BSU.
Pool, Sylvanus - Pvt 9/29/1861. On detached duty as brigade pioneer Aug/1863, near Orange C. H. Aug 1863. D in Jackson Hos Rich, VA, 5/30/1864. Bur there in Hollywood Cem. Hollywood Cem, in Rich, records reads: Pool, S., Company E, 38th GA, D 5/30/1864, marker reads CO E, 38 GA INF.
Poss, William L. - Pvt 5/12/1862. Roll dated 11/4/1864, last on file, shows Absent on furlough given 8/1864. No later record. BSU.
Power, B. D. - Pvt 11/2/1863. In hospital 9/14/1864. No later record. First name unknown. BSU.
Pratt, Elijah - Pvt 9/29/1861. Killed Gaines Mill, VA, 6/27/1862. Listed among seven members of Co. E. as "dead on the field" at the battle of Gaines Mill. BSU.
Raiden, James William - Pvt 5/2/1862. Appt courier for A.Q. M. in 1863. Captured Wilderness, VA 5/6/1864. Released at Elmira, NY, 6/23/1865. Died Oconee Co., 12/24/1894. Buried Watkinsville City Cem, Oconee Co. CSA grave marker reads, CO E, 38 GA INF. Widow, Martha Caroline Campbell Raiden, filed CPA, Oconee Co., 1901.
Robinson, Alexander W. (Alexander "Webb" Webster Robinson) - 3d Cpl 9/29/1861. Captured Waterloo, PA, 7/5/1863. Paroled at Point Lookout, MD, 2/18/1865. Received Boulware & Cox's Wharves, James River, VA, for exchange, Feb. 20-21, 1865. At Camp Lee near Richmond, VA, 2/28/1865. No later record. Died 1907, buried Clifton Cemetery, Fairfax Co., VA.
Robertson, Walter S. - 5th Sgt 9/29/1861. Appt 2d Sgt 4/22/1863. Reduced to ranks 5/1/1863. Appt 2d Sgt 4/6/1864. Appt Sgt Major 4/6/1864. Dangerously W Cedar Creek, 10/19/1864. Admitted to Depot Field Hos Winchester, VA, 10/20/1864, died from hemorrhage and pyaemia, 11/5/1864. Buried Mount Hebron Cem, Stonewall Confed. Cem section, Winchester, VA, Georgia Section, Plot #285.
Scoggins, J. S. (John S. Scoggins) - Pvt 9/29/1861. Died at Camp Mercer, near Savannah, 4/9/1862. CSR contains death claim filed in by widow, Sarah J. Scoggins, of Cobb Co. BSU.
Scoggins, M. C. (Malcom McPherson Scoggins) - Pvt 9/29/1861. Wounded Antietam, 9/17/1862. Det in Richmond, VA, hospital 2/10/1863. Roll of Camp Winder Hospital Richmond, VA dated 10/31/1864, shows him "Attached to hospital as nurse 5/1/1864." No later record. BSU.
Scoggins, William J. - Pvt 9/29/1861. Wounded battle of Antietam, 9/17/1862, gunshot, arm amputated. On wounded furlough. Discharged for wounds, when admitted to General Hos #9, Rich, VA, Mar/1864. CSR contains discharge record. Filed CPA, Oglethorpe Co., 1879. BSU.

Shearer, William S. - Pvt 9/29/1861. Wounded at Antietam, MD, 9/17/1862. Died of wounds 11/20/1862. Buried at Thornrose Cemetery, Staunton, Augusta Co., VA.
Scisson, Larkin R. (or Sisson) Substitute for Oliver H. Arnold, killed 2d Manassas, 8/28/1862. BSU.
Sisson, William D. (William Dyke Sisson) - Pvt 9/29/1861. Appt 5th Sgt Jan/1863; 3d Sgt in 1863. Wounded Petersburg, VA 4/3/1865. Furloughed from General Hos #9, Richmond, VA, 3/17/1865. (B Gilmer Co., GA 2/4/1842. Died Hancock Co., GA 1918) Filed CPA, Hancock Co., 1910. Buried there Lunsford-Wall-Sisson Family Cemetery, CSA Marker reads, CO E, 38 GA INF.
Smith, Charles H. - Pvt 9/29/1861. Discharged prior to 10/31/1862. Discharged disability, in camp near Richmond, VA. CSR contains discharge record. Filed CPA, Clarke Co., 1901, Died 8/9/1919, buried Oconee Hill Cemetery, Oconee Co.
Smith, Doctor H. (or D. C. Smith) (Doctor Simon Smith) - Pvt 5/3/1862. Wounded at Gaines Mill, VA, 6/27/1862. Died of variola in General Hospital, Danville, VA, 11/19/1862. CSR contains full name. Widow, Almeda A. Smith, filed CPA, Oglethorpe Co., 1891. BSU.
Smith, George William - Pvt 5/8/1862. Wounded 2d Manassas, VA, 8/28/1862. Appt 2d Sgt May 1863. Discharged, disability, 11/25/1863. Filed CPA, Oglethorpe Co., 1889. Died 9/30/1896. Buried Clark Cem, Lexington, Oglethorpe Co., Obit in Oglethorpe Echo, 10/2/1896.
Smith, Hay T. - Pvt 9/29/1861. W and C Gettysburg, PA, 7/4/1863. P at PLO, MD, and Trn to City Point, VA for E, 3/16/1864. Rcv there 3/20/1864. Register of Floyd House & Ocmulgee Hos Macon, GA, dated 5/27/1864, shows gunshot wound arm, applied for retirement. Retired 11/19/1864. BSU.
Smith, John T. - Pvt 9/29/1861. Filed CPA, Wilkes Co., 1898. Stated he surrendered at Washington Co., GA after the war ended. J. W. Arnold served witness in CPA and stated, "He served as a sharpshooter and killed as many enemy as any man in the war." BSU.
Smith, John W. - Pvt 4/29/1862. Captured Spotsylvania, VA, 5/12/1864. Died of typhomalarial fever Ft. Delaware, 6/14/1864. Bur Finn's Point National Cem, Salem, NJ, Section CM, Site 2066.
Smith, Manoah - Pvt 4/29/1862. Died Petersburg, VA, 7/8/1862. Widow, Mary E. Smith filed CPA, Oglethorpe Co., 1891. She stated in CPA he died of measles at Lynchburg, VA, Confed. hos. BSU.
Smith, Sion Thomas - Pvt 9/29/1861. Wounded Gaines Mill, 6/27/1862. Died Winder Hospital, Richmond, VA, 8/1/1862. Buried Hollywood Cem, Richmond, VA, bur soldier's section H, lot 179.
Smith, William J. (William James "Willie" Smith) - Pvt 9/29/1861. Appt 1st Sgt 7/23/1862. Killed at Antietam, MD, 9/17/1862. BSU.
Sorrow, Mathew E. - Pvt 9/29/1861. Slighted wounded in head Cedar Creek, reported in Augusta Chronicle, 11/4/1864. Sent to Staunton, VA, hospital, 10/20/1864. Died of pneumonia in General Hospital Farmville, VA, 3/14/1865. BSU.
Taylor, John H. - Pvt 9/29/1861. Sick in hospital Apr/1863. No later record. Post war account by Lt Lester of Company E, states he died at Charlottesville, VA, 9/15/1862. BSU.
Thornton, Samuel A. - Pvt 9/29/1861. Appt 3d Cpl 12/9/1861. Elected Jr. 2d Lt in 1863; 2d Lt 6/14/1863; 1st Lt 6/16/1863. W 2d Manassas, 8/28/1862, W. 5/4/1863 Fred., VA, shot through lung. Surrendered at Appomattox, VA, 4/9/1865. Dentist by profession. BSU.
Tiller, John G. - (John G. "Red" Tiller) Pvt 9/29/1861. Appt 2d Cpl Jan/1863. Wounded in right hip, in attack on Ft. Stedman, Petersburg, VA, 3/25/1865. Captured Richmond, VA, hospital, 4/3/1865, died there of wounds, Jackson Hospital 4/16/1865. Buried there in Hollywood Cemetery.
Tiller, Thomas (Thomas Parson Tiller) - Pvt 5/2/1862. Transferred to Co. H, 6/3/1862, but name does not appear on rolls of said company. Died 5/1/1880. Buried Cedar Creek Bap. Cem, Hart Co.,
Turner, Jesse T. - Pvt 3/10/1862. Died 9/16/1863. Buried in Oakland Cem Atlanta, GA, Section B, Row 7, Grave 19. Grave has a CSA marker, reads Jesse T. Turner Confederate States Army.
Turner, Samuel - Pvt 9/29/1861. K Gaines Mill, VA 6/27/1862. Listed among seven members of Co. E. as "left dead on the field" at battle of Gaines Mill. Bur next day on battlefield by comrades. BSU.
Vaughn, Charles C. (Charles "Charlie C. Vaughn) - Pvt 4/29/1862. Died in Winder Hospital, Richmond, VA, 7/27/1862. Buried there in Hollywood Cemetery. CSR contains death claim filed by widow, Frances Caroline Tiller Vaughn, in Oglethorpe Co.

Waddail, John N. (or Waddell) - Pvt 1/25/1864. C Spotsylvania, VA, 5/12/1864. P Ft. Delaware, Feb/1865. E 3/7/1865. Rcv Boulware & Cox's Wharves, James River, VA, Mar. 10-12, 1865. BSU.
Wall, Benjamin F. (Benjamin Franklin Wall) - Pvt 3/29/1862. Roll dated 11/4/1864, LOF, present. Filed CPA, Oglethorpe Co., 1902, died 2/7/1920. Buried Sardis Bap. Cem, Wilkes Co. CSA marker.
Wall, Julius M. (Julius McIra Wall) - Pvt 9/29/1861. Died Savannah, GA, 6/3/1862. CSR contains a death claim filed by father, J. Thomas Wall, Wilkes Co. BSU.
Walters, Joseph F. (or Walter) - Pvt 10/10/1862. Substitute for John E. Milner. Trn to C. S. Navy 4/6/1864. Appears as 3d Assistant Engineer attached C.S. ram William H. Webb, Lieut. Charles W. Read commanding. C when ram was destroyed 4/24/1865, below New Orleans, La., and later P. BSU.
Webb, Joseph B. (Joseph Benjamin Webb) - Pvt 9/29/1861. W Gaines Mill, VA, 6/27/1862. Discharged, disability, General Hos #19, Rich, VA, 10/24/1862. CSR contains discharge. Filed CPA, Oglethorpe Co., 1899. Died 12/10/1915, bur there Goss/Webb Family Cem, near Collier Bap. Church.
West, James F. - Pvt 9/29/1861. At Chimborazo Hospital, Richmond, VA, pneumonia 4/5/1863, Died 4/24/1863. BSU.
Whitehead, Elijah D. (Elijah Dean Whitehead) - Pvt 5/5/1862. Roll dated 11/4/1864, last on file, shows present. Died 5/31/1895, buried Grey Hill Cemetery, Jackson Co.
Williams, John J. - Pvt 5/2/1862. Died of measles at Richmond, VA, 7/21/1862. Widow, Adaline Williams, filed CPA Madison Co., 1891. Buried Oakwood Cemetery, Richmond, VA.
Wright, James W. - Pvt 9/29/1861. Appt Color Cpl in 1861. Wounded in 1862, wounded at Antietam, 9/17/1862. Roll dated 11/4/1864, last on file, shows present. Captured, date/details unknown, paroled, May/1865, age 38, according to CSR. BSU.
Young, John B. (John Bennett Young) - Pvt 9/29/1861. C Hedgeville, WV, 9/19/1862. E Camp Chase, OH 10/13/1862. Rcv near Vicksburg, Miss. 11/1/1862. C Waterloo, PA, 7/5/1863. E Ft. Delaware, 7/31/1863. C Spotsylvania, VA 5/12/1864. Died of smallpox at Ft. Delaware, 7/31/1864. Bur Finn's Point National Cem, New Jersey. Listed in U.S. Veteran's Admin. records as Pvt John B. Young, U.S. Army.

COMPANY F, THORNTON LINE VOLUNTEERS, HART & ELBERT COUNTIES, GA

*This company was originally known as Co. I and was organized by Cpt John C. Thornton.

Adams, Alfred H. - Pvt 10/15/1861. Captured Gettysburg or Waterloo, PA, Jul/1863. Died of disease Point Lookout, MD, Aug 1, or 26, 1864. Bur Point Lookout Confed. Cem Scotland, St. Mary's Co., MD. Name not on the CSA memorial at cemetery, but many known PLO CSA dead are not listed.
Adams, Harper - Pvt 10/15/1861. Captured Gettysburg, PA, 7/1/1863. Paroled at Point Lookout, MD 2/18/1865. Rcv Boulware & Cox's Wharves, James River, VA for exchange, Feb. 20-21, 1865. BSU.
Adams, H. P. (Hiram P. Adams) - Pvt 8/29/1862. Wounded Fredericksburg, VA 12/13/1862. Died of wounds in General Hospital # 24 (Moore Hos,) Richmond, VA, 1/24/1863. BSU.
Adams, James Joseph (or James Jordan Adams) - Pvt 8/29/1862. Wounded, 1864. Captured near Petersburg, VA, 3/25/1865. Released Point Lookout, MD, 6/23/1865. Died 1905, buried Coldwater United Methodist Church Cemetery, Elbert Co. Tribute and obit in Elberton Star, 12/21/1905.
Adams, Lawrence M. (Laurens M. Adams) - Pvt 10/15/1861. Appt 4th Cpl Feb/1863; 3d Cpl in 1863. Died at Elberton, 11/5/1863. CSR contains death claim filed by father, John M. Adams, Elbert Co. Father lists first name as "Laurens" BSU.
Adams, Tinsley R. - (Tinsley Rucker Adams) Pvt 3/5/1863. C in attack on Ft. Stedman, Petersburg, VA, 3/25/1865. Released PLO, MD 6/23/1865. (Resident GA since 1845) Filed CPA, Elbert Co, 8/13/1919. D 9/22/1919, bur Elmhurst Cem, Elbert Co., CSA marker reads, CO F, 38 GA INF.
Adams, William H. - Pvt 10/15/1861. Appt 4th Sgt May/1862; 3d Sgt Mar/1863; 1st Sgt Jan/1864. C Wilderness, VA 5/6/1864. Released Elmira, NY 6/14/1865. Filed CPA, White Co., 1899. BSU.

Alexander, Philemon W. (Philemon "Phil" Willhite Alexander) - Enlisted as a Pvt in Co. E, 4th Regt GA Inf. 4/28/1861. Discharged, furnished Ulysses M. Robert as substitute, 10/23/1861. Enlisted as Pvt Co. F, 38th Regt 1/23/1862. Appt Quartermaster Sgt 4/1/1862. W & C Gettysburg, PA, 7/3/1863. E City Point, VA 8/28/1863. Appt paymaster clerk in C.S. Navy Apr/1864. P Augusta, 6/3/1865. Filed CPA, Berrien Co., 1908, D 9/8/1909, bur there Fletcher Cem, A.K.A. McMillan-Fletcher Cem.
Almond, Isaac B. - Pvt 10/15/1861. Captured Gettysburg, PA, 7/3/1863. Took OOA to U.S. Govt., enlisted in U.S. service 1/24/1864. On roll for 8/31/1864, reported Died, date unknown." Enlisted in 1st U.S. Vol. Inf. Filed CPA, Elbert Co., 1907. Died 1916, buried there Doves Creek Bap. Cemetery. Obituary in Elberton Star 6/16/1916.
Almond, John A. H. - 5th Sgt 10/16/1861. Died Savannah, GA, 5/14/1862. BSU.
Almond, John Benjamin - Pvt 3/23/1862. W Fred., VA 12/13/1862. CPA shows enlisted as Pvt, Lee's Battalion S Augusta, GA Apr/1865. (D Elbert Co., 11/27/1899) Obit in Elberton Star, 11/23/1899. Bur Elmhurst Cem, Elbert Co. Widow, Sav A. E. Almond, filed CPA, Elbert Co., 1924.
Bailey, John A. - Pvt 5/1/1864. Wounded and captured Cold Harbor, VA 6/3/1864. Released Elmira, NY, 7/11/1865. (Born Bartow Co., 1845. Died Hartwell, 7/2/1910) Filed CPA, Hart Co., 1895. Buried Crossroads Baptist Church Cemetery, Hart Co.
Bentley, John D. (John Davis Bentley) - Pvt 10/15/1861. Wounded 1864. Surrendered at Appomattox, VA, 4/9/1865. BSU.
Bentley, Joseph A. J. - ("Joe") Pvt 10/15/1861. Appt Sgt Aug/1862. Wounded 2d Manassas, 8/28/1862. 2d Sgt Mar/1863; 3d Sgt Aug/1864. Sick in Harrisonburg, VA, hospital, 10/24/1864. Captured near Farmville, VA 4/6/1865. Released Newport News, VA 6/26/1865. Died 1/15/1897, Elbert Co. Obit in Elberton Star 1/15/1897. Obituary states, bur Elberton Cemetery.
Bentley, McAlpin (McAlpin Anslem Bentley) - Pvt 5/10/1862. Wounded at battle of Fredericksburg, VA 12/13/1862. Died 6/6/1864. BSU.
Bentley, R. F. (Robertus Franklin Bentley) - Pvt 5/14/1862. Appt 3d Cpl Jan/1864. Wounded, date and place not given. Roll dated 11/6/1864, last on file, shows present. Died 1871, buried at Wellmaker-Bentley-Thomson Family Cem, Wilkes Co.
Black, John W. - Pvt 5/14/1862. Recruit, enlisting at Savannah, GA. Killed at Second Manassas, 8/28/1862. CSR contains death claim filed by his father, John W. Black, in Elbert Co. BSU.
Bond, Andrew J. (Andrew Jackson Bond) - Pvt 3/10/1862. C and P at Hartwell, GA 5/18/1865. Born 11/10/1837, died 7/10/1909. Bur at Concord United Methodist Church Cemetery, Elbert Co.
Brewer, Columbus (Clemens "Clem" Brewer) - Pvt 3/16/1862. Discharged 6/24/1862. BSU.
Brewer, R. D. W. - Appears on register of Receiving and Wayside Hospital, or General Hospital #9, Richmond, VA, 10/3/1862, sent to Chimborazo Hospital. No later record. First name unknown. BSU.
Broadwell, William T. (William "Billy" Thomas Broadwell) - Pvt 10/15/1861. Appt 4th Cpl Jan/1863. W Wilderness, VA 5/5/1864. Roll dated 11/6/1864, LOF, shows "Home on W frl ; time out unknown." C and P Hartwell, GA 5/18/1865. Filed CPA 10/6/1919. B 1844, D at Abbeville, SC, 10/2/1943, bur there Presbyterian cem.; obit in The Greenville (SC) News, 10/3/1943, page 6A.
Brown, Asa C. (Asa Chandler Brown) - Pvt 5/10/1862. Killed Wilderness, VA 5/5/1864. Widow, Priscilla E. Thornton Brown, filed CPA, Elbert Co., 1891. She stated he was shot through the head and killed instantly. BSU.
Brown, Benager T. (Benager Teasley Brown or Benajah) - 2d Lt 10/15/1861. Elected 1st Lt 6/22/1862. D Staunton, VA 7/27/1862 or 7/29/1862. D of typhoid fever, General Hos Staunton. Bur there Thornrose Cem. CSR contains death claim filed by father, Rolen J. Brown, in Elbert Co.
Brown, Burton T. - Pvt 10/15/1861. Wounded in shoulder, resulting in complete atrophy of muscles of arm from removal of several inches of bone, Fredericksburg, VA, 12/13/1862. Roll dated 11/6/1864; shows "Absent in hos in GA, wounded." unfit for further service. Filed CPA, Hart Co., 1888. Died 9/5/1912, bur Elmhurst Cemetery, Elberton. CSA marker reads PVT 38 GA INF.
Brown, Eppy W. (Eppy White Brown) - Enlisted as a Pvt in Co. C, 16th Regt GA Inf. 7/15/1861. Trn to Co. F, 38th Regt. GA Inf. W by gunshot, right hand, necessitating amputation of three fingers,

Gettysburg, PA, 7/2/1863. Discharged, disability, 10/12/1864. (B Elbert Co., GA 4/23/1836) Filed CPA, Elbert Co., 1895. D 3/25/1925. Bur Rose Hill Cem, Royston.

Brown, John A. J. (John Andrew Jackson Brown) - Pvt 10/15/1861. W in left leg Gaines Mill, VA 6/27/1862. On W frl 10/31/1862. CPA shows trn to 25th Regt GA Inf. 1864. (B Elbert Co. GA, 1843) Filed CPA, Hart Co., 1896. Died 1913 Hart Co., buried there Bethesda Meth. Church, Hart Co. Marker reads CO F, 38 GA INF. Obituary printed in Elberton Star Newspaper 11/25/1913.

Brown, Joshua C. (Joshua Clark Brown) - Pvt Oct/1862. W 1864. D of wounds and chronic diarrhea in Jackson Hos Rich, VA Jul. 1 or 5, 1864. Bur there Hollywood Cem. Marker reads: B 9/21/1818, D 7/5/1864, CO F, 38 GA INF. Widow, Elizabeth Pendleton Brown filed a CPA, Madison Co., 1891.

Brown, William A. D. - Pvt 10/15/1861. Captured Gettysburg, PA, 7/3/1863. Paroled at Point Lookout, MD and trn to Aiken's Landing, VA, for exchange, 2/24/1865, Rcv there 3/10/1865. Filed CPA, Randolph Co., AL. Died 10/08/1901. Buried Bethel East Church Cem, Randolph Co., AL.

Brown, William J. (William Jefferson Brown) - 4th Sgt 10/15/1861. Died of typhoid fever in Belmont Hospital at Lovingston, VA, 8/14/1862. CSR contains death claim filed by wife, Phronia M. J. Teasley Brown. Bur private cemetery at Belmont Plantation, VA. CSA marker reads CO I GA INF.

Burnes, Joseph (or Burns) - Pvt 9/8/1862. Wounded and captured Gettysburg, PA, 7/1/1863. CSR states severe thigh wound at Gettysburg, Paroled DeCamp General Hos, David's Island, NY Harbor, 8/24/1863. Rcv City Point, VA for exchange 8/28/1863. In Columbia, SC hospital, wounded 11/6/1864. NLR. service record dated 2/25/1865 from Gen. Hosp. #1, Columbia, SC states, "Died leaving no effects." Died of pneumonia. Buried at Elmwood Memorial Gardens, Richland Co., SC.

Campbell, James C. - Enlisted as a Pvt in Co. F, 15th Regt GA Inf. 7/15/1861. Trn to Co. F, 38th Regt GA Inf., in E for John H. Hulme, 10/10/1862. Appt 5th Sgt 8/22/1864. C Fisher's Hill, VA 9/22/1864. Released PLO, MD 6/4/1865. D 5/8/1893. Widow, Sarah Ann Higginbotham Campbell filed CPA, Hart Co., 1902. Bur Holly Springs Bapt. Cem, Elbert Co. Obit in Elberton Star, 5/12/1893.

Collins, William - (William E. "Bill" Collins) Pvt 12/10/1862. Deserted 12/10/1863. Died 3/14/1885. Buried Rehoboth Baptist Church Cemetery, Elbert Co.

Cordell, William D. - 8/16/1862. Pvt Enlisted Ruckersville, Elbert Co., GA Discharged for bad health Oct/1862. Born 4/16/1829, died 5/29/1890. Buried Rock Branch Bap. Church Cem, Elbert Co.

Davis, Henry B. - Pvt 10/15/1861. Discharged for tuberculosis 12/16/1861. CSR contains discharge record. B in Habersham Co., GA, 30 years old at discharge. BSU.

Davis, William H. - Pvt 4/23/1862. Died Elberton, GA, 11/11/1863. CSR muster roll dated Oct/1863, lists him as home on sick furlough. BSU.

Denny, J. M. (or Denney) (James M. Denny) - Pvt 2/2/1862. Roll to 8/31/1864, bears remark: "Died, date unknown." CSR states "Died, Elbert Co., GA of disease." Died 5/29/1864. Buried Rehoboth Baptist Church, Elbert Co.

Duncan, Jeptha H. (Jeptha Harris Duncan) - Pvt 10/20/1861. Discharged, furnished substitute, prior to 10/31/1862. Born 6/17/1840, died 4/20/1911. Buried Elmhurst Cemetery, Elberton.

Duncan, J. W. (James Washington Duncan) - Pvt 2/2/1862. Wounded in finger, necessitating amputation, Wilderness, VA 5/6/1864. Wounded Winchester, VA, 9/19/1864. Roll dated 11/6/1864, last on file, shows present. Filed CPA, Elbert Co., 1889. Full name in pension record. BSU.

Dutton, William J. - (John William Dutton) Pvt 2/12/1862. Captured Winchester, VA 9/19/1864. Released Ft. McHenry, MD, 5/5/1865. CSR shows accepted parole after capture; faced court-martial by Federal authorities for violating a "parole of honor." Found guilty and ordered imprisoned at hard labor for the war. D 1928, bur Sardis Bap. Church Cem, Hart Co. Marker reads CO F, 38 GA INF.

Eavenson, John W. (John William Eavenson) - 3d Sgt 10/15/1861. Appt 2d Sgt 6/22/1862. W 12/13/1863 and disabled, Fred., VA. Appt 1st Sgt 3/29/1863. Trn to Co. G, 25th Battn. GA Provost Guard Inf. 1864. Appears on rolls Co. M, 25th Battn. GA Provost Guard Inf. dated 4/16/1864, as 2d Lt., presence or absence not stated. Elected 2d Lt Co. E, 1st Battn. GA Reserve Cavalry 3/23/1864. Roll from Mar. 23 - Oct. 31, 1864, only roll on file, shows absent sick. (B in Elbert Co., 5/28/1840.

D 12/1/1935) Filed CPA, Elbert Co., 1900; Buried Hill Crest Cemetery, Elbert Co. Marker reads CAPT, Co F. 38 GA REGT, C.S.A. Obit in Augusta Chronicle, 12/4/1935, page 2.
Eavenson S. A. (or T. A.) (George A. Eavenson) - Pvt 2/3/1864. Captured Richmond Hospital, 4/3/1865. Sent to POW camp at Newport News, VA, Died there 7/11/1865 of exhaustion and general debility. Buried at Hampton National Cemetery, Hampton, VA. Marker reads, Everton, G. R., GA. CSR lists G. A. Eavenson, resident of Elbert Co.
Eavenson, Thomas M. - Pvt 3/10/1862. W 2d Manassas, VA, 8/28/1862. D from wounds 8/25/1864, or 9/28/1864, Harrisonburg, VA. (B Elbert Co., 1843) Bur there Woodbine Cem, Confed. Section.
Eavenson, Willis Jefferson - Pvt 10/15/1861. Wounded in 1864. At home on sick furlough 11/6/1864. No later record (Born in Elbert Co., GA 4/5/1842) Filed CPA, Elbert Co., 1901. Died 4/17/1919. Buried Concord UMC Church, Elbert Co, Obit in Elberton Star newspaper, 4/18/1919.
Evans, William W. - Pvt 10/15/1861. Absent without leave, 12/1863. No later record.
Fleming, David F. - Pvt 10/15/1861. Died of typhoid fever in General Hospital, Charlottesville, VA 8/26/1862. Buried there in UVA Confederate Cemetery. Soldier's names listed on four bronze plaques on statue. CSR contains death claim filed by father, John P. Fleming, in Elbert Co.
Fleming, David R. (David Reese Fleming) - 4th Cpl 10/15/1861. Discharged, disability, 10/31/1862. CSR contains discharge record. D 2/18/1894 in Elbert Co., bur there Concord Meth. Church Cem.
Fleming, John H. - Pvt 10/15/1861. Discharged, disability, 5/10/1862. Reenlisted in 1863. C Spotsylvania, VA, 5/15/1864. Died of disease Point Lookout, MD, 7/15 or 8/1/1864. Bur PLO Confed. Cem St. Mary's Co., MD, name appears in Baltimore National Cem records as J. R. Fleming.
Fortson, DeLancey (or Dellancy) A. - Pvt 5/14/1862. Trn to Co. I, 15th Regt GA Inf. 1/1/1863. Surrendered Appomattox, VA 4/9/1865. Died 5/26/1914. Buried Dove's Creek Cem, Elbert Co.
Fortson, Eastin L. - Pvt 5/14/1862. Killed at Gettysburg, PA, 7/1/1863. CSR contains death claim filed by father, Haley Fortson, Elbert Co., GA. BSU.
Fortson, George G. - Pvt 5/14/1862. Discharged, furnished substitute, 12/10/1862. Died 7/5/1912, buried at Elmhurst Cemetery, Elbert Co., GA.
Fortson, Jesse W. (Jesse White Fortson) - Pvt 5/10/1862. Died at Staunton, VA, 7/27/1862. Buried there in Thornrose Cemetery. Widow, Sophia Fortson, filed CPA, Elbert Co., 1901. Cenotaph at Doves Creek Baptist Church Cemetery, Elbert Co.
Fortson, Stephen H. (Stephen "Ham" Fortson) - Pvt 5/14/1862. Appears last on roll for Feb/1863. Died 4/12/1916. Buried Falling Creek Bap. Church, Elbert Co. Obit in Elberton Star 4/14/1916.
Fortson, William T. - Pvt 10/15/1861. Appt 5th Cpl 3/1/1862. W Antietam, 9/17/1862. W Fred., 2/13/1862. D of wounds, 12/15/1862. CSR has death claim filed by father, Jesse W. Fortson. BSU
Greenaway, L. A. (Lindsey A. Greenaway) (or Greenway) - Pvt 9/5/1862. Died in Richmond, VA, hospital, 1862. Widow, May A. Greenaway, filed CPA, Elbert Co., 1891. BSU.
Guest, J. B. (John B. Guest) - Pvt 9/5/1862. Roll dated 11/6/1864, LOF, shows him "Absent, Det in Govt. Shoe Shop Augusta, GA Mar/1863." D April/1890. Bur Bethlehem Church, Longstreet District, Elbert Co. Obit in Elberton Star, 5/2/1890. Widow, Martha M. Cox Guest, filed CPA, Hart Co., 1904.
Gulley, R. P. (Robert P. Gulley) - Pvt 5/10/1862. W 2d Manassas, VA 8/28/1862. Det nurse, General Hos Rich, VA Mar/1864. Discharged Jackson Hos Rich, VA, Det to Anderson, SC 11/26/1864. BSU.
Harper, James Willis -1st Cpl 10/15/1861. Wounded in left thigh, Gaines Mill, VA 6/27/1862. Appt 2d Sgt 8/22/1864. Admitted to Jackson Hospital Rich, VA, 3/16/1865, frl for 60 days 3/28/1865. Filed CPA, Hart Co., 1890. Died 6/30/1914, Hart Co., buried there Crossroads Bap. Church Cem. Obit in Hartwell Sun, 7/3/1914, page 1.
Hendricks, W. J. - Pvt 5/22/1864. Captured Cold Harbor, VA 6/3/1864. P Elmira, NY 10/1/1864. E 10/29/1864. D at sea, en-route to Venus Point, Sav River, GA, 11/10/1864. Full name unknown. BSU.

Hickman, Benjamin H. - Pvt 12/14/1861. Wounded and captured Fredericksburg, VA, 12/13/1862. Died of wounds at Douglas Hospital, Washington, D. C. 2/8/1863. Buried at Arlington National Cemetery, VA. Marker reads B. H. Hickman, GA, CSA.
Hickman, Columbus, J. (Columbus Judson Hickman) - Pvt 2/12/1862. Sick in Augusta, GA hospital Apr/1862. Discharged Sav, GA, 5/23/1862. CSR contains discharge record. Died 10/8/1897 in Tuscaloosa, AL. BSU.
Higginbotham, Daniel T. - Pvt 3/26/1862. Sent home with brother, Elijah B. Higginbotham, who lost both legs at Spotsylvania, VA, 5/12/1864; remained until close of war. (Died in Elbert Co., 5/28/1876) Buried Concord United Methodist Church Cemetery, Elbert Co. Widow filed CPA, Elbert Co., 1919. Obit in Elberton Gazette, 6/7/1876.
Higginbotham, Dozier John - Pvt 10/15/1861. Roll for 8/31/1864, shows him "Absent in Major G. W. Lee's Battalion in GA" CPA shows severely wounded in back and hips Gettysburg, PA. Jul/1863; discharged, disability, 1/16/1864. (Born in Elbert Co., GA, 3/28/1840) Filed a CPA in Hart Co., 1904. Died 1919, buried at Liberty Church Cemetery, Madison, Co.
Higginbotham, Elijah B. (Elijah Benson “Bense” Higginbotham) - Pvt 5/14/1862. Wounded in both legs, necessitating amputation below knees, Spotsylvania, VA 5/12/1864. Discharged 1864. Filed CPA Elbert Co., 1879; Died 6/25/1912. Buried Concord Church Cemetery, Elbert Co., CSA grave marker reads, CO F, 38 GA INF. Obit in Elberton Star 6/25/1912.
Higginbotham, John T. - Pvt 10/15/1861. Died at Augusta General Hospital, Augusta, GA, of typhoid fever, 6/13/1862. Buried Holly Springs Baptist Church Cemetery, Elbert Co. Widow, Martha C. Higginbotham filed a CPA, Hart Co., 1891.
Howland, T. J. (Thomas Jefferson Howland) - Pvt 3/17/1862. S Appomattox, VA 4/9/1865. Died 12/22/1926, Jim Hogg Co., TX, Widow, Emma Howland, filed CPA, Jim Hogg Co., #43728. BSU.
Hulme, John H. - Pvt 5/14/1862. Trn to Co. F, l5th Regt GA Inf., in E for James C. Campbell, 10/10/1862. C Dandridge, TN 1/22/1864. D from variola, Rock Island, Ill. 3/24/1864. Bur there in Confed. Cem, plot # 918. Cenotaph located Van's Creek Bap. Cem, Elbert Co., reads 15TH GA INF.
Hulme, J. D. (John Dillard Hulme) - Pvt 9/5/162. On 5/21/1863, received commutation of rations while on sick furlough from Jan 24 - Mar 1863. No later record. Filed CPA, Elbert Co., 1898. Buried at Vans Creek Church Cemetery, Elbert Co. Obituary in Elberton Star, 1/24/1908.
Hunt, B. T. (Benjamin Thornton Hunt) - Pvt 5/14/1862. W Gaines Mill, VA 6/27/1862. D from wounds 7/30/1862. CSR contains death claim filed by widow, Frances E Hunt, in Elbert Co. BSU.
Hunt, Dozier C. (Dozier Calloway Hunt) - Pvt 10/15/1861. C Mine Run, VA 11/28/1863. E PLO, MD, 9/30/1864. D Rich, VA 11/15/1864. Bur there Hollywood Cem, bur Section W, Grave # 155
Hunt, B. A. T. (Beasally A. T. Hunt) - Pvt 8/29/1862. D of smallpox in General Hos Danville, VA 12/1/1862. CSR contains death claim filed by father, Willis Hunt, Elbert Co. First name in CSR. BSU.
Hunt, James Jackson - Pvt Apr/1863. Died of chronic diarrhea at hos in Winchester, VA, Jul/1863. Widow, Mary E. Hunt, filed CPA, Hart Co., 1891. Bur Cokesbury Meth. Church Cem., Hart Co.
Hunt, James W. H. (James Willis Henry Hunt) - Pvt 3/10/1862. Captured Spotsylvania, VA, 5/18/1864. Paroled Elmira, NY, 3/10/1865. Died 7/18/1886. BSU.
Hunt, J. M. B. (Judge Middleton Barrett Hunt) - Pvt 7/4/1863. S Appomattox, VA 4/9/1865. Died. 4/2/1922, Wise Co., TX. Buried there Pleasant Grove Cemetery. Marker reads CO F, 38 GA INF.
Hunt, Reuben S. - (Reuben Smith Hunt) 2d Sgt 10/15/1861. Appt 1st Sgt 6/22/1862. Elected Jr. 2d Lt 3/29/1863. Roll dated 11/6/1864, LOF, shows present. NLR. D 10/9/1912, Elbert Co. BSU.
Hunt, Singleton J. W. - (Singleton Jarrett Waters Hunt) Pvt 3/10/1862. Captured Spotsylvania, VA 5/18/1864. P Elmira, NY, for exchange 3/10/1865. (D Elbert Co., GA 4/2/1914) Filed CPA, Elbert Co., 1901. Bur Hunt Family Cem, Hart Co. Widow, Mary F. Hunt filed CPA, Elbert Co., 1938. BSU.
Hunt, Thomas J. M. - (Thomas Jefferson Monroe Hunt) Pvt 5/10/1862. D Augusta, GA, 5/20/1862. CSR has death claim by father, Willis Hunt, of Elbert Co. Buried Hunt Family Cem, Hart Co.
Hutcherson, Benjamin H. - Pvt 10/15/1861. Appears last on roll for Apr/1863. BSU.
Hutcherson, James M. (or Hutchinson) - Musician, fifer, 10/15/1861. Appt Regt. Musician. K Gaines Mill, VA 6/27/1862. CSR has death claim by father, Joshua Hutcherson, of Elbert Co. BSU.

Hutcherson, John M. - Pvt 10/15/1861. Died 6/21/1862. BSU.
Hutcherson, John M. D. (John McDonald Hutcherson) - Pvt 3/10/1862. Discharged 5/10/1862. Died 4/18/1902, Madison Co., buried there Liberty Church Cemetery.
Jenkins, Elias Pearce - Pvt 9/5/1862. Captured Fisher's Hill, VA, 9/22/1864. Trn to Aiken's Landing, VA for exchange 3/17/1865. (B GA 1/22/1843) Held at Point Lookout prison camp until end of war. Filed CPA, Hart Co., 1901. D 1/19/1926. Buried Bethesda Meth. Church Cemetery, Hart Co., GA
Jordan, F. M. (or Jourdan) (Francis M. Jordan) - Pvt 9/5/1862. Captured and paroled Hartwell, GA 5/18/1865. Died 11/22/1869. Buried Holly Springs Baptist Church Cemetery, Elbert Co. Widow, Mary J. Jordan, filed CPA, Elbert Co., 1891.
King, Thomas W. - Pvt 10/15/1861. Roll dated 11/6/1864, LOF, shows Macon, GA hos. NLR. BSU.
Loftin, A. (or Lofton) (Alexander Lofton) - Pvt 5/11/1862. Roll dated 11/6/1864, LOF, shows AWOL 6/20 - 8/31/1864. NLR. Transferred from Co. I, 59th GA Inf. to Co. F, 38th GA, 5/28/1863. BSU.
Loftin, William D. (or Lofton) - Pvt 9/9/1862. Roll dated 11/6/1864, LOF, present. NLR. BSU.
Maxwell, Benjamin M. (Benjamin "Martin" Maxwell) - Pvt 5/10/1862. Killed 2d Manassas, VA, 8/28/1862. Widow, Drucilla C. Teasley Maxwell, filed CPA, Hart Co., 1891. BSU.
Maxwell, Chandler - Pvt 8/19/1862. W and C Gettysburg, PA, 7/3/1863. E David's Island, NY, Harbor, 9/1/1863. Det Augusta, GA Arsenal Mar/1864. C and P Hartwell, GA 5/18/1865. (Resident of GA since 1844) Filed CPA, Elbert Co., 1919. D 4/25/1926. Bur Concord UMC, Elbert Co.
Maxwell, Jackson O. Jr. (Jackson Oliver Maxwell) 2d Lt 10/15/1861. Elected 2d Lt 6/22/1862; 1st Lt 8/18/1862; Cpt 2/11/1863. W 11/12/18 64. Replaced John C. Thornton when he resigned. Resigned 1/24/1865. D 2/16/1881. Bur Concord United Meth, Elbert Co. Marker reads 38 GA INF, CSA.
Maxwell, William H. (William Hayden "Jadie" Maxwell) - Pvt 3/10/1862. Appt 5th Sgt 3/29/1863; 4th Sgt Jun/1863. Died of chronic diarrhea, Elbert Co., 6/22/1864. Widow, Amelia Maxwell, filed CPA Madison Co., 1891. BSU.
McCurry, James D. - Pvt 10/15/1861. Captured Spotsylvania, VA, 5/20/1864. Released from Elmira, NY, 6/30/1865. BSU.
McCurry, John G., Jr. (John Gordon McCurry Jr.) - Pvt 10/15/1861. Elected Jr. 2d Lt 8/18/1862; 2d Lt 3/29/1863. Roll dated 11/6/1864, LOF, present. NLR. D 12/4/1886. Bur Bio Bap. Church, Hart Co.
McCurry, William E. - Pvt 5/10/1862. Captured at Spotsylvania, VA 5/12/1864. Released from Elmira, NY, 6/30/1865. BSU.
Moss, Benson - Pvt 3/1/1863. Captured Gettysburg, PA, 7/1/1863. Died of disease Point Lookout, MD, 7/1/1864. Buried Point Lookout Confederate Cemetery, Saint Mary's County, MD. Name not listed CSA monument at cemetery, but many names missing.
Moss, Hambleton (Hamilton Moss) - Pvt 5/10/1862. Captured at Gettysburg, PA, 7/1/1863. Released Ft. Delaware, 5/11/1865. Died 9/2/1917, buried Hillcrest Cemetery, Elbert Co., GA.
Motes, W. A. - See Pvt Co. H.
Nelms, James C. - Pvt 10/15/1861. W Fred., VA 12/13/1862. Died of wounds, date not given. BSU.
Nelms, Joshua A. - CSR states he enlisted 10/15/1861. Drummer, discharged 12/16/1861 due to bad health. CSR contains discharge record. Died 4/8/1866. BSU.
Nelms, William - (William S. Nelms) Drummer, 12/14/1862. Surrendered Appomattox, VA 4/9/1865. Died 1/15/1880, Athens. Widow, Harriett J. Nelms, filed CPA, Bibb Co., 1916. BSU.
O'Briant, William (or O'Bryant) (William H. O'Briant) - Pvt 3/25/1862. Discharged, furnished substitute, 9/18/1862. Enlisted in Co. A, 3d Regt GA Militia Jun/1864. CPA shows was detailed to hunt stragglers, Augusta, Mar/1865. Filed CPA. Franklin Co., 1910. Died 11/11/1916, buried Carroll's UMC Cemetery, Franklin Co.
Page, William J. H. - Pvt 10/15/1861. Discharged 12/16/1861. Died 1899. Buried Northview City Cemetery, Hart Co.

Parks, Archibald H. (Archibald Himor Parks) - Pvt 10/15/1861. Roll dated 11/6/1864, LOF, shows him "Absent, teamster, pontoon train. Det by General Lee." Died 11/6/1907. Buried Sardis Bap. Church Cem, Hart Co. Biography published in book titled Memoirs Of Georgia Vol. 1, page 1074.
Partain, Benjamin P. - Pvt 10/15/1861. Wounded Gaines Mill, VA 6/27/1862. Admitted to Jackson Hospital, Richmond, VA, 10/7/1864. Returned to duty 11/9/1864. No later record. BSU.
Partain, B. E. - (Benjamin E. Partain) Pvt 5/10/1862. Died of smallpox 2/6/1863. Widow, Frances E. Gulley Partain, filed CPA, Elbert Co., 1891. She stated he Died Guinea Station, VA. BSU.
Partain, Henry C. (Henry Carter Partain) - Pvt 4/6/1862. Discharged, disability, 6/24/1862. CSR contains discharge record. Born in Anderson District, S.C., 50 years old at discharge. Father of Pvt. Lindsey B. Partain shown below, brother-in-law of Pvt. Wiley Powell, also of Company F. Born 7/8/1811, Died in Hart Co., Georgia, 1864. BSU.
Partain, Henry H. (Henry Hawk Partain) - 3d Cpl 10/15/1861. Appt Color Guard Mar/1862. W in leg Fred., VA 12/13/1862. C Spotsylvania, VA, 5/18/1864. P Elmira, NY 10/11/1864, E 10/29/1864. NLR (B Anderson SC, 3/17/1832) Filed CPA, Hart Co., 1895, D 1917, bur there Sardis Bap. Cem.
Partain, Lindsey B. (Lindsey Bonaparte Partain) - Pvt 4/6/1862. Wounded at Gaines Mill, VA, 6/27/1862. Roll dated 11/6/1864, last on file, shows him present. Born 1/13/1843, Elbert Co., GA. Died 3/22/1921, Elbert Co., buried Rock Branch Baptist Church Cemetery, Elbert County, GA. Son of Pvt. Henry Carter Partain, of same company.
Partain, L. M. - (Lindsey Marshal Partain) Pvt 12/27/1862. Retired by Medical Examining Board Columbia, SC, 9/18/1864. (Born in Anderson Co., SC 6/24/1834) Filed a CPA in Jackson Co., 1898. Died 4/10/1902, buried at Liberty Cemetery, Franklin Co.
Partain, Robert P. - 2d Cpl 10/15/1861. Wounded in 1864. At home on W frl 9/11/1864. C and P Hartwell, GA 5/19/1865. Reported as W Gettysburg, according to Sav Republican 7/30/1863. BSU.
Pearson, George W. - Pvt 5/1/1863. Wounded 6/1864. Captured at Atlanta, GA 9/4/1864. Took oath of allegiance to U.S. Govt. Camp Douglas, IL, enlisted in 6th Regiment U.S. Volunteers, 4/1/1865. Court-martialed and confined at Ft. Laramie, Jan/1866 for stealing a saddle and saddle blanket. BSU.
Pearson, Littleton D. - See Pvt Co. H.
Powell, James L. (James Lewis Powell) - Pvt 10/15/1861, believed to be wounded either Gaines Mill, or Savage Station. The record is not clear. Sent home, Elbert Co., GA and remained there until he died 10/9/1863. CSR contains death claim by mother, Maria Powell, Elbert Co. Buried Rock Branch Baptist Church Cemetery, Elbert Co. Son of Wiley Powell of same company. CSA marker reads PVT 38 GA INF, CO F,
Powell, Wiley - Pvt 7/29/1862. W 2d Manassas, VA 8/28/1862. Died tuberculosis, General Hos #3, Lynchburg, VA 1/24/1864. Bur there in Old City Cem. No. 7, 3d Line, Lot 200 - D Ferguson's Hos. CSR contains a death claim filed by widow, Maria Powell, Elbert Co., Father of William J. Powell and James Lewis Powell, both Co. F. Widow, Mariah Powell filed CPA, Elbert Co., 1893.
Powell, William J. (William Joseph Powell) - Pvt 10/15/1861. Appt 5th Cpl Jun/1863. Killed at Gettysburg, PA, 7/1/1863. CSR contains death claim filed by mother, Maria Powell, of Elbert Co. Son of Wiley Powell, brother of James Lewis Powell shown above. Remains returned to Savannah, GA, after the war and buried there at Laurel Grove Cemetery. CSA marker reads, CO F, 38 GA INF.
Rice, Asa M. - Pvt 5/14/1862. W Jun/1863. At home, W, 11/7/1864. NLR. C, sent to David's Island prison Camp in NY. Filed CPA, Madison Co., 1897. D there 7/5/1905, bur there Bethel Bap. Church.
Roberts, John T. - Pvt 10/15/1861. W Fred., VA 12/13/1862. C Berryville, VA 8/10/1864. D of variola, Elmira, NY 4/23/1865. Bur Woodlawn National Cem, CSA Section, Site 1399, Elmira, NY.
Rucker, William B. (William Barden Rucker) - 1st Sgt 10/15/1861. Elected Jr. 2d Lt 6/22/1862. Died Richmond, VA, 8/1/1862, of typhoid fever, General Hospital 18. Widow, Sarah C. Maxwell Rucker remarried R. J. Brown. She filed CPA, in Elbert Co., 1910. BSU.
Sayer, David W. - Pvt 2/12/1862. Discharged, disability, 2/12/1862. CSR has discharge record. BSU.

Sayer, William O. (William Oscar Sayer) - Pvt 2/12/1862. Captured Winchester, VA, 8/17/1864. Released Elmira, NY, 7/7/1865. (Born Elbert Co., 6/29/1845) Filed CPA, Hart Co., 1906. Died 2/9/1910, buried Burgess City Cemetery, Franklin Co.

Sayer, William T. - Pvt 10/15/1861. Captured Spotsylvania, VA, 5/20/1864. Paroled Elmira, NY, sent to James River, VA, for exchange 3/2/1865. BSU.

Shiflet, W. J. (William J. Shiflet) - Pvt 5/14/1862. K Gettysburg, PA, 7/1/1863. Widow, Frances A. Shiflet filed CPA, Franklin Co., 1891. Cenotaph at Bethany Bap. Church Cem, Hart Co. BSU.

Stamps, James J. (James Johnson Stamps) - Pvt 10/15/1861. C Gettysburg, PA, 7/3/1863. Took OOA to U.S. Govt., enlisted in U.S. service PLO, MD 1/24/1864. Enlisted in Company G, 1st Regt of US Vol. Inf. served through Apr/1866, D 12/31/1928. Bur Oaks Cem, Meeker Co., Minnesota,

Stansell, James - Pvt 11/17/1862. Wounded right arm, Marye's Heights, VA 5/3/1864; right hand, Wilderness, VA, 5/6/1864; right hip, Cold Harbor, VA 6/3/1864. Discharged, disability, 9/10/1864. (B DeKalb Co., GA in 1818) CSR contains discharge record. Filed CPA, Hart Co., 1898. Born in Ireland, Jan/1818. BSU.

Stephenson, Thomas C. (or Stevenson) Pvt 5/10/1862. Killed in battle of 2d Manassas, 8/28/1862. CSR contains a death claim filed in by widow, L. J. Stephenson, Hart Co. BSU.

Teague, William - Pvt 10/20/1861. Roll dated 11/6/1864, LOF, shows him "Absent, Staunton, VA. Had notice from Dr. Marshall that he had detained him until he could get shoes. NLR. BSU.

Teasley, John H. H. (John Henry H. Teasley) - 1st Lt 10/15/1861. Died Savannah, GA hospital 6/19/1862. CSR contains an unusual adoption letter stating his father was John Adams Teasley and he was adopted by William H. Teasley, dated 11/7/1853, signed in Elbert Co. BSU.

Teasley, J. J. (John J. Teasley) - Pvt 8/3/1862. Died Jordon Springs, VA, hospital, 7/5/1863. CSR contains death claim filed by widow, Frances A. Teasley, in Hart Co. Buried at Stonewall Confederate Cemetery, Winchester, VA, marker reads J. Tesley, grave #24.

Terry, Elijah M. (Elijah Monroe Terry) - Pvt 10/15/1861. Roll dated 11/6/1864, last on file, shows present. No later record. Filed a CPA, in Franklin Co., 1905. BSU.

Terry, Joseph C. (Joseph Chandler Terry) - Pvt 12/27/1862. Paroled Farmville, VA, Apr. 11-21, 1865. Died 1891, in Elbert Co. Widow, Lorena Stamps Terry, filed CPA, Elbert Co., 1902. BSU.

Terry, William J. - Pvt 10/15/1861. Wounded Antietam, 9/17/1862. Wounded Fredericksburg, VA, 12/13/1862. Wounded Gettysburg, 7/1/1863. Roll dated 11/6/1864, last on file, shows present. CPA shows surrendered Appomattox, VA, 4/9/1865. (Born Elbert Co., 4/23/1825) Filed CPA, Gilmer Co., 1908. Died 4/12/1909. Widow, Nancy Terry filed CPA, Gilmer Co., 1911. BSU.

Thomason, William G. - (William Gaines Thomason or Thomson) Pvt 10/15/1861. AWOL 10/15-31/1861. NLR D. 11/16/1904. Obit in Elberton Star, 11/17/1904. Bur Holly Springs Cem., Elbert Co.

Thornton, Asa C. (Asa Chandler Thornton) - Pvt 10/15/1861. Roll dated 11/6/1864, LOF, shows "Absent, Det nurse Sept/1863. D 4/13/1888. Obit in Atlanta Constitution, 4/14/1888. Bur Almond - Thornton Cem, Elbert Co. Widow, Dorcas M. Almond Thornton filed CPA, Clarke Co., 1916.

Thornton, Benjamin F. (Benjamin "Benny" Franklin Thornton) - Pvt 10/15/1861. D Rich, VA hos 8/25/1862. Bur there in Hollywood Cem. Cem records show Thornton, M. F., Company F, 38 GA, D 8/7/1862, bur Section C, grave #178. CSR has death claim by father, Fleming Thornton, Elbert Co

Thornton, Depth M. (Jeptha Mercer Thornton) - Pvt 10/15/1861. Wounded 2d Manassas, VA, 8/28/1862. Captured Winchester, VA, 9/19/1864. Paroled at Point Lookout, MD Nov/1864. Rcv Venus Point, Sav River, GA for exchange, 11/15/1864. Filed CPA, Hart Co., 1890. Died 5/12/1907, buried Northview City Cemetery, Hart Co. Obituary in Augusta Chronicle, 5/13/1907.

Thornton, James C. - (James Callaway Thornton) Pvt 10/15/1861. W Fred., VA 12/13/1862. S Appomattox, VA, 4/9/1865. Filed CPA, Elbert Co., 1897. D 11/22/1917. Bur Bio Bap. Cem, Hart Co.

Thornton, James William - Pvt 3/18/1864. Captured Winchester, VA, 9/19/1864. Exchanged Point Lookout, MD 11/1/1864. Rcv Venus Point, Sav River, GA, 11/15/1864. CPA shows enlisted in Reserve Cavalry in 1865, unit disbanded Macon, GA 4/23/1865. (Resident GA since 1845) Died 12/30/1934. Filed CPA, Elbert Co., 1919. Buried Thornton Family Cemetery, Elbert Co.

Thornton, Jesse M. - Pvt 10/15/1861. W 2d Manassas, 8/28/1862. Wounded Fred., VA 12/13/1862. Appt 4th Cpl 8/30/1863. Wounded Monocacy, MD 7/9/1864. Died of wounds same day. BSU.
Thornton, John C. - Cpt 10/15/1861. Wounded Fredericksburg, VA, 12/13/1862. Resigned 2/11/1863. Company commander when company first formed. Resigned for ill health. Elected Cpt of Co. D, 1st Regt GA Reserve Inf. (Fannin's,) 4/29/1864. Died of typhoid fever, Savannah, GA hospital, 10/19/1864.Buried Northview City Cemetery, Hartwell. Widow, Georgia Thornton, filed CPA, Hart Co., 1891. Obit in Daily Constitutionalist, Augusta, 10/26/1864.
Thornton, R. B. (Reuben Benjamin Thornton) - Pvt 5/10/1862. Appt 5th Sgt 2/10/1863; 4th Sgt Mar/1863. Died, chronic bronchitis, Chimborazo Hos Rich, VA 5/29/1863. Bur Oakwood Cem, Rich.
Thornton, Thomas D. - Pvt 10/15/1861. Elected 2d Lt 8/18/1862; 1st Lt 3/29/1863. Surrendered Appomattox, VA, 4/9/1865. (Resident GA since 1839) Filed CPA, Elbert Co., 1919. Died 4/6/1922, buried there at Falling Creek Baptist Cemetery.
Thornton, William M. - (William Marion Thornton) Pvt 10/15/1861. Appt 5th Sgt Jun/1863; 4th Sgt 8/22/1864. Surrendered Appomattox, VA, 4/9/1865. Died 8/2/1900. Buried Elmhurst Cemetery, Elberton. Obituary in Elberton Star, 8/9/1900.
Thornton, William T. - Pvt 9/5/1862. Appt 5th Sgt 11/3/1862. Wounded through right eye Fred., VA 12/13/1862. D of wounds 1/29/1863. Widow, Lucinda Thornton, filed CPA, Elbert Co., 1891. BSU.
Tiner, James A. (James A. Tyner) - Pvt 10/15/1861. Discharged Sav, GA 5/23/1862. Enlisted as a Pvt in Co. B, 24th Regt GA Inf. 3/11/1863. C in 1864 or 1865. Released PLO, MD, 6/20/1865. BSU.
Tiner, Thomas Joshua P. (Thomas Joshua P. Tyner) - Pvt 10/15/1861. Died at home 12/5/1862. CSR contains death claim filed by widow, Mary J. Tyner, in Hart Co. BSU.
Vaughn, C. B. (Charles B. Vaughan) - Pvt 4/6/1864. W in 1864. Roll dated 11/6/1864, LOF, shows "Home on W frl. Time out unknown. Died 3/12/1886, buried Rehoboth Bap. Church Cem, Elbert Co.
Wansley, Thomas Nathaniel (or Warnsley, or Wanslow) - Pvt 9/5/1862. Roll dated 11/6/1864, last on file, shows him "At home on sick furlough. Time out unknown." Filed CPA, Elbert Co. D 2/5/1924, buried at Ruckersville Methodist Church Cemetery, Elbert Co., Georgia.

COMPANY G, BATTEY GUARDS, JEFFERSON COUNTY, GA

*This company was originally known as Co. E. Organized and first organized and led by Captain William H. Battey, of Jefferson County, Georgia.

Adkinson, William F. (William F. Atkinson) - Pvt 10/1/1861.Captured 1862. Exchanged. Captured Fisher's Hill. VA 9/22/1864. Paroled at Point Lookout, MD, 3/17/1865. Filed CPA filed CPA, Burke Co., 1906. Name spelled Atkinson in CPA. BSU.
Anderson, Samuel - Pvt 10/1/1861. Died Staunton, VA 8/4/1862. Bur there in Thornrose Cem.
Anderson, William J. - Pvt 10/1/1861. Died Richmond, VA, 7/28/1862. Bur there Hollywood Cem, section H, grave# 99. CSR contains death claim filed by widow, N. E. Anderson, in Jefferson Co.
Arnold, William C. - Pvt 10/1/1861. Discharged, disability, 7/25/1862. Died of disease, Jefferson Co., 8/23/1862. CRS contains death claim filed by widow, Mary F. Arnold, Jefferson Co. BSU.
Arrington, Hilliary W. - Pvt 10/1/1861. Wounded and captured at Gettysburg, PA, 7/2/1863. Exchanged, killed at Winchester, VA, 9/19/1864. BSU.
Arrington, Joshua P. - Pvt 10/1/1861. Killed Spotsylvania, VA 5/10/1864. Buried there at the Confederate Cemetery, CSA marker reads: CO G, 38 GA INF.
Arrington, Levin C., Jr. (Levin Collins Arrington Jr.) - 3d Cpl 10/1/1861. Discharged 9/13/1862. CSR contains discharge record and death claim filed by father, Levin C. Arrington, Jefferson Co. Died at home of disease, 9/24/1862. BSU.
Arrington, Silas E. - Pvt 10/1/1861. Died Lynchburg, VA, hos 9/6/1862. CSR contains death claim filed by father, Levin C. Arrington Sr., Jefferson Co. BSU.

Arrington, Thomas A. - Pvt 5/1/1862. Died of typhoid fever, Richmond, VA, 8/11/1862. Buried there at Hollywood Cemetery. CSR contains death claim filed by Widow, Dealpha Arrington.
Atwell, Reuben - Pvt 10/1/1861. Killed at Winchester, VA, 6/13/1863. BSU.
Bagget, Seaborn F. - (Seaborn Francis Baggett) Pvt 10/1/1861. Surrendered Appomattox, VA, 4/9/1865. Filed CPA, Emanuel Co., 1901. Died there 3/17/1908. Buried there Canady - Durden Cemetery. Obituary in Macon Telegraph, 3/19/1906.
Barrow, Henry J. - Pvt 10/1/1861. Wounded Antietam, 9/17/1862. Appt 1st Cpl Jul/1863. Admitted Jackson Hos Rich, VA, with chronic diarrhea, 4/25/1864. Died 4/27 or 5/4/1864. Buried in Hollywood Cemetery, Richmond, VA.
Battey, William H. (William Henry Battey) - Cpt 10/1/1861. Killed Antietam, MD 9/17/1862. Cenotaph at Battey-Heard Cem, Jefferson Co. Obit, Augusta Daily Constitutionalist, 12/7/1862. BSU.
Beasley, Benjamin - Pvt 10/1/1861. Appt 4th Cpl in 1862; 1st Sgt May/1864. Surrendered Appomattox, VA, 4/9/1865. Filed CPA, Jefferson Co., 1895. Buried Howard-Beasley Cemetery, Jefferson Co. CSA grave marker reads: CO G, 38 GA INF.
Beasley, Edmund R. - Pvt 10/1/1861. Roll dated 11/4/1864, last on file, shows present. No later record. Died Aug/1871. Widow, Mary E. Beasley, filed CPA, Jefferson Co., 1911. Buried Howard-Beasley Cemetery, Jefferson Co. CSA grave marker reads: CO G, 38 GA INF.
Brinson, John W. (John Wright Brinson) - 1st Lt 10/1/1861. Elected Cpt 9/18/1862. Resigned 1/7/1863. D 4/24/1896, Jefferson Co. Bur there Ways Bap. Cem, marker reads: 38 GA INF CSA.
Brown, Beniah - Pvt 10/1/1861. Appt Hos Steward 2/10/1863. S Appomattox, VA, 4/9/1865. B 1/23/1835, D 8/1/1896, bur Bartow City Cem, Jefferson Co., marker reads: CO G 38 GA INF.
Brown, George C. - Pvt 10/1/1861. Wounded Fred., VA, 12/13/1862. Roll for 11/4/1864, LOF, reads AWOL since 11/1/1864. NLR. B 1/1/1836, D 12/5/1882, bur Bartow City Cem, Jefferson Co.
Brown, James D. (James David Brown) - Pvt 3/2/1862. Died Staunton, VA, 8/9/1862. Bur there Thornrose Cem. CSR contains death claim filed by father, Ebenezer Brown, in Richmond, VA.
Brown, William A. (William Arnold Brown) - Pvt 10/1/1861. Wounded Winchester, VA by bullet in left wrist, 9/19/1864. Absent, wounded, 11/4/1864. No later record. Filed CPA, Jefferson Co., GA in 1889 and filed CPA in Florida, Pension #A04152. Died 2/27/1915 in Marion Co., FL. BSU.
Campbell, Isley - (Isea or Isla) Sick, sent to hos in Richmond, VA, Died of typhoid fever, 6/22/1862. Bur Hollywood Cem, Rich, VA. Widow, Elizabeth Campbell, filed CPA, Emanuel Co., 1891.
Carson, William D. - Pvt 10/1/1861. Died near Richmond, VA 6/29/1862. CSR contains death claim filed by father, Meridith Carson, in Jefferson Co., states he was mortally wounded at battle of Gaines Mill, 6/27/1862, and died on the battlefield, 6/29/1862. BSU.
Chew, John C. - See Pvt Co. C.
Clark, Elimah W. (Elijah W. Clark) - Pvt 5/7/1862. Captured in Jackson Hospital Richmond, VA, 4/3/1865, died there, typhoid fever, 4/28/1865. CRS lists first name. Bur Hollywood Cem, Rich, VA.
Clark, Milton A. (Milton Anthony Clark) - Pvt10/1/1861. Appt 3d Sgt 1863. Wounded in leg, leg amputated, Spotsylvania, VA 5/12/1864. At home, wounded, close of war. Filed CPA in FL, pension #A04691. Died 7/9/1913. Buried Rose Hill Cem, Osceola Co., FL. Marker reads SGT CO G, 38 GA INF, 30 Years a Missionary to the Oklahoma Indians. Obituary in Augusta Chronicle, 7/12/1913.
Connell, Rufus - Pvt 10/1/1861. Appt 3d Cpl in 1862. Captured Petersburg, VA, 4/3/1865. Released Hart's Island, NY Harbor 6/15/1865. Died Mar/1873. Widow, Margaret Connell Smith, filed CPA, McDuffie Co., 1910. BSU.
Daniel, John C. - Pvt 10/1/1861. D in Atlanta, GA (also shown as Jefferson Co., GA) 6/7/1862. BSU.
Daniel, William A. - Pvt 5/1/1862. Discharged 6/20/1862. CSR contains discharge record. BSU.
Dye, David R. - Pvt 10/1/1861. Wounded Spotsylvania, VA 5/12/1864. Died of wounds Lynchburg, VA, 7/12/1864. Name on Linwood Meth. Church Sunday School monument, Ft. Gordon, GA. BSU.
Farmer, George W. - Pvt 10/1/1861. Discharged, disability, 7/27/1862. CSR contains discharge record. Born in Laurens Co., GA, 45 years old at discharge, born abt. 1817. BSU.

Farmer, James J.- Pvt 10/1/1861. Captured near Mine Run, VA, 5/12/1864. Paroled Ft. Delaware and exchanged there 3/7/1865. Died 1882. Buried Ways Baptist Church Cemetery, Jefferson Co.
Farmer, Levin W. (or Leven) - Jr. 2d Lt 10/1/1861. Elected 2d Lt 9/18/1862; Cpt 1/12/1863. Killed Spotsylvania, VA, 5/12/1864. CSR states, "behaved gallantly in battle of Fred." Widow, Alice Brinson Swan, filed CPA in Jefferson Co., 1911. CSR contains marriage record. BSU.
Farmer, Sidney T. - Pvt 3/1/1862. Appt 2d Sgt 7/1/1862. Pvt Nov/1862. Appt Quartermaster Sgt 12/1/1862. Elected Jr. 2d Lt 1/2/1863. Killed Spotsylvania, VA 5/10/1864. Prom to 3rd Lt, 6/1/1863. Bur Confed. Cem, Spotsylvania Courthouse, VA. CSA marker reads, LT, CO G, 38 GA INF.
Farmer, William J. - 2d Sgt 10/1/1861. Appt 1st Sgt 5/27/1862. Elected Jr. 2d Lt 9/24/1862. Killed or captured Fred., VA 12/13/1862. CSR states, "taken prisoner at Fred., not heard from since." Name inscribed on "Linwood Meth. Church Sunday School" monument, Ft. Gordon, GA. BSU.
Farrer, Aguilla M. - Pvt 10/1/1861. Wounded in hip Gettysburg, PA, 7/2/1863. Captured South Mountain, MD 7/5/1863. P Ft. Delaware, 2/1865. Rcv Boulware & Cox's Wharves, James River, VA for E, 3/10-12/1865. Admitted to Wayside, or General Hos #9, Rich, VA 3/14/1865, frl for 30 days 3/15/1865. (B Jefferson Co., 4/10/1831. D Bartow Co., 1917) W Antietam, 9/17/1862. Filed CPA, Bartow Co., 1901. Died 3/10/191 6. Obit states buried Moxley Bap. Church Cem, Jefferson Co.
Farrer, Garvin H. (Gavin Hamilton Farrer) - Pvt 10/1/1861. Wounded Winchester, VA 9/19/1864. Died of wounds Woodstock, VA, 9/23/1864. Listed as color bearer when mortally wounded. On list of casualties for 38th GA, in Augusta Chronicle, 11/11/1864. Buried Massanutten Cem, Woodstock, VA, marker located on Holly Circle in cemetery incorrectly reads C. S. Farrer, 38th VA
Farrer, William Harrison - Pvt 3/1/1862. Caught measles in camp Savannah; died of pneumonia at home while on furlough, 3/20/1862. BSU.
Folks, Green L. - (Or G. L. Faulks) Pvt 10/1/1861. Died of disease in Camp Winder Hos Rich, VA 7/16/1862. Bur Hollywood Cem, Rich, VA, cemetery records show: Faulks, G. L., Co. E 38 GA, D 7/16/1862, bur Section H, Grave #97. CSR has death claim filed by mother, Celia Folks, Jefferson Co.
Freeman, Moses - Pvt 10/1/1861. Died at home, Jefferson Co., 12/30/1861. Bur Freeman Cem, Wrens, Jefferson Co. CSR contains a death claim filed by widow, Louisa Freeman, Jefferson Co.
Futrial, John - (or Feutrial) (John Allen Futral) Pvt 4/28/1862. W Fred., VA 2/13/1862. C near Mine Run, VA 5/12/1864. D bronchitis, Ft. Delaware, 7/3/1864. Bur Finn's Point National Cem, Salem, NJ.
Gaines, John (or Gainous) - Pvt 5/15/1862. Died of diphtheria, General Hos #25, Rich, VA 12/24/1862. Buried Oakwood Cem, Richmond, VA, records read John Gaines, Co. G, 38th GA, CSR contains a death claim filed by father, William Gaines, in Jefferson Co.
Gaines, William (or Gainous) - Pvt 10/1/1861. Discharged, disability, in 1863. BSU.
Gay, Isaac H. - Pvt 5/15/1862. Died from wounds received in the battle of Monocacy, 7/9/1864. Died at Frederick, MD, 7/10/1864. Wound described as "gunshot wound of abdomen." Buried Mount Olivet Cemetery, Grave #244, Frederick, MD. CSA marker reads: CO G 38 GA INF.
Gay, John - Pvt 5/15/1862. Captured near Mine Run (Spotsylvania), VA, 5/12/1864. Paroled at Ft. Delaware, 10/30/1864. Received Venus Point, Savannah River, GA, for exchange, 11/15/1864. No later record. BSU.
Gordon, Alexander - Pvt 10/1/1861. Died in Crumpton's Hospital Lynchburg, VA, 11/17/1862. Buried there in Confederate Cemetery. No. 7, 3d Line, Lot 144.
Gunn, James - 1st Cpl 10/1/1861. Appt 2d Sgt 6/1/1862. Mortally wounded at battle of Gaines Mill, 6/27/1862. Died at Richmond, VA, 7/1/1862. CSR contains death claim filed by father, John Gunn, of Jefferson Co. Buried Oakwood Cemetery, Richmond, VA, records reads Jas Gunn, Co. C, 38th GA.
Gunn, Robert (Robert L. Gunn) - Pvt 10/1/1861. Killed at battle of 2d Manassas. VA, 8/28/1862. CSR contains a death claim filed by father, John Gunn, in Jefferson Co. BSU.
Gunn, William L. - Pvt 10/1/1861. Killed in battle of Gaines Mill, VA, 6/27/1862. BSU.
Hancock, Andrew J. - Pvt 10/1/1861. Wounded and missing 2d Manassas, VA, 8/28/1862. CSR states he was killed at 2d Manassas, 8/28/1862. BSU.

Howard, Jeremiah (Jeremiah H. "Jerry" Howard) - Pvt 5/1/1862. Wounded in 1862. Shot in the wrist/hand while on guard duty at night in Atlanta, GA. Discharged Mar/1863. CSR contains discharge record. Filed CPA Jefferson Co., 1889. Died 10/26/1914, Bulloch Co., buried there East Side Cemetery. Obituary in Statesboro News, 10/27/1914.
Hudson, Hampton J. - Pvt 10/1/1861. Wounded in leg. Roll dated 11/4/1864, LOF, shows AWOL 8/1/1864. NLR. Died 7/19/1909. Widow, Mollie Hudson, filed CPA, Jefferson Co., 1911. BSU.
Hudson, Hezekiah H. - Pvt 10/1/1861. C Wilderness, VA 5/6/1864. Took OOA to U.S. Govt. and enlisted in U.S. service Point Lookout, MD, 6/6/1864. Enlisted Co. B, 1st Regt, U.S. Vol. Inf. BSU.
Hudson, James C. (James Cain Hudson) - 1st Sgt 10/1/1861. Discharged, disability, Atlanta, GA 5/27/1862. Filed CPA, Jefferson Co., 1905. BSU.
Ivey, I. H. (or J. H.) - Pvt 10/1/1861. Died Staunton, VA, 10/28/1862. Buried there Thornrose Cem.
Johnson, Jeremiah J. - Pvt 10/1/1861. Appt Sgt 1862. Died at Richmond, VA, 8/9/1862. Buried there Hollywood Cemetery.
Johnson, Sylvester - Pvt 10/1/1861. Discharged 12/15/1861. CSR contains discharge record. Enlisted Co. K, 5th GA Regt. in 1864 and served until surrender at Greensboro, NC, Apr/1865. Filed CPA, Warren Co., 1898. Born 11/22/1833, Habersham Co., GA. BSU.
Jordan, James A. - Pvt 10/1/1861. Killed at Gettysburg, PA, 7/1/1863. BSU.
Jordan, Thomas J. - Pvt 10/1/1861. W in 1862. W Antietam, 9/171862. D Rich, VA 10/17/1862. Bur there Hollywood Cem. CSR contains death claim filed by father, Thomas G. Jordon, Jefferson Co.
Jordan, William P. (William Parker Jordan) - Pvt 9/6/1862. Detailed as hospital Steward Richmond, VA, 8/5/1863. Detailed as nurse in General Hospital, Camp Winder, Richmond, VA, 11/7/1864. Detailed hospital duty Macon, GA, 2/4/1865. Died 1/7/1894. Buried Laurel Grove Cem., Savannah.
King, James - Pvt 10/1/1861. Captured Farmville, VA, 4/6/1865. Released Newport News, VA, 6/25/1865. BSU.
Lowe, James E. - Pvt in Co. G, 38th Regt GA Inf., Pvt 10/1/1861. Discharged, 6/5/1862, "by reason of chronic complaint rendering him unfit for service." Enlisted as Pvt 9/21/1863, Co. H, 63d Regt GA Inf. In C.S.A. General Hos, No. 11, Charlotte, N. C., 3/12/1865. CSR contains discharge record. Filed CPA, Jefferson Co., 1895. D 2/28/1902, Emanuel Co., bur there Ebenezer Cem.
Martin, Robert A. (Robert Augustus Martin) - Pvt 10/1/1861. Surrendered at Appomattox, VA 4/9/1865. (Borb in Abbeville Co., SC, 9/29/1842) Filed CPA, Lincoln Co., 1904. Died 4/21/1918, buried New Hope Baptist Church Cemetery, Lincoln Co. Capt William Mathews said, "R. A. Martin was in every fight the 38th Georgia was in and was never wounded."
Mathews, William C. - Enlisted as Pvt in Co. E, 1st Regt GA Inf. (Ramsey's) 8/1/1861. Mustered out Augusta, GA 3/18/1862. Enlisted as Pvt, Co. G, 38th Regt GA Inf. 7/1/1862. Appt Ordnance Sgt 7/27/1862. Elected 2d Lt of Co. G, 38th Regt GA Inf. 2/1/1863. Severely wounded Gettysburg, PA, 7/1/1863. Captured Smithburg, MD 7/4/1863. Elected Cpt 5/12/1864. Paroled Johnson's Island, Ohio, forwarded to Point Lookout, MD, for exchange 3/14/1865. Died 9/13/1894, buried Sandersville Old City Cemetery, Washington Co., GA. Tribute in Atlanta Constitution, 9/23/1894.
McCroan, Rhesa H. C. (or McCrone) - Pvt 10/1/1861. Discharged, disability, Sav, GA, 6/5/1862. CSR contains discharge record. B Jefferson Co., GA, 20 years old at discharge. born abt. 1842. BSU.
McGahee, Samuel M. (or McGhee) - Pvt 5/1/1862. W Fred., VA, 12/13/1862. C Spotsylvania, VA 5/19/1864. D of general debility Elmira, NY, 5/4/1865. Bur grave #2759, Woodlawn National Cem, Elmira, Chemung Co., New York. Widow, Millie McTier, filed CPA, Jefferson Co., 1911.
McGahee, William Green (or McGhee) - Pvt 5/1/1862. Deserted 12/29/1864. Took OOA to U.S. Govt. City Point, VA, Trn to Chicago, IL, 1/1/1865. (B in GA) Bur Zoar Meth. Church, Glascock Co.
Murphey, Augustus A. - Pvt 10/1/1861. W Gettysburg, PA, 7/2/1863. Roll dated 11/4/1864, LOF, shows "Absent collecting stragglers in State of GA" Died 5/11/1891, bur Bethany Cem, Jefferson Co.

Patterson, James R. (Samuel R. Patterson) - Pvt 10/1/1861. Died 7/5/1862. Transcription error in name in Henderson's Roster of Georgia Confed. Soldiers. CSR contains death claim filed by father, James Patterson, Jefferson Co. Claim states he was killed in battle at Mechanicsville, VA. BSU.
Peeples, Howell L. (Howell L. Peebles) - Pvt 5/1/1862. Wounded at Wilderness, VA, 5/6/1864. Died of wounds in hospital May 31, or 6/15/1864. BSU.
Peeples, Thomas J. (Thomas Jefferson Peebles) - Pvt 10/1/1861. Appt 4th Sgt Nov/1862. W and C Gettysburg, PA, 7/2/1863. P DeCamp General Hos, David's Island, NY Harbor, Sept/1863. Rcv City Point, VA for E 9/8/1863. C Farmville, VA, 4/6/1865. Released Newport News, VA 6/25/1865. BSU.
Pendry, William John (or John W. Penrow) - Pvt 4/28/1862. S Appomattox, VA 4/9/1865. BSU.
Pennington, William - Pvt 10/1/1861. W Winchester, VA 8/28/1864. D wounds hos 8/30/1864. BSU.
Perdue, John F. - Pvt 10/1/1861. Died Richmond, VA, 7/5/1862. CSR contains death claim filed by father, John Purdue, Jefferson Co. Father stated he died of wounds received in battle. BSU.
Phillips, John P. - Pvt 10/1/1861. Died 11/1/1862. CSR service states, sick in hospital Winchester, VA, Aug/1862. BSU.
Phillips, William J. - Pvt 10/1/1861. Captured Winchester, VA, 9/19/1864. Paroled Point Lookout, MD, and transferred to Aiken's Landing, VA, for exchange, 3/15/1865. BSU.
Pipkin, John G. - Pvt 10/1/1861. Died at home of variola, 10/1/1862. BSU.
Pool, Isaac B. (Isaac Brinson Pool) - Pvt 10/1/1861. Discharged 12/15/1861. Reenlisted 8/25/1862. D 2/18/1863. D 2/17/1863, Widow, Ellen A. Pool, filed CPA, Jefferson Co., 1891. She stated he was sick in the hospital and returned home on furlough, he arrived home and died the next day. BSU.
Pool, Michael W. - Pvt 5/1/1862. Killed Richmond, VA, 6/27/1862. Killed battle of Gaines Mill. CSR contains a death claim filed by widow, Isabelle Pool, Jefferson Co. BSU.
Powell, James P. - Pvt 10/1/1861, 1862. Wounded 2d Manassas, VA, 8/28/1862. Roll for 11/4/1864, last on file, shows him wounded, absent. No later record. BSU.
Prescott, G. W. (George W. Prescott) - Pvt 3/10/1864. Captured Spotsylvania, VA, 5/12/1864. Released Ft. Delaware, 6/16/1865. First name found in CSR. BSU.
Rivers, George Washington - Pvt 10/1/1861. Died of typhoid fever Charlottesville, VA, 8/1/1862. Buried there at UVA Confed. Cemetery. Name listed on memorial at cemetery incorrectly lists unit as 39th GA. CSR contains death claim filed by father, John F. Rivers, of Jefferson Co.
Robbins, Robert C. (Robert Chauncey Robbins) - Pvt 5/16/1862. Died of disease, near Richmond, VA, 7/7/1862. Cenotaph located at Battey-Heard Cem, Louisville, Jefferson Co., Death notice in Constitutionalist, (Augusta, GA) 7/19/1862, page 2. Buried Battey-Heard Cem, Jefferson Co.
Rodgers, Alexander A. - Pvt 10/1/1861. Killed at Spotsylvania, VA, 5/12/1864. BSU.
Rodgers, Edwin M.- Pvt 3/26/1862. D Sav, 5/15/1862. CSR contains death claim filed by father, Allen Rodgers. Bur Jackson's Bap. Church Cem, Washington Co. Marker reads, Co. G, 38 GA INF.
Rodgers, George A. - Pvt 6/24/1862. Killed at Gaines Mill, VA, 6/27/1862. CSR contains a death claim filed by mother, Harriet Rogers, in Jefferson Co. BSU.
Rodgers, James W. (or Rogers) - Pvt 10/1/1861. Wounded Gaines Mill, VA 6/27/1862. Died of wounds 7/5/1862. CSR contains a death claim filed by father Eleanah Rogers, in Jefferson Co. BSU.
Rodgers, John S. - Pvt 10/1/1861. Appt 1st Cpl Jun/1862. Pvt Jul/1863. Roll dated 11/4/1864, last on file, shows present. No later record. BSU.
Sessions, Robert J. - Pvt 10/1/1861. Died Pope Hill, GA 9/25 or 10/8/1863. BSU.
Shirley, A. - (Aaron Shirley) Pvt May/1864. Captured Cedar Creek, VA, 10/19/1864. Released Point Lookout, MD, 6/19/1865. BSU.
Shirley, Marshall H. - 3d Sgt 10/1/1861. Transferred to Co. H, 22d Regt GA Inf. Mar/1862. Paroled Farmville, VA, hospital, Apr. 11-21, 1865. BSU.
Smith, Amory S. - (Armory Solon Smith) Pvt 11/21/1861. Appt Acting Ordnance Sgt S Appomattox, VA 4/9/1865. Filed CPA, Jefferson Co., 1919, D 7/3/1922, bur there Bethany Cemetery.
Smith, David T. - 4th Cpl 10/1/1861. CSR states Wounded at Antietam, 9/17/1862. Captured Farmville, VA, 4/6/1865. Released Newport News, VA, 6/25/1865. BSU.

Smith, James (James M. Smith) - Pvt 10/1/1861. Died 1/1/1862. Died Savannah, buried Resaca Confed. Cem, Gordon Co. CSR contains death claim filed by mother, Francis Smith, in Jefferson Co.
Smith, James M. - Pvt 3/1/1862. Killed at battle of Gaines Mill, VA, 6/27/1862. BSU.
Smith, Jordan - Pvt 10/1/1861. Killed at battle of 2d Manassas, VA, 8/28/1862. BSU.
Smith, Leroy - Pvt 10/1/1861. Appt 1st Sgt 9/24/1862. Wounded Spotsylvania, VA, 5/12/1864. Absent, wounded, 9/13/1864. No later record. BSU.
Smith, Luther C. - Pvt 3/1/1862. Wounded at 2d Manassas, 8/28/1862. Wounded Winchester, VA 9/19/1864. On wounded furlough 11/4/1864. No later record. BSU.
Smith, Patrick Henry - ("Pat") Pvt 10/1/1861. Appt 4th Sgt 7/16/1862; 2d Sgt 11/1862. C Spotsylvania, VA, 5/12/1864. Roll dated 11/41864, LOF, shows Absent POW. NLR. W 2d Manassas, 8/28/1862. Confined Ft. Delaware, Feb/1865. Died 12/9/1891, buried Midville City Cem, Burke Co.
Smith, Richard - Pvt 10/1/1861. Wounded at 2d Manassas, 8/28/1862. Captured Gettysburg, PA, 7/2/1863. Died of disease Ft. at Delaware, 1/28/1864. Buried Finn's Point National Cem, Salem, NJ.
Smith, Weems F. - Pvt 10/1/1861. Killed at 2d Manassas, VA, 8/28/1862. CSR contains a death claim filed by mother, Nancy L. Smith, of Jefferson Co. BSU.
Stewart, Beverly L. - Pvt 10/1/1861. Died at Mount Jackson, VA, 12/15/1862. Buried at Our Soldiers Cem, Mount Jackson, Shenandoah Co., VA.
Stewart, James B. - Pvt 10/1/1861. Surrendered at Appomattox, VA, 4/9/1865. Buried Mt. Moriah United Methodist Church Cemetery, Jefferson Co., CSA marker reads CO G, 38 GA INF.
Stewart, Levin Collins (Levin "Coot" Collins Stewart) - Pvt 3/1/1862. W 1862, Wounded 2d Manassas, 8/28/1862. Captured Wilderness, VA 5/12/1864. Paroled at Ft. Delaware, Feb/1865. Rcv Boulware & Cox's Wharves, James River, VA, for exchange, Mar. 10-12, 1865. Filed CPA, Jefferson Co., 1904. D 1931, buried Mount Moriah Cemetery, Mathews, Jefferson Co.
Stewart, Robert - Pvt 10/1/1861. K 2d Manassas, VA, 9/1/1862. Killed battle of Ox Ford, (Chantilly) VA, 8/31/1862. CSR contains a death claim filed by father, Eli A. Stewart, of Jefferson Co. BSU.
Stewart, Thomas J. - Pvt10/1/1861. Appt 3d Sgt 8/9/1862. K Fred., VA 12/13/1862. CSR contains death claim by father, Eli A. Stewart, Jefferson Co. Bur Mount Jackson Cem in Shenandoah Co., VA.
Stewart, Thomas P. - Pvt 10/1/1861. Died of typhoid fever, General Hospital, Gordonsville, VA, 8/11/1862. Died at Gordonsville Receiving Hos (Exchange Hotel). The Exchange Hotel still stands in Gordonsville and is a museum. The Confed. dead, initially bur near the hospital, were later moved to Maplewood Cem. Mass CSA grave at Maplewood Cemetery, but soldier's names are not listed.
Stewart, William - Pvt 10/1/1861. Appt 2d Cpl in 1862. Wounded at 2d Manassas, 8/28/1862. Detailed in hospital on account of disability in 1863. Captured Harper's Ferry, VA, 7/10/1864. Released Elmira, NY, 5/29/1865. Filed CPA, Jefferson Co., 1906. BSU.
Swann, Henry J. (Henry Jackson Swann) - Pvt 10/1/1861. Roll for Jan.-Feb 1862, shows discharged, date not given. Died 11/14/1861, bur Ebenezer Associate Reform Presbyterian Cem, Jefferson Co.
Swann, Jabez N. - 4th Sgt 10/1/1861. D Rich, VA, 7/16/1862. Bur Hollywood Cem, Rich, VA, Cem records show I. M. Swan, Co., G, 3rd GA, D 7/17/1862. No record of I. M. Swan in the 3rd GA Regt.
Swann, Joshua K. (Joshua Key Swann) - Pvt 5/15/1862. C and P in 1862. Det in hos 8/1/1864. D in Jackson Hos Rich, VA, 2/8/1865. Bur there in Hollywood Cem. Grave marker appears in opening scene of documentary titled "Death and the Civil War," released in 2012 by film maker Ric Burns.
Swann, Thomas E. - 5th Sgt 10/1/1861. Wounded at 2d Manassas, 8/28/1862. Roll dated 11/4/1864, LOF, shows "Absent Detailed with Major Harman 10/28/1864." No later record. Filed CPA, Jefferson Co., 1899. Died 10/4/1899, buried Ways Baptist Church, Jefferson Co.
Teems, Daniel - (or Teem) Pvt Enlisted at Conscript Camp GA during May/1864. Captured battle of Monocacy, MD, 7/9/1864. Sent to USA Hos Baltimore, MD, W in the right thigh, slightly, by a Minnie ball. Died in captivity, 10/29/1864. Buried Loudon Park National Cem, Baltimore City, MD .

Thompson, Andrew L. - Pvt 3/1/1862. Died Richmond, VA, Jul 6 or 16, 1862. CSR contains death claim filed by father, John Thompson, in Jefferson Co. BSU.
Thompson, Edmund - Pvt 10/1/1861. Died Sav, GA, 1/1/1862. Capt C. Hanleiter diary entry reads, "Witnessed the funeral ceremonies of a member of Capt Battey's Company who D last night" 1/2/1862. CSR has death claim by wife, Diannah Thompson, Jefferson Co. D of disease. BSU.
Thompson, John Jesse - Pvt 8/12/1862. Wounded Fredericksburg, VA, 12/13/1862. Gunshot wound to arm, discharged for wound, 2/13/1864. CSR has discharge record. Filed CPA, McDuffie Co., 1889. Born in Jefferson Co., GA, resident of GA since birth, 8/11/1847, 17 years old at discharge. BSU.
Thompson, William A. - Pvt 10/1/1861. Killed Spotsylvania, VA, 5/18/1864. BSU.
Vaughn, Isaac C. - 2d Lt 10/1/1861. Appears on list of wounded at Antietam, 9/17/1862. Elected 1st Lt 9/18/1862. Captured Spotsylvania, VA, 5/12/1864. Released Ft. Delaware, 6/16/1865. Died 9/23/1883, buried Carswell Family Cemetery, Wrens, Jefferson Co.
Way, George M. T. - 2d Cpl 10/1/1861. Pvt in 1862. Captured Spotsylvania, VA, 5/19/1864. Died at Elmira, NY, 7/25/1864. Buried Woodlawn National Cemetery, Grave #2850, Elmira, NY.
Way, William F. - Pvt 10/1/1861. Discharged for opacity of cornea, General Hospital, Farmville, VA, 4/18/1863. CSR contains discharge record. Filed CPA, Jefferson Co., 1902. Died 5/27/1907, buried Moxley Cemetery, Emanuel Co.
Weeks, Jesse W. - Pvt 10/1/1861. Wounded Winchester, VA, 9/19/1864. At home on wounded frl close of war. Filed CPA Jefferson Co., 1907. D 10/3/1912. Bur Mount Moriah Cem, Jefferson Co.
Whitehead, Robert - Pvt 10/1/1861. Discharged for general ill health and chronic rheumatism Gordonsville, VA, 7/27/1862. CSR contains discharge record. Filed CPA, Jefferson Co., 1895. Died 3/1/1901, in Jefferson Co. BSU.
Willoughby, George T. - Pvt 10/1/1861. Died Middleburg, VA, 9/16/1862. Wounded 2d Manassas, 8/28/1862. CSR contains a death claim filed by widow, Martha Willoughby, Jefferson Co. BSU.
Willoughby, John B. - Pvt 10/1/1861. W in abdomen, Gettysburg, PA, 7/2/1863, D there of wounds, at General Hos, 8/3/1863. Remains returned to Sav, GA. Bur Laurel Grove Cem. CSA grave marker.
Wise, Chelsea S. (Chesley Simmons Wise) - Pvt 11/24/1861. Wounded at 2d Manassas, VA, 8/28/1862. Discharged Mar/1863. Died 10/2/1900, at Aiken, SC. Buried at Magnolia Cem, Augusta, GA. Mary Elizabeth Daniel Wise, filed CPA, Richmond Co., 1912. BSU.
Wright, William A. (William Anderson Wright) - Pvt 5/13/1862. Wounded 2d Manassas, 8/28/1862. Captured Sailor's Creek, VA, 4/6/1865. Released Lincoln USA General Hos, DC, 6/12/1865. Died 1934, buried Greenwood City Cemetery, Parker Co. TX. Maker reads: CO. G 38 GA INF CSA.
Young, Jacob F. - Pvt 5/1/1862. Roll dated 11/4/1864, LOF, shows present. No later record. BSU.
Young, James W. - Pvt 5/10/1862. Died Rich, VA, 12/24/1862, typhoid fever. Bur Hollywood Cem, cem records show: Lt D. W. Young, Company C, 38th D 12/12/1862, buried in Section Q, grave 12.
Young, Jesse - Pvt 10/1/1861. Paroled Farmville, VA, Apr. 11-21, 1865. Died 1868, Emanuel Co. Widow, Sylvira Young, filed a CPA, in Jefferson Co., 1915. BSU.
Young, Leroy B. - Pvt 10/1/1861. Died, 4/11/1862. BSU.
Young, Noah - Pvt 10/1/1861. Deserted. Took OOA to U.S. Govt. City Point, VA, 1/1/1865. Filed CPA, Columbia Co., 1890. D 9/10/1903. Widow, Sarah Young filed CPA, Columbia Co., 1904. BSU.
Young, Thomas J. - Pvt 5/15/1862. Died near Rich, VA 7/22/1862. Buried there in Hollywood Cem. Hollywood Cemetery records show: Young, J. J., Company G, 38 GA, D 7/22/1862, CSR contains death claim filed by widow, Jane B. Young, Jefferson Co. Claim states he died of disease.

*MUSTER ROLL OF COMPANY H, GOSHEN BLUES, ELBERT COUNTY GA

*This company was also known as Old Co. B. and was organized by Capt. Robert P. Eberhart.

Adams, Hiram Gaines - Pvt. 10/15/1861. App 2d Corp. Oct/1862. Captured Gettysburg, PA, 7/3/1863. Sent to Ft Delaware, DE 7/7/1863. Trans to Point Lookout, MD, and exchanged there 2/10/1865. Rcv Cox's Landing, James River, VA 2/21/1865. Died 9/4/1910. Buried Bowman City Cemetery, Elbert Co. Marker reads, Born 10/14/1839. Obit in Elberton Star, 9/6/1910.
Adams, James R. (James Robert Adams) - Pvt. 3/1/1864. C Spotsylvania, VA 5/19/1864. P at PLO MD 1864. Rcv Venus Point, Sav River, GA for E 11/15/1864. NLR. Died 1886. Bur Vaughan Family Cem, Elbert Co. Marker reads: CO H 38 GA INF. Widow, Sarah Adams, filed CPA, Elbert Co., 1901.
Anderson, J. M. (James M. V. Anderson) - Pvt. 2/9/1862. Killed Gaines Mill, VA, 6/27/1862. CSR states he D. 6/30/1862. CSR has death claim filed by father, James E. Anderson, Elbert Co., GA BSU
Anderson, W. A. (William Adrian "Aid" Anderson) - Pvt. 10/15/1861. Captured Spotsylvania, VA, 5/12/1864. P Ft Delaware for E Mar/1865. Rcv Boulware R Cox's Wharves, James River, VA, Mar. 10-12, 1865. Died. 1883, buried Union Cemetery, Mitchell Co, GA. Widow, Permelia Oglesby Anderson, filed CPA, Elbert Co., 1901.
Andrews, A. J. - Pvt. 10/15/1861. App 5th Sgt. Jan/1862. Wounded 2d Manassas, VA, 8/28/1862. Killed at the battle of the Wilderness, VA, 5/5/1864. BSU.
Andrews, J. O. (John O. Andrews) - 3d Corp. 10/15/1861. Elected Jr. 2d Lt. 7/22/1862; 2d Lt. 9/22/1862; 1st Lt. 7/2/1863. D Locust Grove, VA, 4/4/1864. Buried Fredericksburg, VA, Confederate Cemetery. CSA marker reads. "J. O. Andrews, GA." Bur Section 14, Row 19, Grave 1.
Andrews, R. S. (Robert Stanford Andrew) - Pvt 9/10/1862. Died Ladies' Relief Hospital Lynchburg, VA, 12/25/1862. Buried there Confederate Cemetery. CSR has death claim by wife Eliza Andrews.
Andrews, W. T. (William Tate Andrew) - 1st Lt. 10/15/1861. Resigned 6/1/1862. Resigned due to poor health. Died 9/2/1887, buried Stinchcomb United Methodist Church, Elbert Co.
Barnes, Thomas P. - Pvt. Listed in record of effects of deceased soldier turned over to C.S. Quartermaster dated 1864. Buried Old City Cem, Lynchburg, VA, Lot 197, Row 1, grave #10.
Bennett, A. H. M. (Achilles "Killis" H. M. Bennett) - 4th Corp. 10/15/1861. App 2d Corp. Dec/1863. W Spotsylvania, VA 5/12/1864. C & P Greenville, SC 5/23/1865. Filed CPA, Clarke Co., 1890. BSU.
Berryman, C. T. (Charles Tabor Berryman) - Pvt. 3/5/1864. Captured Spotsylvania, VA, 5/12/1864. (Place of capture also given as Petersburg, VA, 3/25/1865) Released from Lincoln USA General Hospital, Washington, D. C., 6/12/1865. BSU.
Bond, E. F. - Pvt. 3/1/1864. Died of pneumonia, Jackson Hospital Richmond, VA, June 9 or 11, 1864. Buried there at Hollywood Cem, Section U, grave #184. Full name unknown.
Bond, E. M. (Ephrioditus "Eppy" M. Bond) - 2d Corp. 10/15/1861. W, in hand 2d Manassas, VA, 8/28/1862. Roll for 8/31/1864, shows "Sent up application to be retired. Approved and returned to applicant." D 10/2/1914. Widow, Nancy Bond, filed CPA, Elbert Co., 1919. Bur Stinchcomb United Meth. Church, Elbert Co. Marker reads CORP, CO H 38, GA INF. Obit in Elberton Star 10/6/1914.
Bond, Henry M. - Pvt. 3/1/1862. Died Savannah, GA 4/7/1862. CSR contains death claim filed by widow, Eliza Jane Bond, Elbert Co. BSU.
Bond, James W. - Pvt. 3/1/1862. C Farmville, VA 4/6/1865. Released Newport News, VA, 6/24/1865. Filed CPA Elbert Co., 1897. Died 1910. Buried Stinchcomb United Methodist Church, Elbert Co. Marker reads: CO H, 38 GA INF CSA. Obituary in Elberton Star newspaper, 3/8/1910.
Bond, John W. - Pvt. 3/1/1862. Died 9/3/1862. Buried Thornrose Cemetery, Staunton, VA. CSR contains death claim filed by father, Henry W. Bond, of Elbert Co.
Bond, Nelson C. - Pvt. 3/1/1862. Died Grove Hospital in Nelson Co., VA, 8/17/1862. CSR contains death claim filed by widow, Lucy A. Bond, of Elbert Co., GA. BSU.
Bond, T. H. M. (Thomas H. M. Bond) - Pvt. 3/1/1864. Wounded Wilderness, VA, 5/5/1864. Absent, Wounded, 11/6/1864. No later record. BSU.
Bond, William E. M. - Pvt. 10/15/1861. Died 6/1/1862, Savannah, GA. Death recorded in the diary of Col. Cornelius Hanleiter. Remains packed in charcoal and sent home by freight train. CSR contains death claim filed by mother, Arminda Bond, of Elbert Co. BSU.

Booth, Absalom - Pvt. 10/15/1861. Wounded Gaines Mill, VA, 6/27/1862. Roll dated 11/6/1864, LOF shows him "absent with leave since 8/26/1864." Captured and paroled Andersonville, SC 5/23/1865. Filed CPA, Elbert Co., 1889. BSU.

Booth, D. - Pvt. 10/1/1861. No later record. Full name unknown. BSU.

Booth, Gabriel H. - Pvt. 10/15/1861. Killed Gaines Mill, VA, 6/27/1862. CSR contains death claim filed by father, Victor E. Booth, of Elbert Co. BSU.

Booth, James C. (James Carter Booth) - Pvt. 10/15/1861. Roll for 11/6/1864, LOF, shows present. (B Elbert Co., 7/18/1833) Filed CPA Madison Co., 1907. Bur Jones Chapel Meth Church, Madison Co. Born 7/18/1833, died 1918. CSA grave marker reads: CO H, 38 GA INF.

Booth, John A. - Pvt. Wounded Sharpsburg (Antietam), MD, 9/17/1862. Filed CPA, Cobb Co, 1889. Capt. Eberhart, Co. Commander, attested to his service in CPA. CSR not found. Died 4/7/1924, Buried Maloney Springs Cemetery, Cobb Co. Obituary printed in Cobb County Times, 5/8/1924.

Booth, John C. (John Callaway or Caswell Booth) - Pvt. 9/2/1862. Captured Winchester, VA 9/19/1864. Released Point Lookout, MD 6/9/1865. Died 7/30/1894. Buried Rehoboth Baptist Church Cemetery. Elbert Co. Widow, F. M. Booth, filed CPA, Elbert Co., 1910.

Booth, John W. - Pvt. 3/1/1862. Wounded left leg, necessitating amputation below knee, Fisher's Hill, VA 9/25/1864. Roll dated 11/6/1864, LOF shows "Absent, W, since 9/25/1864." Filed CPA, Elbert Co., 1879. D 5/29/1916. Bur Deep Creek Baptist Church, Elbert Co. Widow, Eliza P. Booth, filed CPA, Elbert Co., 1916. CSA marker reads CO H, 38 GA INF. Obit in Elberton Star, 5/30/1916.

Booth, John W. G. - Drum Major 10/15/1861. Trans to Co. H. On special service, G. W. Lee's Battn. in GA, disabled, 11/6/1864. C Macon, GA, Apr/1865. Wounded Antietam, MD, 9/17/1862. BSU.

Booth, J. L. C. (James L. C. Booth) - Pvt.10/15/1861. Wounded Antietam, MD, 9/17/1862. Roll dated 11/6/1864, last on file, shows present. Captured and paroled at Anderson, SC, 5/23/1865. BSU.

Booth, Samuel S. (Samuel Stephen Booth) - Pvt. 2/27/1862. Wounded Gaines Mill, VA, 6/27/1862. Roll for 8/31/1864, shows absent, sick Columbia, SC hospital. Sent away from regiment, sick, 5/20/1864." No later record. BSU.

Booth, W. A. (William Absalom Booth) - Bass Drummer in regimental band, 10/15/1861. Trans to Co. H. App Sgt., wounded and disabled Fredericksburg, VA 12/13/1862. On special service with G. W. Lee's Battalion in GA, disabled, 8/31/1864. C Macon, GA Apr/1865. Died 2/15/1897, in Cobb Co., buried there Gresham Cemetery. Widow, Carrie R. Booth, filed a CPA, Cobb Co., 1920.

Bowers, A. M. (Asa Marion Bowers) - Pvt. 3/7/1864. Wounded Winchester, VA 9/19/1864. Roll dated 11/6/1864, LOF shows "Absent, wounded." No later record. (B Coweta Co., GA 2/5/1837) Filed CPA, Hart Co., 1901. Died 11/27/1926. Buried Milshoals Bapt, Church Cem, Madison Co.

Bowers, John H. (John Harris Bowers) - 3d Sgt. 10/15/1861. Wounded Fredericksburg, VA 12/13/1862. Elected 2d Sgt. Dec/1863. Surrendered Appomattox, VA, 4/9/1865. D 8/3/1909. BSU.

Bowers, Lloyd S. - Pvt. 6/27/1862. At home, sick, 7/28 - 8/31/1864. NLR. BSU.

Brown, Francis M. - Pvt. Died at Richmond, VA, of typhoid fever 10/14/1862. CSR contains a death claim filed by widow, Martha E. Brown, of DeKalb Co. BSU.

Brown, D. S. (David S. Brown) - Pvt. 3/1/1864. Surrendered at Appomattox, VA, 4/9/1865. Died 4/16/1902, buried Rehoboth Baptist Church Cemetery, Elbert Co.

Brown, Ira W. - Pvt. 10/15/1861, discharged for tuberculosis, 10/31/1861. BSU.

Brown, John D. (John Duncan Brown) - Pvt. 3/1/1862. Wounded Fredericksburg, VA, 12/13/1862. Surrendered Appomattox, VA, 4/9/1865. Died 9/29/1912. Obit in Elberton Star, 10/1/1912. Bur Holly Springs Baptist Church Cem, Elbert Co. Widow, Martha A. Brown, filed CPA, Elbert Co., 1913.

Brown, J. B. (Jesse B. Brown) - Pvt. Mar/1864. Died of chronic diarrhea in hospital of 2d Corps, A.N.V., Orange Court House, VA, Apr. 5, or 15, 1864. BSU.

Brown, J. H. - Pvt. 10/15/1861. Discharged 12/16/1861. Reenlisted 9/10/1862. Died 9/26/1862. Buried at Old City Cem, Lynchburg, VA, No. 2, 5th Line, Lot 130. Died at Langhorne's Hospital. Burial location should read No.2, 5th Line, Lot 180. First name unknown.

Brown, Jesse M. (Jesse Marshall Brown) - Pvt. 3/1/1864. Wounded left arm Spotsylvania, VA, 5/12/1864. Absent without leave Aug 15 - 31, 1864. CPA shows on wounded furlough until end of war. (B Elbert Co., 10/24/1844. Died Covington, GA, 6/8/1922) Filed CPA, Elbert Co., 1906. Buried Covington Mill Cem, Newton Co. CSA grave marker reads: PVT CO H, 38 GA INF.
Brown, Milton R. (Middleton R. Brown) - Pvt. 9/10/1862. D American Hotel Hospital, Staunton, VA 12/14/1862. Bur there Thornrose Cem. CSR has death claim by widow, Sarah J. Brown, of Elbert Co.
Brown, S. H. - Pvt. 9/10/1862. Died 1862. Full name unknown. BSU.
Brown, W. J. (William J. Brown) - Pvt. 8/29/1862. Roll dated 11/6/1864, LOF, shows him AWL on special service since 8/1/1863." Died 4/8/1907. Buried Rehoboth Baptist Church Cem., Elbert Co.
Burden, J. N. (or I. N.) (Isaac Nelson Burden) - Pvt. 10/15/1861. D Elbert Co., GA 11/16/1862. CSR has death claim by father, Nelson Burden, Elbert Co. Bur Holly Springs Bapt. Cem, Elbert Co.
Burden, W. W. (Willis W. Burden) - Pvt. 3/1/1862. Wounded at the battle of Gaines Mill, VA, 6/27/1862. Absent, sick, Aug/1863. No later record. BSU.
Butler, G. W. (George W. Butler) - Pvt. 3/1/1862. App 1st Corp. Mar/1862. Killed Gaines Mill, VA 6/27/1862. CSR contains death claim filed by widow, Barbary J. Butler, Elbert Co., GA She states he died Richmond, VA, hospital, 7/3/1862. Likely mortally wounded at Gaines Mill. BSU.
Butler, John F. - Pvt. 7/1/1862. Wounded at battle of 2d Manassas, VA, 8/28/1862. Roll dated 11/6/1864, last on file, shows present. Captured and paroled Greenville, SC 5/23/1865. BSU.
Butler, T. N. (Thomas Newton Butler) - Pvt. 10/15/1861. Wounded 2d Manassas, VA, 8/28/1862; Chancellorsville, VA 5/3/1863. C Cedar Creek, VA, 10/19/1864. P at PLO, MD Mar/1865. Trans to Aiken's Landing, VA for exchange, 3/17/1865. Rcv Boulware's & Cox's Wharves, James River, VA 3/19/1865. Filed CPA, Elbert Co., 1914. Died 7/19/1924. Buried Rehoboth Baptist Church, Elbert Co.
Campbell, Jeptha E. - Pvt. 3/1/1862. Wounded Fredericksburg, VA, 12/2/1862. App Sgt. in 1864. Wounded in left arm and left lung Cold Harbor, VA, 6/1/1864. Surrendered Appomattox, VA, 4/9/1865. (Born GA 4/23/1842. Died Elbert Co., 6/9/1919) Filed CPA, Elbert Co., 1888. Widow, Georgia Campbell filed CPA, Elbert Co., 1938. Buried Elmhurst Cemetery, Elbert Co.
Campbell, W. G. (William G. Campbell) - Pvt. 3/1/1863. S Appomattox, VA, 4/9/1865. BSU.
Carlton, Henry C. - Pvt. 9/10/1862. D Rich, VA, 12/21/1862. Bur in Old City Cem, Conf. Section, Lynchburg, VA, 12/22/1862. No. 8, 1st Line, Lot 123. D Langhorne's Hospital. CSR contains an Elbert Co., GA court document dated 1849, appointing Benjamin Winn legal guardian of Henry C. Carlton and siblings, they being orphans. Record contains death claim filed by Mr. Winn.
Carruth, B. O. (Benjamin C. Carrouth) - Pvt. 9/10/1862. Died Richmond, VA, 11/15/1862. BSU.
Carruth, F. A. (Francis Asbury Carrouth) - Pvt. 9/10/1862. Captured Spotsylvania, VA, 5/12/1864. Paroled Elmira, NY, sent to James River, VA, for exchange, 3/10/1865. (Born Franklin Co., 1839. Died Madison Co., 1904) Filed CPA, Madison Co., 1901. BSU.
Carruth, G. A. (Gideon Asbury Carruth) - Pvt. 10/15/1861. S Appomattox, VA 4/9/1865. Filed CPA. Cherokee Co., 1902. B 7/4/1838, D 1/7/1912. Bur Carmel Bapt. Chur Cem, Woodstock, Cherokee Co. Marker reads, "Enlisted in the Conf. Army 10/13/1861, Co. H, 38th GA REGT in Elbert Co., GA."
Carruth, J. A. (Joseph Adam Carrouth) - Pvt. 10/15/1861. S Appomattox, VA 4/9/1865. BSU.
Carruth, J. E. (James Elbert Carrouth) - Pvt. 2/1/1864. Surrendered Appomattox, VA 4/9/1865. Listed in Appomattox roster as wagoner. Filed CPA, Madison Co., 1922. Died 10/30/1923. Buried Old Fork Cemetery, Madison Co., CSA grave marker.
Carruth, Thomas Sanford (Carrouth) - Pvt. 3/1/1862. Wounded in Seven Days' Fight near Richmond, VA, Jun/1862. Died of wounds in Elbert Co., 8/7/1862. CSR contains death claim filed by mother, Harriet Carrouth, of Elbert Co. BSU.
Cash, S. J. (Seaborn J. Cash) - Pvt. 10/15/1861. Died Sav, GA, 11/19/1861. Died at Camp Bartow, Savannah, GA, 11/29/1861, of typhoid fever. His remains returned to Elbert Co. Diary of Capt Cornelius Hanleiter contains an account of his death. BSU.

Christian, B. B. (Benjamin B. Christian) - Pvt. 10/15/1861. Wounded and captured Winchester, VA, 9/19/1864. Died of wounds at U.S.A. Post Hospital, Fort McHenry, MD, 5/15/1865. Buried at Loudon Park National Cemetery, Baltimore, MD.
Christian, C. F. (Charles F. Christian) - Pvt. 2/10/1864. Wounded Fisher's Hill, VA, 9/25/1864. Roll dated 11/6/1864, LOF, shows "Absent, wounded since 9/25/1864." Captured and paroled at Hartwell, 5/20/1865. Filed CPA, Elbert Co., 1900. Died 8/12/1905, buried Rehoboth Baptist Church, Elbert Co.
Christian, John W. - Pvt. 10/15/1861. Wounded, right hand, second finger amputated, Antietam, MD, 9/17/1862. Absent with leave, wounded, 11/6/1864. Captured and paroled Hartwell, GA, 5/19/1865. Filed CPA, Madison Co., 1891. (Born in GA, 3/13/1844. Died Elbert Co., 2/5/1909) Buried Mill Shoal Baptist Church Cemetery, Madison Co. CSA marker reads, CO H, 38 GA INF.
Christian, M. M. - Pvt. 8/29/1862. Roll dated 11/6/1864, LOF, shows him, "AWOL since 7/15/1863." Union records show "M. Christian, Pvt., Co. K, 8th Regiment GA Inf. (name not found on said rolls,) C Gettysburg, PA, 7/4/1863 ; that Luther M. Christian, Co. E, 38th Regiment GA Inf. (name not found on said rolls,), a Conf. deserter, was sent to City Point, VA 2/26/1865. Luther M. Christian D at USA Post Hospital, Cairo, Illinois, on 3/22/1864, and bur Cairo. BSU.
Christian, William R. - Pvt. 3/1/1862. Died of diarrhea in Ladies' Relief Hospital Lynchburg, VA 11/26/1862. Buried there in Old City Cemetery, Confederate Section, No. 3, 1st Line, Lot 100.
Colvard, John U. (John Usry Colvard) - Pvt. 3/1/1862. W and disabled Gaines Mill, VA 6/27/1862. Disabled. On special service in 2d Corps Hospital 3/1/1863 - 11/4/1864.C High Bridge, VA 4/6/1865. Released Newport News, VA 6/26/1865. Filed CPA, Elbert Co., 1890. B 5/13/1841. Died 5/18/1914, buried Holly Springs Baptist Church Cem., Elbert Co. Obit in Elberton Star newspaper, 5/22/1914.
Crook, James P. - Pvt. 10/15/1861. Died at 2d GA Hospital Richmond, VA, 9/8/1862, of typhoid fever. Buried at Oakwood Cemetery, Richmond, VA, Division B, Row 4.
Crook, William R. (William Riley Crook) - Pvt. 3/1/1862. Discharged for disability, 5/5/1862. Died 1900. Buried Stinchcomb Church Cemetery, Elbert Co. Obituary in Elberton Star, 6/14/1900.
Daniel, James Cicero - 1st Lt. 3/4/1865. See Pvt. Co. E.
David, H. C. (Henry Clay David) - Pvt. 10/23/1863. Roll dated 11/6/1864, last on file, shows present. NLR. Enlisted at Macon, GA, trans from Jackson Hos, Rich, VA, to Columbia, SC., 5/28/1864. BSU.
Deadwyler, Henry Robinson - 1st Sgt. 10/15/1861. Elected Capt. 2/16/1864. C Spotsylvania C. H., VA, 5/12/1864. Released from Ft. Delaware 6/16/1865. B 11/12/1838, D 4/20/1885, Elbert Co. Bur Deadwyler Family Cem, Elbert Co. Marker reads: CAPT, CO H 38 GA, Immortal 600. Obituary in Atlanta Constitution, 4/22/1885. Widow, Cornelia W. Deadwyler, filed CPA, Madison Co., 1920.
Deloach, John A. - CSR shows him listed as a patient the University of Virginia Hospital, Charlottesville, VA. Died there of typhoid fever 7/1/1862, buried there Confederate Cemetery. Listed on the Confederate memorial at cemetery as D, J. A. 38th GA.
Dickerson, R. J. (Robert J. Dickerson) - Pvt. 3/9/1863. Captured Gettysburg, PA, 7/3/1863. Died of measles at Ft Delaware, 8/21/1863. First name in CSR, notation in record states "absent, sick, left in the hands of the enemy." Buried Finn's Point National Cem, Salem, NJ.
Dickerson, T. S. (Thomas S. Dickerson) - Pvt. 10/15/1861. Deserted 4/14/1864. Took oath of allegiance to U.S. Govt. and furnished transportation to Philadelphia, PA, 4/21/1864. Widow, Emily McConnel Dickerson filed CPA, Elbert Co., 1919. Pension application was disapproved. BSU.
Dixon, J. S. (John Samuel Dixon) - Pvt. 2/27/1862. Captured Harper's Ferry, VA, 7/8/1864. Paroled Elmira, NY, and trans for exchange 3/10/1865. Filed CPA, Sumter, FL, 1907, pension #A04343. Born in Elberton, GA., 2/14/1836, moved to FL 1856. Died 12/14/1916. Buried Dixon Cem, Lake Co., FL.
Eberhart, F. P. (Frances P. Eberhart) - Pvt. 10/15/1861. Discharged at Savannah, 12/16/1861. Died 8/8/1870. Buried Antioch Baptist Church, Elbert Co.
Eberhart, James E. - Pvt. Conscript, 9/10/1862. Discharged Richmond, VA, 1/30/1863 for glaucoma. CSR mixed with records of John G. Eberhart below. BSU.

Eberhart, John G. - Pvt. 9/10/1862. Discharged on account of glaucoma, from Chimborazo Hospital #2, Rich, VA, 2/7/1863. D. on way home. No record of glaucoma found for John G. Eberhart in CSR, discharged from hos 2/7/1863 after suffering from typhoid fever. CSRs mixed with records of James E. Eberhart shown above. BSU.
Eberhart, John W. (John Washington Eberhart) - Pvt. 3/1/1862. App 2d Corp. Mar/1862; 1st Corp. Oct/1862. C Spotsylvania, VA 5/12/1864. P Ft Delaware, Feb/1865. Rcv Boulware & Cox's Wharves, James River, VA for E, Mar. 10 - 12, 1865. (B in Hall Co., 5/23/1843) Filed CPA, Paulding Co., 1902. D 4/22/1926. Bur Poplar Cem, Paulding Co. CSA marker reads: CO H, 38 GA REGT.
Eberhart, Robert Patton - Capt. 10/15/1861. W in right lung Fred., VA, 12/13/1862. Resigned 2/16/1864. (B in GA, 10/13/1834) Participated in the Gen. William Walker expedition to Nicaragua, 1850's. Filed CPA, Fulton Co., 1889. Born 10/13/1834, died 1/17/1907, Atlanta, buried there Westview Cemetery. Obit in Atlanta Constitution newspaper, 1/21/1907, page 6. Tribute in Confederate Veteran Magazine, July 1907. Name spelled "Eberhardt" on grave marker.
Eberhart, William C. - Pvt. 3/1/1862. Captured Spotsylvania, VA, 5/12/1864. Released Elmira, NY, 7/7/1865. CSR states D chronic diarrhea 8/14/65, at U.S. Hosp Sav, GA, bur there Laurel Grove Cem.
Edwards, D. W. - (Daniel W. Edwards) Pvt. 10/15/1861. Discharged 12/16/1861. Died 6/17/1900, Buried Mount Bethel Cemetery, Marietta, Cobb Co.
Escoe, William T. (or Eascoe) - Pvt. 1/18/1862. Discharged Savannah, GA 5/23/1862. BSU.
Gaines, M. C. (Marion C. Gaines) - Pvt. 10/15/1861. Died of typhoid fever in General Hospital at Staunton, VA 7/27/1862. Buried there in Thornrose Cemetery. CSR contains death claim filed by father, George Gaines, of Elbert Co.
Ginn, Gaines W. (Gaines Washington Ginn) - Pvt. 10/15/1861. App 4th Corp. 3/1862; 3d Sgt 12/1863. Wounded Spotsylvania, VA, 5/12/1864. Wounded 12/6/1864. No later record. Filed CPA, Elbert Co., 1888. Died 5/1/1917, buried Rehoboth Baptist Church Cemetery, Elbert Co.
Ginn, James A. (James Alfred Ginn) - Pvt. 6/27/1862. Killed 2d Manassas, VA, 8/28/1862. Born 1829, cenotaph at Hillcrest Cemetery, Bowman, Elbert Co. CSR contains a death claim filed by widow, Julieann Ginn, Elbert Co. Widow filed a CPA, Elbert Co., 1891. BSU.
Ginn, J. W. (Jeptha W. Ginn) - Pvt. 10/15/1861. Discharged, disability, 6/2/1862. BSU.
Ginn, S. S. (Singleton Satterwhite Ginn) - Pvt. 6/27/1862. W Seven Days' Fight near Richmond, VA in 1862. S Appomattox, VA, 4/9/1865. D 9/20/1895. Bur Holly Springs Bapt Church Cem, Elbert Co.
Gulley, Allen Y. - Pvt. 3/1/1862. Wounded, shot through neck and captured Spotsylvania, VA, 5/12/1864. Exchanged at Ft Delaware, 3/7/1865. Received Boulware & Cox's wharves, James River, VA, Mar. 10-12, 1865. Filed CPA, Elbert Co., 1895. BSU.
Gulley, James W. A. (James William A. Gulley) - Pvt. 3/1/1862. Captured Spotsylvania, VA, 5/12/1864. Paroled and exchanged Ft Delaware, 3/7/1865. Rcv Boulware & Cox's Wharves, James River, VA, Mar. 10-12, 1865. Died Conf. Soldiers' Home, Atlanta, 6/12/1930. Filed CPA, Madison Co., 1905. Buried Old Fork Baptist Church Cem, Madison Co. CSA marker reads CO H 38 GA INF.
Gulley, R. W. (Robert W. Gulley) - Pvt. 10/15/1861. Deserted 9/25/1864. Went to New Creek, W. VA, where he took OOA to U.S. Govt. at New Creek WV, Oct/1864. BSU.
Hall, Beverly W. - Pvt. 10/15/1861. App 3d Corp. in 1862. W Wilderness, VA, 5/5/1864. D of wounds 6/9/1864. Bur Hollywood Cemetery Richmond, VA, bur Section U, grave #581.
Hall, Charles Woodson - Pvt. 1/18/1862. Appt 1st Sgt. Jul/1863. Captured and paroled Greensville, SC, 5/24/1865. Filed CPA, Elbert Co., 1902. Died 9/24/1907. Buried Rehoboth Baptist Church Cemetery, Elbert Co. Obituary in Elberton Star, 9/24/1907.
Hall, F. M. (Francis Marion Hall) - Pvt. 9/10/1862. AWOL May 5 - Nov. 6, 1864. CPA shows he left command, ill with fever, 4/1/1865. Sent to Richmond, VA, hospital, furloughed home for 30 days, Apr/1865. (Resident GA since 1845) Filed a CPA, Elbert Co., 1920. Buried Hillcrest Cem, Elbert Co.

Hall, James C. - (James Chambers Hall) 2d Lt. 10/15/1861. Elected 1st Lt. 6/1/1862. Died at Warwick House Hospital Lynchburg, VA, 8/16/1862, of typhoid fever. Buried at Hall-Conwell-Fortson Cemetery, Elbert Co., Tribute in The Elberton Star newspaper, 3/23/1889.
Hall, Lindsey E. (or Lindsay) - Pvt. 9/10/1862. D Fred., VA, 4/15/1863. Bur in Conf. Cem Spotsylvania, VA, marker reads, 38 GA INF. Widow, Mary E. Hall filed CPA, Elbert Co., 1891.
Hall, Robert Pendleton, Jr. ("Petch") - Pvt. 3/12/1864. CPA shows on detail to guard wagon train 4/9/1865. (Born in Elbert Co., 1845) Filed a CPA in Madison Co., 1908. Died 8/18/1911, buried at Rehoboth Baptist Church Cemetery, Elbert Co.
Hall, Simeon - Pvt. 1/5/1862. AWOL Jul 28 - Nov 6, 1864. NLR. B 3/29/1831, D 2/12/1911, bur Town Creek Bapt. Cem., DeKalb Co., AL. Widow, Marietta Hall, filed CPA DeKalb Co., AL, 1912.
Hall, William H. - Pvt. 1/18/1862. C Winchester, VA, 9/19/1864. Released PLO, 6/28/1865. BSU.
Hall, William S. - Enlisted as a Pvt. in Co. F, 15th Regiment GA Inf. 7/15/1861. App 2d Sgt. 1/18/1862. Trans to Co. H, 38th Regt. GA Inf., 1/13/1863. Captured Gettysburg, PA, 7/1/1863. Roll dated 11/6/1864, last on file, shows "Absent, a prisoners of war since 7/1/1863." NLR. BSU.
Hammond, C. H. - Pvt. 10/15/61. Appears last on roll Feb/1862. Died in service. BSU.
Hansard, James B. - Pvt. 2/27/1862. Appears last on roll 10/31/1862. BSU.
Hansard, Jeptha R. - (Jeptha Riley Hansard) Pvt. 3/1/1862. Captured Spotsylvania, VA, May 12 or 20, 1864. Released Elmira, NY, 6/30/1865. Filed CPA, Elbert Co., 1910. Died 5/3/1922, Buried Harmony Baptist Church Cemetery, Elbert Co.
Hansard, P. H. - (Patrick Henry Hansard) Pvt. 11/18/1863. Captured Harper's Ferry, VA 7/8/1864. Paroled Elmira, NY, 10/11/1864. Exchanged 10/29/1864. Admitted to Ocmulgee Hospital Macon, GA, 11/15/1864, trans 11/18/1864, place not stated. No later record. Filed CPA, Clarke Co., 1889. Died 1/9/1897. Buried Town Creek Baptist Church Cemetery, DeKalb Co., AL.
Hansard, William J. - (William James Hansard) Pvt. 2/27/1862. Killed near Mechanicsville, VA 6/1/1864. BSU.
Harbin, Jasper L. - Pvt. 3/1/1862. App 3d Corp. Oct/1862; Wounded at 2nd Manassas. Regimental Ensign, 4/26/1864. Elected Lt. Captured Cold Harbor, VA, 6/1/1864. Exchanged. Admitted to Stuart Hospital Rich, VA 4/2/1865, returned to duty same day. Regtal color bearer. Resident NY 1886. BSU.
Harbin, William M. - (William "Marion" Harbin) Pvt. 2/19/1862. Discharged, under-age, Nov/1862. Born 10/22/1846. CSR contains discharge certificate, 16 years discharge, 90 pounds, and four feet eight inches tall. Filed a CPA, DeSoto Co., FL, 1918. BSU.
Hendrix, Charles W. (or Hendrick) (Charles Woot Hendricks) - Pvt. 10/15/1861. Wounded, third finger right hand amputated, hand rendered useless, Chancellorsville, VA, 5/3/1863. Wounded Wilderness, VA 5/6/1864. (Born Elbert Co., 2/2/1834. Died 7/20/1928) Filed CPA, Madison Co., 1893. Buried Hill Crest Cemetery, Elbert Co. CSA marker.
Hendrix, Elijah G. (Elijah G. Hendrick) - Pvt. 3/1/1862. Discharged, disability, Savannah, GA, 5/23/1862. CSR contains discharge record. Died 12/22/1917, Elbert Co. Widow, Nancy Tucker Hendricks filed CPA, Elbert Co.1919. BSU.
Hendrix, Francis M. (Francis Marion Hendrick) - Pvt. 3/1/1862. Captured Spotsylvania, VA 5/12/1864. E 3/7/1865. Admitted Receiving and Wayside Hospital (General Hospital #9) Rich, VA 3/10/1865. Sent to Camp Lee near Rich, VA, 3/11/1865. At home, W frl close of war. (Resident GA since 1829) Filed CPA, Elbert Co., 1919. Died 12/14/1922. Bur Hillcrest Cem, Elbert Co., GA
Hendrix, John C. (or Hendrick) - Pvt. 3/4/1862. Died Savannah, GA, May 10 or 11, 1862. CSR contains a death claim filed by mother, Margarete Hendrick, Elbert Co. BSU.
Hendrix, Stephen L. (or Hendrick) - Pvt. 3/1/1862. Died of pneumonia in hospital of 2d Army Corps, ANV., Orange Court House, VA, Apr. 7 or 15, 1864. BSU.
Hendrix, William C. (or Hendrick) - Pvt. 1/18/1862. Wounded Fredericksburg, VA, 12/13/1862. Absent without leave, May 6 - Nov 6, 1864. BSU.
Hewell, Josiah T. (Josiah Tuck Hewell) - Pvt. 10/15/1861. Wounded in leg, Antietam, MD 9/17/1862. Absent without leave, 5/1/1863 - 11/6/1864. Filed CPA, Oglethorpe Co., 1890. Died there 12/21/1896. Buried Sardis Baptist Church Cemetery, Wilkes Co. Widow, Caroline Hewell filed CPA Oglethorpe Co., 1897. Obit in Oglethorpe Echo newspaper, 12/25/1896.

Hewell, Philip W. - Pvt. 9/10/62. Captured Gettysburg, PA, 7/3/1863. Died of smallpox at Fort Delaware, 11/16/1863. Buried at Finn's Point National Cemetery, Salem, NJ.
Jones, J. E. S. - (John E. S. Jones) Pvt. 9/19/1862. Died Mount Jackson, VA, 12/31/1862, buried "Our Soldiers Cem," Mt. Jackson, Shenandoah Co., VA. CSR states he was a substitute for William O'Braint. CSR contains a death claim filed by widow, Fanny E. Jones, Elbert Co.
Jones, Martin - Pvt. 3/18/1863. Appears last on roll 4/30/1863. Filed CPA, Franklin Co., 1901. Stated in CPA trans to 15th GA Regt., Co. F and surrendered Appomattox C.H. 4/9/1865. Died 1905. Buried Old Fellowship Meth. Church Cemetery, Vanna, Hart Co., GA
King, F. M. (Francis "Frank" Marion King) Pvt. 4/18/1864. Admitted to General Hospital Charlottesville, VA, 7/26/1864, and Trans to Lynchburg, VA same date. Roll LOF, 11/6/1864, shows "Absent, sick." NLR. Filed CPA, Cobb Co., 1904. Full name in CPA. B in Elbert Co, 4/1/1843. BSU.
King, J. H. - (James H. King) Pvt. 10/15/1861. Discharged prior to 8/31/1864. CSR contains discharge record. Born 11/5/1818, died 4/1/1887. Buried Holly Springs Baptist Church Cemetery, Paulding Co. Widow, Rachel King, filed a CPA, in Cobb Co., 1896.
King, John L. W. - Pvt. 10/15/1861. W Fred., VA 12/13/1862. Roll dated 11/6/1864, last on file, shows present. CPA shows on detail with wagon train near Appomattox, VA, 4/9/1865. (Died 11/12/1895) Widow, Sarah E. Denney King filed CPA, Madison Co., 1911. BSU.
King, William B. (William Benjamin King) - Pvt. 9/10/1862. Transferred to Co. I, 15th Regiment GA Inf. 7/1863. Deserted near Knoxville, TN Jan/1864. Filed a CPA in TN, pension file # 1623904. Died 10/15/1924. Buried Oak Grove Church Cemetery, Monroe Co, TN.
Kirby, William W. - Pvt. 10/16/1861. Roll dated 11/6/1864, LOF, shows present. NLR. D 11/3/1906. Bur Holly Springs Bapt Church, Elbert Co. Widow, Rachel Kirby, filed CPA, Franklin Co., 1910.
Mann, E. J. (Edwin Jackson "Jack" Mann) - Pvt. 10/15/1861. Wounded and disabled Wilderness, VA 5/5/1864. Surrendered Appomattox, VA 4/9/1865. Filed CPA, Elbert Co., 1901. Died 3/15/1909. Buried Deadwyler Family Cemetery, Elbert Co. Obituary in Elberton Star, 3/16/1909.
Mann, George H. (George Henry Mann) - Pvt. 3/28/1864. Wounded and captured Monocacy, MD 7/9/1864. Paroled Ft McHenry, MD and Trans to Point Lookout, MD 9/27/1864. Exchanged 9/30/1864. C and P Hartwell, 5/18/1865. (B Elbert Co., 12/6/1846) Filed CPA, Madison Co., 1906. Died 10/22/1924, bur Union Bapt. Church, Madison Co. CSA grave marker reads, CO H, 38 GA INF.
Martin Eli - Pvt. At Gen. Hosp Farmville, VA 8/21/1862. Returned to duty 10/15/1862. NLR. BSU.
Martin, R. H. - Pvt. 10/15/1861. Roll dated 11/6/1864, last on file, shows absent without leave since June/1862. Full name unknown. BSU.
Maxwell, Charles W. (Charles Woodson Maxwell) - Pvt. 3/1/1862. Discharged, disability, Sav, GA 5/23/1862. Filed CPA, Elbert Co., 1900. Died 4/20/1934. Bur Deep Creek Bapt. Church, Elbert Co.
Mobley Isaac M. - Enlisted Pvt, Co. I, 15th Regt. GA Inf. 7/15/1861. Trans Co. H, 38th Regiment GA Inf. 1/1/1863. Roll 11/6/1864, last on file, shows present. (Born Oglethorpe Co., 8/29/1838) Filed CPA, Madison Co., 1899. BSU.
Moon, J. S. (James S. Moon) - Sgt. 3/1/1862. Issued furlough for 24 days, 2/4/1865. BSU.
Moon, John S. - Enlisted as a Pvt. in Co. I, 15th Regt. GA Inf. 7/15/1861. Discharged for disability, at Richmond, VA, 10/25/1861. Enlisted as Pvt., Co. H, 38th GA, 3/1/1862. Appointed 5th Sgt. Dec/1863. Wounded by gunshot, Oct/1864. Died 10/1/1864. Buried at Old City Cemetery, Lynchburg, VA. Widow, Martha Moon, filed a CPA, in Madison Co., 1901.
Moon, William P. (William Peter Moon) - Enlisted as Pvt., Co. I, 15th Regiment GA Inf. 7/15/1861. Discharged, disability, 10/25/1861. Enlisted as Pvt., Co. H, 38th Regt. GA Inf. 3/1/1862. W 2d Manassas, VA 8/28/1862. Died of wounds 9/2/1862. CSR contains death claim filed by father, James B. Moon, Elbert Co. BSU.
Moon, W. G. (William Gable Moon) - Pvt. 9/10/1862. Captured Winchester, VA 9/19/1864. Exchanged Point Lookout, MD 3/15/1865. Rcv Boulware & Cox's Wharves, James River, VA, 3/18/1865. Filed CPA, Elbert Co., 1899. Died 10/11/1918. Buried Rehoboth Baptist Cemetery, Elbert Co. CSA grave marker reads CO H, 38 GA INF. Obituary in Elberton Star, 10/18/1918.

Moore, J. W. (John W. Moore) - Pvt. 10/15/1861. App Musician (Drummer) in 1861. W Gaines Mill, VA 6/27/1862. W and C Gettysburg, PA, 7/5/1863, severe thigh W, sent to NY, P DeCamp General Hospital, David's Island, NY Harbor, Sep/1863. Rcv City Point, VA, for E 9/8/1863. C Spotsylvania, VA 5/23/1864. Sent to PLO, MD, then Elmira, NY, released 6/30/1865. First name in CSR. BSU.

Moore, Thomas A. - Pvt. 10/15/1861. App 4th Corp. Oct/1862. W Fredericksburg, VA 12/13/1862. C Gettysburg, PA, 7/5/1863. P DeCamp General Hospital, David's Island, NY Harbor, Sep/1863. Rcv City Point, VA 9/16/1863. App 3d Corp. in 1864. S Appomattox, VA, 4/9/1865. Filed CPA, Elbert Co., 1901. D 1907, bur Moore Family Cem, Elbert Co. CSA marker reads, CO H, 38 GA VOL INF.

Moore, William M. - Pvt. 10/15/1861. Appears last on roll Apr/1863. Trans to Co F 15th GA, captured Dandridge, TN 1/22/1864. Transferred from Rock Island, IL, for exchange 3/2/1865. Filed CPA, Elbert Co., 1902. D 5/9/1907, Franklin Co. Buried Canon City Cemetery, Franklin Co.

Motes, W. A. (William A. Motes) - Pvt. 9/10/1862. Trans to Co. F, 38th Regiment GA Inf. Sep/1862. Captured Gettysburg, PA, 7/3/1863. Released Ft Delaware, 6/16/1865. Filed CPA, Hart Co., 1897. Died 9/5/1897, buried Coldwater United Methodist Church. Marker reads, CO F, 38 REG GA INF.

Nelms, Hiram A. - Pvt. Conscript, 9/10/1862. Died of typhoid fever Dec. 15 or 18, 1862. CSR contains death claim filed by widow, Sarah S. Nelms, Elbert Co. BSU.

Oglesby, Abda - 4th Sgt 10/15/1861. Captured Spotsylvania, VA, 5/12/1864. Paroled Ft Delaware, sent to Aiken's Landing, VA, for exchange 9/18/1864. Rcv Varina, VA, 9/22/1864. Admitted Jackson Hospital Rich, VA, abscess on arm, 9/22/1864, furloughed 30 days, 9/26/1864. Enlisted as Pvt., Co. H, 2d Battalion GA Militia Cavalry 10/1/1864. App 1st Lt., Acting Adj. S Macon, GA 4/22/1865. (B Elbert Co., 6/9/1836, died there 2/4/1907) Widow, Mary L. Oglesby filed CPA, Elbert Co., 1908. Buried there Elmhurst City Cemetery. Obituary in Elberton Star newspaper, 2/7/1907.

Oglesby, John Jr. - 2d Lt. 10/15/1861 Elected 2d Lt. 7/22/1862; 1st Lt. 9/22/1862. Severely wounded Gaines Mill; mortally wounded 7/1/1863, at Gettysburg, died there next day. CSR contains death claim filed by widow, Margaret P. Oglesby, of Elbert Co. BSU.

Oglesby, William (William M. "Shuck" Oglesby) - Pvt. 10/15/1861. Wounded Fredericksburg, VA 12/13/1862. Roll dated 11/6/1864, last on file shows "absent with leave, special service in Lee's Battalion, in Georgia since 12/13/1862." Filed CPA, Elbert Co., 1896, died there 3/4/1917, buried Antioch Baptist Church. CSA marker. Obituary in Elberton Star newspaper, 3/6/1917.

Oglesby, W. B. (William Baird Oglesby) - Pvt. 10/15/1861. Wounded Fredericksburg, VA 12/13/1862. Roll dated 11/6/1864, LOF shows present. NLR. D 1884. Bur Antioch Baptist, Elbert Co.

Owens, S. - or T. Owens, Pvt. Listed on a register of Wayside Receiving Hospital or General Hos # 9, Rich, VA, Sept. 1864, D there of acute dysentery 9/24/1864. Buried Oakwood Cem, Richmond, VA

Parham, Jeptha I. - Pvt. 10/15/1861. Captured Spotsylvania, VA, 5/12/1864. Exchanged at Ft Delaware, DE 3/7/1865. At home on furlough close of war. Filed CPA, Madison Co., 1918. (Died Madison Co., 1924) Died 4/13/1924, buried Oak Grove Baptist Church Cemetery in Madison Co.

Parham, John W. - Pvt. 3/1/1862. Killed 2d, Manassas VA, 8/28/1862. CSR contains death claim filed by widow, Sarah F. Parham in Elbert Co. BSU.

Parham, William S. - Pvt. 1/29/1864. Roll dated 11/6/1864, LOF, shows present. (B Madison Co., 1842) Filed CPA, Madison Co., 1908. Died 1924 Madison Co., buried there Vineyard Creek Cem.

Parks, E. O. (Elbert O. Parks) - Pvt. 8/29/1862. Wounded and captured Chancellorsville, VA 5/3/1863. Died Post Hospital Ft Delaware, 11/8/1864. Bur Finn's Point National Cem, Salem, NJ.

Payton, Moses (or Peyton) - Pvt. 3/1/1864. Wounded Fredericksburg, VA, 12/13/1862. CSR shows AWOL from Mar/1863 - Nov/1864. Captured and paroled Hartwell, GA 5/18/1865.BSU.

Pearson, Littleton D. - Pvt. 2/12/1862. Trans to Co. F, Jan/1863. Captured Rapidan Station, VA 11/11/1863. Paroled at Point Lookout, MD 2/18/1865. Rcv Boulware A Cox's Wharves, James River, VA, Feb. 20-21, 1865. Widow, Nancy Pearson, filed CPA, Elbert Co., 1891. She said he

never returned home. Buried Confederate Hill, Loudon Park National Cem, Baltimore, MD, Plot D-12.
Peyton, John G. (or Payton) - Pvt. Aug/1862. Died 10/7/1862, of measles. Buried Thornrose Cemetery, Staunton, VA. Widow, Elizabeth C. Peyton filed CPA, Elbert Co., 1891.
Power, J. M. (John Mankin Power) - Pvt. 10/15/1861. Captured and paroled Hartwell, GA, 5/18/1865. D 6/11/1888, Habersham Co. Widow, Amanda Power, filed CPA, White Co., 1901. BSU.
Pulliam, James P. - Pvt. 3/1/1862. Died typhoid pneumonia, CSA General Hospital Charlottesville, VA Jun 30 or 7/1/1862. Buried there at University of Virginia, Confederate Cemetery. Marker lists J. Pulline. CRS contains death claim filed by father, Matthew Pulliam, in Elbert Co., GA
Pulliam, Nathan B. - Pvt. 1/5/1862. Discharged, disability, 1/14/1862. Enlisted as Pvt. in Co. I, 15th Regiment GA Inf. 3/26/1863. Substitute for Elijah B. Norman. Discharged, furnished Elijah B. Norman as substitute, 8/24/1863. CSR contains discharge certificate. Died 6/15/1865. Widow, Dallas Pulliam filed CPA, Elbert Co., 1891. BSU.
Pulliam, William (William Thomas "Honey" Pulliam) - Pvt. 10/15/1861. Wounded 2d Manassas, VA, 8/28/1862. C and P Hartwell, GA 5/19/1865. Filed CPA, Elbert Co., 1889. Died 6/14/1905. Buried Stinchcomb United Meth. Church, Elbert Co., CSA marker reads: CO H 38, GA INF.
Pulliam, Willis- Pvt. 2/9/1862. Discharged prior to 11/6/1864. CSR contains discharge certificate, born Elbert Co., GA, age 46 at time of discharge, born abt. 1818. BSU.
Rice, J. D. (James David Rice) - Pvt. 10/15/1861. Discharged 12/3/1861. B 11/17/1843. CSR contains discharge record. Died 2/12/1909. Buried Blacks Creek Baptist Church Cemetery, Madison Co.
Rice, William J. (William Jasper Rice) - Pvt. 9/10/1864. Captured Fisher's Hill, VA, 9/22/1864. Paroled at Point Lookout, MD, transferred to Aiken's Landing, VA, for exchange 3/17/1865. Received Boulware & Cox's Wharves, James River, VA, for exchange, 3/19/1865. (Resident GA since 1831) Filed CPA, Elbert Co., 1920, Died 4/16/1920, buried there Hillcrest Cemetery.
Roberts, H. W. (Henry W. Roberts) - Pvt. 3/1/1862. Deserted, received Provost Marshal General's Office, Washington, D. C., from Army of the James, 2/24/1865. Took oath of allegiance to U.S. Govt., furnished transport to Columbus, OH. BSU.
Roberts, O. M. (Olive M. Roberts) - Pvt. 6/20/1862. Discharged for deafness 7/27/1862. CRS contains discharge record. Filed CPA, Elbert Co., 1888. Buried there Stinchcomb United Methodist.
Roberts, William G. - Pvt. 3/1/1862. Captured Gettysburg, PA, 7/2/1863. Exchanged City Point, VA, 4/30/1864. Roll dated 11/6/1864, last on file, shows absent without leave since 7/11/1864. BSU.
Roberts, Wiley T. - Pvt. 3/1/1862. Killed Fredericksburg, VA, 12/13/1862. Widow, Frances Roberts filed CPA, Madison Co., 1891. BSU.
Rousey, Archibald Alexander - Pvt. 10/15/1861. W Gettysburg, PA, 7/2/1863. C Berryville, VA, 8/10/1864. Released Elmira, NY, 6/23/1865. D Oct/1881, bur Bethlehem UMC, Elbert Co. CSA marker reads: CO H 38th GA INF. Widow, Eliza J. Vassor Rousey, filed CPA, Elbert Co., 1901.
Rousey, J. (James Rousey) - Pvt. 10/15/1861. Killed at Gaines Mill, VA, 6/27/1862. CSR contains death claim filed by mother, Martha Rousey, Elbert Co. BSU.
Rousey, Mitchell G. (Mitchell Gaines Rousey) - Pvt. 9/10/1862. Wounded Fredericksburg, VA, 12/13/1862. Died typhoid fever, General Hospital #3, Lynchburg, VA, 10/23/1863. "Sent to Lexington, GA" Body sent to Lexington, GA by rail, buried Stinchcomb Meth. Church. CSA marker reads, CO H, 38 GA INF. Widow, Malinda E. Bond Rousey, filed CPA, Elbert Co., 1910.
Rousey, Woodson - Pvt. Admitted Jackson Hospital, Rich, VA, 3/22/1865. C 4/3/1865 in Rich, VA, hos. Admitted prison hos Newport News, VA, 5/28/1865, died 6/17/1865, chronic diarrhea. Bur at P. West's farm, Newport News, VA. Remains disinterred in 1890 and reburied Greenlawn Cem, Newport News, VA. Names of the soldiers on Conf. memorial at Cem, but his name is missing.
Sanders, Charles H. - Pvt. 10/15/1861. Wounded in leg, necessitating amputation below knee, and captured Gettysburg, PA, 7/1/1863. P 9/27/1863. At home, wounded, close of war. Filed CPA, Elbert Co. Bur Stinchcomb United Meth. Church Cemetery, Elbert Co., marker reads: CO H 38, GA INF.

Sanders, Jacob E. - Pvt. 10/15/1861. Severely wounded in abdomen at Gettysburg, PA, 7/1/1863, died of wounds, 7/6/1863. CSR has death claim filed by widow, Mary A. Sanders, of Elbert Co. She filed a CPA, Elbert Co., 1891. BSU.
Sanders, J. J. - (John James Sanders) Pvt. 9/10/1862. Severely wounded Marye's Heights, VA, 5/3/1863. Roll dated 11/6/1864, last on file, reads absent with leave, home since 5/8/1863. Sent in papers to be retired." Died 11/16/1867, buried Stinchcomb Methodist Church Cemetery, Elbert Co.
Sanders, Richard - Pvt. 3/1/1862. Wounded Gaines Mill, VA, 6/27/1862. Captured Spotsylvania, VA, 5/12/1864. Roll dated 11/6/1864, LOF shows "Absent, POW since 5/12/1864." NLR. BSU.
Saunders, Peter M. (or Sanders) - 1st Corp. 10/15/1861. D typhoid fever, General Hospital #3, Lynchburg, VA Dec 1 or Aug/1863. Bur there Conf. Cem. CSR has death claim by James H. Sanders.
Scales, J. S. (John S. Scales) - Pvt. 10/15/1861. Discharged, disability, 7/27/1862. CSR contains discharge record. D 1/30/1888, bur Ruhamah Methodist Church Cem, Anderson Co., SC.
Settle, James S. - Pvt. 3/1/1862. D dysentery, General Hosp. Mount Jackson, VA 12/8/1862. Bur Our Soldiers Cem, Mt. Jackson, Shenandoah Co., VA. Widow, Sena Settle, filed CPA, Forsyth Co., 1891.
Seymour, Charles Marion - Pvt. 9/10/1862. W Chancellorsville, VA, 5/3/1863. Roll dated 11/6/1864, LOF shows "Absent, sick since 5/13/1864, Harrisonburg, VA, hospital." D Oct/1901, Elbert Co. Bur Antioch Church Cem, Elbert Co. Obituary in Elberton Star, 10/24/1901.
Seymour, J. A. (or Seamore) (Jesse A. Seymour) - Pvt. 9/10/1862. W and C Gettysburg, PA, 7/3/1863. D of smallpox, Ft Delaware, 11/22/1863. Bur Finn's Point National Cem, Salem, NJ.
Seymour, J. G. (Jonathan Greene Seymour) - Pvt. 10/15/1861. Surrendered at Appomattox, VA, 4/9/1865. (Resident of GA since 1839) Filed CPA, Elbert Co., 1919. Died 4/13/1923, buried Dewy Rose Baptist Church Cemetery, Elbert Co.
Seymour, Marshall M. - Enlisted as Pvt., Co. I, 15th Regiment GA Inf. 3/4/1862. Transferred to Co. H, 38th Reg. GA Inf. Jul/1863. Absent without leave, 10/1/1864. Filed CPA, Elbert Co., 1897. Died 1/27/1903. Buried Deep Creek Baptist Church Cemetery, Elbert Co.
Smith, F. O. - Pvt. 10/15/1861. Appears last on roll 4/1862. Other records show he died service. Full name unknown. BSU.
Smith, John A. - Pvt. 10/15/1861. W Fred., VA, 12/13/1862. Died from wounds Mount Jackson, VA, 9/4/1864. Bur Our Soldiers Cem, Shenandoah Co., VA. Mary E. Smith filed CPA, Elbert Co., 1891.
Smith, P. H. - (Pleasant "Ples" Holly Smith) Pvt. 3/1/1862. W and C Gettysburg, PA, 7/2/1863. P DeCamp General Hospital, David's Island, NY Harbor, date not stated. Roll dated 11/6/1864, LOF, shows present. C and P Anderson, SC, 5/3/1865. D 4/19/1910, Dallas Co., AR, bur there Liberty Cem.
Smith, Sandford B. - Pvt. 3/1/1862. W, left arm permanently disabled, Spotsylvania, VA, 5/12/1864. (Resident GA since 10/4/1833, D Taliaferro Co., 1902.) Filed CPA, Taliaferro Co., 1890. BSU.
Smith, W. R. H. (William R. H. Smith) - Pvt. 10/15/1861. D Petersburg, VA, 8/20/1862. Sick in 3rd GA Hos, Rich, VA, fever, 7/15/1862. CRS has death claim by wife, Parthena Smith, Elbert Co. BSU.
Tiller, Thomas - See Pvt. Co. E.
Tucker, Lewis - Pvt. 2/9/1862. Discharged, disability, 6/5/1862. BSU.
Vasser, George L. (or Vassar) - Enlisted as Pvt., Co. F, 15th Regt. GA Inf. 7/15/1861. Transferred to Co. H, 38th Regt. 7/15/1863. Wounded 1864. Roll dated 11/6/1864, last on file shows absent with leave, since 9/1/1864, Staunton, VA, hospital, wounded. BSU.
Vassar, John J. (or Vasser) - 2d Sgt. 10/15/1861. App 1st Sgt. 7/22/1862. Elected Jr. 2d Lt. 7/18/1863, 1st Lt. Dec/1863. Wounded Wilderness, VA, 5/6/1864, died two days later in a field hospital, at Staunton, VA, 5/9/1864. He was an orphan, and had only one brother, Lucius, to mourn his loss. Tribute in Elberton Star, 3/30/1889. Bur Thornrose Cem, Staunton, VA.

Vaughan, Alexander (or Vaughn) - Pvt. 10/16/1861. Appointed Hospital Steward 1862. Wounded at Gettysburg, PA, 7/1/1863, died there of wounds, 7/17/1863, while in the hands of the enemy. (Born 12/22/1841) Widow, Nancy Vaughn, filed CPA, Franklin Co., 1891. BSU.
Vaughan, A. W. (or Vaughn) (Alexander Wilkins Vaughn) - Pvt. 9/10/1862. W Fred, VA, 12/13/1862. Wounded and captured Gettysburg, PA, 7/2/1863. P DeCamp General Hospital, David's Island, NY, Harbor, 1863. Rcv City Point, VA, 9/16/1863. Captured and paroled Hartwell, GA 5/19/1865. Filed CPA Elbert Co., 1898. Died 1/29/1905. Buried Vaughan Cemetery, Elbert Co., GA.
Vaughan, Jacob D. (or Vaughn) (Jacob David Vaughan) - Pvt. 10/15/1861. Died Savannah, GA, 5/31/1862. Remains packed in charcoal, sent home by train. Buried Vaughan Family Cem, Elbert Co. CSA grave marker. Widow, Martha J. Hewell Vaughan filed CPA, Elbert Co., 1891. She said he got a congestive chill and died a few days later.
Vaughan, J. H. (or Vaughn) (Isaiah H. Vaughan) - Pvt. 3/1/1862. Died Charlottesville, VA, 7/1/1862. CSR also shows died 6/27/1862. CSR contains death claim filed by widow, Peninah Vaughan, in Elbert Co. BSU.
Vaughan, M. B. (Miles B. Vaughan) - Pvt. 10/15/1861. Killed Fredericksburg, VA, 12/13/1862. Widow, Susan Ann Vaughan, filed CPA, Oglethorpe Co., 1891. She stated in the application he was shot in head and killed instantly, 12/13/1862. BSU.
Vaughan, William W. (or Vaughn) - Pvt. 3/1/1862. Absent, sick, Aug/1863. Died in service. BSU.
Vaughan, W. H. (or Vaughn) (William H. Vaughn) - Pvt. 3/1/1862. Killed Gaines Mill, VA, 6/27/1862. CSR contains death claim filed by mother, Kessiah Vaughn, Madison Co. BSU.
Webb, Andrew J. - (Andrew Jackson Webb) Pvt. 10/15/1861. Trans to Co. F, 15th Regt. GA Inf. 9/15/1862; to Co. I, 1/14/1863. Surrendered Appomattox, VA, 4/9/1865. Filed CPA, Clarke Co., 1901. Died 4/29/1917. Buried Antioch Baptist Church, Elbert Co. Obit in Elberton Star, 5/1/1917.
Webb, G. W. (George W. Webb) - Pvt. 3/1/1862. Died Richmond, VA, hospital Jul 2 or 6, 1863. CSR contains a claim death filed by widow, Sarah. M. Webb, Elbert Co., GA BSU.
White, Garnett - Pvt. 3/1/1862. Captured in Loudon Co., VA, 7/16/1864. Died U.S. General Hospital Elmira, NY, 7/27/1865. Buried grave #2859, Woodlawn National Cem, Elmira, NY.
White, Robert - Pvt. 10/15/1861. Captured Spotsylvania, VA, 5/12/1864. Paroled Ft Delaware, Feb/1865. E 3/7/1865. Rcv at Boulware & Cox's Wharves, James River, VA, 3/10/1865. Died 5/8/1888 in Elbert Co. Widow, Lucinda M. Colvard White, filed CPA, Elbert Co., 1911.BSU.
White, S. J. (Stephen J. White) - Pvt. 10/15/1861. Died at Savannah, GA 5/23/1862. Buried Hillcrest Cem, Elbert Co., CSR contains letter filed in Elbert Co., GA, court appointing the administrators of his estate as John N. Moore and Elizabeth F. White.
Wilhite, P. A. (Philip A. Wilhite) - Pvt. 10/15/1861. Discharged 12/16/1861. Enlisted in Company I, 66th GA Regt. as Pvt. 9/1/1863. Reported missing since retreat from Missionary Ridge, TN, 11/25/1863. Buried Elmhurst Cemetery, Elberton. CSA marker reads: CO I. 66 GA INF.

COMPANY I, DAWSON FARMERS, DAWSON COUNTY, GA.

*This company was also known as New Co. I, Old Co. N, and New Co. G. Date of enlistment given also as 5/15/1862. The company was first organized by Captain Elijah Blackburn. Company was known as Company N during the battle of Fred., 12/13/1862.

Anderson, James P. (or Andrews) - Pvt 5/6/1862. C Oct/1862. E in 1862. Deserted 1863. BSU.
Anderson, R. N. - Pvt (Robert Newton Anderson) 5/15/1862. Deserted in 1863. Appears on register of Wayside Hospital or General Hos # 9, Richmond, VA; typhoid fever, died 4/19/1863, did not desert. Buried Oakwood Cemetery, Richmond, VA; cemetery records show: R. M. Anderson, 38th GA, Co. H. Widow, Eliza C. Anderson, filed CPA, Dawson Co., 1891.
Andrews, John F. - Pvt 5/6/1862. Wounded Gaines Mill, 6/27/1862. Deserted subsequent to 1/3/1863. Filed CPA, Cherokee Co., 1889. CPA said totally disabled by wounds, never AWOL. BSU.
Andrews, Samuel E. - Pvt 5/6/1862. Captured in 1862, paroled 1862. Appt 3d Cpl Aug/1863; 1st Cpl in 1864. Sick in Staunton, VA, hospital, Nov/1864. Died of disease, 1864. BSU.

Bell, Augustus C. - ("Gus") Jr. 2d Lt 10/24/1862. See Pvt Co. B.
Bennett, Hiram P. - Pvt 5/6/1862. D 7/8/1862. Possibly misidentified as R. Bennett in CSR records. Records of R. Bennett of 38th GA, Co. N, states, D 7/5/1862 of typhoid fever, Charlottesville, VA, bur there UVA Confed. Cem. Name of R. Bennett on CSA memorial cem, no individual grave maker.
Blackburn, Elijah - Cpt 5/6/1862. Died typhoid pneumonia, Gordonsville, VA, hos 9/2/1862. Died at Exchange Hotel Wayside Receiving Hos. Believed buried there in unmarked grave at Maplewood Cem. CSR has death claim by widow, Eliza Blackburn, Dawson Co. Exchange Hotel is a museum.
Blackburn, Henry W. Pvt 5/6/1862. Died of smallpox, VA, 3/4/1863. CSR contains death claim filed by father, Jesse W. Blackburn, Dawson Co. BSU.
Blackstock, Richard (Richard R. Blackstock) - Pvt 5/6/1862. Captured Rockville, MD 7/14/1864. Paroled Elmira, NY, 3/14/1865. Filed CPA, Pickens Co., 1887. B 1835, D 12/10/1906, buried Four Mile Bap. Church Cem, Pickens Co., GA. Widow, Mary Blackstock filed CPA, Pickens Co., 1907.
Britton, Jackson - Pvt 4/30/1862, enlisted Atlanta. Appears on a company muster roll for May 1 to Oct 31 1862. Died in hospital, Richmond, VA, 7/31/1862. BSU.
Brooks, Aaron A. - Pvt 5/6/1862. Killed 2d Manassas, VA, 8/28/1862. BSU.
Brooks, Moses A. (Moses Aaron Brook) - Pvt 5/6/1862. Died of disease Orange Court House, VA 5/6/1864. BSU.
Burns, Cornelius - Pvt 5/6/1862. Wounded and captured Monocacy, MD, 7/9/1864. Paroled at Point Lookout, MD, 1864. Exchanged Venus Point, Sav River, GA, 11/15/1864. No later record. Filed CPA, Scott, AR, 1894. Filed CPA in OK. Buried Lower Cemetery, Le Flore Co., OK.
Burns, George W.- See Pvt, Co. L.
Callis, John W. - Pvt 5/6/1862. Sick in hos Jan/1863. Died, smallpox. Buried Old City Cem, CSA Section, Lynchburg, VA 1/2/1863. Buried in unmarked grave in Yankee Square Section of cemetery. Archaeologists began excavation of this section in 2014; his grave may one day be identified and his grave marked. Widow, Kisiah Callas filed CPA, Dawson Co., 1891. She stated he was accidently shot in foot by a comrade, sent to the hospital, contracted smallpox and died.
Carlyle, John - Pvt 6/13/1862. Died of fever at Shady Grove Hospital, Staunton, VA, 8/7/1862. Buried at Thornrose Cemetery, Staunton, VA, Name listed in cemetery records as J. Carlisle, 38th GA, Co. B. CSR contains a death claim filed by his widow, Caroline Carlisle.
Carlyle, W. Henry - Pvt 5/6/1862. Roll dated 11/7/1864, last on file, shows "Sick in hospital, Atlanta, GA when last heard from in 12/1862." Filed CPA, Gwinnett Co., 1894. BSU.
Castleberry, Benjamin W. - (Benjamin William Castleberry) Pvt 5/6/1862. C near Petersburg, VA, 3/25/1865. Released PLO, MD, 6/26/1865. (B GA, 3/4/1843. Died Forsyth Co., 1/5/1931) Filed CPA, Forsyth Co., 1911, buried there Salem Bap. Church Cem. CSA marker reads PVT CO I, 38 GA INF.
Castleberry, Timothy R. - Pvt 5/6/1862. Captured Harrisonburg, VA 9/25/1864. Released Point Lookout, MD 6/26/1865. (Born Forsyth Co., 5/20/1841) Filed CPA, Forsyth Co., 1903. BSU.
Collier, Zada A. (Zadith Anderson Collier) - Pvt 5/6/1862. Wounded at 2d Manassas, VA 8/28/1862. Died of smallpox at Guinea Station, VA, Mar/1863. CSR contains death claim filed by mother, Elizabeth Collier, Dawson Co. BSU.
Croy, William J. - (William Joshua "WJ" Croy) Pvt 5/6/1862. Wounded in head 2d Manassas, VA 8/28/1862. C Chambersburg, PA 7/5/1863. Rcv Ft. Delaware, 8/19/1863. Roll dated 11/7/1864, LOF, shows him "Absent, POW." No later record. (Resident GA since 1843) Filed CPA, Dawson Co., 1890. Died 2/19/1920. Buried Funston Baptist Church Cemetery, Funston, Colquitt Co.
Darnell, Elias - Pvt 5/6/1862. W Malvern Hill, VA 7/1/1862. Appt 2d Sgt Oct/1862. W and disabled 1864. Roll dated 11/7/1864, LOF, shows "AWOL, Co. Surveyor, Dawson Co., GA, since Jan/1864." NLR. Purportedly joined the Yankee guerillas of the North GA Mountains, 1st GA Infantry (USA), and served as Cpt. This unit was never accepted into U.S. Army. Died after 1894. BSU.
Dufrees, Elias (or Dufraes, or Dupree) - Pvt 5/6/1862. Wounded at battle of 2d Manassas, VA, 8/28/1862. Died of wounds 10/27/1862. BSU.
Edwards, Sylvester - Pvt 5/6/1862. Discharged, disability, 7/22/1862. D General Hos #8, Rich, VA, 7/24/1862. CSR contains discharge record. B in Greenville, SC, 33 years old at discharge. BSU.

Ellis, Bailey M. - Pvt 5/6/1862. D of disease, Chimborazo Hos Rich, VA 6/9/1863. Bur Oakwood Cem, Rich, VA. Cem records incorrectly list unit as 38th AL. CSR states D of erysipelas, an acute streptococcus bacterial infection. Widow, Margaret M. Martin Ellis, filed CPA Forsyth Co., 1891.
Forbes, William M. - Pvt 5/6/1862. Killed Gaines Mill, VA, 6/27/1862. CSR contains a death claim filed by widow, Nancy E. Forbes, Lumpkin Co. Widow filed CPA, Dawson Co., 1891. BSU.
Gary, J. W. - Listed on Regtal return dated May/1862, states on 60 day frl until 6/6/1862. NLR. BSU.
Godfrey, William L.- 5th Sgt 5/6/1862. W 2d Manassas, VA, 8/28/1862. Rcv pay Atlanta, 1/3/1863. Roll dated 11/7/1864, LOF, shows him "AWOL since 12/1862; unknown where. D May/1865. Widow, Caroline Godfrey filed CPA, Dawson Co., 1891. Bur Bethel Meth. Church, Dawson Co.
Goswick, John W. - Enlisted as a Pvt in Co. H, 1st Regt GA Inf. (Ramsey's,) 7/21/1861. Mustered out Augusta, GA 3/18/1862. Elected Jr. 2d Lt of Co. I, 38th Regt GA Inf. 5/6/1862; 2d Lt 9/2/1862; 1st Lt 9/2/1862. C Fred., VA 12/13/1862. E. 12/17/1862. Elected Cpt 2/25/1864. W, right shoulder Spotsylvania, VA 5/12/1864. W near Canton, GA 9/19/1864. CPA states home on W frl close of war. (B Lumpkin, GA 9/11/1839) Filed CPA, Milton Co., 1901. D 3/27/1922. Bur Alpharetta Cem, Fulton Co. Death notice and funeral invitation printed in Atlanta Constitution, 3/29/1922.
Green, Bishop S. - Pvt 5/6/1862. Died of disease Staunton, VA, 11/9/1862. Buried there Thornrose Cem. Widow, M. Elizabeth Green, filed CPA, Hall Co., 1891.
Green, Monroe M.- (Madison Monroe Green) 1st Cpl 5/6/1862. Died of rheumatism of heart at Howard's' Grove Hospital Richmond, VA, 11/16/1862. Buried Oakwood Cemetery, Richmond, VA. CSR contains a death claim filed by father, John Green.
Grogan, W. D. - (William D. Grogan) - Pvt 5/6/1862. D of smallpox Guinea Station, VA Mar/1863. CSR has death claim filed by widow, Tabitha Grogan. Widow filed CPA, Dawson Co., 1891. BSU.
Hicks, Henry W. (or Hix) - Pvt 5/6/1862. Wounded Gaines Mill, VA, 6/27/1862. Killed at Monocacy, MD, 7/9/1864. BSU.
Hill, Andrew H. (Andrew Hudlow Hill)- 3rd Sgt 5/6/1862. Wounded Gaines Mill, VA 6/27/1862; Antietam, MD 9/17/1862. Elected Jr. 2d Lt 9/25/1862; 2d Lt 10/24/1862. Resigned due to wounds 2/13/1863. Died 8/4/1911, buried Hill City Baptist Church Cemetery, Gordon Co.
Hill, Starling Emanuel - Pvt 5/1/1864. Captured Fisher's Hill, VA 9/22/1864. Sent to Point Lookout, MD, 10/1/1864. No later record. CSR states died in hospital #5104, Point Lookout Prisoner of War Camp, of chronic diarrhea. Buried at Point Lookout Cemetery, Scotland, St. Mary's Co., MD.
Hill, William A.- 2d Lt 5/6/1862. Elected 1st Lt 9/2/1862; Cpt 9/2/1862. D near Guinea Station, VA 1/23/1863. Prom to Cpt, assumed command of company. Bur Fred. Confed. Cem, Fredericksburg.
Holtzclaw, John - Pvt 5/6/1862. Discharged 6/6/1862. D 3/2/1902, bur Crossroads Bap. Church Cem, Forsyth Co. Widow, Elizabeth Holtzclaw, filed CPA Forsyth Co., 1912.
Jay, John W. (John Wesley Jay) - Pvt 5/6/1862. W and disabled Gettysburg, PA, 7/2/1863. Appt 4th Cpl Jan/1864. Roll dated 11/7/1864, LOF, shows "Absent home on W frl since June 1864." Filed CPA, Newton Co., 1905. D 2/28/1919, bur Nazareth United Meth. Cem, Bartow Co.
Jay, Robert H. - Enlisted as a Pvt in Co. H, 1st Regt GA Inf. (Ramsey's,) 7/24/1861. Mustered out Augusta, GA 3/18/1862. Enlisted as Pvt in Co. I, 38th Regt GA Inf. 5/15/1862. Appt 3rd Sgt 9/25/1862. W and C Gettysburg, PA, 7/4/1863. P De Camp General Hos, David's Island, NY Harbor, 8/24/1863. E City Point, VA 8/28/1863. Appt 2d Sgt May/1864. W, permanently disabled Monocacy, MD 7/9/1864. At home W frl end of war. D 7/7/1931, Desoto Parish LA, bur there Keatchie Cem.
Johnson, Samuel C. (Samuel Caraway Johnson)- Enlisted as a Pvt, Co. L, 38th Regt GA Inf. 3/4/1862. Elected 1st Lt of I, 5/6/1862; Cpt 9/2/1862. Resigned, disability, 9/2/1862. B 10/10/1831, D 10/2/1870, bur Old Salem Meth. Cem, Dawson Co. Obit in Atlanta Constitution, 10/6/1870, page 2.
Kellis, John W. - Pvt 5/6/1862. W Cedar Mountain, VA 8/9/1862. Died of wounds in 1862. BSU.
Kelly, John - Pvt 5/6/1862. Roll dated 11/7/1864, last on file, shows "absent without leave since Aug/1863, Deputy Sheriff, Dawson Co." Died 9/28/1908 Buried Bethel Baptist Church Cemetery, Dawson Co. Widow, Frances L. Kelly, filed CPA Dawson Co., 1908.

Kelly, William - 2d Sgt 5/6/1862. Died of disease, 7/18/1862. BSU.
Lawrence, James A. - Pvt 5/6/1862. Died of variola, (smallpox) in General Hos Danville, VA, 12/16/1862. CSR contains a death claim filed by mother, Rachel Lawrence, Dawson Co. BSU.
Loggins, J. A. (or E. A.) (Ervin Alexander Loggins) - Pvt 5/6/1862. W Spotsylvania, VA 5/12/1864. W, disabled Cold Harbor, VA 6/1/1864. At home on W frl 11/7/1864. Filed CPA, Jackson Co., 1890. D there 2/17/1926. Bur Cabin Creek Cem, Jackson Co. CSA marker reads: Co I 38 GA INF.
Loggins, Thomas A. - Pvt 5/6/1862. Wounded Malvern Hill, VA, 7/1/1862. Deserted at Forsyth Co. Jan/1864. Took oath of allegiance to U.S. Govt. at Chattanooga, TN, 4/7/1864. BSU.
Loggins, Timothy S. - Pvt 5/6/1862. Killed at Gaines Mill, VA, 6/27/1862. BSU.
Martin, Absalom C. - (Absalom B. Martin) Pvt 5/6/1862. Appt 2d Cpl in 1864. W Winchester, VA 9/19/1864. S Appomattox, VA 4/9/1865. Filed CPA, Forsyth Co., 1895. D. 4/23/1914, bur Hopewell Cem, Cumming, Forsyth Co. Widow, Nancy A. Martin, filed CPA, Forsyth Co., 1919.
Martin, Eli T.- 2d Cpl 5/6/1862. Pvt Dec/1862. Appt 3rd Cpl Jan/1864. S Appomattox, VA 4/9/1865. Filed CPA Dawson Co., 1904. D 6/19/1927. Bur Bethel United Meth. Church Cem, Dawson Co.
Martin, John M.- Enlisted as a Pvt in Co. H, 1st Regt GA Inf. (Ramsey's,) 3/18/1861. Mustered out Augusta, GA, 3/18/1862. Enlisted as Pvt in Co. I, 38th Regt GA Inf. 5/15/1862. Captured Spotsylvania, VA 5/20/1864. Transferred from Point Lookout, MD to Elmira, NY, 7/6/1864, died there, chronic diarrhea 10/24/1864. Buried at Woodlawn National Cem, Elmira, NY.
Martin, Thomas M. - Pvt 5/6/1862. Killed at Gaines Mill, VA, 6/27/1862. CSR contains a death claim filed by father, Toleston G. Martin, Lumpkin Co. BSU.
Martin, Van Buren - Enlisted as a Pvt in Co. H, 1st Regt GA Inf. (Ramsey's,) 7/24/1861. Mustered out Augusta, GA 3/18/1862. Enlisted as a Pvt in Co. I, 38th Regt. GA Inf. 5/15/1862. W Malvern Hill, VA 7/1/1862. Captured Fredericksburg, VA, 12/13/1862, paroled 12/17/1862. Appt 3d Sgt Aug/1863. Wounded, disabled and captured Spotsylvania, VA, 5/10/1864. E 8/31/1864. CPA shows home on wounded furlough 1864, to close of war. Georgia Confederate pension record not found. BSU.
Mashburn, John W. (or Mashbourne) (John Wesley Mashburn) - Pvt 5/6/1862. Appt 1st Sgt 7/8/1862. C near Mine Run, VA 5/12/1864. P Ft. Delaware, DE. Feb/1865, E 3/7/1865. D 5/15/1865. Widow, Mary M. Mashburn, filed a PA in Forsyth Co., 1891. She stated he died of chronic diarrhea, 5/15/1865, soon after returning home. Bur Ebenezer Cem, Forsyth Co. Marker reads: 38 GA INF.
McAfee, William Hamilton - Enlisted as a Pvt in Co. E, Phillips' Legion GA Inf. 7/1/1861. Elected Jr. 2d Lt of Co. I, 22d Regt GA Inf. 10/20/1862; Cpt of Co. I, 38th Regt GA Inf. 5/22/1863. Resigned to become Chaplain of the 22d Regt GA Inf. 2/25/1864. Surrendered at Appomattox, VA, 4/9/1865. Died 11/25/1915. Buried Mount Hope Cemetery, Lumpkin Co. Served in Georgia Senate for 2 terms after the war, Obituary in Atlanta Constitution 11/30/1915.
McDonald, Lewis W.- 3d Cpl 5/6/1862. Appt 2d Cpl Mar/1863. Deserted in 1863. BSU.
McDonald, W. L. (Washington Lafayette McDonald) - Pvt 5/6/1862. Appt 6th Sgt Oct/1862. W Fred., VA 12/18/1862. Appt 4th Sgt 1864. C Spotsylvania, VA 5/12/1864. P Ft. Delaware, DE Feb/1865. E 3/7/1865. D 4/10/1881. Widow, S. J. McDonald, filed CPA, Lamar Co., AL, 1899.
Millsaps, P. J. (or Isaiah) (Peter Isaiah Millsaps) - Pvt 5/6/1862. Killed at Spotsylvania, VA, 5/12/1864. Widow, Faralena Milsepa, filed CPA, Dawson Co., 1891. BSU.
Payne, B. E.- (Benjamin E. Payne Jr.) Pvt 5/15/1862. Captured at Falling Waters, MD, 7/4/1863. Exchanged at Point Lookout, MD 2/18/1865. At Camp Lee near Richmond, VA 2/21/1865 Died 1908. Widow, Louisa J. Payne, filed CPA Cherokee Co., 1911. BSU.
Payne, Samuel H. (Samuel Houseworth Payne) - Pvt 5/15/1862. W Fred., VA 12/13/1862. C Gettysburg, PA, 7/2/1863. E 7/31/1863. Roll dated 11/7/1864, LOF, shows him AWOL since Oct/1863. Filed CPA, Dawson Co., 1905. Died 5/16/1917. Bur Juno Bap. Church Cem, Dawson Co.
Payne, Zebulon C. - Pvt 5/15/1862. Wounded and disabled Cedar Mountain, VA, 8/9/1862. Admitted to Jackson Hospital, Richmond, VA, rheumatism, 5/18/1864, frl for 60 days 12/15/1864. Filed CPA, Dawson Co., 1894. Died 5/23/1915, buried Juno Bap. Church Cemetery, Dawson Co.

Perry, James G. - Pvt 5/15/1862. Accidentally shot and wounded in 1862. Discharged 6/17/1862. CSR contains discharge record. Born in Franklin, GA, 26 yrs old at discharge. Filed CPA, Lumpkin Co., 1906. Died Mar/1907. Widow, Elizabeth Perry, filed CPA, Lumpkin Co., 1910. BSU.
Potts, Montraville Patton - Pvt Conscripted 4/14/1864, Deserted to enemy Bowling Green, VA, 5/31/1864. Stated to Federal officials he was a Union man and had avoided conscription by hiding in the woods, desired to take the OOA and go to TN. Sent to Elmira Prison Camp, NY, shot by guard after late night disturbance. Recovered from wound, D there of chronic diarrhea 4/10/1865. BSU. Probably bur under the name of POTTER, R H, bur at Woodlawn National Cem, D 4/9/1865.
Purdy, Thomas N. (Thomas Newton Purdy) - Pvt 5/6/1862. W Fred., VA 12/13/1862. Deserted Mar/1864. Took OOA to U.S. Govt. Chattanooga, TN And released 3/16/1864. BSU.
Reese, Jonas - Pvt 5/6/1862. Appt 4th Sgt 8/1862; 2d Sgt 1862. K Fred., VA 12/13/1862. BSU.
Richardson, John C. - 1st Sgt 5/6/1862. "Rcv pay Atlanta, GA on 1/3/1863, to 1st of 11/1862." Roll dated 11/7/1864, LOF, shows "AWOL since Jun/1862." D 4/9/1915. Bur Alta Vista Cem., Hall Co.
Robinson, George R. - Pvt 5/6/1862. Discharged, having been elected to civil office, 2/26/1864. Elected Sheriff of Dawson Co., prior to discharge. BSU.
Rudicil, Emanuel - Pvt 5/6/1862. Discharged 5/26/1862. BSU.
Simmemon, D. W. (David W. Simmemon) - Pvt 5/15/1862. Deserted. Rcv from Provost Marshal General, Army of Potomac, 3/26/1865, and sent to Colonel T. Ingraham, P. M. G., as a Confed. deserter 3/29/1865. D 9/10/1895. Bur Bethel Bap. Church Cem, Dawson Co.
Sluder, W. R. - (William R. Sluder) Pvt 5/6/1862. Accidentally W 6/26/1862. C Spotsylvania, VA 5/12/1864. P Ft. Delaware, Feb/1865. E Boulware & Cox's Wharves, James River, VA, 5/10/1865. (D Dawsonville, 11/17/1917) Filed CPA, Dawson Co., 1911. Bur there Harmony Bap. Church Cem.
Smith, Joseph - Pvt 5/15/1862. Discharged, disability, 10/5/1862. CSR contains discharge record. Enlisted as a Pvt Co. A, 11th Regt GA Militia Cavalry in 1864. Surrendered in Dawson Co., 1865. Filed CPA, Hall Co., 1910. BSU.
Smith, Solomon J. (Solomon Jasper Smith) - Pvt 5/6/1862. Enlisted in G. W. Lee's command in Atlanta, GA Feb/1864. CPA shows discharged, disability, Feb/1864, enlisted in Findley's Cavalry 8/1/1864, and discharged Macon, GA 4/15/1865. (B GA in 1844) Filed CPA, Forsyth Co., 1917. D 7/2/1920. Bur Salem Bap. Church Cem, Forsyth Co. Marker reads, Co I, 38 GA INF.
Swilling, James M. - Pvt 5/6/1862. Appt 4th Sgt Oct/1862. Reduced to ranks Aug/1863. Roll dated 11/7/1864, last on file, shows "absent without leave since 7/1/1863." Died 10/23/1888. Buried Jenkins Swillings Cemetery, Logan Co., AR.
Tatum, F. Daniel - (Francis Daniel Tatum) 4th Cpl 5/6/1862. W Antietam, MD 9/17/1862. Roll dated 11/7/1864, LOF, shows "Absent, Det in C. S. Laboratory Apr/1863." D 4/26/1919, bur New Hope United Meth. Cem, Gainesville, Forsyth Co. Widow, Mary J. Tatum, filed CPA, Forsyth Co., 1919.
Tatum, Van Buren - Pvt 5/6/1862. W in arm, necessitating amputation, 2d Manassas, VA 8/28/1862. Roll dated 11/7/1864, LOF, shows "Absent, Enrolling Officer, Pickens Co., GA since 7/1863." Filed CPA, Gordon Co., 1902. Died 12/10/1924. Buried Bayou Scie Church, Belmont, Sabine Co., LA.
Terry, J. D. - Pvt 5/15/1862. Died 7/3/1862. BSU.
Tescunier, John - (John Moses Tesseneer) Pvt 6/19/1862. Died 9/1/1862. CSR contains death claim filed by widow, Mary Tesseneer, Dawson Co. BSU.
Thackston, James Martin - Pvt 5/6/1862. Died of typhoid fever 7/28/1862. Buried UVA Confed. Cemetery, Charlottesville, VA Soldier's names listed on bronze plaques on memorial at the cemetery.
Thomas, Andrew J. - Pvt 5/6/1862. Discharged on account of chronic rheumatism Gordonsville, VA 7/22/1862. CSR contains discharge record. Filed CPA, Dawson Co., 1897. BSU.
Thompson, W. G. - Pvt 5/15/1862. Wounded Gaines Mill, VA 6/27/1862. Deserted in 1863. BSU.
Tumlin, Lewis - Pvt 5/16/1862. Wounded at 2d Manassas, VA, 8/30/1862. Discharged, disability, 10/28/1862. CSR contains discharge record. BSU.
Turner, T. W. - Pvt 5/15/1862. Deserted prior to 8/21/1864. BSU.

Wallis, David - Pvt 5/15/1862. Appointed 1st Cpl 12/1862. Roll dated 11/7/1864, last on file, bears remark: "Gone to the Western Army Feb/1864." BSU.
Wallis, David J. - (David James Wallis) 4th Sgt 5/6/1862. W 2d Manassas, VA 8/28/1862. D from wounds 10/8/1862. CSR contains death claim filed by widow, Sarah C. Wallis, Dawson Co. BSU.
Wallis, Harvey C. - (Harvey Collins Wallis) Pvt 5/15/1862. W through thigh and permanently disabled, Gaines Mill, VA 6/27/1862. Roll dated 11/7/1864, LOF, shows on unlimited frl. (B GA 10/12/1840) Filed CPA, Forsyth Co., 1890. D 9/13/1924, buried Hopewell Cemetery, Forsyth Co.
Weatherby, G. - (or G. Weatherbite) (George Weatherby) Pvt Captured Bowling Green, KY, 6/1/1864. D of chronic diarrhea Elmira, NY, 9/25/1864. Bur Grave #368, Woodlawn National Cem.
Westbrook, John R. - (John Reece Westbrook) Pvt 5/15/1862. Appt 4th Cpl Aug/1863; 5th Sgt in 1864. Sick in Lynchburg, VA, hospital Oct/1864. No later record. Died 11/11/1895, buried New Hope United Meth. Church Cem, Forsyth Co. Obituary in The Baptist Leader, 1/3/1896.
Wilson, Richard V. (Richard Verable Wilson) - Musician 5/6/1862. Captured at High Bridge, VA, 4/6/1865. Released Newport News, VA, 6/25/1865. Born 11/16/1829. Filed CPA, Marshall Co., AL, 1907, pension #4572. BSU.

MUSTER ROLL OF 2d COMPANY I, IRWIN INVINCIBLES, HENRY CO. ALABAMA

This company was first organized and led by Capt Henry L. Jones. Many members of this company came from Henry Co., Alabama. The town of Franklin, Alabama was the meeting place where the company was organized and was located near the Georgia border with Alabama. In Dec/1861, 3d Co. E, 25th Georgia Inf., this company ("Irwin Invincibles," afterwards "Henry Light Infantry"), was ordered from West Virginia to Georgia by the Secretary of War. It arrived in Georgia 1/12/1862, and went into camp near Savannah, Georgia and was assigned to the Company E, 25th Georgia Infantry. Many members of this company were transferred to form 2d Co. I, 38th Georgia Infantry on May 2nd, 1862. While assigned to the 38th Georgia, this Company saw some of the most horrific battles of the war, including in the battles of Gaines Mill, 2nd Manassas, Antietam, and Fredericksburg, plus several smaller affairs. On 3/1/1863, it was again transferred and became 2d Co. A, 60th Georgia Infantry. It was written the men of the 38th Georgia and this company both vigorously protested the transfer of the company to the 60th Georgia, as they had become attached to each other and had forged deep bonds of camaraderie upon numerous fields of battle. Finally, this company was transferred for the last time and became Company K, 61st AL. Inf., 4/11/1864. The company served with the 38th Georgia from May 1862, until March 1st, 1863.

Anderson, John H. - Pvt 8/8/1861. Trn from 3d Co. E, 25th GA Inf. to 2d Co. I, 38th GA Inf. 5/2/1862; to 2d Co. A, 60th GA Inf. 3/1/1863; to Co. K, 61st AL. Inf. 4/11/1864. Captured Cedar Creek, VA, 10/18/1864. Paroled at Point Lookout, MD, Mar/1865. Transferred to Aiken's Landing, VA, 3/28/1865. Received Boulware & Cox's Wharves, James River, VA, 3/30/1865. BSU.
Applewhite, George W. - Pvt 3/14/1862. Transferred from 3d Co. E, 25th GA Inf. to 2d Co. I, 38th GA Inf. 5/2/1862. Killed at 2d Manassas, VA, 8/28/1862. BSU.
Applewhite, John (John Barnes Applewhite) - Pvt 7/16/1862. In Richmond, VA, hospital Feb/1863. No later record. Died 1906, buried Green Hill Cemetery, Titus Co., TX.
Arnold, Abner - Pvt 4/21/1862. Transferred to 2d Co. A, 60th GA Inf. 3/1/1863; to Co. K, 61st AL Inf. 4/11/1864. Missing in action at battle of Winchester, VA, 9/19/1864. BSU.
Barnes, John W. - Pvt 5/10/1862. Wounded Antietam, 9/17/1862. Wounded Fredericksburg, 12/13/1862. Appt 1st Cpl in 1863. Trn to 2d Co. A, 60th GA Inf. 3/1/1863; to Co. K, 61st Regt AL. Inf. 4/11/1864. W 1864. Captured Ft. Stedman, VA 3/25/1865. Released PLO, MD, 6/9/1865. BSU.
Barnes, Thomas H. - Pvt 8/8/1861. Trn from 3d Co. E, 25th GA Inf. to 2d Co. I, 38th GA Inf. 5/2/1862. Wounded Antietam, MD, 9/17/1862, sent to Winchester, VA, hos. Appt 2d Cpl in 1862. Trn to 2d Co. A, 60th GA Inf. 3/1/1863. At home, Wounded furlough 1/15/1864. Transferred to Co. K, 61st AL. Inf. 4/11/1864. Absent, sick, Eufaula, AL hospital, Aug/1864. No later record. BSU.
Barnes, William R. - Pvt 8/8/1861. Transferred from 3d Co. E, 25th GA Inf. to 2d Co. I, 38th GA Inf. 5/2/1862. Died in Staunton, VA, hospital 1/2/1863. BSU.

Bodiford, Elias P.- Pvt 8/8/1861. Trn from 3d Co. E, 25th GA Inf. to 2d Co. I, 38th GA Inf. 5/2/1862; to 2d Co. A, 60th GA Inf. 3/1/1863; to Co. K, 61st AL Inf. 4/11/1864. C Petersburg, VA 4/2/1865. Released PLO, MD, 6/9/1865. Filed CPA, Geneva Co., AL. Died 2/17/1917. BSU.
Bond, Abel - Pvt 5/10/1862. W Fred., VA 12/13/1862. Trn to 2d Co. A, 60th GA 3/1/1863; to Co. K, 61st AL Inf. 4/11/1864. Being unfit for field service, Det to report as guard to Major Harris, Macon, GA, 3/14/1864. Sick, Ft. Gaines, GA hos 7/8/1864. NLR. Filed CPA, Etowah Co., AL. D 1/18/1921, bur Collinsville Cem, DeKalb Co., AL Marker reads: Born 12/15/1830, PVT CO I, 61 AL INF.
Bond, George W. - Pvt 1/12/1862. Transferred from 3d Co. E, 25th GA Inf. to 2d Co. I, 38th GA Inf. 5/2/1862. Wounded at Gaines Mill, VA, 6/27/1862. Died of wounds July 13 or 16, 1862. CSR contains death claim filed by mother, Jane Bond, Henry Co., AL. BSU.
Bray, Henry C. - Pvt 8/8/1861. Trn from 3d Co. E, 25th GA Inf. to 2d Co. I, 38th GA 5/2/1862; to 2d Co. A, 60th GA 3/1/1863; to Co. K, 61st AL. Inf. 4/11/1864. C Petersburg, VA, 4/8/1865. Released PLO, MD, 6/24/1865. D 10/4/1932, Covington Co., AL, bur Weeds Bap. Cem, Crenshaw Co., AL.
Bray, Louis D. - Pvt 8/8/1861. Trn from 3d Co. E, 25th GA Inf. to 2d Co. I, 38th GA Inf. 5/2/1862.; to 2d Co. A, 60th GA Inf. 3/1/1863; to Co. K, 61st AL. Inf. 4/11/1864. Sick Ft. Gaines, GA, 7/8/1864. NLR. B 3/28/1840, D 8/19/1912, buried Bogart Baptist Church Cemetery, Oconee Co., GA.
Britton, Jackson - Pvt 4/20/1862. Died in Richmond, VA, hos 7/31/1862. Bur Oakwood Cem, Rich, VA. Cem records show J. J. Britton, no unit or state listed. Buried in Division A, Row M, Grave 64.
Brown, Andrew J. - Pvt 1/12/1862. Transferred from 3d Co. E, 25th GA Inf. to 2d Co. I, 38th GA Inf. 5/2/1862. Wounded at Ox Ford, VA, 8/31/1862. Wounded 9/27/1862. Transferred to 2d Co. A, 60th GA Inf. 3/1/1863. Deserted to enemy 2/19/1864. BSU.
Bruce, Thomas J. - Pvt 5/10/1862. Died Richmond, VA, Nov 10th or 18th, 1862. Bur there Hollywood Cem, CSA marker. CSR has death claim filed by father, George P. Bruce, Henry Co., AL.
Campbell, Reuben M. (Reuben Marian Campbell) - Pvt 8/8/1861. Trn from 3d Co. E, 25th GA Inf. to 2d Co. I, 38th GA Inf. and elected Jr. 2d Lt 5/2/1862. Elected 1st Lt in 1862. Trn to 2d Co. A, 60th GA Inf. as 1st Lt 3/1/1863; to Co. K, 61st AL Inf. as 1st Lt 4/11/1864. C Petersburg, VA 4/2/1865. Released Johnson's Island, OH 6/18/1865. Filed CPA, Richmond Co., 1911. He stated his company was C Petersburg, VA 4/1/1865. D 1910, bur Magnolia Cem, Augusta, Richmond Co., GA.
Chumney, James C. (James Claybourne Chumney) - Pvt 3/14/1862. Trn from 3d Co. E, 25th GA Inf. to 2d Co. I, 38th GA Inf. 5/2/1862; to 2d Co. A, 60th GA Inf. 3/1/1863; to Co. K 61st AL. Inf. 4/11/1864. W in 1864. Paroled Winchester, VA, 4/17/1865. Filed CPA, Henry Co., AL. B 9/17/1833. D 3/14/1921. Buried Rocky Grove Cem, Jackson Co., FL. Marker reads PVT, CO K, 61 AL INF.
Chumney, John T. - (John Tyler Chumney) Pvt 8/8/1861. Transferred from 3d Co. E, 25th GA Inf. to 2d Co. I, 38th GA Inf. 5/2/1862; to 2d Co. A, 60th GA Inf. 3/1/1863; to Co. K, 61st AL Inf. 4/11/1864. C Ft. Stedman VA 3/25/1865. Released Point Lookout, 5/13/1865. Died 8/5/1925, buried Hopewell Cemetery, Freestone Co., TX.
Chumney, Richard E. - Pvt 8/8/1861. Transferred from 3d Co. E, 25th GA Inf. to 2d Co. I, 38th GA Inf. 5/2/1862. Wounded Gaines Mill, VA, 6/27/1862. Transferred to 2d Co. A, 60th GA Inf. 3/1/1863; to Co. E, 61st AL. Inf. 4/11/1864. Roll dated Aug/1864, last on file, shows present. No later record. Filed CPA, #34004, Henry Co., AL. Died 1899. BSU.
Chumney, William - (William T. Chumney) Pvt 8/8/1861. Transferred from 3d Co. E, 25th GA Inf. to 2d Co. I, 38th GA Inf. 5/2/1862. In Rich, VA hos Feb/1863. NLR. D Feb/1863. BSU.
Clark, Anderson P. - Pvt 5/10/1862. In Danville, VA, hospital, 12/1/1862. Disease listed as chronic nephritis, NLR. B 4/19/1831, D 2/5/1906. Buried Lee's Chapel Cemetery, Geneva Co. AL.
Clark, C. S. - Pvt 4/16/1862. Trn from 3d Co. E, 25th GA Inf. to 2d Co. I, 38th GA Inf. 5/2/1862. Roll for 11/12/1862, last on which borne, shows him present. W, 2d Manassas, 8/28/1862. BSU.

Clark, Drayton - Pvt 4/16/1862. Trn from 3d Co. E, 25th GA Inf. to 2d Co. I, 38th GA Inf. 5/2/1862; to 2d Co. A, 60th GA Inf. 3/1/1863; to Co. K, 61st AL. Inf. 4/11/1864. Captured Spotsylvania, VA, May 28 or 6/10/1864. Released Elmira, NY, 6/11/1865. Died 6/5/1904, Geneva Co., AL. Widow, Amanda D. Clarke filed CPA, Houston Co., AL, pension file #34309. BSU.
Clark, Mathew - Pvt 8/8/1861. Trn from 3d Co. E, 25th GA Inf. to 2d Co. I, 38th GA Inf. and Appt 2d Cpl 5/2/1862. Pvt in 1863. Transferred to 2d Co. A, 60th GA Inf. 3/1/1863; to Co. K, 61st AL. Inf. 4/11/1864. Wounded Wilderness, 5/5/1864. Surrendered at Appomattox, VA, 4/9/1865. BSU.
Cooley, Michael - Pvt 5/10/1862. W Fred., VA 12/13/1862. Roll for Jan. - Feb. 1863, last on which borne, shows "home on frl." NLR. B 2/6/1836, D 7/30/1907. Bur Big Creek UMC Cem, Houston, AL.
Cooper, Daniel - Pvt 4/24/1862. Trn from 3d Co. E, 25th GA Inf. to 2d Co. I, 38th GA Inf. 5/2/1862. W Fred., VA 12/13/1862. Trn to 2d Co. A, 60th GA Inf. 3/1/1863; to Co. K, 61st AL. Inf. 4/11/1864. Captured, date and place not given. Released Point Lookout, MD 6/10/1865. Died 3/8/1908. BSU.
Cooper, Reuben - Pvt 5/10/1862. Died of chronic diarrhea, 7/23/1862 at Ladies' Relief Hospital, Lynchburg, VA. Body was packed for his brother, Stephen Cooper, but body was never shipped and his brother, Stephen was killed 11/20/1863. Buried Lynchburg, VA, Old City Cemetery, CSA section, Lot 162, Grave #3, Row #5.
Cooper, Stephen - Pvt 5/10/1862. Trn to 2d Co. A, 60th GA Inf. 3/1/1863. Killed 11/20/1863. BSU.
Davis, Thomas G. - Pvt 5/10/1862. Transferred to 2d Co. A, 60th GA Inf. 3/1/1863; to Co. K, 61st AL. Inf. 4/11/1864. C Macon, GA Apr. 20-21, 1865. D 11/9/1917. Widow, Prudence Davis, filed a CPA, #3210, in Henry Co., AL. Dec/1917. BSU.
Deal, William A. - Pvt 8/8/1861. Trn from 3d Co. E, 25th GA Inf. to 2d Co. I, 38th GA Inf. 5/2/1862. Sick in Lynchburg, VA hos Feb/1863. Trn to 2d Co. A, 60th GA Inf. 3/1/1863. Wounded Mine Run, VA 11/27/1863. Trn to Co. K, 61st AL. Inf. 4/11/1864. Appt 2d Cpl 6/1/1864. Sick in Pratt's Hos Lynchburg, VA 8/1864. NLR. D 2/11/1924, Dale Co., AL. Bur there Newton Cem, grave unmarked.
Etheridge, Isaac R. - (or J. R. Etheridge) Pvt 8/8/1861. Transferred from 3d Co. E, 25th GA Inf. to 2d Co. I, 38th GA Inf. 5/2/1862. Discharged 6/5/1862. BSU.
Fleming, John - Pvt 8/8/1861. Trn from 3d Co. E, 25th GA Inf. to 2d Co. I, 38th GA Inf. 5/2/1862. K 2d Manassas, VA 8/28/1862. Widow, Elizabeth M. Fleming filed CPA, Geneva Co., AL. BSU.
Fortson, William M. - Pvt 8/8/1861. Trn from 3d Co. E, 25th GA Inf. to 2d Co. I, 38th GA Inf. 5/2/1862; to 2d Co. A, 60th GA Inf. 3/1/1863; to Co. K, 61st AL. Inf. 4/11/1864. Appt Color Cpl. Surrendered Appomattox, VA, 4/9/1865. Died 11/2/1908. Buried Bay Springs Baptist Church Cemetery, Houston Co., AL. Death notice in Dothan Eagle newspaper, Dothan, AL, 11/10/1908.
Glover, Alexander (Alexander "Alex" McAlister Glover) - Pvt 3/2/1862. Discharged 12/31/1862. Died 4/1/1910. Widow, Edna Marrow Glover filed CPA, Montague Co., TX. BSU.
Grace, James L. - Pvt 5/10/1862. Discharged, disability, Danville, VA 10/22/1862. CSR contains discharge record. Filed CPA, Geneva Co., AL. D 5/7/1906. Bur Hurricane United Meth. Cem, Geneva Co., AL. Widow, Mattie E. Grace filed CPA, Geneva Co., AL., 4/25/1914, pension #33235.
Grantham, John D. - Pvt 8/8/1861. Transferred from 3d Co. E, 25th GA Inf. to 2d Co. I, 38th GA Inf. 5/2/1862. Elected 1st Lt 6/28/1862; Cpt in 1862. Transferred to 2d Co. A, 60th GA Inf. 3/1/1863; to Co. K, 61st GA Inf. 4/11/1864. Wounded 1864. Admitted to General Hos #4, Richmond, VA, wounded in left thigh, 8/6/1864, furloughed 8/30/1864. Died 4/15/1915, buried Marion Cemetery, Perry Co., AL. Widow, Issie Grantham, filed CPA, Perry Co. AL, 1916
Gregory, John J. (John Jacob Gregory) - Pvt 8/8/1861. Trn from 3d Co. E, 25th GA Inf. to 2d Co. I, 38th GA Inf. 5/2/1862. Killed 2d Manassas, VA 8/28/1862. CSR contains death claim by mother, Maomi Quartlebum, Henry Co., AL. Record identifies Jacob as son from her first marriage. BSU.
Griffith, Franklin O. - Pvt 8/8/1861. Trn from 3d Co. E, 25th GA Inf. to 2d Co. I, 38th GA Inf. 5/2/1862 and Appt 2d Sgt 5/2/1862; to 2d Co. A, 60th GA Inf. 3/1/1863. C Gettysburg, PA, Jul 1 - 4, 1863. D DeCamp General Hos, David's Island, NY Harbor 8/20/1863. Buried Cypress Hills National Cem, Brooklyn, NY, name listed as F. O. Griffiths, bur in Section 1, Site 802.

Grimsley, Harmon A. (Harmon Augustus Grimsley) - 4th Cpl 8/8/1861. Trn from 3d Co. E, 25th GA Inf. to 2d Co. I, 38th GA Inf. and Appt 5th Sgt 5/2/1862. Wounded Fredericksburg, VA 12/13/1862. Pvt 12/31/1862. Trn to 2d Co. A, 60th GA Inf. 3/1/1863. Discharged, disability, from General Hospital #l, Danville, VA, 8/1/1863. Moved to Texas after the war and changed his name to W. G. Winslow. Died 9/9/1922. Buried Oakwood Cem, Walker Co., TX. Marker reads: W. G. Winslow. Descendants say Winslow was mother's maiden name.
Hall, James C. - 3rd. Sgt 8/8/1861. Transferred from 3d Co. E, 25th GA Inf. to 2d Co. I, 38th GA Inf. 5/2/1862. Killed at 2d Manassas, VA, 8/28/1862. BSU.
Hawkins, Jefferson - Pvt 1/18/1862. Trn from 3d Co. E, 25th GA Inf. to 2d Co. I, 38th GA Inf. 5/2/1862. D of pneumonia in Rich, VA, hos 12/2/1862. CSR contains death claim by father, Josiah Hawkins, in Henry Co., AL. Bur Hollywood Cem, Rich, VA. Cem records read: JNO. Hawkins.
Hawkins, John A. - Pvt 3/1/1862. Trn from 3d Co. E, 25th GA Inf. to 2d Co. I, 38th GA Inf. 5/2/1862. Wounded and captured Antietam, MD 9/17/1862. Died of wounds in U.S.A. General Hospital, Frederick City, MD, 10/28/1862. Buried Mount Olivet Cem, Frederick, MD. Grave #150. Marker reads: Hawkins, John W. Pvt Co A, 38th GA
Hawkins, Thomas - Pvt 1/12/1862. Transferred to 2d Co. A, 60th GA Inf. 3/1/1863; to Co. K, 61st AL. Inf. 4/11/1864. Absent without leave, 9/10/1864. NLR. BSU.
Haymond, John (or Hammond) - Pvt 5/10/1862. Transferred to 2d Co. A, 60th GA Inf. 3/1/1863. Died of pneumonia at Chimborazo Hospital #2, Richmond, VA, 4/2/1863. BSU.
Henderson, William H. - Pvt 8/8/1861. Transferred from 3d Co. E, 25th GA Inf. to 2d Co. I, 38th GA Inf. 5/2/1862; to 2d Co. A, 60th GA Inf. 3/1/1863; to Co. K, 61st AL. Inf. 4/11/1864. Appt Sgt. Paroled Lynchburg, VA, 4/13/1865. Wounded at 2d Manassas, 8/28/1862. BSU.
Hicks, Crawford (Or Crawford Hix) - Pvt 5/10/1862. Killed at Gaines Mill, VA, 6/27/1862. CSR contains death claim by widow, Sarah Hix, Henry Co., AL. BSU.
Hicks, Harbin, D. (Harbin David Hicks) Pvt 4/3/1862. Trn from 3d Co. E, 25th GA Inf. to 2d Co. I, 38th GA Inf. and Appt 4th Cpl 5/2/1862; to 2d Co. A, 60th GA Inf. 3/1/1863; to Co. K, 61st AL. Inf. 4/11/1864. C Ft. Stedman, VA 3/25/1865. Released PLO, MD 6/13/1865. D 12/26/1914. Bur Otter Creek Cem, Holmes Co., FL. Widow, Cyntha Somelin Hicks filed CPA, #A04121. Holmes Co., FL.
Hicks, James J. (James Jeremiah Hicks) - 3d Cpl 8/8/1861. Trn from 3d Co. E, 25th GA Inf. to 2d Co. I, 38th GA Inf. 5/2/1862; to 2d Co. A, 60th GA Inf. 3/1/1863. W Mine Run, VA 11/27/1863. Trn to Co. K, 61st AL. Inf. 4/11/1864. C Frederick City, MD, 7/9/1864. E PLO, MD 2/13/1865. Rcv Cox's Landing, James River, VA Feb. 14/15/1865. NLR. Filed CPA, Holmes Co., FL, 1900, PA #A06412. D 1925, bur St John's Cem, Holmes Co., FL.
Hicks, John D. - Pvt 4/3/1862. Transferred from 3d Co. E, 25th GA to 2d Co. I, 38th GA 5/2/1862; to 2d Co. A, 60th GA 3/1/1863. Absent without leave, 1/15/1864. Transferred to Co. K, 61st AL. Inf. 4/11/1864. Absent without leave, 8/17/1864. BSU.
Hodges, M. Green - Pvt 5/10/1862. D in Staunton, VA hos 7/10/1862. Bur there Thornrose Cem. CSR contains death claim by widow, Elizabeth Hodges, Henry Co., AL. D of typhoid fever.
Holmes, Joel H. - Pvt 8/8/1861. Trn from 3d Co. E, 25th GA Inf. to 2d Co. I, 38th GA Inf. 5/2/1862; to 2d Co. A, 60th GA Inf. 3/1/1863; to Co. K, 61st AL. Inf. 4/11/1864. Captured Winchester, VA 9/19/1864. Paroled at Point Lookout, MD for exchange 3/15/1865. Received Boulware & Cox's Wharves, James River, VA, 3/18/1865. No later record. Died 5/22/1907 Buried Shortville Cemetery, Henry Co., AL. Widow, Cynthia Holmes, filed a CPA, #2511, Henry Co., AL, 1920.
Hughes, John - Pvt 5/10/1862. Deserted 7/1/1862. BSU.
Irwin, A. B. (Andrew Berry or Barry Irwin) -1st Lt 5/2/1862. Elected Cpt 6/27/1862. D Rich, VA 7/3/1862. Bur there in Hollywood Cem, Section C, grave #151. Cem records list, Irving, A. B., Company I, 38 GA, D 8/20/1862. Cenotaph at Warthen Family Cem, Washington Co., GA, reads B Washington Co. 3/4/1838, D Rich, VA 7/24/1862, Cpt Irwin Invincibles, 38 GA, Wright's Legion.
Jones, Abner E. - From the Adjutant General: "The records show that, Abner E. Jones, Pvt, Co. K, 61st AL. Inf., formerly Co. I 38th Georgia Inf., enlisted Aug/1862, Franklin, GA, no record of capture or parole. Wounded 2d Manassas, 8/28/1862. Resigned 1/7/1863, as 3rd Lieut. Died 11/26/1878. Widow, Frances A. Jones filed a CPA, #28219, Coffee Co., AL, 1904. BSU.

Jones, James E. - 2d Lt., Jr. 2d Lt 8/8/1861. Transferred from 3d Co. E, 25th GA Inf. to 2d Co. I, 38th GA Inf. and elected 2d Lt 5/2/1862; Cpt in 1862. Wounded 2d Manassas, 8/28/1862. Killed in battle of Fredericksburg, shot in the head leading his company in a charge, 12/13/1862. BSU.
Jones, Henry L. - 1st Lt 8/8/1861. Transferred from 3d Co. E, 25th GA Inf. to 2d Co. I, 38th GA Inf. and elected Cpt 5/2/1862. D 6/27/1862. K in the battle of Gaines Mill, 6/27/1862. BSU.
Jones, William T. - Pvt 9/29/1862. Transferred to 2d Co. A, 60th GA Inf. 3/1/1863; to Co. K, 61st AL. Inf. 4/11/1864. Roll for Aug. /1864, LOF, shows AWOL since 8/12/1863. NLR. BSU.
Kirkland, H. E. (Henry E. Kirkland) - Pvt 5/10/1862. Transferred to 2d Co. A, 60th GA Inf. 3/1/1863; to Co. K, 61st AL. Inf. 4/11/1864. Killed at Winchester, VA 9/19/1864. Widow, Martha Kirkland, filed CPA, Henry Co., AL, 06/22/1889. BSU.
Kirkland, William (William "Bill" Kirkland) - Pvt 5/10/1862. Transferred to 2d Co. A, 60th GA Inf. 3/1/1863; to Co. K, 61st AL. Inf. 4/11/1864. C Petersburg, VA, 4/2/1865. Released PLO, MD 6/14/1865. D 1/29/1883 in Marianna, Jackson, FL. Widow, Sarah Ann Kirkland, filed CPA, Henry Co., AL, 1903. She stated he was "Captured by Yankees while carrying water." BSU.
Kirkland, William H. - Pvt 5/10/1862. Transferred to 2d Co. A, 60th GA Inf. 3/1/1863; to Co. K, 61st AL. Inf. 4/11/1864. Killed 7/18/1864. BSU.
Kirkland, William R. - Pvt 3/25/1862. Transferred from 3d Co. E, 25th GA Inf. to 2d Co. I, 38th GA Inf. 5/2/1862. Discharged 6/23/1862. Records show was admitted to CSA. General Hos Farmville, VA, with chronic diarrhea, 5/1/1864, furloughed 5/31/1864. No later record. BSU.
Kirkland, William W. (William Wilson Kirkland) - 1st Cpl 5/2/1862. Appt 5th Sgt 12/31/1862. Trn to 2d Co. A, 60th GA Inf. 3/1/1863; to Co. K, 61st AL. Inf. 4/11/1864. W in 1864. C Petersburg, VA 4/2/1865. Released PLO, MD 6/14/1865. D 9/8/1911, Henry Co., AL, bur there Union Springs Cem.
Maley, John G. - Pvt 5/10/1862. Wounded at Antietam, 9/17/1862. Transferred to 2d Co. A, 60th GA Inf. 3/1/1863; to Co. K, 61st AL Inf. 4/11/1864. Died at home 9/22/1864. Widow, L. A. Maley, filed CPA, Henry Co., AL, 5/19/1891. BSU.
Mason, George Lee - Pvt 5/10/1862. Roll for 2/28/1863, last on file, shows him AWOL. No later record. Died 11/21/1885. First buried in 1885, Young Family Cemetery, Washington Co., GA, remains moved and re-buried beside his wife in the City Cemetery, Swainsboro, Emanuel Co., GA.
McGauley, Patrick L. (or McCarley) - Pvt 8/8/1861. Transferred from 3d Co. E, 25th GA Inf. to 2d Co. I, 38th GA Inf. 5/2/1862 and Appt 1st Sgt 5/2/1862. Wounded at 2d Manassas, 8/28/1862. Transferred to 2d Co. A, 60th GA Inf. 3/1/1863. Deserted to enemy 2/19/1864. BSU.
McMath, H. - Pvt 1/13/1862. Trn from 3d Co. E, 25th GA Inf. to 2d Co. I, 38th GA Inf. 5/2/1862. Discharged from service General Hospital, Huguenot Springs, VA, 9/19/1862. BSU.
Miller, John G. - Pvt 8/8/1861. Trn from 3d Co. E, 25th GA Inf. to 2d Co. I, 38th GA Inf. 5/2/1862; to 2d Co. A, 60th GA Inf. 3/1/1863; C Waterloo, PA 7/5/1863. Trn to Co. K, 61st AL. Inf. 4/11/1864. P at PLO, MD 2/18/1865, and E there 2/28/1865. NLR. Died 4/11/1900, buried Howard Grove Primitive Bap. Cem, Houston Co., AL. Widow, Jane Miller, filed CPA, Houston, Co., AL, 1907.
Miller, William Henry H. - Pvt 8/8/1861. Trn from 3d Co. E, 25th GA Inf. to 2d Co. I, 38th GA Inf. 5/2/1862. W 2d Manassas, VA 8/28/1862. Trn to 2d Co. A, 60th GA Inf. 3/1/1863. C Mine Run, VA 11/27/1863. Released in 1864. Transferred to Co. K, 61st AL. Inf. 4/11/1864. Captured at Ft. Stedman, VA, 3/25/1865. Released Point Lookout, MD 5/14/1865. Filed CPA, 1914, Houston Co., AL, pension #34173. D 3/6/1925. Buried Pilgrim's Rest Baptist Cemetery #2, Houston Co., AL.
Nichols, Joel N. (or James M.) - Pvt 1/12/1862. Transferred from 3d Co. E, 25th GA Inf. to 2d Co. I, 38th GA Inf. 5/2/1862; to 2d Co. A, 60th GA Inf. 3/1/1863; to Co. K, 61st AL. Inf. 4/11/1864. Sick in Lynchburg, VA, hospital 8/31/1864. No later record. BSU.
Palmore, John (or Palmer) - Pvt 8/8/1861. Trn from 3d Co. E, 25th GA Inf. to 2d Co. I, 38th GA Inf. 5/2/1862. Wounded 2d Manassas, VA, 8/28/1862. Trn to 2d Co. A, 60th GA Inf. 3/1/1863. Deserted to enemy, 2/19/1864. BSU.
Perkins, John J. - Pvt 8/8/1861. Trn from 3d Co. E, 25th GA Inf. to 2d Co. I, 38th GA Inf. 5/2/1862; to 2d Co. A, 60th GA Inf. 3/1/1863; to Co. K, 61st AL. Inf. 4/11/1864. Appt 3rd. Sgt in 1864. C Frederick City, MD, 7/9/1864. Paroled and exchanged PLO, MD, 3/14/1865. Received Boulware & Cox's Wharves, James River, VA, 3/16/1865. No later record. Filed CPA, Holmes Co.,

FL, 1907. Died 1/30/1928, buried Red Hill Cem., Holmes Co., FL. CSA marker reads 61st ALA, CO. K.
Peterman, James T. - Pvt 5/10/1862. Trn to 2d Co. A, 60th GA Inf. 3/1/1863. C Mine Run, VA 11/29/1863. Trn to Co. K, 61st AL. Inf. 4/11/1864. Released Elmira, NY 6/14/1865. BSU.
Peterman, J. B. - (James B. Peterman) Pvt 5/10/1862. Roll for 2/28/1863, LOF, shows present. NLR. Widow, Arian Bray Peterman, filed a CPA in Oglethorpe Co., GA, 1891. She stated he was K instantly by a grape shot through the head, fired by a Federal battery, during the battle for Maryre's Height, 5/4/1863. Service claimed in pension record as a member of Co. A, 60th GA Inf. BSU.
Peterman, Thomas E. (Thomas Edward Peterman) - Pvt 5/10/1862. W 2d Manassas, VA 8/28/1862. Transferred to 2d Co. A, 60th GA Inf. 3/1/1863. Unfit field service due to leg wound, Det as guard in Ordnance Dept. Macon, GA 3/4/1864. Trn to Co. K, 61st AL. Inf. 4/11/1864. Absent, sick Eufaula, AL 8/31/1864. NLR. Filed CPA Houston Co., AL., 1917, D 2/24/1923, bur there Dothan City Cem.
Pitman, Jasper N. (or Pittman) - Pvt 8/8/1861. Trn from 3d Co. E, 25th GA Inf. to 2d Co. I, 38th GA Inf. 5/2/1862 and Appt 4th Sgt Wounded Fredericksburg, VA, 12/13/1862, gunshot to joint of right arm. Trn to 2d Co. A, 60th GA Inf. 3/1/1863. Transferred to Co. K, 61st AL. Inf. 4/11/1864. Retired 9/16/1864. Twenty years old, as of 3/1/1863, according to CSR. BSU.
Pitman, Robert M. (or Pittman) - Pvt 3/14/1862. Trn from 3d Co. E, 25th GA Inf. to 2d Co. I, 38th GA Inf. 5/2/1862; to 2d Co. A, 60th GA Inf. 3/1/1863. W and C Mine Run, VA 11/27/1863. Trn to 1st Division General Hos, Alexandria, VA, 12/4/1863, D there of wounds 12/25/1863. First bur there in military cem, moved to Christ Church Cem, Alexandria, VA. Marker lists R. Pittman, 60th GA
Pitman, W. H. (or Pittman) - Pvt 8/8/1861. BSU.
Polson, Walter P. - 2d Cpl 8/8/1861. Transferred from 3d Co. E, 25th GA Inf. to 2d Co. I, 38th GA Inf. 5/2/1862; to 2d Co. A, 60th GA Inf. 3/1/1863. Roll for 1/15/1864, LOF, shows present. Does not appear on rolls of Co. K, 61st AL Inf. BSU.
Porter, James M. - Pvt 4/21/1862. Trn from 3d Co. E, 25th GA Inf. to 2d Co. I, 38th GA Inf. and Appt 3rd. Cpl 5/2/1862. Trn to 2d Co. A, 60th GA Inf. 3/1/1863; to Co. K, 61st AL. Inf. 4/11/1864. D Mount Jackson, VA 9/10/1864. Bur Our Soldiers Cem., Mount Jackson, Shenandoah Co., VA.
Register, Irwin - Pvt 8/8/1861. Trn from 3d Co. E, 25th GA Inf. to 2d Co. I, 38th GA Inf. 5/2/1862. Killed Cold Harbor, VA 6/1/1862. Should read "Killed Gaines Mill, 6/27/1862," according to CSR. Widow, Eliza Register filed CPA, Henry Co., AL, 1887. BSU.
Register, John - Pvt 5/10/1862. Died in Frederick City, MD hos 9/16/1862. Bur Mount Olivet Cem, Grave #2, Frederick, MD. A modern CSA marker reads: JOHN S. REGISTER, PVT CO A, 6 AL INF, D 9/16/1862. The old original grave marker, reads JNO REGISTER, DALE CO. AL.
Roland, Henry M. - Pvt 8/8/1861 Trn from 3d Co. E, 25th GA Inf. to 2d Co. I, 38th GA Inf. 5/2/1862; to 2d Co. A, 60th GA Inf. 3/1/1863; to Co. K, 61st AL. Inf. 4/11/1864. Captured Cedar Creek, VA 10/19/1864. E Point Lookout, MD, 2/13/1865. NLR. W 2d Manassas, 8/28/1862. BSU.
Saunders, Joseph H. - Pvt 8/8/1861. Trn from 3d Co. E, 25th GA Inf. to 2d Co. I, 38th GA Inf. 5/2/1862; to 2d Co. A, 60th GA Inf. 3/1/1863; to Co. K, 61st AL. Inf. 4/11/1864. Appt 2nd Sgt 1864. Roll for 8/31/1864, Last on file, shows present. No later record. BSU.
Sellers, Riley - Pvt 5/10/1862. Discharged 7/22/1862. CSR contains discharge record. BSU.
Smith, G. W. (George W. Smith) - Pvt 5/10/1862. Wounded 2d Manassas, 8/28/1862. Transferred to 2d Co. A, 60th GA Inf. 3/1/1863; to Co. K, 61st AL. Inf. 4/11/1864. Roll for 8/31/1864, last on file, shows absent, sick. BSU.
Smith, Henry M. - Pvt 5/10/1862. Trn to 2d Co. A, 60th GA Inf. 3/1/1863; to Co. K, 61st AL. Inf. 4/11/1864. Roll for 8/31/1864, Last on file, shows AWOL. No later record. Filed CPA, Geneva Co., AL, 1916. Died 5/28/1921, buried New Hope Primitive Baptist Church Cemetery, Houston Co., AL.
Smith, J. W. (John W. Smith) - Pvt 5/10/1862. W Fred., VA 12/13/1862. Admitted to Chimborazo Hos #2, Rich, VA, with gunshot wound in left arm, developed smallpox and was trn to Howard's Grove, Rich, VA, 2/12/1863. Died there in General Hos of variola and gangrene 2/16/1863. BSU.
Smith, Nathan W. - Pvt 1/13/1862. Trn from 3d Co. E, 25th GA Inf. to 2d Co. I, 38th GA Inf. 5/2/1862; to 2d Co. A, 60th GA Inf. 3/1/1863. Died of tuberculosis, Fair View Hospital, Lexington, VA, 9/2/1863. BSU.

Smith, N. E. (Nathan E. Smith) - Pvt 1/20/1862. Transferred to 2d Co. A, 60th GA Inf. 3/1/1863. W in 1863. Trn to Co. K, 61st AL. Inf. 4/11/1864. Roll for 8/31/1864, LOF, shows AWOL. NLR. BSU.
Smith, S. J. - Pvt 4/2/1862. Discharged 6/5/1862. BSU.
Smith, W. J. (William Jackson Smith) - Pvt 5/10/1862. Trn to 2d Co. A, 60th GA Inf. 3/1/1863; to Co. K, 61st AL. Inf. 4/11/1864. Sick in Macon, GA hos 8/31/1864. NLR. Filed CPA, Houston Co., AL, 1926. Bur Oak Hill Cem, Autauga Co., AL. Marker incorrectly lists regt. as CO I, 31ST GA.
Starling, Isaac - Pvt 8/28/1861. Trn from 3d Co. E, 25th GA Inf. to 2d Co. I, 38th GA Inf. 5/2/1862. W Fred., VA 12/13/1862. W and C battle of Fred., 12/13/1862. D. Old Capital Prison USA hos, D.C., gunshot wound of side 1/1/1863. Buried Piney Grove Primitive Bap. Church Cem, Henry Co., AL.
Strickland, John C. (John Chambers Strickland) - Pvt 4/12/1862. Trn from 3d Co. E, 25th GA Inf. to 2d Co. I, 38th GA Inf. 5/2/1862; to 2d Co. A, 60th GA Inf. 3/1/1863; to Co. K, 61st AL. Inf. 4/11/1864. C Petersburg, VA 4/2/1865. Released PLO, MD 6/17/1865. D. 1882, bur Old Zion Church, Henry Co., AL. Widow, S. M. Strickland, filed CPA, Henry Co., AL, 1897.
Strickland, William - Pvt 8/8/1861. Trn from 3d Co. E, 25th GA Inf. to 2d Co. I, 38th GA Inf. 5/2/1862; to 2d Co. A, 60th GA Inf. 3/1/1863; to Co. K, 61st AL. Inf. 4/11/1864. Captured Petersburg, VA 4/2/1865. Released at Point Lookout, MD, 6/17/1865. BSU.
Taylor, Richard - (Richard C. Taylor) Pvt 8/8/1861. Trn from 3d Co. E, 25th GA Inf. to 2d Co. I, 38th GA Inf. 5/2/1862; to 2d Co. A, 60th GA Inf. 3/1/1863; to Co. K, 61st AL Inf. 4/11/1864. Roll for 8/31/1864, last on file, shows present. No later record. Buried Ramah Church Cemetery, Wilkinson Co. Grave marker reads, CO. I, 38 GA INF.
Thorp, W. A. (or Tharpe) (William A. Tharp) - Pvt 8/8/1861. Transferred from 3d Co. E, 25th GA Inf. to 2d Co. I, 38th GA Inf. 5/2/1862. Discharged 6/5/1862. BSU.
Varnum, Colin S. (Colen Shaw Varnum) - Pvt 8/8/1861. Appt 1st Sgt Trn from 3d Co. E, 25th GA Inf. to 2d Co. I, 38th GA Inf. 5/2/1862 and Appt 3d. Sgt; to 2d Co. A, 60th GA Inf. and Appt 2d Sgt 3/1/1863; to Co. K 61st AL Inf. 4/11/1864. Captured Petersburg, VA 4/2/1865. Released Point Lookout, MD, 7/21/1865. D 5/27/1905. Buried Piney Grove Primitive Baptist Church, Henry Co., AL. Widow, J. C. Varnum filed CPA, Houston Co., AL.
Warren, Reddick P. - Pvt 3/1/1862. Trn from 3d Co. E, 25th GA Inf. to 2d Co. I, 38th GA Inf. 5/2/1862; to 2d Co. A, 60th GA Inf. 3/1/1863; to Co. K, 61st AL. Inf. 4/11/1864. C Fisher's Hill, VA 9/22/1864. Took OOA to U.S. Govt. PLO, MD, enlisted in U.S. service in 2nd U.S. Vol., (Union). Mustered out of Union service 6/18/1866, Ft. Leavenworth, KS. BSU.
Whiddon, Edward - Pvt 5/10/1862. Died of typhoid fever in Charlottesville, VA hos 7/18/1862. Bur there in Confed. Cem. Bur UVA Confed. Cem, Charlottesville, Albemarle Co., VA Soldier's names listed on four bronze plaques. CSR has death claim filed by widow, Sarah Whiddon, Henry Co., AL.
Whiddon, James - Pvt 8/8/1861. Trn from 3d Co. E, 25th GA Inf. to 2d Co. I, 38th GA Inf. 5/2/1862; to 2d Co. A, 60th GA Inf. 3/1/1863; to Co. K, 61st AL. Inf. 4/11/1864. W in 1864. Appt Sgt. S Appomattox, VA 4/9/1865. D 10/17/1900, bur Newville Bap. Church, Newville, AL.
Whiddon, Joel - Pvt 5/10/1862. Died of chronic diarrhea Camp Lee near Richmond, VA 10/16/1862. Buried at Hollywood Cem, Richmond, VA, Cemetery records show "Whitden," no first name listed, Co. I, 27th GA, died 10/20/1862. This is likely Joel Whiddon, as there is not a soldier named "Whitden" or similar name, on the roster of the 27th GA.
Whiddon, Lott - Pvt 4/17/1862. Transferred from 3d Co. E, 25th GA Inf. to 2d Co. I, 38th GA Inf. 5/2/1862; to 2d Co. A. 60th GA Inf. 3/1/1863; to Co. K, 61st AL. Inf. 4/11/1864. Appt Cpl in 1864. Killed in battle of Spotsylvania Court House, 5/10/1864. BSU.
Whiddon, William N. (or William A.) - Pvt 1/13/1862. Transferred from 3d Co. E, 25th GA Inf. to 2d Co. I, 38th GA Inf. 5/2/1862. Died 8/31/1862. BSU.
Williams, Benjamin N. - Pvt 8/8/1861. Trn from 3d Co. E, 25th GA Inf. to 2d Co. I, 38th GA Inf. 5/2/1862. Wounded 2d Manassas, VA, 8/28/1862. Died of wounds 9/11/1862. BSU.

COMPANY K, BARTOW AVENGERS OR DeKALB AND FULTON BARTOW AVENGERS DEKALB COUNTY, GA.

This company was also known as Co. G, and New Co. B. It was first organized by Captain William Wright. Captain William C. Mathews recalled, "Company K was one of the best fighting companies in the Regt. We sometimes called them the "orphans" because in every fight they would lose their commander and a Sergeant would command them. This always fell on Sergeant F. L. Hudgins, and he was always equal to the occasion, no matter what it was. Wounded several times severely, he always got back in time for the next fight. He commanded the company for the last six months of the war and it has no better commander than he. He was a truly brave and gallant soldier."

Adams, Enos - Pvt 9/26/1861. AWOL 10/15/1863 - 8/31/1864. Filed a CPA, DeKalb Co., 1889, died abt. 1904. Buried Nancy Creek Primitive Baptist Church Cemetery, Chamblee, DeKalb Co. Widow, Elizabeth Adams filed CPA. 1905.
Adams, John - (John R. Adams) Pvt 3/8/1862. Trn 10/24/1862. Died of disease in camp. BSU.
Akers, John H. - Pvt 9/26/1861. W 2d Manassas, VA 8/28/1862. D from wounds 9/4/1862. CSR shows death claim filed by Nancy Akers, but the actual death claim not in CSR. BSU.
Almand, A. W. (or Almond) (or A. W. Allman) - Pvt 3/8/1862. Wounded at Gettysburg, PA, 7/2/1863. Killed near Cedar Creek, VA, 10/19/1864. BSU.
Anderson, James Lewis - Pvt 9/26/1861. W in 1862. Appt 3d Cpl June/1863; 2d Cpl 7/2/1863. Wounded in 1864, Spotsylvania Court House. Appears on roll of men employed in Anderson's Div Army of Tennessee, roll not dated, as paroled Greensboro, NC, 1865. Died 6/22/1906, buried New Hope Church Cemetery, Marietta. Obit in Marietta Journal, page 4, 6/28/1906.
Armstrong, James J. - Pvt 9/26/1861. Discharged, disability, Savannah, GA 6/6/1862. (Born abt. 1813) Discharge record in CSR. BSU.
Austin, Richard - Pvt. Died of variola (smallpox) age 23, on 1/12/1863, buried at Huguenot Springs Confederate Cemetery, Powhatan Co., VA.
Autry, Isaac W. (Isaac Washington Autry) - Pvt 9/26/1861. C Petersburg, VA 3/25/1865. Released PLO, MD 6/9/1865. C in battle of Fred., 12/13/1862, P for E 12/17/1862, according to CSR. BSU.
Autry, William Allen - (or Awtrey) Pvt 3/8/1862. Wounded in right leg and permanently disabled, Winchester, VA, 9/19/1864. Paroled Albany, GA 5/20/1865. (Born in GA 10/24/1843) Filed CPA, Fulton Co., 1899. BSU.
Bailey, Luke R. - Pvt 9/26/1861. Transferred to Co. C, Cobb's Legion GA Inf. 7/1/1862. Wounded, date and place not given. On wounded furlough 1/30/1865. BSU.
Ball, James E. - Pvt 9/26/1861. Appt 4th Cpl 6/20/1862. Severely wounded in chest and captured Gettysburg, PA, 7/2/1863. Died of wounds, hospital in or near Gettysburg, PA, 7/10/1863. BSU.
Bayne, Hendley V. - Pvt 9/26/1861. W & C Wilderness, VA, 5/6/1864. Released Elmira, NY 6/19/1865. Widow, Ella E. Cook Bayne, filed CPA, Fulton Co., 1920. D 1/23/1906. Bur Oakland Cem, Section 8, Lot 360, Grave 6, Fulton Co. Obit in Atlanta Constitution, 1/24/1906, page 3.
Bowman, Robert (or Boman) - Pvt 9/26/1861. Discharged on account of chronic rheumatism Charlottesville, VA July 22 or 27, 1862. Discharge record in CSR. BSU.
Bowman, William M. (or Boman) - Pvt 5/15/1862. Sick Atlanta, GA 10/31/1862, and died there 11/23/1862. Buried Oakland Cemetery, Atlanta.
Boyd, John B. - Pvt 9/26/1861. C Winchester, VA 9/19/1864, or Fisher's Hill, VA 9/22/1864. P West's Bldgs. USA General Hos Baltimore, MD, and sent to James River, VA for E, Feb/1865. CSR states captured at Fisher's Hill, 9/22/1864. D 7/9/1904, bur Midway United Meth. Church, Fulton Co. Widow, Isabella A. Barrett Boyd, filed CPA, DeKalb Co., 1904. Marriage license in CPA.
Brisentine, William H. (or Brisendin) - (William H. Brisendine) Pvt 9/26/1861. Captured Jericho Ford, VA 5/24/1864. Paroled at PLO, MD 2/13/1865. Rcv Cox's Landing, James River, VA for E, Feb.14 or 15, 1865, frl 40 days. Receiving and Wayside Hos, or General Hos #9, Rich, VA 2/21/1865. Filed CPA, Cherokee Co., AL. B 4/11/1843, D 2/10/1909, bur Black Oak Cem, Cherokee Co., AL.
Brooks, William A. - (William Andrew Brooks) Pvt 9/26/1861. W in head, date and place not given. Surrendered at Appomattox, VA 4/9/1865. (Born in SC, 1842) Filed CPA, Milton Co., 1901.

B 6/20/1844, D 8/26/1926. Buried Ebenezer Primitive Baptist. Church Cemetery, Roswell, Fulton Co.
Brown, Allen - Pvt 9/26/1861. Sick in hos Aug 9 - Nov 7, 1864. S Appomattox, VA 4/9/1865. (D 2/15/1935. Bur in Antioch Bap. Church Yard, DeKalb Co. Age 93 years) Filed CPA, DeKalb Co., 1912. Noted in obituary as the last survivor of Co. K, 38th GA Regt.
Brown, Killis - Pvt 3/8/1862. W White Oak Swamp, VA 6/13/1864. C Harrisonburg, VA 9/25/1863. P at PLO, MD and transferred to Aiken's Landing, VA for E, 3/17/1865. BSU.
Brown, Lewis (Lewis A. Brown) - Pvt 9/26/1861. S Appomattox, VA, 4/9/1865. (B in GA) Filed CPA, Fulton Co., 1920. B 11/29/1838, D 4/15/1926, in Fulton Co., bur there College Park Cem.
Burdett, Humphrey M. (or Berdett) - Pvt 9/26/1861. Shot through right foot and hip Chancellorsville, VA 5/3/1863. Appt 3d Sgt Aug/1862. AWOL May 10 - Nov 7, 1864. CPA shows totally disabled by wounds. (B GA 1/8/1822) W 2d Manassas, 8/28/1862. Filed CPA, Fulton Co., 1887. B 1/8/1822, D 1895. Buried Sandy Springs United Meth. Church Cemetery, Fulton Co.
Burdett, John S. (John Sylvester Burdett) - Pvt 9/26/1861. C Cedar Creek, VA 10/19/1864. Released PLO, MD 5/12/1865. Filed CPA, Chambers Co., AL. Born 1839, died 1916, buried Bethel United Methodist Church Cemetery, Troup County. CSA marker reads: CO K 38 GA INF.
Caudle, John H. W. H. - See Pvt. Co. M.
Chandler, James E. - Pvt 9/26/1861. Appt 2d Cpl 4/30/1862. Killed Antietam, MD 9/17/1862. Capt William C. Mathews wrote, "Cpl James E. Chandler commanded the company Sharpsburg and D bravely in the thickest of the fight." Death claim by Hester Chandler, in CSR. DeKalb Co. BSU.
Chandler, John W. - Pvt 9/26/1861. Killed 2d Manassas, VA, 8/30/1862. BSU.
Chandler, William B. - Pvt 3/8/1862. Died of disease in camp 5/2/1863. Grave marker at Old Macedonia Cem, DeKalb Co., GA, reads: Rev. W. B. Chandler, CO K 38 GA INF.
Childress, Joseph H. - Pvt 9/26/1861. Roll dated 11/7/1864, LOF, shows absent, sick, unknown where, since 7/12/1864. CPA shows on sick frl Aug/1864, could not reach command. Reported to Colonel Findley. P Kingston, GA 5/12/1865. (B TN 1841) Filed CPA, Fulton Co., 1903. BSU.
Childress, William A. (William Anderson Childress) - Pvt 9/26/1861. W in left hip Seven Days Fight near Rich, VA 6/20/1862. W in left shoulder (back) and right leg (ankle) Cedar Creek. VA 10/19/1864. Absent, W, 11/7/1864. (B GA 1831) Filed CPA, Fulton Co., 1895. D 10/1/1909, bur Nancy Creek Primitive Bap. Church Cem, DeKalb Co. Obit in Atlanta Constitution, 10/2/1909.
Collier, James S. - Pvt 9/26/1861. D typhoid fever Charlottesville, VA, 6/29/1862. Bur there in UVA Confed. Cem., Charlottesville, VA. CSR contains death claim by father, John Collier, DeKalb Co.
Collier, John E. J. - Pvt 9/26/1861. Transferred to Co. D, 8/15/1864. Wounded near Newtown, VA, 7/17/1864 and sent to hospital. Received pay Richmond, VA, 9/3/1864. Filed CPA, Hempstead Co., AR 8/4/1904, Died 6/7/1908, buried there Ayers Cemetery.
Connor, B. G. - Pvt 5/15/1862. Deserted 6/3/1862.
Cook, George G. - Pvt 9/26/1861. On sick frl Dec/1861. Enlisted as Pvt Co. A, 1st Regt GA State Troops (Howell's Company,) 8/14/1862. Appt 3d Cpl 4/1/1864. Severely wounded in neck, Atlanta, GA 7/22/1864. Detailed on W.K.A. Railroad. (Born in GA 8/18/1837) Filed CPA, Fulton Co., 1899. Died 2/9/1906, buried at Cross Roads Cemetery, beside Cross Roads Primitive Baptist Church, Sandy Springs. Obituary in Atlanta Constitution, 2/10/1906.
Cook, James R. - Pvt 5/15/1862. Died Staunton, VA, 7/24/1862. Buried there Thornrose Cemetery. CSR contains death claim filed by widow, Cynthia Cook. Widow filed CPA, Polk Co., 1891. She stated he took fever during the march to Richmond and was taken to Staunton Hospital, were he died.
Cordell, J. H. - Pvt 9/26/1861. Wounded in 1862. At home, wounded, 10/31/1862. absent without leave Dec/1862 - Aug/1863. Roll for 9/10/1863, shows him "dropped by orders BSU.
Cowan, John Jordan - Pvt 3/20/1862. Captured at the Wilderness, VA, 5/6/1864. Released Elmira, NY 6/16/1865. Died 1907, at Newton. BSU.
Cowan, John S. - Pvt 3/20/1862. Died 6/20/1862. BSU.

Cowan, Zach J. - (Zachariah J. Cowan) Pvt 3/20/1862. S Appomattox, VA, 4/9/1865. B 12/10/1839, D 10/29/1897, buried Harden Cem, Rockdale Co. Widow, S. C. Cowan, filed CPA, Turner Co., 1911.
Daniel, Jesse F. - (Jesse Fletcher Daniel) Pvt 9/26/1861. Appears last on roll Apr/1862. D 1908, bur Prays Mill Bap. Church Cem, Douglas Co. Widow, Jane T. Daniel filed CPA, Douglas Co., 1911.
Dorris, John M. - (or John M. Dowis) Pvt 9/26/1861. Appt Color Cpl Jan/1862. Killed Gaines Mill, VA, 6/27/1862. CSR contains a death claim filed by widow, Dianna Dowes, Fulton Co. BSU.
Ellis, William H. - Pvt 5/15/1862. Died of measles in general Hospital #16, Rich, VA, 6/14/1863. Buried Oakwood Cemetery, Richmond, VA. Widow, Nancy E. Ellis, filed CPA, Fulton Co., 1891.
Ennis, John - Pvt 9/26/1861. Discharged 5/15/1862. BSU.
Ennis, William - Pvt 9/26/1861. Wounded in 1862. BSU.
Farr, David N. (Or David N. Fair) - Pvt 9/26/1861. Wounded Gaines Mill, VA, 6/27/1862. Roll dated 11/7/1864, LOF, shows "Absent, sick hos since 6/3/1864." Filed CPA, Marshall Co., AL. Buried New Prospect Cem, Marshall Co., AL. CSA marker reads, Born 1824, Died 1905, CO K, 38 GA INF.
Fletcher, Richard H. - Pvt. Transferred from 31st GA Regt. Elected Cpt 12/12/1864. Admitted to Stuart Hospital Richmond, VA, 2/9/1865, and frl for 30 days 3/29/1865. Killed 5/11/1869. Buried Cedar Hill Cemetery, Dawson, Terrell Co. Murdered by gunman on highway, story of death in Macon Telegraph, 5/21/1869. Follow-up article 7/30/1869, describes lynching of the killer.
Gazaway, Francis M. (Francis Marion Gasaway) - 1st Cpl 9/26/1861. Killed 2d Manassas, VA, 8/28/1862. Widow, Mary A. E. Gazaway filed CPA, Campbell Co., 1891. BSU.
Gazaway, James H. (James Henry Gazaway) - 3d Cpl 9/26/1861. Appt 2d Sgt, 5/14/1863. Captured Gettysburg, PA, 7/1/1863. P at PLO, MD, for E 10/11/1864. Rcv Boulware & Cox's Wharves, James River, VA, 10/15/1864. Admitted to Jackson Hos Rich, VA 10/16/1864, frl for 60 days, 11/6/1864. (B GA 1840) Filed CPA, DeKalb Co., 1920, D 8/24/1933, bur there Wesley Chapel United Meth. Cem.
Gazaway, John (John Gordon Gasaway) - Pvt 3/8/1862. Discharged 7/27/1862. Discharge record in Confederate Service Record. BSU.
Gazaway, William (William Marion Gasaway) Pvt 9/26/1861. Wounded and disabled in 1862 by gunshot. Wounded 2d Manassas, 8/28/1862. Captured near Lithonia, GA, 7/26/1864. Took OOA to U.S. Govt., enlisted in U.S. service, Camp Chase, OH and transferred to Chicago, IL 3/20/1865. Enlisted in 6th U.S. Vol. 3/16/1865. Died 12/18/1925, Bates Co, MO. BSU.
Gentry, Asbury M. - Pvt 9/26/1861. D Sav, GA 12/31/1861. Cause of death not stated in CSR. BSU.
Gober, Julius J. - 1st Lt 9/26/1861. Died from typhoid fever in General Hospital at Staunton, VA, 7/26/1862. Buried there in Thornrose Cemetery.
Goodwin, Charles H. (Charles Harris Goodwin) - Pvt 9/26/1861. Wounded Gaines Mill, VA, 6/27/1862. Died of wounds in DeKalb Co., GA, July 25, or 8/4/1862. CSR contains a death claim filed by father, Harris Goodwin, Fulton Co. Buried at Goodwin Family Cemetery, DeKalb Co.
Goodwin, Gustin E. - 2d Lt 9/26/1861. Elected 1st Lt 7/26/1862. Wounded at 2d Manassas, VA 8/28/1862. Died of wounds 9/7/1862. CSR states he was in command of company when wounded. CSR contains death claim filed by widow, DeKalb Co. BSU.
Goodwin, William F. (William Franklin Goodwin) - Pvt 4/28/1862. W Gaines Mill, VA 6/27/1862. Elected Jr. 2d Lt 12/1/1862. K Gettysburg, PA, 7/2/1863. CSR states he was K on 7/1/1863. BSU.
Goss, William T. (William Thomas Goss) - Enlisted as a Pvt in Co. D, 25th Battalion, GA Provost Guard Inf. 8/3/1863. On roll for July - 8/1863, LOF, presence or absence not stated. Enlisted as a Pvt in Co. K, 38th Regt GA Inf. 3/7/1864. Captured Winchester, VA, 9/19/1864. P Baltimore, MD and sent to James River, VA for E Feb/1865. (Born in DeKalb Co., GA 12/17/1847) Filed a CPA, Fulton Co., 1902. Died 3/15/1931. Bur Byron City Cemetery, Peach Co., marker reads: CO K, 38 GA INF.
Grogan, Gideon F. - Pvt 3/8/1862. Killed at Antietam, MD 9/17/1862. CSR contains a death claim filed by father, Bartholmew Grogan, Fulton Co. BSU.

Grogan, Jerry H. - Pvt 9/26/1861. Wounded, disabled in 1862. Detailed for special service 1864. Absent, sick, 9/12/1864. BSU.
Grogan, Joseph D. - Pvt 3/8/1862. Killed at Antietam, MD, 9/17/1862. CSR contains death claim filed by father, Bartholmew Grogan, in Fulton Co. BSU.
Grogan, Josiah T. - Pvt 9/26/1861. W left side Antietam, MD 9/17/1862. Trn to Co. D, 8/15/1864. S Appomattox, VA 4/9/1865. (B 8/22/1833) Filed CPA, Fulton Co., 1891. D abt. 3/1/1917. BSU.
Guess, Franklin L. (Franklin "Frank" L. Guess) - Pvt 9/26/1861. Trn to Co. B, 9th Battalion, GA Light Artillery 4/21/1862. Appears last on roll for June 1862. Filed CPA, DeKalb Co., 1903, under the name "Francis L. Guest. D 1912, buried at Sardis United Meth. Church Cem, Atlanta.
Hammond, Joshua T. - See Pvt, Co. M. BSU.
Head, Henry Tandy - Pvt 9/26/1861. Trn to Co. C, Cobb's Legion GA Inf. in 1863. S Appomattox, VA 4/9/1865. D 4/16/1898. Bur Rock Spring Presbyterian Cem, Fulton Co.
Henry, Jesse L. - Pvt 9/26/1861. Killed at Gaines Mill, VA 6/27/1862. Confederate Service Record contains a death claim filed by mother, Nancy A. Henry. BSU.
Henry, William Robert - 2d Sgt 9/26/1861. Elected 1st Lt 11/1862. Wounded in left leg, necessitating amputation below knee, Fredericksburg, VA, 12/13/1862. Resigned Oct. 6 or 16, 1863. (Died Ellenwood, GA 4/12/1909) Filed CPA, Clayton Co., 1889. Buried Tanners Cemetery, Clayton Co. Obit in Atlanta Georgian & News, 4/16/1909, page 8.
Hildebrand, William B. H. - Pvt 11/16/1861. Wounded in 1863. Captured Shepherdstown, WV, 7/1/1863. Paroled at Point Lookout, MD 2/18/1865. Died abt. 1900. Buried Old Mount Paran Baptist. Church Cemetery, Fulton Co., marker reads: William B. Hilderbrand, CO K, 38 GA INF.
Hornbuckle, H. H. (Horton H. Hornbuckle) - Pvt 5/15/1862. Wounded Gaines Mill, VA, 6/27/1862. Died of pneumonia in General Hospital, Charlottesville, VA, 2/6/1863. Buried there UVA Confederate Cemetery. Soldier's names listed on four bronze plaques on memorial.
Hudgins, Francis L. (Francis Lee Hudgins) - 4th Sgt 9/26/61. Seriously wounded Malvern Hill, VA, in arm. 7/1/62; severely wounded Gettysburg, PA, 7/1/63. Appt 1st Sgt 12/17/64. Surrendered Appomattox, VA, 4/9/65. B DeKalb Co., 10/17/1842, Filed CPA, DeKalb Co., 1895. Died at Confederate. Soldiers' Home in Atlanta, 11/27/1928. Bur Prospect Meth. Church Cem, Chamblee.
Johnston, John S. - 1st Sgt 9/26/1861. Killed Gaines Mill, VA, 6/27/1862. CSR contains death claim filed by mother, Sarah Johnson, in Laurens District, SC. BSU.
Jones, Charles S. W. - Pvt 9/26/1861. W, Rich, VA, 1864. D of wounds, Rich, VA Hos 5/18/1864. Bur there in Hollywood Cem. Cem records list Jones, C. S., Company K, 38 GA, D 5/28/1864.
Jones, James H. - Pvt 3/20/1862. W in 1862. C Wilderness, VA 5/6/1864. P Ft. Delaware, Mar/1865. (B Anderson Co., SC. 1837) W Antietam, 9/17/1862. Filed CPA, Fulton Co., 1901. BSU.
Jones, John W. - Pvt 9/26/1861. Wounded in 1863. Appt 5th Sgt Sept/1863. Wounded and captured Winchester, VA, 9/19/1864. (Place of capture also shown Harrisonburg, VA 9/25/1864) P at Point Lookout, MD, Nov/1864. Received Venus Point, Sav River, GA, for E, 11/15/1864. BSU.
Jones, Jordan M. (Jordan Marion Jones) - Pvt 9/26/1861. Roll dated 11/7/1864, last on file, shows "Absent, sick, not known where, since 4/1/1864." Paroled Albany, 5/25/1865. Filed CPA, DeKalb Co., 1921. Died 6/28/1925, buried Macedonia Cemetery, DeKalb Co.
Jones, J. H. (John H. Jones) - Pvt 11/15/1861. Wounded 1863. Captured Wilderness, VA, 5/6/1864. Paroled Ft. Delaware, Feb/1865. Received Boulware & Cox's Wharves, James River, VA, exchanged, Mar 10-12, 1865. Filed CPA, Fulton Co., 1901. BSU.
Jones, Robert F. - Pvt 3/8/1862. K Gaines Mill, VA 6/27/1862. Marker Old Macedonia Cem, DeKalb Co., reads, "Rev Robert F. Jones, D 6/28/1862, "Nobly he fell while fighting for liberty," unknown if a cenotaph, or grave. Widow, Mary L. Jones, filed a CPA, Fulton Co., 1891.
Jones, R. D. F. (Robert Daniel Francis Jones) - Pvt 3/8/1862. Surrendered Appomattox, VA 4/9/1865. Captured Fredericksburg, 12/13/1862, exchanged 12/17/1862. Died 3/18/1896. Buried Mount Zion West Bap. Church Cem, Haralson Co. Filed CPA Cleburne Co., AL, 1893. Stated in CPA, Wounded right hand at Malvern Hill. Descendants say died Cleburne Co., AL.
Kelly, John F. - (Or John T. Kelly) Pvt 9/26/1861. Appt 2d Sgt Nov/1862. Pvt Mar/1863. Discharged, over-age, Way Hos, Lynchburg, VA, 10/12/1864. Discharge record in CSR. BSU.
Kelly, John H. - Pvt 9/26/1861. Deserted 4/7/1865. BSU.

Kelly, J. T. - (James Thomas Kelly) Pvt 3/7/1864. Absent, sick, 8/9/1864. Admitted to CSA. General Hos Charlottesville, VA with chronic diarrhea 9/1/1864, frl for 60 days 9/11/1864. Filed CPA, McLennan Co., TX, 1913. D. 9/27/1923. Bur there Gholson Cem. Marker reads 38 GA INF.
Lee, A. J. - Pvt 3/4/1862. Appears last on roll for Feb/1863. Discharged. Full name and BSU.
Lee, George - Pvt 9/26/1861. Died of disease in camp 4/22/1863. CSR contains death claim filed by sister, Susan E. Vaughan, DeKalb Co. Buried Hollywood Cemetery, Richmond, VA, cemetery records read, L. Lee, died 6/22/1863, Co. K, 38th GA
Little, William J. - Pvt 9/26/1861. W by gunshot, date and place not given. C near Petersburg, VA 3/26/1865. Released PLO, MD, 6/29/1865. Filed CPA, Paulding Co., 1898. D 10/12/1916. Bur Pine Log Cem, Paulding Co., Widow, Francis C. Thompson, filed CPA, Paulding Co., 1920.
Maddox, Joseph A. (or Mattox) - 3d Sgt 9/26/1861. Appt 1st Sgt 6/28/1862. W Wilderness, VA, 5/5/1864. D of pneumonia in General Receiving Hos (Charity Hos) Gordonsville, VA 12/17/1864. D the Gordonsville Receiving Hos (E Hotel). The Confed. dead, initially bur near the hos, were later moved to Maplewood cem. The E Hotel is still and today is a museum. There is a mass grave at Maplewood Cem where all the CSA soldiers are bur, but no markers identify the men by name.
Mangham, Wiley - (or Wiley Manghon) Pvt 9/26/1861. Appears last on roll for 4/30/1864. BSU.
Martin, Benjamin S. - (Benjamin Stephen Martin) Pvt 9/26/1861. Trn from Co A, Jan/1862. W in arm, necessitating amputation below elbow, Spotsylvania, VA 5/12/1864. Absent, W 8/31/1864, to end of war. Filed CPA, Logan Co., AR, 1892, D 8/12/1899, bur there Liberty Church, CSA marker.
McClain, Berryrnan S. - (Berryman S. McClain) Pvt 9/26/1861. Discharged, disability, 4/10/1862. D in camp, typhoid, 4/11/1862. CSR has death claim by father, William McClain, DeKalb Co. BSU.
McGuire, William R. - Pvt 9/26/1861. W, skull fractured, Gaines Mill, VA 6/27/1862; in right arm 2d Manassas, VA 8/28/1862. C Gettysburg, PA, 7/3/1863. E PLO, MD, 5/3/1864. Rcv Aiken's Landing, VA 5/8/1864.(B GA 9/26/1841) Filed CPA, Milton Co., 1889. D 10/8/1909. Bur Mount Carmel Primitive Bap. Church Cem, Douglas Co.
Mitchell, Elisha J. - Pvt 5/15/1862. Roll dated 11/7/1864, LOF, shows "Absent, sick in Greensboro, GA hospital since 10/15/1862." Buried Bethel Community Cem, Stone Mountain, DeKalb Co. Marker reads Born 4/15/1840, Died 8/27/1920.
Mitchell, James R. - Pvt 9/26/1861. Killed Fredericksburg, VA, 12/13/1862. CSR contains a death claim filed by father, John E. Mitchell, DeKalb Co. BSU.
Morgan, William A. (William Abram Morgan) - Pvt Jan/1862. W in 1863. AWOL 6/10/1864 - 8/31/1864. Filed CPA, DeKalb Co., 1914. D 12/1/1917. Bur Sandy Springs Meth. Cem, Fulton Co.
Morris, Enoch H. C. - 5th Sgt 9/26/1861. Elected Jr. 2d Lt 7/28/1862. K 2d Manassas, VA, 8/28/1862. CSR contains death claim filed by father, Enoch Morris, Fulton Co. BSU.
Nash, John W. - Pvt 5/15/1862. Severely wounded Fredericksburg, VA, 12/13/1862. Died of wounds in Richmond, VA, hospital, 1/2/1863. BSU.
Owens, William L. - Pvt 9/26/1861. Appt 1st Cpl 8/28/1862. AWOL 3/8/1864 - 8/31/1864. D 12/6/1889. Bur Tabernacle Bap. Church Cem, Aiken Co., SC, CSA marker reads CO K, 38 GA INF.
Phillips, John W. - Pvt 5/15/1862. Killed at Gaines Mill, VA, 6/27/1862. Widow, Emaline H. Phillips, filed CPA, Fulton Co., 1891. BSU.
Pruett, Jerry J. - Pvt 9/26/1861. Discharged 6/6/1862. BSU.
Richardson, Daniel B. - Pvt 9/26/1861. Died Hanover Junction hos, of typhoid fever, 2/3/1863. BSU.
Richardson, Josephus S. - Pvt 9/26/1861. Killed at Gaines Mill, VA, 6/27/1862. CSR contains a death claim filed by father, William Richardson, DeKalb Co. BSU.
Richardson, William M. - Pvt 9/26/1861. W, left arm permanently disabled, 2d Manassas, VA 8/28/1862. W Lynchburg, VA 5/9/1864. Absent, W, 8/31/1864. (Resident of GA since 1840) Filed CPA, DeKalb Co., 1888. D 4/2/1906. Bur Jones Memorial Meth. Cem, Forest Park, Clayton Co.
Robinson, George F. - Enlisted, Pvt in Co. E, 4th Regt GA Inf. 4/28/1861. Appt Adjutant of the 38th Regt GA Inf. May/1863. Elected Cpt of Co. E, 8/13/1863. Retired to Invalid Corps 12/12/1864. BSU.
Seals, Benjamin A. - Pvt 3/25/1862. Surrendered Appomattox, VA, 4/9/1865. Died 12/10/1902, in Fulton Co, buried there Sardis United Methodist Church Cemetery, Buckhead.

Smith, William A. - Pvt 9/26/1861. Appt 2d Cpl 9/17/1862. D in Lynchburg, VA Hos Mar. or 5/1863, and bur there in Old City Cem, CSA Section, No. 3, 2d Line, Lot 184. D Reid's Hospital.
Stowers, Asbury M. (Asbury W. Stowers) - Pvt 9/26/1861. Captured Martinsburg, VA, 7/23/1863. Released Ft. Delaware, 6/15/1865. BSU.
Stubbs, George W. - Jr. 2d Lt 9/26/1861. Elected 2d Lt 7/26/1862; 1st Lt 9/8/1862. Promoted to Cpt. Killed near Winchester, VA, 7/24/1864. Buried Mount Hebron Cemetery, Stonewall Confederate Cemetery, Winchester, VA.
Swiney, James M. (or J. M. Summey, or J. M. Sweeney) - Pvt 9/26/1861. Wounded in left side and lung, Gaines Mill, VA, 6/27/1862. Appt 4th Cpl 7/10/1863. S Appomattox, VA, 4/9/1865. (Resident GA since Mar 1834) Filed CPA, Fulton Co., 1888. Died 9/20/1899. Fulton Co. Buried Clifton Methodist Cemetery, DeKalb Co. Widow, Nancy C. Swiney, filed CPA, Fulton Co., 1901.
Swiney, S. J. J. (Sampson J. J. Swiney) - Pvt 9/26/1861. Appt 2d Cpl 12/31/1862. Killed at Winchester, VA, 6/13/1863. BSU.
Thomas, Thomas Jefferson - See Pvt Co. A
Thompson, John B. - Pvt 3/8/1862. Discharged on account of youth and feeble development, Lynchburg, VA, 10/13/1862. CSR has discharge record. Died 1919, buried Oakland Cem, Atlanta. Obit in Atlanta Constitution, 2/14/1919. Widow, Malinda L. Thompson, filed CPA, Fulton Co., 1920.
Tony, Charles W. (or Toney) - Musician. Transferred to Co. K. Roll dated 11/7/1864, shows "absent without leave since 8/1/1863." Listed as drummer in CSR. Union records show captured at Macon, GA, Apr/1865. Filed CPA, DeKalb Co., 1900. Died 4/6/1915, Fulton Co. BSU.
Tony, James (or Toney) - Fifer 9/26/1861. Discharged on account of hernia, 10/15/1862. CSR contains discharge record, 58 yrs old at discharge. BSU.
Tweedle, M. J. (Marion J. Tweedell) - Pvt 5/15/1862. Appt 3d Sgt Jan/1864. W Winchester, VA, 9/19/1864. In hos, W, close of war. Filed CPA, DeKalb Co., 1894. D 8/27/1907, DeKalb Co. BSU.
Vaughn, J. S. (Joel S. Vaughan) - Pvt 5/15/1862. Severely wounded (multiple gunshot wounds) Gaines Mill, VA 6/27/1862. Discharged on account of wounds Mar/1863. Filed CPA, DeKalb Co., 1890. Died 8/16/1897, in AL. BSU.
Vaughn, Robert L. - Pvt 9/26/1861. Discharged prior to Jan/1862. Post war roster reads "Died at Savannah." BSU.
Vaughn, W. T. (William "Bill" T. Vaughn) - Pvt 3/25/1862. S Appomattox, VA, 4/9/1865. Had both hands blow off after the war while blasting near Atlanta." D 2/10/1887, bur Prospect Cem, Roswell Junction, Fulton Co. Death notice in Atlanta Constitution newspaper 2/11/1887.
Victory, Middleton - Pvt 9/26/1861. Discharged, for old age and disability, 12/31/1861. BSU.
Wade, Enoch D. - Pvt 9/26/1861. Surrendered Appomattox, VA, 4/9/1865. Filed CPA, Chattooga Co., 1888. BSU.
Wade, F. M. (Franklin "Frank" M. Wade) - Pvt 3/4/1864. Sick in Lynchburg, VA, hospital June 18-8/31, 1864. CPA shows left command on sick furlough, Nov/1864. (Born DeKalb Co., Oct/1844) Filed CPA, Fulton Co., 1902. BSU.
Wade, George W. (George Washington Wade) - Musician 9/26/1861. Surrendered Appomattox, VA 4/9/1865. Listed as a Drummer in CSR. Filed CPA, Marshall Co. AL, 1901. Died 4/7/1906, buried Union Cemetery, Marshall Co., AL.
Walker, James M. - 4th Cpl 9/26/1861. Discharged 6/20/1862. BSU.
Ward, William O. - 2d Cpl 9/26/1861. Died in 2d GA Hos, GA Apr. 28 or 30, 1862, of typhoid fever. Augusta Chronicle dated 5/1/1862. Buried Magnolia Cem, Confederate Square, Augusta, GA.
Wheeler, Amos - Pvt 9/26/1861. Killed Spotsylvania, VA, 5/12/1864. BSU.
Wiggins, Ellis W. - Pvt 5/10/1862. K Antietam, MD, 9/17/1862. CSR has death claim by widow, Leanna P. Wiggins, DeKalb Co. Widow, R. B. Guffin (or Griffin) filed CPA, Fulton Co., 1920.
Wiggins, George W. - Pvt 3/8/1862. W in 1863. C Wilderness, VA 5/6/1864. Released Elmira, NY, 6/30/1865. D 10/31/1905. Widow, Unisa J. McWilliams Wiggins, filed CPA, 1920, DeKalb Co. BSU.
Wiggins, Jackson C. - Pvt 9/26/1861. Elected 2d Lt 10/24/1862. K Cold Harbor, VA 6/1/1864. BSU.

Wiggins, James M. - Pvt 5/8/1862. Appt Cpl. Surrendered Appomattox, VA, 4/9/1865. (Born DeKalb Co., 12/28/1842. Died at Confederate Soldiers Home, Atlanta, 3/19/1931) Filed CPA, Fulton Co., 1925. Buried Clifton Church Cemetery, Atlanta.
Wiggins, M. O. (Mathew O. Wiggins) - Pvt 3/8/1862. Wounded in shoulder, reported in Augusta Chronicle, 11/4/1863, and disabled, Cedar Creek, VA, 10/19/1864. Absent, wounded, 11/7/1864. Filed CPA, DeKalb Co., 1890. (Born Henry Co., 5/10/1843. Died Confederate Soldiers' Home, Atlanta, 2/8/1926. Buried Macedonia Baptist Church, DeKalb Co)
Wiggins, R. W. (Robert W. Wiggins) - Pvt 3/8/1862. Killed Ft. Stedman, VA, 3/25/1865. Capt William C. Mathews recalled, "Robt. Wiggins was conspicuous for gallantry in every engagement and was killed at Ft. Hare." Ft. Hare was also known as Ft. Stedman. BSU.
Wilkins, James B. - Pvt 9/26/1861. Appt 3d Cpl 7/2/1863. C near Flat Rock, GA, 7/27/1864. E Boulware's Wharf, James River, VA, 3/27/1865. Admitted to Jackson Hos Rich, VA with chronic diarrhea 3/28/1865, turned over to Provost Marshal, 4/14/1865. D 7/31/1895. Bur Martin's Chapel United Meth. Church, Gwinnett Co. Widow, Mary Malinda Wilkins filed CPA, Gwinnett Co., 1911.
Wilson, Aaron Jordan - Pvt 9/26/1861. W Gaines Mill, VA 6/27/1862. D of wounds 6/30/1862. CSR contains a death claim filed by father, William A. Wilson, DeKalb Co. BSU.
Wilson, Benjamin L. - Pvt 9/26/1861. Killed Fredericksburg., VA, 12/13/1862. CSR record states "Killed at Fredericksburg, VA, 5/4/1863, during battle of Chancellorsville. CSR has death claim filed by father, William A. Wilson, DeKalb Co. BSU.
Wilson, J. H. - Pvt 5/6/1862. Killed at Gettysburg, PA, 7/2/1863. BSU.
Wood, William B. (or William R.) - Pvt 9/26/1861. Absent, sick, 8/20/1864 - 11/7/1864. NLR BSU.
Wright, William - Cpt 9/26/1861. Resigned, disability, 9/13/1862. Resignation accepted 10/13/1862. Died 3/28/1889. Buried William Wright Family Cemetery, DeKalb Co. Obit in Atlanta Constitution, page 17, 3/31/1889. Widow, Emma Gentry Wright, filed CPA, Fulton Co., 1915.
Wright, William A. (William Aron Wright) - Pvt 9/26/1861. Appt 5th Sgt Aug 1862. Wounded 2d Manassas, VA, 8/30/1862. Captured Cold Harbor, VA, 6/1/1864. Released Elmira, NY, 6/16/1865. (Born in Henry Co., 1/15/1846) Filed CPA, DeKalb Co., 1909. Died there 11/23/1925, buried Greenwood Cemetery, Fulton Co.

COMPANY L, CHESTATEE ARTILLERY, FORSYTH AND DAWSON COUNTIES, GA

This company was first raised and organized by Captain Thomas H. Bomar. It was commissioned Chestatee Artillery 9/30/1861, by order of War Dept. and was ordered by Wright's Legion to report to Camp Kirkpatrick, Dawson Co., GA It was mustered into Confederate State service as a Light Artillery Company, 10/9/1861. It was originally designated Co. H, and was later known as the New Co. N. The Chestatee Artillery under Capt Bomar served for some time near Charleston, S.C., taking a prominent part in the defense of Battery Bee and Morris Island. On 5/5/1864, the company was ordered to re-join the 38th GA Infantry Regt and serve as infantry, in the Army of Northern Virginia, where it did good service to the end. Note: Chestatee is a small town in northeastern Forsyth Co., Georgia, and many members of this company came from Forsyth Co.

Andoe, Theodore M. (or Theobold M.) - Pvt 10/13/1861. Appt 1st Cpl Oct/1862. Wounded Cold Harbor, VA, 6/1/1864. Died from wounds 9/4/1864. BSU.
Arthur, James J - (or James S. Arthur) Pvt 2/18/1863 Enlisted Atlanta, Captured Winchester 9/19/1864, sent to Point Lookout. Took OOA and joined U.S. service, 10/17/1864. Joined Company D, 2nd US Volunteers, stationed at Ft. Leavenworth, KS in 1866. BSU.
Baxter, Jasper Joseph - Pvt 11/13/1861. Died 1/14/1863. BSU.
Bell, Hugh L. - Pvt 10/13/1861. Sick in Springfield Hospital, Aug/1862. Discharged 1862. BSU.
Bennett, Elijah B. - Pvt 10/13/1861. Surrendered Appomattox, VA, 4/9/1865. BSU.
Bennett, J. Masser - (Jessie Messiah Bennett) Pvt 2/1/1863. Absent, sick, May 18 - Aug 31, 1864. Died 3/5/1904. Buried Phillips Family Cemetery, Forsyth Co, remains relocated in 1950's, to Shady Grove Baptist Church Cemetery, Forsyth Co.

Bennett, William J. - Pvt 5/10/1862. Wounded and captured Monocacy, MD, 7/9/1864. Exchanged at Point Lookout, MD, 10/30/1864. Filed CPA, Dawson Co., 1888. Born 12/21/1844, died 4/30/1892. Buried Crossroads Baptist Church Cemetery, Mars Hill Crossroads, Forsyth Co.
Bolton, James S. (James Samuel Bolton) - Pvt 10/13/1861. Surrendered Appomattox, VA, 4/9/1865. Born 5/21/1835, died 5/15/1888, buried Coal Mountain Baptist Cemetery, Forsyth Co. Widow, Elender C. Martin Bolton, filed a CPA in Forsyth Co., 1920.
Bomar, Thomas H. - (Thomas Hayne Bomar) Cpt in Co L, 10/13/1861. Elected Major 7/2/1863. C Cedar Creek, VA, 10/19/1864. Released Ft. Delaware, 7/24/1865. Commanded the 38th Regt at Battle of Monocacy. Died 3/13/1927, buried Evergreen Almeda Cemetery, El Paso, TX. CSR has 211 pages.
Booker, O. P. (or O. B.) (Obediah P. Booker) - Pvt 1/28/1863. S Appomattox, VA, 4/9/1865. BSU.
Bowen, Francis Marion (or Borin) (Francis M. Boren) - Pvt 2/3/1863. W and C Monocacy, MD 7/9/1864. P at PLO, MD 10/30/1864. Rcv Venus Point, Sav River, GA 11/15/1864. Filed CPA Forsyth Co., 1894. B. 8/27/1842, Coweta Co., GA. Name spelled as "Boren" in CPA. BSU.
Boyd, John A. - Pvt 5/10/1862. Sick in hos 8/25 - 31, 1864. D 7/9/1904, DeKalb Co. Bur Midway United Meth. Church, Barrow Co. Widow, Isabella Barrett Boyd filed CPA, Fulton Co., 1904.
Boyd, Robert - (Robert Asbury Boyd) Pvt 5/10/1862. Furloughed home for sickness. AWOL May 1 - Aug 31, 1864. Filed CPA, DeKalb Co., AL. Died, buried Ft. Payne Cemetery, DeKalb Co., AL.
Brice, Daniel Pinkney - Pvt 12/15/1861. Roll dated 9/15/1864, last on file, shows present. Died in 1864. CSR shows he was wounded and admitted to Petersburg Hospital on 3/25/1865, died Mar. 27, probably mortally wounded in the attack on Ft. Stedman, 3/25/1865. BSU.
Brice, John R. (John Ratliff Brice) - Pvt 10/13/1861. Appt 7th Cpl in 1863; 4th Cpl in 1864. C Winchester VA 9/19/1864. Released PLO, MD, 5/12/1865. (Resident GA since 1841) Filed CPA, Hall Co., 1910. D 9/25/1935. Buried Alta Vista Cemetery, Gainesville. Marker reads: 38 GA INF.
Brice, Thomas J. - Pvt 9/1/1862. Absent without leave May 14 - Aug 31, 1864. Died 1/17/1922. Buried Liberty Baptist Church, Forsyth Co.
Brittain, John C. - (or Britton) Enlisted as Pvt Co. A, Phillips' Legion GA Cavalry 9/25/1861. Trn to Co. L, 38th Regt. 4/30/1863. Roll dated 8/31/1863, shows sick in camp. CPA show wounded near Richmond, VA, and furloughed home Sept/1863. Elected Sheriff while on furlough (Died Gainesville, GA, 1912) Filed CPA, Lumpkin Co., 1904. Buried Alta Vista Cemetery, Gainesville.
Brown, John C. - Pvt 5/15/1862. Absent without leave Feb/1863. BSU.
Burns, George W. (George Washington Burns) - Enlisted as Pvt, Co. H, 1st Regt GA Inf. (Ramsey's) 7/24/1861. Mustered out Augusta, GA 3/18/1862. Enlisted as Pvt in Co. L, 38th Regt GA Inf. 5/15/1862. Trn to Co. I, in 1862. Wounded Gaines Mill, VA 6/27/1862. Captured Spotsylvania, VA 5/12/1864. P at Point Lookout, MD, Feb/1865. Rcv Boulware & Cox's Wharves, James River, VA, for E, 3/12/1865. At home on frl end of war. Filed CPA Lumpkin Co., 1898. Died 2/19/1924, buried Mt. Hope Cemetery, Dahlonega. Widow, Harriet Burns, filed CPA Lumpkin Co., 1924.
Campbell, George S. - Pvt 10/13/1861. On special service June 26 - Sept 15. CSR contains discharge record. BSU.
Campbell, Seaborn - Artificer (mechanic) 10/13/1861. Appears last on roll for Dec/1861. BSU.
Cannon, Moses C. - (Moses Columbus Cannon) 6th Sgt 10/13/1861. Appt 5th Sgt Jan/1862; 4th Sgt 1864. AWOL Apr. 1 - Sept. 15, 1864. Died 8/24/1901. Buried Liberty Bap. Church Cem, Forsyth Co.,
Cantrell, Samuel - Pvt May/1863. Roll dated 9/15/1864, LOF, shows "Absent, sick in Mount Vernon, VA, hospital since 7/20/1864." Discharged Oct/1864. BSU.
Cantrell, William F. - Pvt 5/10/1862. C Milford Station, VA, 5/21/1864. P at PLO, MD and trn to Aiken's Landing, VA, for E 3/17/1865. Rcv Boulware & Cox's Wharves, James River, VA, 3/19/1865. Filed CPA, Dawson Co., 1898. D 7/16/1912, bur Liberty Bap. Church Cem, Forsyth Co.
Carver, William - Pvt 8/23/1862. Roll dated 9/15/1864, LOF, shows sick frl, Aug 1-31, 1864. BSU.
Charlton, James R. - Pvt 5/10/1862. Deserted 6/14/1862. BSU.
Cheek, Sylvester - Pvt 6/13/1863. Absent without leave, Mar. 1 - 9/15/1864. B 6/3/1845, Died 1/22/1938. Buried Pleasant Ridge Cemetery, or Patton Cemetery, Marion Co., AR.

Childers, Abraham O. - Pvt 2/1/1863. Roll dated 9/15/1864, LOF, shows on frl Aug. 18 -31, 1863. CPA shows home, sick frl to end of war. (B Laurens, SC 1831) Filed CPA, Gordon Co., 1898. B 1831, in SC. Bur Sugar Valley Bap. Chur Cem, Gordon Co. CSA marker reads: Co N 38 GA INF.
Childers, John, Sr. - 3d Sgt 10/13/1861. Roll dated 9/15/1864, LOF, shows present. BSU.
Childers, John, Jr. - Pvt 10/10/1861. Received by Army of Potomac, a Confederate deserter, 3/26/1865. Sent to City Point, VA, 3/27/1865. BSU.
Childers, Samuel - Pvt 10/13/1861. Discharged, disability, right arm useless, near New Market, VA, 10/9/1864. CSR contains discharge record. BSU.
Childers, William - Pvt 10/13/1861. Discharged 8/14/1862. CSR has discharge, 64 years old. BSU.
Childers, William Jackson - Pvt 10/13/1861. W through both hips and C Cedar Creek, VA 10/19/1864. P at PLO, MD, E 2/10/1865. At Camp Lee near Rich, VA 2/27/1865. (Resident GA, 1848) Filed CPA, Forsyth Co., 1888. D 10/12/1931, bur there Salem Bap. Church. B 3/30/1844,
Clarke, David (David Clark) - Pvt 10/13/1861. Discharged 5/25/1862. CSR contains discharge record, 54 years old. Bur Concord Bap. Church, Forsyth Co., GA
Cogburn, George W. - Pvt 3/6/1862. On sick frl 1864. D 1864. BSU.
Condon, W. D. (William D. Condon) - Enlisted as Pvt in Croft's Battery GA Light Artillery 3/11/1862. Transferred to Co. L, 38th Regt GA Inf., 1/1/1863. Roll dated 9/15/1864, last on file, shows present. Died 1/29/1908, buried Woodlawn Cemetery, Marion Co., FL. Widow, Anna Dukes Condon, filed CPA, Marion Co., FL, 1909.
Cowen, William A. (or Cowan) - Pvt, 9/26/1861. Died Savannah, GA, no further record. BSU.
Crane, Solomon F. - Pvt 10/13/1861. Elected Jr. 2d Lt 1/1/1864; 2d Lt 2/20/1864. Killed at Cold Harbor, VA 6/1/1864. BSU.
Crane, William H. (William Harper Crane) - Enlisted Co. N, 2nd NC Battalion, trans to 38th GA 1863. Pvt 1/3/1863. Captured Cedar Creek, VA 10/19/1864. Paroled at Point Lookout, MD, transferred to Aiken's Landing, VA, for exchange 3/17/1865. Rcv Boulware & Cox's Wharves, James River, VA, 3/19/1865. Filed CPA, Hall Co., 1900. D 1/4/1927 Hall Co. BSU.
Crawford, William A. (William Asbury Crawford) - Pvt 10/13/1861. AWOL June 17 - 8/31/1863. B 12/23/1840, died 11/9/1926. Buried at Juno Baptist Church Cemetery, Dawson Co.
Crow, Isaac - Pvt 10/13/1861. Roll dated 9/15/1864, LOF, shows AWOL since 8/12/1863. (B Campbell Co., GA 1832. Born 6/4/1832, died 1/6/1908, Morgan Co.) Filed CPA, Forsyth Co., 1905. Buried Sugar Creek Baptist Church Cemetery, Morgan Co.
Dacus, James - Pvt Aug/1862. Discharged by civil authority, 11/12/1862. BSU.
Dacus, James A. (James Authur Dacus) - Pvt 10/10/1863. C Winchester, VA, 8/12/1864. Released Camp Chase, OH, 5/13/1865. D 11/17/1915. Filed CPA, McMinn Co., TN. Pension # S12909. Bur Cedar Grove Cem, McMinn Co., TN. Twin of William R. Dacus of same company.
Dacus, John (John W. Dacus) - Pvt 10/13/1861. D Sav, GA 7/8/1862. CSR contains death claim filed by father, Arthur Dacus, Forsyth Co. Bur Laurel Grove Cem, Lot 892, Sav, marker reads GA ARTY.
Dacus, Paschal Hawkins ("Hawk") - Pvt 5/10/1862. C Fisher's Hill, VA 9/19/1864. D PLO, MD 3/10/1865. Bur PLO Confed. Cem, MD, Widow, Sarah Dacus, filed CPA, Forsyth Co., 1911.
Dacus, William Ryal (Riley W. Dacus) - Pvt 10/13/1861. Discharged by civil authority Sav, 11/12/1862. D 11/20/1906. Bur Cedar Grove Cem, McMinn Co., TN. Marker reads Riley W. Dacus.
Daniel H. Thomas - Pvt 5/10/1862. Roll dated 11/15/1864, last on file, shows absent without leave since 5/14/1864. BSU.
Daniel, J. W. - (James W. Daniel) Pvt 2/1/1863. Deserted. Took OOA to U.S. Govt. Chattanooga, TN. 4/7/1864. D 5/27/1902. Bur New Hope Meth. Church, Forsyth Co.
Dean, Edward C. - Pvt 11/1/1862. Det Hos Steward 1862. Transferred to Co. L, 25th Battn. GA Provost Guard Inf. Elected 1st Lt of Co. 5th Regt GA Reserve Inf. S Greensboro, NC 4/26/1865. D 7/10/1909 Norcross, Gwinnett Co. Widow, E. C. Dean, filed CPA, Fulton Co., 1919. BSU.
Dean, Thomas W.(Thomas Webster Dean) - Pvt 10/12/1861. Appt 2d Sgt Jan/1862. Roll for 8/21/1862, shows present, B 11/18/1843, D 10/19/1874. Bur Peachtree Memorial Cem, Gwinnett Co.

Dobbs, W. C. (William Calvin Dobbs) Pvt 9/8/1862. W in 1864. AWOL May 1 - Aug. 31, 1864. Other records show paid Atlanta, 5/11/1864. D Forsyth Co., 1900. BSU.
Dunlap, Tyler - Pvt 2/4/1863. Roll dated 9/15/1864, LOF, shows present. BSU.
Dunn, James - Pvt 5/10/1862. Deserted 6/14/1862. BSU.
Elliott, W. H. (William Hampton Elliott) - Pvt 5/13/1863. Sick in Mount Jackson, VA hos 8/31/1864. CPA shows home ill with chills, fever, bronchitis, end of war. (B GA 3/5/1845) Filed CPA, Dawson Co. D 10/14/1921. Bur Dawsonville Cem, Dawson Co. CSA marker reads: CO N 38 GA INF.
Elrod, Charles M.- Pvt 2/1/1863. Wounded Cold Harbor, VA, 6/1/1864. Roll 9/15/1864, last on file, shows absent, sick. BSU.
Fisher, Frederick - Musician 4/29/1862. Transferred to Co. D, 32d Regt GA Inf. 11/1/1862. Det Regtal bugler. BSU.
Freeman, John - Pvt 10/13/1861. Captured Winchester, VA, 9/19/1864. BSU.
Garrett, Andrew Jackson - Pvt 10/13/1861. Appt 6th Cpl 11/1862; 3d Cpl 1864. W & C Frederick, MD, 7/10/1864. P at PLO, MD, 1864. Rcv Venus Point, Sav River, GA for E, 11/15/1864. (B Forsyth Co., 8/30/1842) Filed CPA, Hall Co., 1905. D 7/5/1926, bur Alta Vista Cem, Hall Co. GA.
Garrett, Benjamin Franklin - Pvt 3/1/1863. Deserted. Rcv by Provost Marshal General, Army of Potomac, Confed. deserter, date not given. Sent to Cpt Potter, Provost Marshal, City Point, VA, 3/27/1865. Filed CPA, Dawson Co., 1889. D 6/13/1905, bur Alta Vista Cem, Gainesville. Widow, Samantha Garrett, filed CPA, Hall Co., 1906.
Garrett, George W. - Pvt 12/24/1861. K Cold Harbor, VA 6/1/1864. BSU.
Garrett, Hosea - Pvt 3/3/1862. Roll dated 11/15/1864, LOF, shows "AWOL since 4/1/1864. Left on sick frl ." On sick frl end of war. (B Franklin Co. 1835) Filed CPA, Hall Co., 1897. D aft. 1903. BSU.
Gibson, John B. - Pvt 5/10/1862. W Cold Harbor, VA 6/1/1864. frl for 30 days from Jackson Hos Rich, VA 6/15/1864. Roll dated 11/15/1864, LOF, shows absent, sick since 8/31/1864. Bio in History & Bio. Record of North & West TX. D 3/19/1918, Montague Co., TX, bur there Salona Cem.
Green, John (John William Green or Greene) - 2d Sgt 10/13/1861. Appt 3d Sgt 2/15/1863. Captured Ft. Stedman, VA, 4/25/1865. Released Point Lookout, MD, 6/27/1865. Died 11/27/1880, buried Beech Grove Cemetery, Greene Co., AR. Marker reads: SGT CO N, 38 GA INF. Widow, Manerva Greene, filed CPA, Greene Co., AR, 1913.
Green, Roger - Pvt 10/13/1861. Discharged 1/8/1862. CSR has discharge record, 63 years old. BSU.
Hall, William P. - Pvt 11/8/1863. C Guinea Station, VA 5/21/1864. At PLO, MD, 8/31/1864. D 1864. Bur PLO Confed. Cem., MD. Not listed on CSA monument at cem, but many names missing.
Hardeman, A. J. (Andrew Jackson Hardeman) - Pvt 8/30/1862. Roll dated 9/15/1864, LOF, shows AWOL since 2/13/1864. D 5/18/1912, bur Salem Bap. Church Cem, Forsyth Co. Widow, Norey Annie Hutchinson (or Hutchins) Hardeman, filed CPA, DeKalb Co., 1925.
Hardin, B. F. - (Richard Taulbert Hardin) Pvt 9/20/1862. Roll dated 9/15/1864, LOF, shows AWOL since 2/1/1864. Union POW records state Richard T. Hardin took OOA to U.S. Govt., Chattanooga, TN 4/7/1865. Filed a CPA, Dawson Co., 1901. Bur there Bethel United Meth. Church. D 7/24/1911.
Harris, David J.- Pvt 2/7/1863. Died Hardeville, SC, hospital, Oct/1863. BSU.
Henderson, Ezekiel L. - Pvt 10/13/1861. W Cold Harbor, VA 6/1/1864. Roll dated 9/15/1864, LOF, present. (B Forsyth Co., 7/22/1842) Filed CPA, Hall Co., 1903. Bur there Zion Hill Bapt. Church.
Henderson, R. Daniel (Richard Daniel Henderson) - Pvt 5/10/1862. D Sav, GA 6/23/1862. D of fever. CSR contains death claim filed by widow, Julia Anna Henderson, Forsyth Co. BSU.
Hendrix, John C. (John Chappell Hendrix) - 2d Lt 10/13/1861. Elected 2d Lt 4/19/1862. Resigned Jan/1863. Dropped from rolls Aug/1863, General Order #96, A. & I.G.O., for being absent without leave Jan 19 - Aug/1863. Left camp by order General Mercer, Special Order #19, 1/19/1863. Died 9/10/1917. Buried Oakland Cemetery, Atlanta.

Hendrix, Prater - Pvt 12/18/1861. Discharged for injured shoulder/maimed hand 6/6/1862. Died 9/13/1893, Dawson Co., buried there Bethel United Methodist Church Cemetery. Widow, Sara M. Hendrix filed CPA, Dawson Co., 1901.
Hendrix, William - 1st Sgt 10/13/1861. Elected Jr. Lt 4/19/1862; 2d Lt Aug/1863; 1st Lt 1864; Cpt 6/1/1864. Wounded Cold Harbor, VA, 6/1/1864. Surrendered Appomattox, VA, 4/9/1865. BSU.
Hubbard, Early - Pvt 10/1/1862. Wounded Lynchburg, VA, 6/20/1864. Roll 9/15/1864, last on file, shows absent, wounded. BSU.
Hubbard, Joseph O. - Pvt 10/13/1861. Roll on 9/15/1864, last one file, shows absent without leave, in North Georgia since 6/1/1864. BSU.
Hubbard, Swancey - Pvt 10/13/1861. Discharged 8/22/1862. CSR contains discharge. D 7/11/1897, Stephenville, Erath Co., TX. D by suicide; article in Ft. Worth Morning Register, page 1, 7/14/1897. Bur there East End Cem. Widow, Melinda A. Hood Hubbard filed CPA, Erath Co., TX, 1899.
Hubbard, William W. (William Wesley Hubbard) - Pvt 10/1/1862. C Petersburg, VA, 4/3/1865. Released Hart's Island, NY Harbor, 6/15/1865. D. 2/13/1898, Dawson Co., bur Mount Vernon Bap. Church, Dawson Co. Widow, Cintha E. Hubbard, filed CPA, Dawson Co., 1911.
Hutchins, Drury B. - Pvt 8/30/1862. Roll dated 9/15/1864, LOF, shows AWOL since 3/1/1864. D 12/24/1891. Bur Salem Bap. Cem., Forsyth Co. Nancy J. Hutchins, widow, filed CPA, Hall Co., 1910.
Johnson, Samuel C. - Pvt 3/4/1862. Elected 1st Lt 5/6/1862; Cpt 9/2/1862. Resigned, disability, 9/2/1862. CSR contains resignation letter, resigned due to poor health. BSU.
Jones, Joseph L. - Pvt 3/3/1862. Wounded in thigh & captured Winchester, VA. 9/19/1864. Paroled at Point Lookout, MD and exchanged 10/30/1864. Received Venus Point, Savannah, GA, 11/15/1864. CPA shows furloughed home and unable for further duty. (Born Spartanburg Co., SC Jun/1831) Filed CPA, Fulton Co., 1895. Died 10/29/1911, buried Salem Baptist Church Cemetery, Forsyth Co.
Jordan, Harden (Harden A. Jordan) - 5th Sgt 10/13/1861. Appt 4th Sgt Jan/1862. Det guidon 2/15/1863. Captured Winchester, VA, 9/19/1864. E 10/30/1864. Filed CPA, Forsyth Co., 1890. CPA states he died 1904. Buried Mt. Zion Cemetery, Forsyth Co., maker reads: SGT CO N 38 GA INF.
Kitchens, S. B. (Stephen B. Kitchens) - Pvt 2/1/1863. Roll 9/15/1864, last on file, shows absent without leave since 3/14/1864. BSU.
Lane, James F. - Pvt 5/10/1862. Appt 6th Cpl Nov/1862; 3d Sgt 1864. Roll 9/15/1864, last on file, shows present. No later record. BSU.
Light, Pleasant Green - Pvt 3/1/1862. Appt 1st Cpl 1864. Roll dated 9/15/1864, last on file, shows present. No later record. Died 6/24/1889, Greene Co., AR. Buried there Mt. Zion Cemetery.
Loggins, Irwin - Pvt 5/10/1862. Wounded, arm disabled, in VA, 1864. BSU.
Maloney, William M. - Pvt 5/10/1862. Deserted 6/14/1862. BSU.
Martin, James H. - Pvt 2/1/1863. Roll dated 9/15/1864, LOF, shows AWOL since 3/1/1864. NLR. Bur Wolffork Bap. Church Cem, Rabun Co., Marker reads: CO G 38 GA INF, CSA, no dates.
Mason, Bluford - Pvt 7/1/1862. Deserted at Cold Harbor, VA, 5/6/1864. Entered US Lines at Altoona 7/26/1864. Took oath of allegiance to U.S. Govt. at Louisville, KY, 8/3/1864. Sent from Department of Cumberland for release north of Ohio River, 8/4/1864. BSU.
Mason, Merrell - Pvt 3/4/1862. Roll dated 9/15/1864, LOF, shows sick Lynchburg, VA, since Jun/1864. At CSA General Hos, Charlottesville VA, 4/25/1864 returned to duty 9/28/1864. BSU.
Massingill, Bailey D. - (Bailey David Massingill) Pvt 11/1/1861. Roll dated 9/15/1864, LOF, shows sick in hos since 5/22/1864. Filed CPA, Floyd Co., 1910. (D 9/12/1912) Bur Salmon Cem, Floyd Co.
McDaniel, E. B. (Egbert Baldwin McDaniel) - Pvt 4/30/1864. Captured Cold Harbor, VA, 6/3/1864. Paroled Elmira, NY, sent to James River, VA for exchange 3/14/1865. Received Boulware & Cox's Wharves, James River, VA, Mar. 18-21, 1865. Died 7/30/1894. (Buried at Oakland Cemetery Atlanta) Obit in Atlanta Constitution, 7/31/1894.
McDaniel, Ira O. (Ira Oliver McDaniel Jr.) - Pvt 5/10/1862. Detailed in Commissary Dept. Savannah, GA in 1862. Elected Jr. 2d Lt 2/20/1864 ; 1st Lt 6/1/1864. Captured Frederick, MD, 10/26/1864. Exchanged 10/31/1864. Rcv Venus Point, Sav River, 11/15/1864. (Buried Oakland Cemetery, Atlanta) Died 1/17/1907. Obituary in Atlanta Constitution, 1/19/1907.

McDaniel, Phillip S. (Phillip Sanders McDaniel) - Pvt 11/21/1861. Appt 2d Sgt 1863. Killed at Monocacy, MD, 7/9/1864. BSU.
McDaniel, R. S. - Pvt Jan/1862. Appointed Ordnance Sgt 1862. Full name unknown. BSU.
McKinney, James O. - 1st Cpl 10/13/1861. Reduced to ranks Oct/1862. Appt 6th Cpl 1864. Captured Swift Creek, VA, 4/3/1865. Released Hart's Island, NY Harbor, 6/15/1865. BSU.
McKinney, William L. - Pvt 2/1/1863. Captured Winchester, VA, 9/19/1864. Took oath of allegiance to U.S. Govt., enlisted in US service 10/17/1864. Enlisted Company D, 2nd U.S. Volunteer Infantry. Died 1928. Buried at Pleasant Hill Baptist Church Cemetery, Hall Co.
Mendez, F. Augustus - Musician 12/29/1861. Trn to Regimental band 5/13/1862; Trn to Maxwell's Battery GA Light Artillery. Roll for Jun/1864, shows present. Listed in CSR as a Bugler. BSU.
Miller, John - Pvt 2/1/1863. Captured Milford Station, VA, 5/21/1864. Paroled at Point Lookout, MD, Trn to Aiken's' Landing, VA and E, 9/15/1864. Rcv Varina, VA 9/22/1864. D Sept/1864. BSU.
Monk, Robert - Pvt 7/14/1863. Roll dated 9/15/1864, LOF, shows present. NLR. BSU.
Mooney, Robert A. - 4th Sgt 10/13/1861. Appt 3d Sgt Jan/1862. C, details not noted. Took OOA to U.S. Govt. Chattanooga, TN 4/7/1864. (B Hall Co., 1834) Filed CPA, Gwinnett Co., 1906. BSU.
Mooney, James H. - Pvt 12/20/1861. Wounded in foot 1862. Discharged by civil authority, on writ of habeas corpus, 11/23/1862. BSU.
Mooney, William Oliver - Pvt 10/13/1861. Died at Savannah, GA 12/16/1862. CSR contains death claim filed by mother Levnia Mooney, Forsyth Co. BSU.
Morgan, Henry S. - Pvt May/1862. Admitted to Charlottesville, VA hospital 9/1/1864. Transferred to Lynchburg, VA, 9/27/1864. Captured Petersburg, VA, 3/25/1865. P 1865. Filed CPA, Forsyth Co., 1903. Died 1/30/1923, bur Penia Baptist Church Cemetery, Crisp Co.
Morgan, Simeon - Pvt 5/10/1862. Killed at Cold Harbor, VA, 6/1/1864. CSR states killed at Turkey Ridge, Cold Harbor. BSU.
Morgan, William R. - Pvt 10/13/1861. Died 7/8/1862. CSR contains death claim filed by widow, Rachel Morgan, DeKalb Co. Buried Laurel Grove Cem, Lot 858, Sav, Marker reads, Chestatee Arty.
Murphy, Michael - Pvt 5/10/1862. Deserted 6/14/1862. BSU.
Neal, William F. - Pvt 8/22/1862. P Farmville, VA, Apr. 11-21, 1865. BSU.
Norrell, John - 4th Cpl 10/13/1861. Appt 1st Sgt 4/19/1862. Roll dated 9/15/1864, LOF, shows AWOL since 4/1/1864. On sick frl. D 2/9/1892, bur Coal Mountain Bap. Church Cem, Forsyth Co.
Oliver, Lindsey D. - Pvt 9/1/1862. Admitted to C. S. A. General Military Hos #4, Wilmington, NC, with pulmonary bronchitis, 2/12/1865. Sent to General Hos #6, 2/20/1865. BSU.
O'Shields, David - Pvt 12/20/1861. Surrendered at Appomattox, VA, 4/9/1865. Died 1885, buried Harmony Baptist Church Cemetery, Dawson Co.
O'Shields, Hiram - Pvt 8/31/1862. Captured Milford Station, VA 5/21/1864. Paroled at Point Lookout, MD, and transferred to Aiken's Landing, VA, for exchange 9/18/1864. Rcv Varina, VA, 9/22/1864. Admitted Jackson Hospital Richmond, VA, for pneumonia 9/22/1864, furloughed 30 days 9/26/1864. (Resident GA since 8/15/1845) Filed CPA, Cobb Co., 1888. BSU.
O'Shields, Pinkney - Pvt 10/13/1861. Roll for 8/31/1863, shows him present. No later record. BSU.
Owen, A. J. - Pvt 8/30/1862. Sick in Charleston hospital Farmville, VA. Died there 4/9/1865. Buried there, specific grave site unknown. BSU.
Owen, Francis Marion - Pvt 3/10/1862. C Winchester, VA, 9/19/1864. Arrived PLO, MD, 11/18/1864. Admitted to Jackson Hos Rich, VA, 2/16/1865, frl for 60 days 3/9/1865. (B Forsyth Co., 10/14/1841) Filed CPA, Hall Co., 1903. D 5/3/1920, bur Salem Bap. Church, Chestatee, Forsyth Co.
Owen, George W. - Pvt 12/26/1861. Roll dated 9/15/1864, LOF, shows AWOL since 4/1/1864. NLR. Filed CPA, Forsyth Co., 1919. D 4/21/1931, bur Salem Bap. Church, Chestatee, Forsyth Co.
Owen, Isaac Newton - Pvt 3/15/1862. Wounded Hatcher's Run, VA, 2/5/1865. Widow, Ellen Owen Massengale, filed CPA, Jackson Co., 1916. She stated in CPA he died March/1865. BSU.
Owen, Jesse P. - Pvt 10/13/1861. C Battle of North Anna, VA 5/23/1864. D of chronic diarrhea Elmira, NY, 8/23/1864. Bur Woodlawn National Cem, Elmira NY, listed as J. P. Owens, GA Art.
Owen, John H. (John Henderson Owen) - Pvt 5/30/1862. Roll. 8/31/1863, shows present. No later record. Discharged for heart dropsy, 8/26/1863. Died 9/20/1887. Buried Salem Baptist Church Cemetery, Forsyth Co. Widow, Fannie Owen, filed CPA, Forsyth Co., 1896.

Owen, Thomas A. - Pvt 5/10/1862. Captured at Cold Harbor, VA, 6/3/1864. Paroled at Point Lookout, MD and transferred to Aiken's Landing, VA for exchange 9/18/1864. Died 1864. BSU.
Owen, William A. - Pvt 10/13/1861. Roll for 8/31/1863, shows present. D 1864. BSU.
Owen, William Christopher - Pvt 5/10/1862. AWOL, 1862. In arrest Sullivan's Island, SC 8/30/1864. NLR. Filed CPA, Forsyth Co., 1889. D 11/30/1887. Bur Salem Bap. Chur, Forsyth Co.
Owen, William Jasper - Pvt 10/13/1861. Killed at Cold Harbor, VA, 6/1/1864. BSU.
Parting, Reuben (or Partin) - Pvt 2/1/1863. Captured Frederick, MD, 7/10/1864. Paroled as an invalid POW, Ft. McHenry, MD, 1864. Trn to Point Lookout, MD 9/27/1864. Exchanged 9/30/1864. At Camp Lee near Rich, VA, 10/10/1864. NLR. CSR states W Monocacy, shot in the face. BSU.
Patterson, Enoch - 5th Cpl 10/13/1861. Wounded, left leg permanently disabled, Cold Harbor, VA, 6/1/1864. At home on wounded furlough Jun/1864, to end of war. Filed CPA, Forsyth Co., 1912. Died 2/19/1929, buried there New Hope Methodist Church Cemetery, Forsyth Co.
Patterson, Hiram - Pvt 8/30/1862. Captured Winchester, VA 9/19/1864. Paroled at Point Lookout, MD, and trn to Aiken's Landing, VA, for exchange, 3/15/1865. Rcv Boulware & Cox's Wharves, James River, VA, 3/18/1865. (B GA) Filed CPA, Forsyth Co., 1911. Died 5/26/1921. Buried New Hope Methodist Church, Forsyth Co. Obit in Forsyth Co. News, 9/22/1921.
Patterson, John D. (John D. Bettis) - Pvt 10/13/1861. Roll dated 9/15/1864, LOF, shows AWOL since 6/15/1864. Filed CPA, Dawson Co., 1907. CPA shows name as John D. Patterson, but changed his name to John D. Bettis. Unknown date of name change. Buried Salem Baptist Church Cemetery, Forsyth Co. Grave marker reads: John D. Bettis, Born 9/5/1838, Died 12/24/1924.
Patterson, Joshua (Joshua B. Patterson) - Pvt 8/30/1862. Appt artificer 1862. Det mechanic Nov/1862. Roll dated 9/15/1864, LOF, shows present. D 3/6/1893. Bur New Hope United Meth. Church Cem, Forsyth Co. Widow, Darkus E. Cogburn Patterson, filed CPA, Forsyth Co., 1901.
Patterson, Samuel - Pvt 1/25/1863. AWOL May 14th - Sept. 4th, 1863. CPA shows accidentally W Dec/1864, frl home. Unable to rejoin command. (B Hall Co., 8/18/1826) Filed CPA, Hall Co., 1901. D there in 1916. Bur Chestatee Bapt. Church, Hall Co.
Perry, Charles T. - Pvt 8/26/1862. Captured Harrisonburg, VA, 9/25/1864. Released Ft. McHenry, MD, 5/10/1865. BSU.
Perry, W. H. (William Horton Perry) - Pvt 5/10/1862. Appt Musician Dec/1862. Died at Richmond, VA, 3/4/1865. Buried there Hollywood Cemetery, listed in CSR as Bugler.
Perryman, D. A. - Pvt 3/6/1863. Absent, sick, 6/19 - 8/31/1864. S Appomattox, VA 4/9/1865. BSU.
Pharr, J. T. (John Thomas Pharr) - Enlisted as a Pvt in Co. D, 1st Regt GA Regulars 3/1/1861. Trn to Maxwell's Battery. GA Light Artillery. 8/1862; to Co. L, 38th Regt GA Inf. 4/26/1864. Deserted, Petersburg, VA 3/23/1865. Rcv Washington, D. C., a Confed. deserter, 3/29/1865, took OOA to U.S. Govt., furnished trans to Harrisburg, PA 4/9/1865. Buried Louden Park Cemetery, Baltimore, MD.
Phillips, Edward W.- Pvt 11/1/1861. Appears last on roll for Feb/1863. BSU.
Phillips, James M. - Pvt 2/4/1863. Reported slightly W, head Cedar Creek, 10/19/1864. C, date and place not given. Released Harts Island, NY Harbor, 6/15/1865. Bur Haw Creek Bapt., Forsyth Co.
Pigeon, J. F. (or Pigean) - (Joseph F. Pegean) Pvt 5/10/1862. Elected Jr. 2d Lt 8/26/1863. Deserted. Took OOA to U.S. Govt. Chattanooga, TN 4/7/1864. CSR contains documents concerning the arrest of Lt J. F. Pigeon as a Yankee spy and ordering him be turned over to Gen. Winder, Provost Marshal Rich, VA. Record also shows he was charged with forging furloughs for other soldiers. BSU.
Porter, H. W. - Pvt 9/8/1862. D of chronic diarrhea, General Hos Staunton, VA, 9/18/1864. BSU.
Porter, Joseph - Pvt 1/27/1863. On sick frl 30 days from 4/15/1863. Filed, CPA, Forsyth Co., 1899. D Jun/1910. Bur there Salem Bapt. Cem. Widow, Elizabeth Porter, filed CPA, Forsyth Co., 1911.
Prater, A. P. (Alexander Preston Prater) - Pvt 12/1/1862. Deserted. Took oath of allegiance to U.S. Govt. Chattanooga, TN 4/7/1864. Died abt. 1882, Collin Co., TX. BSU.
Prater, Benjamin G. - Pvt 8/26/1862. Union records show deserted from Co. F, this Regt Charleston, SC 5/15/1864. C in Hall Co., 1864. Took OOA to U.S. Govt. Louisville, KY, released to remain north of Ohio River during war, 9/26/1864. D 11/17/1924. Bur Dayton Cem, Dayton, Washington Co.

Robinson, J. H. (Jolly H. Robinson) - Pvt 10/9/1862. Appears last on roll for Feb/1863. Transferred to Co. A, Phillips GA Legion, according to CSR. Deserted 1864, C in Hall Co., took the OOA, ordered to remain of the Ohio River. D 7/8/1917, bur Chelsea Cem, Rogers Co., OK.
Sanford, Raymond B. - Pvt 10/13/1861. Appt 6th Cpl Sep/1862. Discharged by civil authority 12/3/1862. Died 6/18/1864, buried Liberty Baptist Chur Cemetery, Dawson Co., CSA grave marker.
Sanford, Truman H. - 1st Lt 10/13/1861. Resigned 3/22/1862. Filed CPA, Forsyth Co., 1900. Died 11/24/1902. Buried Alta Vista Cemetery, Hall Co.
Satterfield, John A. - Pvt 10/13/1861. Roll dated 9/15/1864, LOF, shows AWOL from 7/1/1864. Filed CPA, Pickens Co., GA, 1899. D 9/8/1902. Bur Corinth Bap. Church Cem, Pickens Co. Marker inscribed Co. N 38 GA INF. Widow, Harriet Satterfield, filed CPA, Pickens Co., 1902.
Shiflet, James Monroe (or Shiftlet) - Pvt 9/25/1862. Captured at Ft. Stedman, VA, 3/25/1865. Released Point Lookout, MD, 6/19/1865. J. M. Shiflett filed CPA, Floyd Co., 1899. Died 1/31/1918, buried Wax Cemetery, Silver Creek, Floyd Co.
Singleton, Pyremus W. (Pyremus Whitfield Singleton) - Pvt 10/13/1861. Sick in camp 9/21/1863. CPA shows on sick frl 1/11/1864. Unable to rejoin command. (B Forsyth Co., 1842) Filed CPA, Hall Co., 1905. D 6/21/1920, bur New Hope Meth. Church, Forsyth Co.
Smith, Clark W. - Pvt 2/8/1863. Roll dated 9/15/1864, LOF, shows AWOL since 2/1/1864. BSU.
Smith, Harrison G. - 3d Cpl 10/13/1861. Appt 2d Cpl 1864. Captured Middletown, VA, 8/12/1864. Paroled Elmira, NY, sent to James River, VA, for exchange, 3/14/1865. Received Boulware & Cox's Wharves, James River, VA, Mar. 18 - 21, 1865. BSU.
Smith, Richard R.- Pvt 1/28/1863. Roll dated 9/15/1864, LOF, shows present. NLR. BSU.
Smith, William M. - Pvt 8/30/1862. Roll dated 9/15/1864, LOF, shows AWOL since 1/15/1864. D. 12/9/1894, bur New Hope Meth., Forsyth Co. Widow, Sarah A. Smith filed CPA, Gwinnett Co., 1911.
Stephens, John R. - (John Roberson Stephens) Pvt 10/13/1861. Absent without leave 10/12/1864. Died 6/9/1904, buried Nix-Stephens Family Cemetery, Union Hill, Cherokee Co.
Stephens, Lewis A. (Lewis Asberry Stephens) - Pvt 3/4/1862. Appointed blacksmith; artificer. Absent with leave 10/12/1864. No later record. Died 3/9/1912. Buried Bethel Methodist Church Cemetery, Dawson Co. Widow, Nancy A. Bruce Stephens filed CPA in Dawson Co., 1913.
Stovall, George W. (George Wilkes Stovall) - Pvt 5/10/1862. C Baker's Creek, MS, 5/16/1863. AWOL Aug. 7-31, 1863. CPA shows discharged from Ft. Delaware, Jul/1864, home close of war. (Resident GA since 1838) Filed CPA, Forsyth Co., 1920. D 7/22/1930. Bur Alta Vista Cem, Hall Co.
Stripland, W. B. - (William Bailey Stripland) Pvt 5/10/1862. Roll for 9/15/1864, LOF, shows him AWOL since 6/1/1864. NLR. Filed CPA, Forsyth Co., 1908. D 12/15/1924. Bur Concord Bap. Church, Forsyth Co. CSA marker reads CO N 38 GA INF. Obit in Forsyth Co. News, 12/24/1924.
Stroud, Willis W. - Pvt 10/13/1861. Discharged on account of ill health and old age, 5/25/1862. CSR contains discharge record, age 53. Died 6/12/1888, buried at Mt. Zion Cemetery, Forsyth Co., marker reads: CO L 38 GA INF, Widow, Nancy Ann Stroud, filed CPA in 1891.
Tatum, Elisha - Pvt 3/3/1862. Wounded in 1864. Roll dated 9/15/1864, LOF, shows on wounded furlough since 7/20/1864. Died 11/4/1900. Bur Crossroads Bap. Church Cem, Forsyth Co.
Tatum, Horatio R. - 2d Cpl 10/13/1861. Appt 4th Sgt 1864. Surrounded Appomattox, VA, 4/9/1865. Died 2/9/1908, Dawson Co. Buried there at the Dawsonville City Cemetery.
Tatum, Moses - Pvt 12/20/1861. Transferred to Croft's Battery GA Light Artillery Jan/1863. Surrendered Citronelle, AL. 5/4/1865. Paroled Meridian, MS, 5/11/1865. (Born in SC 3/10/1830) Filed CPA, Hall Co., 1895. Died there 1906. BSU.
Tatum, Silas E. - Pvt 3/3/1862. C near Washington, D.C., 7/13/1864. D, pneumonia Elmira, NY 5/20/1865. Bur there Woodlawn National Cem, Grave #2941, Elmira, NY.
Taylor, C. P. (Caswell P. Taylor) - Pvt 10/15/1863. Absent, sick Lynchburg, VA, Jul/1864. BSU.
Taylor, George T. - Pvt 10/13/1861. Roll dated 9/15/1864, LOF, sick frl since 9/1/1864. NLR. BSU.
Taylor, James S. - (James Samuel Taylor) Pvt 1/28/1863. On sick furlough, 10/5/1864. Died 3/2/1888, buried Liberty Baptist Church Cemetery, Dawson Co.

Taylor, John M. - (John Marion Taylor) Pvt 5/10/1862. AWOL Jan. 1- Aug. 31, 1864. CPA shows at home on sick frl Feb/1864, to end of war. (B Lumpkin Co., 1835) Filed CPA, Hall Co., 1899. D 1910. Bur Liberty Bap. Church Cem, Dawsonville, Dawson Co. Marker reads CO N 38 GA INF.
Taylor, Mulchy M. (Moses Mulchy Taylor) - Pvt 4/30/1863. On special service with Provost Guard 9/15/1864. Surrendered Appomattox, VA, 4/9/1865. BSU.
Taylor, Samuel (Samuel James Taylor) - Pvt 1/28/1863. On sick furlough 8/29/1864. Died 1864. Buried Concord Baptist Church Cemetery, Forsyth Co.
Taylor, Samuel E.- 2d Lt 10/13/1861. Elected 1st Lt 4/19/1862. Sick in camp 9/21/1863. Resigned Oct/1863. Elected to the General Assembly, Dawson Co. and not allowed to hold a commission and be an elected official. D 2/27/1864. Bur Liberty Bap. Church Cem, Dawsonville, Dawson Co.
Taylor, William H. (William Henry Taylor) - Pvt 10/13/1861. Appt Cpl. S Appomattox, VA, 4/9/1865. Filed CPA, Dawson Co., 1900. D 11/25/1914, bur Liberty Bap. Church Cem, Dawson Co.
Touchstone, B. R. (Benjamin Russell Touchstone) - Pvt Mar/1862. C Middleton, VA, 8/12/1864. Released Elmira, NY, 7/7/1865. (B in Campbell Co., 6/24/1841) Filed CPA, 1905, Richmond Co., D there 1/30/1924. Bur there Westview Cem, CSA marker reads: Co N 38 GA INF.
Tuck, Bennett D. - Pvt 2/2/1862. Discharged Sav, GA 5/25/1862. CSR has discharge record. BSU.
Watson, George W. - Pvt 10/13/1861. S Appomattox, VA 4/9/1865. Filed CPA, Cherokee Co., 1905. Bur Field's Chapel Cem, Cherokee Co. CSA marker reads: CO N 38 GA INF CSA.
Watson, Harrison (Harrison Judson Watson) - Pvt 10/13/1861. Appt 5th Sgt 1864. Des 3/22/1865. Took OOA to U.S. Govt. Washington, D.C., 4/9/1865, furnished transportation to Harrisburg, PA. D Oct/1918, Bur St. Mary's Episcopal Cem, Baltimore, MD. Marker reads, SGT, 38 GA INF.
Watson, Rich (or Richard Watson) - Pvt 3/1/1862. S Appomattox, VA 4/9/1865. Filed CPA, Van Buren Co., AR 1907, veteran #12990. D 10/4/1912. BSU.
Westbrook, Samuel - Pvt 8/30/1862. Wounded Cold Harbor, VA 6/1/1864. On wounded furlough 8/31/1864. Died 9/1/1901, buried Cumming City Cemetery, Forsyth Co.
Westbrook, William R. - Pvt 1/28/1863. W, left arm, necessitating amputation above elbow, Cold Harbor, VA, 6/1/1864. Absent, W, 8/31/1864, to close of war. (Resident GA since 10/10/1844) Filed CPA, Forsyth Co., 1880. D 6/23/1899, bur New Hope Meth. Church Cem, Forsyth Co.
Wetherford, Alfred L. (Alfred L. Weatherford) - Pvt 10/13/1861. C Harrisonburg, VA, 9/25/1864. P at PLO, MD, and E 10/31/1864. Rcv Venus Point, Savannah River, GA, for E, 11/15/1864. Filed CPA, St. Francis Co., AR, 1901, veteran #9384. D abt. 1912, bur Hughes Cem, St. Francis Co., AR.
Whitmire, Elias E. (Elias Earl Whitmire) - Pvt 12/17/1861. C Winchester, VA, 9/19/1864. Released PLO, MD, May 12 -14, 1865. (B GA in 5/9/1835) Filed CPA, Forsyth Co., 1911. D 10/30/1917, Forsyth Co., bur there Salem Bap. Church Cem. Marker vandalized, broken in half.
Whitmire, Elisha M. (Elisha Murphy Whitmire) - Pvt 10/13/1861. Appt 8th Cpl Sept/1862. Roll dated 9/16/1864, last on file, shows absent without leave since 2/1/1864. Died 5/4/1895, buried Piney Grove Bap. Church Cemetery, DeKalb Co., AL, marker reads: CO N, 38 GA INF. Widow, Mary J. Whitmire, filed CPA, DeKalb Co., AL, 1899.
Whitmire, George C. (George Christopher Whitmire) - Pvt 8/30/1862. Captured Strasburg, VA 8/19/1864. Released PLO, MD, 6/22/1865. (B in Forsyth Co., GA in 1844) Filed CPA, Forsyth Co., 1902. D 7/19/1924. Bur Lebanon Meth. Cem, Cherokee Co., marker reads: CO L, 38 GA INF.
Whitmire, James Christopher - Pvt 12/20/1861. Absent, sick, 9/2/1864. Died in war. NLR. BSU.
Whitmire, John A - Pvt 8/20/1862. Appt bugler in 1863. Wounded Weldon's Railroad, VA, 8/19/1864. Absent, wounded, 9/17/1864. CPA shows home on wounded frl end of war. (B in Forsyth Co., 6/16/1842. D Hall Co., 1920) Filed CPA, Hall Co. Buried Alta Vista Cem, Gainesville, Hall Co.
Whitmire, Radcliffe Boone - Pvt 2/20/1862. Roll dated 8/31/1863, shows "Present, sick." Died in service. Died 9/4/1863, in SC. BSU.
Whitmire, Samuel R. (Samuel Reece Whitmire) - Pvt 2/1/1863. Wounded in 1864. AWOL Mar. - Aug 31, 1864. CPA shows enlisted Co. A, 11th Regt GA Militia Cavalry in 1864. S Kingston, GA 5/12/1865. (B Hall Co., abt. 1849) Filed CPA, Hall Co., Died 4/8/1921. Bur Calvary Cem, Hall Co.
Whitmire, W. Reeves - (William Reeves Whitmire) Pvt 5/10/1862. Sick in hos 10/11/1864. Died 7/28/1916. Buried Mount Hope Cem, Dahlonega, Lumpkin Co. Marker reads CO H, 38 GA INF.

Wilson, Robert T. - Pvt 5/10/1862. Appt 5th Cpl Jan/1864. Roll dated 9/15/1864, last on file, shows absent, sick. Filed CPA, Chattooga Co., 1900. BSU.
Wofford, John D. (Or Wafford) - Pvt 10/13/1861. Captured Fisher's Hill, VA, 9/21/1864. Paroled at Point Lookout, MD, transferred to Aiken's Landing, VA, for exchange 3/17/1865. BSU.
Wood, Joseph P. - Guidon 10/13/1861. AWOL Mar. 15 - Sept. 21, 1863. Died 10/11/1876, buried Crossroads Bap. Church Cem, Forsyth Co. Widow, Mary E. Wood, filed CPA, Forsyth Co., 1891.
Wood, William H. (William Hasley Wood) - Pvt 5/10/1862. C Winchester, VA, 9/19/1864. P at PLO, MD, trn to Aiken's Landing, VA, for E 3/15/1865. Filed CPA, Fulton Co., 1891. D 1908. BSU.
Worley, James T. - Pvt 3/16/1863. Muster roll dated Aug/1863 reports, "absent, sick McPersonsville, SC since 6/13/1863. Took OOA to U.S. govt. 5/3/1864. Died 12/14/1894. Buried Four Mile Creek Bapt. Church, Pickens Co.

*MUSTER ROLL OF COMPANY M, JOE THOMPSON ARTILLERY, FULTON CO.

This Co. was originally known as Co. M, 38th Georgia Volunteer Infantry Regiment. This Company of men was transferred from the 38th Georgia Infantry Regiment, when the regiment departed Georgia for Virginia, June 8th, 1862. This company became Cpt Hanleiter's Co. GA Light Artillery (Joe Thompson Artillery) 6/10/1862 and was assigned to the Army Of Tennessee C. S. A. The focus of this roster only includes men who joined the Co. before 6/10/1862, since afterwards it was detached from the 38th Georgia Regiment and never rejoined the regiment. First commanded by Lt Col George W. Lee, Capt Cornelius R. Hanleiter assumed command 6/10/1862 and led the Company until the end of the war.

Ables, Andrew J. - Pvt 9/26/1861. Forcibly expelled 3/7/1862. On 3/5/1862. Forcibly expelled for accidently shooting and killing a fellow soldier of his Company. Story found in Diary of Colonel Cornelius Hanleiter. BSU.
Adams, Edmond Raymond - Pvt 4/25/1862. Appt 8th Cpl Oct/1862; 7th Cpl Mar/1863; Pvt Jun/1863; 5th Cpl Mar/1864; Pvt Oct/1864. Surrendered Greensboro, NC, 4/26/1865. Filed CPA, in Morgan Co., 1903. Died 1/28/1936, buried there Rehobeth Cemetery. Marker reads: Born 11/5/1846.
Armistead, William J. - Pvt 11/15/1861. In arrest by sentence of Court Martial 4/7/1862. Dishonorably discharged, Special Order #17, 6/13/1862. (B Walton Co., 1833) Filed CPA, Fulton Co., 1897. Stated he served 4 years and surrendered with the western army in Apr/1865. Buried Hollywood Cemetery, Atlanta, Fulton Co. Obit in Atlanta Constitution, 3/25/1913.
Bailey, Joseph E. (Joseph Edward Bailey) - Pvt 9/26/1861. Appt 6th Sgt Oct/1861. Appt Quartermaster Sgt Jun/1862; 1st Sgt Sep/1862. Transferred to Army Northern VA, 8/19/1864; Brook Art., SC, Appt Cpl, Surrendered Appomattox, VA, 4/9/1865. Filed CPA, Cherokee Co., TX, 1905. Buried West Hill Cemetery, Grayson Co., TX.
Boring, George W. - Pvt Jun/1862. Substitute for Henry Bradford Perry. On extra duty, carpenter for Beaulieu Battery, Oct/1862 - Oct/1864. BSU.
Caldwell, John W. C. - (John Wesley Caldwell) Pvt 2/24/1862. Roll for 10/31/1864, LOF, shows present. NLR. Widow, Amanda Caldwell filed CPA, Pike Co., 1891. CPA states K Aversboro, NC, 3/15/1865. BSU.
Carlton, Spencer (Spencer M. Carlton) - Pvt 1/10/1862. Surrendered Greensboro, NC, 4/26/1865. Filed CPA, Cleburne Co., AL, bur there Crossroads Bethsaida Cemetery. B 11/3/1840, D 2/15/1921.
Cash, John - (John Hamilton. Cash) Pvt 5/15/1862. Roll for 10/31/1864, LOF, present. CPA shows left command Raleigh, NC, on special detail, Apr/1865. (B GA, 4/20/1837) D Fulton Co. GA, 3/24/1923. Bur Mount Gilead Meth., Fulton Co. Widow, Catherine Cash filed CPA, Fulton Co.
Caudle, John H. W. H. - Pvt 9/26/1861. Transferred to Co. K, 5/5/1862. No later record. BSU.
Center, William - Pvt 11/15/1861. Discharged, disability, 6/6/1862. Discharged due to old age, according to the diary of Colonel Cornelius Hanleiter. Born about 1805 in SC, died 9/14/1863, buried Oakland Cemetery, Atlanta, Fulton Co., GA. Buried in potter's field section, no grave marker.
Chapman, John M. - Pvt 9/26/1861. Sick in Fulton Co., GA, Dec/1862. NLR. BSU.

Connor, Edward - Pvt 9/26/1861. On detached service 1863 - Apr/1865. Paroled Greensboro, NC, 4/26/1865, which time he was absent on duty at Yadkin River Bridge. BSU.
Craven, Elijah J. - 2d Lt 9/26/1861. Elected 1st Lt 10/25/1861. Resigned. 10/20/1862. Elected Cpt of Co. C, 63d Regt GA Inf. 11/29/1862. Furloughed for 15 days 2/12/1864. Absent, sick, Aug. 1 - 25, 1864. No later record. Born 9/9/1832. Died 10/1/1881. Buried Oak Wood Cemetery, Grayson Co., TX. Widow, Mary McCamie Craven, filed CPA, Grayson, TX, 1899.
Craven, Newton S. - 1st Sgt 9/26/1861. Transferred to Co. B, 18th Battalion, GA Inf. in 1862, to Co. C, 63d Regt GA Inf. and elected 2d Lt, 12/1/1862. Absent, sick, 9/2/1864. (Born in GA) Died 5/26/1875. Widow, Lottie E. Craven, filed CPA. Grayson, TX, in 1899. BSU.
Craven, Thomas A. S. - Pvt 2/24/1862. Died of pneumonia, Fulton Co., 6/20/1862. BSU.
Daniel, James E. - Pvt 9/26/1861. Died of pneumonia Pembroke, GA, 4/17/1862. BSU.
Daniel, Jesse F. (Jesse Fletcher Daniel) - Pvt 9/26/1861. Surrendered Greensboro, NC, 4/26/1865. Died 10/13/1907, buried at Prays Mill Baptist Church Cemetery, Douglas Co., GA.
Daniel, John T. (John Thomas Daniel) - Pvt 2/24/1862. Appt 4th Cpl Mar/1864; 3d Cpl Aug/1864. Roll 10/31/1864, last on file, present. Filed CPA, Cullman Co., AL. Died 09/14/1908, buried Mt. Joy Baptist Church Cemetery, Cullman Co., AL.
Dansby, William D. (William Durham Dansby) - Pvt 9/26/1861. Appt Color Cpl Jan/1862. Surrendered Greensboro, NC, 4/26/1865. Died in Troup Co., GA, 8/26/1876. Widow, Martha B. Dansby, filed CPA, Troup Co., 1901. BSU.
Davenport, William H. - Pvt 9/26/1861. Died in Savannah, GA hos 6/21/1862. BSU.
Defoor, James A. (James Abner Defoor) - 3d Sgt 9/26/1861. Appt 2d Sgt Sept/1862. Pvt 1863. On extra duty in Quartermaster Dept. in 1864. Sick in hospital, 10/26/1864. Born 1839, died 1912. Buried Oak Lawn Cemetery, Atlanta. Obituary in Atlanta Constitution, 3/16/1912, page 5.
Devlin, Terrence - Pvt, 11/13/1861. Trn to Co. D, 1st Regt GA Regulars 5/26/1862; to Maxwell's Batt. GA Light Art. Aug/1862. Deserted from Camp Houston, near Savannah, GA, 8/22/1862 BSU.
Douglas, William A. - Pvt 9/26/1861. Appt 7th Cpl Jun/1862; 3d Cpl Sept 1862; 2d Cpl Oct/1862; 5th Sgt Jun/1863; Pvt Nov/1863. Roll 10/31/1864, last on file, shows present, No later record. BSU.
Englett, Daniel B. - Pvt 9/26/1861. Appt 6th Cpl 10/1861; 5th Cpl Jan/1862. Pvt Sept/1862. S Greensboro, NC, 4/26/1865. Born. 10/18/ 1842, D 8/7/1877. Bur Oakland Cemetery, Fulton Co.
Englett, James J. - Pvt 5/15/1862. Surrendered Greensboro, NC, 4/26/1865. Born in DeKalb Co, 2/18/1837. Filed CPA, Fulton Co., 1897. Died 5/22/1905. Buried Decatur Cemetery, DeKalb Co. Widow, Nancy J. Whitlock Englett, filed CPA, Fulton Co., 1905.
Englett, Pulaski W. (Pulaski Washington Englett) - Pvt 9/26/1861. Died Savannah, GA, 5/30/1862. Reported by Capt Hanleiter as dying of typhoid & catajara (SIC) fever. Funeral account in Cpt C. Hanleiter's diary. Born 4/21/1839. CSR contains a death claim filed by father, Benjamin Englett, Fulton Co. Buried Decatur Cemetery, DeKalb Co.
Etheridge, Zachariah - Pvt 9/26/1861. On special duty Atlanta, GA, Oct/1862. Discharged 10/1/1864. (Born in Wake Co., NC, Aug/1817) Filed CPA, Hart Co., 1897.
Frost, John - Pvt 9/26/1861. Appt Sgt Oct/1861. Appt 6th Cpl Oct/1862, 5th Cpl Jun/1863, 3d Cpl Jun/1863, 2d Cpl Nov/1863, 3d Sgt Mar/1864. Des 8/6/1864. Bur Forest Park City Cem, Clayton Co.
Giles, John T. - 1st Cpl 9/26/1861. 1st Cpl 6/10/1862. Appt 4th Sgt Sep/1862; 3d Sgt 6/1863. Pvt Mar/1864. Surrendered Greensboro, NC, 4/26/1865. BSU.
Hammond, Abner - Enlisted as a Pvt in Co. C, 21st Regt GA Inf. 6/26/1861. Discharged, disability, camp near Centreville, VA, 10/5/1861. Enlisted as Pvt, Co. M, 38th Regt GA Inf. 1/24/1862. Discharged, disability, 6/6/1862. CSR contains discharge record, 64 years old at discharge. BSU.
Hammond, Joshua T. - (Joshua "Josh" T. Hammond) Enlisted as a Pvt in Co. C, 21st Regt GA Inf. 6/26/1861. Discharged, disability, near Centreville, VA, 10/5/1861. Enlisted as Pvt, Co. M, 38th Regt GA Inf. 11/15/1861. Transferred to Co. K, 6/2/1862. Killed at Antietam, MD, 9/17/1862. CSR contains a death claim filed by father, Abner Hammond, Clayton Co. BSU.
Hanleiter, Cornelius R. (Cornelius Redding Hanleiter) - 1st Lt 9/26/1861. Elected Cpt 10/25/1861. Roll for 10/31/1864, LOF, present. NLR (B in Sav, GA) Owned several newspapers in GA before, after the war. Filed CPA, Fulton Co., 1895. Died 4/19/1897, Atlanta. Bur Oak Hill Cem, Spalding Co.

Hanleiter, William Robertson - Enlisted 4/10/1861 Atlanta, in Co. A., 1st GA Regulars, elected 2nd Lt, Jo Thompson Art., 11/19/1862, transferred from VA, to Savannah, arrived 1/15/1863. Surrendered 4/26/1865, Greensboro, NC, with Johnston's army. Died 2/14/1916, buried Oak Hill Cemetery, Spalding Co. Obit in Atlanta Constitution, 2/15/1916.
Holbrook, James G. - Pvt 9/26/1861. Discharged, disability, 5/6/1862. Enlisted as a Pvt in Cpt Hanleiter's Co. GA Light Art. (Jo Thompson Art.) 9/26/1862. Roll for 10/31/1864, last on file, shows present. Home, wounded furlough Mar/1865 to end of war. (Born in GA, 2/6/1837) Filed CPA, Fulton Co., 1900. D 6/10/1901. Buried Ben Hill United Meth. Church Cem Fulton Co., GA CSA marker incorrectly reads: CAPT RITTER'S GA LIGHT ART.
Holbrook, William Milton - 2d Cpl 9/26/1861. Appt 2d Sgt 11/15/1861. Pvt Jun/1862. Roll dated 10/31/1864, last on file, shows present. No later record. Died 8/29/1872. Widow, Martha A. Holbrook Morris, filed CPA, Fulton Co., 1916. BSU.
Holliday, John W. - 2d Sgt 9/26/1861. Discharged on account of chronic disease, 11/15/1861. Enlisted as a Pvt in Cpt Hanleiter's Co. GA Light Art. (Jo Thompson Art.) 3/25/1863. Surrendered Greensboro, NC, 4/26/1865. BSU.
Holmes, Augustus M. - Pvt 9/26/1861. Appt 4th Cpl Sept/1862; 5th Sgt Oct/1862; 4th Sgt Jun/1863. Died 3/14/1864. BSU.
Hornsby, James H. - Pvt 12/28/1861. Surrendered Greensboro, NC, 4/26/1865. Filed CPA, Morgan Co., AL. Died 8/19/1906, buried Old Canaan Cemetery, Cullman Co., AL.
Hornsby, William Gilbert - Pvt 11/10/1861. Roll for 10/31/1864, LOF, shows present. Filed CPA, Fulton Co., 1895. Died 11/4/1902. Widow, Julian Hornsby filed CPA, Fulton Co., 1903. BSU.
Horton, William R. - Pvt 9/26/1861. Des 3/15/1864. Joined from desertion, 4/8/1864. NLR. BSU.
Hudson, Joseph (or Hutson) - 4th Cpl 9/26/1861. Pvt Jan/1862. Appears last on roll Apr/1863. BSU.
Hudson, William (or Hutson) - Pvt 2/24/1862. Surrendered Greensboro, NC, 4/26/1865. Filed CPA, Oconee Co., 1911. BSU.
Johnson, Benjamin J. - Pvt 11/15/1861. Died of typhoid fever, 6/1/1862. Buried at Laurel Grove Cemetery, Savannah, GA, marker incorrectly identifies unit as 37th GA. Detailed account of his death and burial included in Capt Hanleiter's diary.
Joice, William H. - Pvt 9/26/1861. Appt 6th Cpl Jan/1862; 2d Cpl Jun/1862. 2d Cpl 6/10/1862. Appt 5th Sgt Sep/1862. Det with Major Millen, 6/18/1864. Dropped, supposed to be voluntarily, within the enemy's lines, 10/17/1864. BSU.
Lasseter, John L. (or John L. Lassiter) - Pvt 9/26/1861. Discharged, on account of chronic disease of hip joint, Atlanta, 11/20/1862. CSR shows enlisted in 3rd GA State Reserves in 1863, served at Andersonville Prison. Filed a CPA, Newton Co., 1895. D 1897. BSU.
Law, Aaron - See Lowe, Aaron.
Lawrence, James B. - Pvt 1/26/1862. Appt 8th Cpl Jan/1862. Det Atlanta, Oct/1862. Appt 1st Sgt Jun/1864. Paroled Charlotte, NC, 5/3/1865. BSU.
Lee, Barney Drewry - Pvt 9/26/1861. Appt Sgt Major in 1861. On special service Atlanta, GA, with Cpt G. W. Lee, 10/1862. Transferred to Co. A, 25th Battalion GA Provost Guard Inf. Lee's Battalion) 11/25/1862. No later record. (Died Atlanta, GA 2/14/1891) Obit in Atlanta Constitution, 2/16/1891. Widow, Emma J. Lee, filed CPA, Fulton Co., 1911. BSU.
Lee, George Washington - Pvt 9/26/1861. Elected Lt Colonel 10/11/1861; Colonel 2/18/1862. Resigned 7/1862. Appt Maj, to rank from 5/2/1863, (from State of GA,) to report to Atlanta, GA. Assigned to the 25th Battn. GA Provost Guard Infantry (Atlanta, GA) Maj, date not stated. Elected Lt Colonel. (This Battn. was disbanded 6/24/1864. Elected Lt Col of the 3d Battn. GA State Guards (Atlanta Fire Battn.) date not given. Resigned from the 38th GA in Jul./1862 due to ill health. D 4/3/1879, Rome, buried there in Myrtle Hill Cemetery, Obit in Macon Telegraph 4/15/1879.
Littleton, Benjamin F. - Pvt 9/26/1861. Appears last on roll for Feb/1863. BSU.
Long, William H. - Pvt 9/26/1861. Appt 7th Cpl Oct/1862; 6th Cpl, June 1863. Det in Quartermaster Dept. Oct/1863. Roll for 10/31/1864, last on file, shows present. NLR. BSU.
Lothridge, Thomas - Pvt 11/15/1861. Roll for 10/31/1864, LOF, shows present. NLR. BSU.

Lowe, Aaron (or Law) - Pvt 2/24/1862. Sick in Garrison Hos 10/29/1864. NLR. Filed CPA, Cobb Co., 1911. Widow, Nancy Sewell Lowe, filed CPA, Cobb Co., 1928. D 4/21/1928, bur Milford Bap. Chur Cem, Cobb Co. Marker reads CO M, 38 GA INF. Obit in Marietta Journal, page 7, 4/26/1928.
Marlow, John - Pvt 2/24/1862. Surrendered Greensboro, NC, 4/26/1865. BSU.
McDaniel, Aaron - Pvt 9/26/1861. Discharged, disability, 3/15/1862. Discharged for severe chest cough, possible consumption, according to CSR. BSU.
McDaniel, Green Berry - Pvt 2/24/1862. Appt 5th Cpl Oct/1862; Pvt Jun/1863. Det blacksmith 10/1863. Roll 10/31/1864, LOF, present, artificer. NLR (B SC) Filed CPA, Fulton Co., 1899. BSU.
McDaniel, John W. - Pvt 2/24/1862. Roll for 10/31/1864, LOF, shows present. NLR. BSU.
McDaniel, William P. - Jr. 2d Lt 9/26/1861. Elected 2d Lt 10/25/1861. Resigned 10/15/1862. BSU.
McKennie, Samuel R. (or McKemie) - Pvt 9/26/1861. Roll, Jun/1863, shows present. NLR. BSU.
McKensie, William M. (or McKenzie) - 4th Sgt 9/26/1861. 4th Sgt 6/10/1862. Pvt in 1863. Trn to Co. C, 63d Regt GA Inf., 1863. Appt 3d Sgt Aug/1863. Furloughed, 2/17/1864. NLR. BSU.
McNabb, David A. - Pvt 9/26/1861. Discharged on account of ulcerated leg, 2/15/1862. Injured his leg chasing a chicken according to CSR. CSR contains discharge. Died 3/5/1918, Atlanta, GA, buried there at Hollywood Cemetery. Obit in Atlanta Constitution, 3/6/1918.
McNamara, Dan F. (Dan Frank McNamara) - Pvt 3/30/1862. Appt 8th Cpl June 1862. Roll for 6/30/1862, shows present. Deserted to the enemy blockade off the coast of GA, on 8/30/1862, Picked by U.S. Steamer Conemaugh off Skidaway Island, while in a small boat. Transported to New York City aboard the U.S. Steamer Massachusetts. BSU.
Mitchell, Horace M. - Pvt 9/26/1861. Appt Commissary in 1861. No later record. BSU.
Moore, John - Pvt 9/26/1861. Appears last on roll for 6/30/1862. BSU.
Moore, William P. - Pvt 2/24/1862. Attached to Co. F, 22d Battalion GA Heavy Art. S Greensboro, NC, 4/26/1865. (B Jasper Co., 1833) Filed CPA, Fulton Co., 1900. D Fulton Co. 2/11/1908. BSU.
Morgan, Matthew R. - Pvt 9/26/1861. Admitted to C.S.A. General Hospital #3, Greensboro, NC Apr/1865. BSU.
Morgan, William M. - Pvt 11/13/1861. Det in Quartermaster Dept. June/1864. Roll 10/31/1864, LOF, shows Absent, Det indefinitely, nurse, Guyton Hos. BSU.
Parr, Benjamin W. - Pvt 9/26/1861. Appt 5th Cpl 10/1861; 4th Cpl Jan/1862. Pvt, 6/10/1862; trans to Co. A, 25th Battalion GA Provost Guard Inf., 11/25/1862; to Co. K, 5th Regt GA Inf., 1863. S Greensboro, NC, 4/26/1865. D 5/8/1899. Buried at Oakland Cemetery, Atlanta, obit in Atlanta Constitution 5/8/1899.
Parr, Lewis J. – Major, Cpt in Co. M, 9/26/1861. Elected Maj 10/11/1861; Lt Col 2/14/1862. Lt Col Parr accompanied the 38th GA when it left GA for VA. Wounded at Gaines Mill, 6/27/1862, arm shot off near the shoulder. Resigned 7/15/1862, or 12/13/1862. Filed CPA, FL, pension #A03222, 1907. Died in Osceola, FL, 4/15/1908, buried there Mount Peace Cemetery.
Payne, Raymond B. - Pvt 9/26/1861. Died from accidental gunshot wound by another soldier at Skidaway Island, GA, 3/6/1862. Full account in the diary of Colonel Cornelius Hanleiter. BSU.
Perry, Henry Bradford - Pvt 2/26/1862. Discharged, furnished George W. Boring as substitute, Jun/1862. Born 5/4/1830, died 10/31/1884. Buried at Perry Family Cemetery, Newton Co.
Pinion, Sanford V. - Pvt 9/26/1861. Appt 6th Cpl Jun/1862. Appt 2d Cpl Sep/1862. Dropped, supposed to be voluntarily within the enemy's lines, 10/17/1864. (Born in Greenville, SC, 1842) Filed CPA, Fulton Co., 1900. Buried Oakland Cemetery, Fulton Co.
Ransone, Benjamin J. (or Benjamin F) - Pvt 11/15/1861. Discharged, disease of spine, Beaulieu, near Savannah, GA, 11/1/1862. BSU.
Robbins, Algernon S. (Algernon Sidney Robbins Jr.) - 3d Cpl 9/26/1861. 3d Cpl 6/10/1862. Appt 1st Cpl Sep/1862. Pvt Jun/1863. Roll for 10/31/1864, last on file, shows present. Died 1897, buried Oakland Cemetery, Fulton Co. Widow, M. M. Robbins, filed CPA, Fulton Co., 1919.
Robbins, James W. (James Wiley Robbins) - Pvt 9/26/1861. Appt carpenter, 10/1862. Admitted to CSA General Hos Q11, Charlotte, NC, with chronic diarrhea 4/1/1865. Died 4/17/1873. Widow, Mary Ann Robbins, filed CPA, 1891,Fulton Co. BSU.
Robbins, John W. (John Wesley Robbins) - Pvt 9/26/1861. Discharged Beaulieu near Sav, GA 11/13/1862. Reenlisted 3/20/1863. Det in Quartermaster Dept. Jun/1864. Admitted General Hos #11,

Charlotte, NC, 4/1/1865. Died 9/28/1913. Buried Oakland Cemetery, Fulton Co. Widow, Carion Robbins, filed CPA, Fulton Co.

Robbins, Joseph P. (Joseph Pliny Robbins) - Pvt 9/26/1861. Appt 7th Cpl June 1863. S Greensboro, NC, 4/26/1865. (Resident GA, 1843) Filed CPA Fulton Co, 1911. D 9/27/1931. Bur Mt. Gilead UMC.

Robbins, Querlis W. - Pvt 9/26/1861. Appt 6th Cpl Sept/1862; 4th Cpl Oct/1862; 2d Cpl Jun/1863; 1st Cpl Nov/1863. Pvt Mar/1864. Surrendered Greensboro, NC, 4/26/1865. BSU.

Roberts, William E. - Pvt 9/26/1861. Detailed in Quartermaster Dept. 10/1/1862. Roll for 10/31/1864, last on file, shows present. No later record. BSU.

Roberts, Willis R. - Pvt 9/26/1861. Transferred to Captain Hanleiter's Company Ga. Light Artillery (Jo Thompson Artillery,) 6/10/1862 . D 3/25/1864. BSU.

Rodgers, James W. - Pvt Jan/1864, CPA stated he served until 4/23/1865, when he was surrendered, Camden SC. Filed a CPA, DeKalb Co., 1907. Died 5/7/1924. Buried at Decatur Cemetery, DeKalb Co. Obit in Atlanta Constitution newspaper, 5/8/1924.

Shaw, Augustus (Augustus "Gus" A. Shaw) - Pvt 9/26/1861. Elected Jr. 2d Lt 10/25/1861. Appt Acting Adjutant 5/2/1862. Elected 1st Lt 11/8/1862. S Greensboro, NC, 4/26/1865. Accompanied 38th GA Regt to VA as Adjutant. Captured Gaines Mill, 6/27/1862, Exchanged Aug/1862. Died 11/5/1905. Bur Westview Cem, Fulton Co. Widow, Flora Trout Shaw filed CPA, 1920. Fulton Co.

Shaw, Samuel H. - Pvt 2/18/1862. Appt 3d Cpl Mar/1864. 2d Cpl 8/1864. Roll for 10/31/1864, LOF, shows present. NLR. D 10/9/1889. Bur Westview Cem, Fulton Co.

Sherling, Marion Hamilton - Pvt 2/24/1862. Roll for 10/31/1864, LOF, shows present. NLR. BSU.

Simril, Robert E. - (Robert "Bob" Eaton Simril) Drummer, 9/26/1861. Surrendered Greensboro, NC, 4/26/1865. Filed CPA Coweta Co., 1897. Died 6/13/1901, Newman, buried there Oak Hill Cem. CSA marker reads, Hanleiter's Co, Thompson's Light Artillery.

Smith, Joseph B. - Pvt 9/26/1861. Deserted 2/9/1864. Filed CPA, Hall Co., 1910. BSU.

Steele, John H., Jr. - Pvt 10/1/1861. Transferred to Co. C, 6/1/1862. Surrendered Appomattox, VA, 4/9/1865. CSR states missing Gettysburg, July/1863, later shown as a paroled prisoner. BSU.

Stevenson, William H. - Pvt 9/26/1861. Appt bugler Jun/1863. Roll for 10/31/1864, last on file, shows present. No later record. BSU.

Teague, Isaac M. - Pvt 9/26/1861. Transferred to Cpt Daniel's Co. Hurt Guards, 10/2/1861. (Not found on rolls of Cpt Daniel's Co. GA Vols., CSA. Hurt Guards not identified.) BSU.

Thrash, William Henry - 5th Sgt 9/26/1861. 2d Sgt 6/10/1862. On detail duty as Ordnance Sgt with Beaulieu's Battery Sept/1862 - Jun/1863. Trn to Co. D, 2d Engineer Corps 9/9/1863. Roll for Nov - Dec 1863, shows present, artificer. No later record. BSU.

Trainer, Thomas C. - Pvt 9/26/1861. Appt 5th Sgt 10/1861. Appt Quartermaster Sgt Jan/1863. Roll dated 10/1/1864, shows "Absent, Det with Cook & Bro., Special Orders #55, A. & I. O., Rich, VA, 3/7/1864." NLR. BSU.

Waits, Andrew M. - Pvt 9/26/1861. Union POW records show he deserted to Federal lines 12/21/1864 at Savannah, Georgia. Died 1893. Widow, Mary Frances Waits filed CPA in Fulton Co., 1901. Buried at Hollywood Cemetery, Fulton Co.

Waits, James M. - Pvt 9/26/1861. Roll for 10/31/1864, shows present. CPA shows sick, Raleigh, NC, hos close of war. (Born in Campbell Co., GA, 1838) Filed CPA, Fulton Co., 1904. Buried at Bucky James Cemetery, Franklin Co., GA.

Westerfield, Peter (or Westerfeld) - Pvt 2/12/1862. Transferred to Cpt McAlpin's Co. Engineer Corps, 9/9/1863, but name is not borne on the roll of this Co., dated 12/31/1863. Printed a notice in Atlanta Constitution 6/29/1890 seeking his Cpt to attest to his honorable service. Name in notice listed as Wustefield. BSU.

Whiting, Samuel J. - Pvt 1/26/1862. Appt 7th Cpl Jan/1862; 4th Cpl Jun/1862. Deserted to the enemy blockade off the coast of GA, 8/30/1862, while on garrison duty with Beaulien Battery, on the Vernon River. Picked by the U.S. Steamer Conemaugh off Skidaway Island, while in a small boat. Transported to NY on U.S. Steamer Massachusetts. BSU.

Williams, Thomas H. - Pvt 3/4/1862. Appt A.Q.M. assigned to Cobb's Legion 5/7/1863. NLR. BSU.

Wilson, William A. (William Arnold Wilson) - Pvt 9/26/1861. Appt 5th Cpl Sept/1862; 3d Cpl Oct/1862; 1st Cpl June 1863. Pvt 11/1863. Appt 2d Sgt Mar/1864. On detail duty with hos train Apr/1865. (Died 1904) Widow, Sarah E, Wilson, filed CPA, Fulton Co., 1920. Buried Westview Cem, Fulton Co.
Wooten, Daniel B. (Daniel Bartholomew Wooten) - Pvt 9/26/1861. Deserted 5/18/1864. Took oath of allegiance to U.S. Government at Nashville, TN 2/26/1865. Born in GA, 4/20/1822. Died 10/7/1904, Nashville, TN, buried there at Mount Olivet Cemetery.

END NOTES

Chapter 1 **The Beginning – 1861**

[1] Gray, Mary A. H., Life in Dixie During the War, Atlanta, Ga., Foote and Davies, 1901 page 24
[2] Hanleiter, C. R., History of the Confederate Veterans Association of Fulton County, Georgia (1890) The Recollections of Captain C. R. Hanleiter, March 2nd, 1890.
[3] Lester, George H., The War Record of the Tom Cobb Infantry 38th Georgia, Co E, Oglethorpe Echo Newspaper, April 10th, 1879
[4] Ibid.
[5] Henderson, Lillian, Henderson's Roster of Georgia Confederate Soldiers, Compiled by Lillian Henderson, National Archives

Chapter 2 **To Savannah!**

[1] Hanleiter, C. R., History of the Confederate Veterans Association of Fulton County, Georgia (1890) The recollections of Captain C. R. Hanleiter, March 2nd, 1890.
[2] Lester, George H., The War Record of the Tom Cobb Infantry, 38th Georgia, Company E, Extracts from the Oglethorpe Echo Newspaper (April 10th, 1879)
[3] Hanleiter, C. R., The Hanleiter Diary, The Atlanta Historical Society, Entry from 19 Nov 1861
[4] Hanleiter, C. R., The Hanleiter War Diary, Atlanta Historical Society, page 18, entry for Nov 29th, 1861
[5] Henderson, Lillian, Henderson's Roster of Georgia Confederate Soldiers, National Archives
[6] Lester, George H., The War Record of the Tom Cobb Infantry, 38th Georgia, Company E, Oglethorpe Echo Newspaper (April 10th, 1879)
[7] Georgia State Archives, Roster of the Confederate soldiers of Georgia, 1861-1865 compiled by Lillian Henderson
[8] Hanleiter, C. R., The Hanleiter War Diary, Atlanta Historical Society, entry for Mar. 5th, 1862.
[9] Lieutenant George H. Lester, The War Record of the Tom Cobb Infantry, 38th Georgia, Company E, Extracts from the Oglethorpe Echo Newspaper (April 10th, 1879)

Chapter 3 **On To Virginia!**

[1] The Hanleiter War Diary, Atlanta Historical Society, entry for June 6th, 1862
[2] History of the Confederate Veterans Association of Fulton County, Georgia (1890) The recollections of Captain C. R. Hanleiter, March 2nd, 1890.
[3] Georgia State Archives, Microfilm Reel 2759, McCleskey Collection
[4] G. W. Nichols, A Soldier's Story Of His Regiment (6ist Georgia) And Incidentally Of The Lawton-Gordon-Evans Brigade Army Northern Virginia, 1898, pages 38-40

Chapter 4 **The Battle of Gaines Mill - June 27th, 1862**

[1] Nichols, G.W., A Soldier's Story of His Regiment, (61st Georgia) and Incidentally the of Lawton-Gordon- Evans Brigade, Army of Northern Virginia, 1898, Jesup, Georgia, page 47
[2] Evans Led the Charge, Macon Telegraph Newspaper, Jan 30th, 1894
[3] Ibid.
[4] From the Report of Brig. General Alexander R. Lawton, C. S. Army, commanding Fourth Brigade, Second Division (Jackson's), of the battle of Gaines' Mill, Dated July 28th, 1862
[5] Evans Led the Charge, Macon Telegraph Newspaper, Jan 30th, 1894
[6] From the Report of Brig. General Alexander R. Lawton, C. S. Army, commanding Fourth Brigade, Second Division (Jackson's), of the battle of Gaines' Mill, Dated July 28th, 1862
[7] From the Report of Brig. General Alexander R. Lawton, C. S. Army, commanding Fourth Brigade, Second Division (Jackson's), of the battle of Gaines' Mill, Dated July 28th, 1862
[8] Ibid.
[9] Report of Captain William H. Battey, Thirty-eight Georgia Infantry, of the battles of Gaines' Mill and Malvern Hill, Jul 27th, 1862.
[10] Evans Led the Charge, Macon Telegraph Newspaper, Jan 30th, 1894
[11] Ibid.
[12] Lieutenant George H. Lester, The War Record of the Tom Cobb Infantry, 38th Ga., Oglethorpe Echo Newspaper (April 10th, 1879)
[13] From the Report of Brig. General Alexander R. Lawton, C. S. Army, commanding Fourth Brigade, Second Division (Jackson's), of the battle of Gaines' Mill, dated July 28th, 1862
[14] Evans Led the Charge, Macon Telegraph Newspaper, Jan 30th, 1894, L.A.M.
[15] Nancy Hanks was a famous racehorse in the early 1890's.
[16] Evans Led the Charge, Macon Telegraph Newspaper, Jan 30th, 1894
[17] Report of Captain William H. Battey, Thirty-eight Georgia Infantry, of the battles of Gaines' Mill and Malvern Hill, Jul 27th, 1862.
[18] Evans Led the Charge, Macon Telegraph Newspaper, Jan 30th, 1894
[19] Report of Captain William H. Battey, Thirty-eight Georgia Infantry, of the battles of Gaines' Mill and Malvern Hill, Jul 27th, 1862.
[20] Report of Colonel William S. H. Baylor, 5th Va. Inf., Jul 9th, 1862.
[21] Dabney, Robert Lewis, 1820-98, Life & campaigns of Lieut.-Gen. Thomas J. *Jackson* (Stonewall *Jackson*), 1866, P. 455.
[22] Evans Led the Charge, Macon Telegraph Newspaper, Jan 30th, 1894
[23] The Savannah Republican, July 19th, 1862, Page 1, Lawton's Brigade in the Battles Before Richmond, E. P. Lawton, and Savannah Republican, Page 2, July 18th, 1862, Capt William H. Battey. Report of Captain William H. Battey, Thirty-eight Georgia Infantry, of the battles of Gaines' Mill and Malvern Hill, Jul 27th, 1862.
[24] Evans Led the Charge, Macon Telegraph Newspaper, Jan 30th, 1894
[25] Lieutenant George H. Lester, The War Record of the Tom Cobb Infantry, 38th Ga., Oglethorpe Echo Newspaper (April 10th, 1879)
[26] Ibid.
[27] Macon Telegraph newspaper article, 1900, Wept Over the Wrong Man, Daniels
[28] The War Record of the Tom Cobb Infantry 38th Georgia, Co E, By George H. Lester, Oglethorpe Echo Newspaper, April 10th, 1879
[29] Lawton, Alexander, Report of Brig. General Alexander R. Lawton, C. S. Army, commanding Fourth Brigade, Second Division (Jackson's), of the battle of Gaines' Mill, Dated July 28th, 1862
[30] The War Record of the Tom Cobb Infantry 38th Georgia, Co E, By George H. Lester, Oglethorpe Echo Newspaper, April 10th, 1879
[31] Macon Telegraph newspaper article, 1900, Wept Over the Wrong Man
[32] Frederick, Gilbert, A Record Of The Military Service Of The fifty-Seventh New York State volunteer Infantry

[33] Ibid.
[34] Lieutenant George H. Lester, The War Record of the Tom Cobb Infantry, 38th Ga., Oglethorpe Echo Newspaper (April 10th, 1879)
[35] Oakwood Cemetery Records of the Confederate Dead, Transcribed 1991, City of Richmond
[36] Nichols, G.W., A Soldier's Story of His Regiment, (61st Georgia) and Incidently the of Lawton-Gordon- Evans Brigade, Army of Northern Virginia, 1898, Jesup, Georgia, page 47
[37] Roster of the Confederate soldiers of Georgia, 1861-1865 compiled by Lillian Henderson
[38] Georgia State Archives, Microfilm Reel 2759, McCleskey Collection
[39] Georgia. State Division of Confederate Pensions and Records, (1959 - 1964). Roster of the Confederate soldiers of Georgia, 1861-1865. [Hapeville, Ga.: Longina & Porter. and Confederate Pension Applications, Georgia Confederate Pension Office, RG 58-1-1, Georgia Archives.

Chapter 5 **The 2nd Manassas Campaign, July - Aug 1862**

[1] Mathews, W. C., Wrights Legion, The Sunny South, January 10th, 1891
[2] Letter courtesy of Ms. Joan Bond, descendant of G. W. Gaddy, of San Antonio, TX.
[3] Mathews, W. C., Wrights Legion, The Sunny South, January 10th, 1891
[4] Georgia. State Division of Confederate Pensions and Records. Roster of the Confederate Soldiers of Georgia, 1861-1865. [Hapeville, Ga.: Longina & Porter, 1959-1964
[5] Neese, George Michael, Three years in the Confederate horse artillery, The Neale Publishing Co., 1911, page 90
[6] Mathews, W. C., Wrights Legion, The Sunny South, January 10th, 1891
[7] Lester, George H., War Record of the Tom Cobb Infantry, Company E, 38th Georgia Regiment, Oglethorpe Echo newspaper, April 10th, 1879
[8] Mathews, W. C., Wrights Legion, The Sunny South, January 10th, 1891
[9] Bill Arp was the pen name of Charles Henry Smith, a noted Georgia politician and humorist whose column ran in the Atlanta Constitution for many years after the war.
[10] Lester, George H., The War Record of the Tom Cobb Infantry, 38th Ga., Oglethorpe Echo Newspaper (April 10th, 1879)
[11] Ibid.
[12] Nichols, G. W., A Soldier's Story of His Regiment (61st Georgia), and incidentally of the Lawton-Gordon-Evans brigade, Army Northern Viginia, (Jesup Georgia, 1898) P.80
[13] Lester,George H., War Record of the Tom Cobb Infantry, Company E, 38th Georgia Regiment, Oglethorpe Echo newspaper, April 10th, 1879
[14] Ibid.
[15] Abott, John Willis, The Story of Our Army for Young Americans: From Colonial Days to Present Times, Dodd, Mead, 1914, page 393
[16] Sheeran, James B., Confederate Chaplain: A War Journal, Bruce Publishing Company, 1960, page 78
[17] Lester, George H., War Record of the Tom Cobb Infantry, Company E, 38th Georgia Regiment, Oglethorpe Echo newspaper, April 10th, 1879
[18] Ibid.
[19] Mathews, W. C., Wrights Legion, The Sunny South, January 10th, 1891
[20] Hunter McGuire, Speech at dedication of Jackson Memorial Hall at VMI, The Confederate cause and conduct in the war between the states: as set forth in the reports of the History Committee of the Grand Camp, C.V., of Virginia, and other Confederate papers, L. H. Jenkins, 1907, page 187.
[21] Lester, George H., War Record of the Tom Cobb Infantry, Company E, 38th Georgia Regiment, Oglethorpe Echo newspaper, April 10th, 1879
[22] Braswell, Samuel H, Confederate Pension Applications, Georgia Confederate Pension Office, RG 58-1-1, Georgia Archives.
[23] Sunny South, Sep. 19, 1896 -- page 10, A Telltale Dispatch

[24] Smith, Abram P., History of the Seventy-Sixth Regiment New York Volunteers: What it Endured, Truair, Smith and Miles, printers, Syracuse, 1867 - New York (State) pages 119 - 120
[25] Lester, George H., War Record of the Tom Cobb Infantry, Company E, 38th Georgia
[26] National Archives, M836, Confederate States Army Causalities: Lists and Narrative Reports, 1861 - 1865, Casualty Report of 38th Georgia Regiment for 2nd Manassas and Ox Ford, Va.
[27] Henderson's Roster of Georgia Confederate Soldiers; National Archives, M836, Confederate States Army Causalities: Lists and Narrative Reports, 1861 - 1865, Casualty Report of 38th Georgia Regiment for 2nd Manassas and Ox Ford, Va.
[28] National Archives, M836, Confederate States Army Causalities: Lists and Narrative Reports, 1861 - 1865, Casualty Report of 38th Georgia Regiment for 2nd Manassas and Ox Ford, Va.
[29] Lester, George H., The War Record of the Tom Cobb Infantry, 38th Ga., Oglethorpe Echo Newspaper (April 10th, 1879)

Chapter 6 **The Bloodiest Day – Antietam, September 17th, 1862**

[1] Mathews, W. C., Wrights Legion, The Sunny South, January 10th, 1891
[2] Lowe, John T, Report of Major J. T. Lowe, O.R.-- Series I—Volume XIX/1 [S# 27]
[3] Hood, John Bell, Advance and retreat: Personal experiences in the United States and Confederate States armies, published for the Hood Orphan Memorial Fund [by] G. T. Beauregard, 1880, pages 41-42
[4] Lowe, John T, Report of Major J. T. Lowe, O.R.- Series I-Volume XIX/1 [S# 27]
[5] Dawes, Rufus Robinson, Service with the Sixth Wisconsin volunteers, E.R. Alderman & Sons, 1890, P. 87
[6] Mathews, W. C., Wrights Legion, The Sunny South, January 10th, 1891
[7] Early, Jubal, Report of General Jubal Early, O.R.-- Series I—Volume XIX/1 [S# 27]
[8] Bradwell, Isaac, Confederate Veteran, October 1921, 29:378-80
[9] Lowe, John T, Report of Major J. T. Lowe, O.R.- Series I-Volume XIX/1 [S# 27]
[10] Bradwell, Isaac, Confederate Veteran, October 1921, 29:378-80
[11] Ayres, Ezra, Ezra Ayres Carman Papers, Library of Congress, Box 14, Reel 1, No. 15, "The Battle of the Union Right and Confederate Left, Daybreak to 7:30 a.m
[12] Lowe, John T, Report of Major J. T. Lowe, O.R.- Series I-Volume XIX/1 Pages 975 - 976
[13] Jackson, Thomas, Report of Lieutenant General Thomas J. Jackson, O.R.-- Series I—Volume XIX/1 [S# 27]
[14] Mathews, W. C., Wrights Legion, The Sunny South, January 10th, 1891
[15] Pension Applications, Georgia Confederate Pension Office, RG 58-1-1, Georgia Archives; Roster of the Confederate soldiers of Georgia, 1861-1865 compiled by Lillian Henderson
[16] The National Archives, Mirofilm roll: "M-266 Compiled Service Records of Confederate Soldiers Who Served in Organizations from Georgia - Thirty Eighth Infantry, Rolls 433 through 441
[17] Hooker, Joseph, Report of General Joseph Hooker Battle Report, Official Records, Series 1, Vol 19, Part 1 (Antietam - Serial 27) , Pages 213 - 219
[18] Bradwell, Isaac, Confederate Veteran, October 1921, 29:378-80
[19] Ibid.
[20] R.W.D, Sketches and Anecdotes, The Sunny South, Feb 11th, 1882, Page 5
[21] Ayres, Ezra, Ezra Ayres Carman Papers, Library of Congress, Box 14, Reel 1, No. 15, "The Battle of the Union Right and Confederate Left, Daybreak to 7:30 a.m
[22] Confederate Pension Applications, Georgia Confederate Pension Office, RG 58-1-1, Georgia State Archives.
[23] Ayres, Ezra, Ezra Ayres Carman Papers, Library of Congress, Box 14, Reel 1, No. 15, "The Battle of the Union Right and Confederate Left, Daybreak to 7:30 a.m
[24] Confederate Pension Applications, Georgia Confederate Pension Office, RG 58-1-1, Georgia State Archives

[25] Ayres, Ezra, Ezra Ayres Carman Papers, Library of Congress, Box 14, Reel 1, No. 15, "The Battle of the Union Right and Confederate Left, Daybreak to 7:30 a.m
[26] US War Department, The War of the Rebellion: a Compilation of the Official Records of the Union and Confederate Armies (OR), 128 vols., Washington DC: US Government Printing Office, 1880-1901, Vol. 51/Part1 (Ser #107), pp. 139-140
[27] Ayres, Ezra, Ezra Ayres Carman Papers, Library of Congress, Box 14, Reel 1, No. 15, "The Battle of the Union Right and Confederate Left, Daybreak to 7:30 a.m."
[28] Ayres, Ezra, Ezra Ayres Carman Papers, Library of Congress, Box 14, Reel 1, No. 15, "The Battle of the Union Right and Confederate Left, Daybreak to 7:30 a.m
[29] Lester, George H., War Record of the Tom Cobb Infantry, Company E, 38th Georgia Regiment, , Oglethorpe Echo newspaper, April 10th, 1879
[30] Ayres, Ezra, Ezra Ayres Carman Papers, Library of Congress, Box 14, Reel 1, No. 15, "The Battle of the Union Right and Confederate Left, Daybreak to 7:30 a.m."
[31] Dawes, Rufus Robinson, Service with the Sixth Wisconsin Volunteers, E.R. Alderman & Sons, 1890, P. 90
[32] Ayres, Ezra, Ezra Ayres Carman Papers, Library of Congress, Box 14, Reel 1, No. 15, "The Battle of the Union Right and Confederate Left, Daybreak to 7:30 a.m."
[33] Lester, George H., War Record of the Tom Cobb Infantry, Company E, 38th Georgia Regiment, Oglethorpe Echo newspaper, April 10th, 1879
[34] Lowe, John T., Report of Major J. T. Lowe, O.R.-- Series I—Volume XIX/1 [S# 27]
[35] Bradwell, Isaac, Confederate Veteran, October 1921, 29:378-80
[36] Confederate Pension Applications, Georgia Confederate Pension Office, RG 58-1-1, Georgia State Archives
[37] Ibid.
[38] Ibid.
[39] Ibid.
[40] Early, Jubal, Report of General Jubal Early, O.R.-- Series I - Volume XIX/1 [S# 27]
[41] Dawes, Rufus Robinson, Service with the Sixth Wisconsin volunteers, E.R. Alderman & Sons, 1890, P. 90
[42] Baxter, John, In Virginia and Maryland In 1862, Atlanta Journal, March 24th, 1902, P. 5
[43] Lowe, John T, Report of Major J. T. Lowe, O.R.- Series I-Volume XIX/1 Pages 975 - 976
[44] Jackson, Thomas, Report of Lieutenant General Thomas J. Jackson, O.R.-- Series I - Volume XIX/1 [S# 27]
[45] Ibid.
[46] Hood, John B., Report of General John B. Hood, O.R.-- Series I--Volume XIX/1 [S# 27]
[47] Rufus Robinson Dawes, Service with the Sixth Wisconsin volunteers, E.R. Alderman & Sons, 1890, P. 92
[48] Daffan, Katie Litty, My Father as I Remember Him (Houston: Gray and Dillaye, 1908, page 81
[49] Gordon, John B., Reminiscences of the Civil War, New York: Charles Scribner's Sons, 1904, P. 89
[50] Lowe, John T, Report of Major J. T. Lowe, O.R.-- Series I - Volume XIX/1 [S# 27]
[51] Lester, George H., War Record of the Tom Cobb Infantry, Company E, 38th Georgia Regiment, , Oglethorpe Echo newspaper, April 10th, 1879
[52] The National Archives, Confederate States Army Casualty List and Narrative Reports, Ewell's Division, Aug 22 - 20 Sept, 1862, M836, Roll 0007,
[53] Mathews, W. C., Wrights Legion, The Sunny South, January 10th, 1891
[54] Bradwell, Isaac, Confederate Veteran, October 1921, 29:378-80
[55] Mathews, W. C., Wrights Legion, The Sunny South, January 10th, 1891
[56] Isaac Bradwell, Confederate Veteran, October 1921, 29:378-80
[57] Henderson's Roster of Georgia Confederate Soldiers; National Archives, M836; CS Army Causalities: Lists and Narrative Reports, 1861 - 1865, Casualty Report of 38th Georgia Regiment for Antietam

Chapter 7 **The Battle of Shepherdstown, or Boteler's Ford September 19 & 20th, 1862**

[1] The War of the Rebellion: v.1-53 [serial no. 1-111] BGen William Pendleton's Official Report of September 24, 1862, US Govt Printing Office, 1887, Report 210, page 829
[2] General Jubal Early, OR, Series 1, Volume XIX, Numbers 269. Report of Brigadier General Jubal A. Early, C. S. Army, commanding Ewell's division, of operations September 3-27, Page 965-973
[3] Lieutenant General Thomas J. Jackson, O.R.-- Series I -Volume XIX/1 [S# 27]
[4] Official Records, Series 1, Vol. 19, Part 1, Page 832, Numbers 210. Report of Brigadier General William N. Pendleton, C. S. Army, Chief of Artillery, of operations August 20-September 24. Pages 829-835
[5] John B. Isler, Report of October 14, 1862 on skirmish at Shepherdstown Sept 20, 1862, Official Records Series 1, Vol 19, Part 1 (Antietam - Serial 27) , Pages 344 - 345
[6] Mathews, W. C., Wrights Legion, The Sunny South, January 10th, 1891, page 5
[7] Portrait Monthly: Containing sketches of departed heroes, and prominent personages of the present time, interesting stores, etc, Vol 1, T. B. Leggett and Co, 1864, Page 187
[8] The Rebellion Record Vol 6 A Diary of American Events, D. Van Norstrum Publisher, 1873, page 30
[9] Official Records, Series 1, Vol. 19, Part 1, Page 832, Numbers 210. Report of Brigadier General William N. Pendleton, C. S. Army, Chief of Artillery, of operations August 20-September 24,.pages 829-835
[10] Unidentified Union Soldier, Second District Regiment, Baltimore Sun Newspaper, Oct 10th, 1862
[11] John B. Isler, Report of October 14, 1862 on skirmish at Shepherdstown Sept 20, 1862, Official Records: Series 1, Vol 19, Part 1 (Antietam - Serial 27), Pages 344 - 345
[12] The War of the Rebellion: v.1-53 [serial no. 1-111] , Report of John-Fitz Porter, US Army, Report 84, page US Govt Printing Office, 1887, page 340
[13] Gen A. P. Hill, OR, Battle report for Shepherdstown, Series 1, Vol 19
[14] NARA M266. Compiled service records of Confederate soldiers from Georgia units,
[15] General Jubal Early, OR, Series 1, Volume XIX, Numbers 269. Report of Brigadier General Jubal A. Early, C. S. Army, commanding Ewell's division, of operations September 3-27, Page 965-973
[16] Mathews, W. C., Wrights Legion, The Sunny South, January 10th, 1891, page 5

Chapter 8 **The Battle of Fredericksburg, Sept - Dec 1862**

[1] Lester, George H., The war Record of the Tom Cobb Infantry 38th Georgia, Co E, Oglethorpe Echo, April 10th, 1879
[2] Blain, James S., Condition of Troops in Virginia, Savannah Republican, Dec 4th, 1862
[3] Guillaume, Army Correspondent Near the Battlefield, Sandersville (Ga.) Central Georgian, Jan 14, 1863
[4] Lester, George H., The war Record of the Tom Cobb Infantry 38th Georgia, Co E, Oglethorpe Echo, April 10th, 1879
[5] Early, Jubal, Lieutenant General Jubal Anderson Early, C.S.A.:Autobiographical Sketch and Narrative of the War Between the States, J.B. Lippincott, 1912, page 167
[6] Letter from Ben, Charleston Daily Courier, December 30, 1862
[7] Lester, George H., The war Record of the Tom Cobb Infantry 38th Georgia, Co E, Oglethorpe Echo, April 10th, 1879
[8] Evans, Clement, Report of Col Clement Evans, OR, Battle report, Chap. XXXIII, page 671
[9] Ibid.
[10] Stiles, Robert, Four Years With Marse Robert, Neal Publishing Co., 1904, page 135
[11] Evans, Clement, Report of Col Clement Evans, OR, Battle report, Chapt. XXXIII, page 671

[12] Booth, Carrie R., CSA Pension Record of widow, Ms. Carrie R. Booth, filed in Cobb Co., Ga. 1919, National Archives
[13] Thornton, Lucinda, Widow's CSA pension application, Confederate Pension Applications, Georgia Confederate Pension Office, RG 58-1-1, Georgia State Archives
[14] Vaughn, Susan Ann, Widow's CSA pension application, Confederate Pension Applications, Georgia Confederate Pension Office, RG 58-1-1, Georgia State Archives
[15] R. Henry, CSA pension application, Georgia Confederate Pension Office, RG 58-1-1, Georgia State Archives
[16] Lester, George H., The war Record of the Tom Cobb Infantry 38th Georgia, Co E, Oglethorpe Echo, April 10th, 1879
[17] Guillaume, Army Correspondent Near the Battlefield, Sandersville (Ga.) Central Georgian, Jan 14, 1863
[18] Letter from Ben, Charleston Daily Courier, December 30, 1862
[19] Lawton's Brigade at the Battle of Fredericksburg, Augusta Chronicle, January 15th, 1863
[20] Guillaume, Army Correspondent Near the Battlefield, Sandersville (Ga.) Central Georgian, Jan 14, 1863
[21] Unknown Confederate Soldier, Lawton's Brigade at the Battle of Fredericksburg, Augusta Chronicle, January 15th, 1863
[22] Guillaume, Army Correspondent Near the Battlefield, Sandersville (Ga.) Central Georgian, Jan 14, 1863
[23] Evans, Clement, Report of Col Clement Evans, OR, Battle report, Chapt. XXXIII, page 671
[24] Lester, George H., The war Record of the Tom Cobb Infantry 38th Georgia, Co E, Oglethorpe Echo, April 10th, 1879
[25] Evans, Clement, Report of Col Clement Evans, OR, Battle report, Chapt. XXXIII, page 671
[26] Unknown Confederate Soldier, Lawton's Brigade at the Battle of Fredericksburg, Augusta Chronicle, Jan 15th, 1863; Guillaume, Army Correspondent Near the Battlefield, Sandersville (Ga.) Central Georgian, Jan 14, 1863.
[27] Army Correspondent of the Savannah Republican, Lawton's Brigade at Fredericksburg, The Macon Telegraph, Jan 3rd, 1863, page 3.
[28] Nichols, G. W., A Soldier's Story of His Regiment (61st Georgia), and incidentally of the Lawton-Gordon-Evans brigade, Army Norhtern Viginia, (Jesup Georgia, 1898) page 64
[29] Confederate Veteran, Volume 30, S.A. Cunningham, 1922, The Georgia Brigade at Fredericksburg, by Isaac Bradwell, page 18-19
[30] Evans, Clement, Report of Col Clement Evans, OR, Battle report, Chapt. XXXIII, page 671
[31] Guillaume, Army Correspondent Near the Battlefield, Sandersville (Ga.) Central Georgian, Jan 14, 1863, Page 2.
[32] Sparks, David S., Inside Lincoln's Army: The Diary of Marsena Rudolph Patrick, Provost Marshall General, Army of the Potomac, Marsena Rudolph Patrick, T. Yoseloff, 1964, page 193
[33] Mathews, W. C., Wrights Legion, The Sunny South, January 10th, 1891

Chapter 9 **Chancellorsville Campaign, Jan - May 1863**

[1] Lester, George H., The War Record of the Tom Cobb Infantry, 38th Ga., Oglethorpe Echo Newspaper (April 10th, 1879)
[2] Stiles, Robert, Four Years With Marse Robert, Neal Publishing Co., 1904, page 157-158
[3] Harrison, Tip, Christmas in War Time, The Anderson intelligencer., Anderson Court House, S.C., January 20, 1897, Page 4
[4] A Snow Battle During the War, The Weekly Constitution, Atlanta, Ga. December 21, 1886
[5] Harrison, Tip, Christmas in War Time, The Anderson Intelligencer, Anderson Court House, S.C., January 20, 1897, Page 4
[6] Stiles, Robert, Four Years With Marse Robert, Neal Publishing Co., 1904, page 158

[7] Harrison, Tip, Christmas in War Time, The Anderson intelligencer., Anderson Court House, S.C., January 20, 1897, Page 4
[8] A Snow Battle During the War, The Weekly Constitution, Atlanta, Ga. December 21, 1886
[9] Unknown 38th Ga. Soldier, Thirty Eighth Georgia Regiment, Augusta, Jan 9th, 1863, Page 2
[10] The National Archives, Mirofilm roll: "M-266 Compiled Service Records of Confederate Soldiers Who Served in Organizations from Georgia - Thirty Eighth Infantry, Rolls 433 through 441, Record of John Brinson
[11] The National Archives, Mirofilm roll: "M-266 Compiled Service Records of Confederate Soldiers Who Served in Organizations from Georgia - Thirty Eighth Infantry, Rolls 433 through 441, Capt William McLeod
[12] Lester, George H., The War Record of the Tom Cobb Infantry, 38th Ga., Oglethorpe Echo Newspaper (April 10th, 1879)
[13] Mathews, W. C., Wrights Legion, The Sunny South, January 10th, 1891
[14] Baxter, John, Letter from John Baxter to Emily Hardman, Mar. 8th, 1863, From Camp at Fredericksburg, Va. owned by author
[15] Official Records of the Union and Confederate Armies, Series, 1, Chapt. 27, Chap. XXXIX, Confederate Correspondence, p. 865
[16] Gordon, John B., Reminiscences of the Civil War, New York: Charles Scribner's Sons, 1904, page 92-93
[17] Ibid, page 95.
[18] McCleskey, Francis C., Letter to mother, April 18th, 1863, Georgia State Archives
[19] Sedgwick, John, General Sedgwick's official report, May 15th, 1863, OR
[20] Hudgins, Francis L., With The 38th Georgia, Confederate Veteran Magazine, 1918.
[21] Bradwell, Isaac, Confederate Veteran, Oct 1915, 23, 446-447
[22] Hudgins, Francis L., With The 38th Georgia, Confederate Veteran Magazine, 1918.
[23] Bradwell, I. G., Confederate Veteran, October 1915, 23: 446-447
[24] Bache, Richard, Meade, Life of General George Gordon Meade, H.T. Coates & co., 1897, P. 260 commander of the Army of the Potomac
[25] O.R.- Series I-Volume XXV/1 (General Order #47)
[26] Sedgwick, John, General Sedgwick's official report, May 15th, 1863, OR
[27] Mathews, William C., The Central Georgian, June 3rd, 1863
[28] Ibid.
[29] Early, Jubal, Lieutenant General Jubal Anderson Early, C.S.A., Autobiographical sketch and narrative of the war between the states, J.B. Lippincott Company, 1912, P. 218
[30] Hudgins, Francis L., With the 38th Georgia, Confederate Veteran Magazine, 1918.
[31] Mathews, William C., The Central Georgian, June 3rd, 1863
[32] Gordon, John B., Reminiscences of the Civil War, New York: Charles Scribner's Sons, 1904, page 100
[33] Mathews, William C., The Central Georgian, June 3rd, 1863
[34] Gordon, John B., Reminiscences of the Civil War, New York: Charles Scribner's Sons, 1904, page 100-101
[35] V.A.S.P., Letter From The Army in Virginia, Savannah Republican Newspaper, June 18th, 1863, P. 2
[36] Mathews, William C., The Central Georgian, June 3rd, 1863
[37] Hudgins, Francis L., With the 38th Georgia, Confederate Veteran Magazine, 1918
[38] Gordon, John B., Reminiscences of the Civil War, New York: Charles Scribner's Sons, 1904, page 101
[39] Mathews, William C., The Central Georgian, June 3rd, 1863
[40] Nichols, George W., G. W. Nichols, A Soldier's Story of His Regiment (61st Georgia), and incidentally of the Lawton-Gordon-Evans brigade, Army Northern Virginia, (Jesup Georgia, 1898)
[41] Mathews, William C., The Central Georgian, June 3rd, 1863
[42] Hudgins, Francis L., With the 38th Georgia, Confederate Veteran Magazine, 1918
[43] Snow, William Parker, Lee and his generals, Richardson & Company, 1867, P. 96

[44] Hudgins, Francis L., With the 38th Georgia, Confederate Veteran Magazine, 1918

Chapter 10 **The Battle of 2nd Winchester, June, 1863**

[1] Bradwell, Isaac G., Confederate veteran, Volume 30, (S.A. Cunningham 1922) 330.
[2] Nichols, G. W., A Soldier's Story of His Regiment (61st Georgia), and incidentally of the Lawton-Gordon-Evans brigade, Army northern Virginia (Jesup Georgia, 1898) 116-117
[3] Bradwell, Isaac G., Confederate veteran, Volume 30, (S.A. Cunningham 1922) 330.
[4] Milroy, Robert, Letter from Robert Milroy to Mary Milroy, Jan 18th, 1863, Milroy papers
[5] G. W. Nichols, A Soldier's Story of His Regiment (61st Georgia), and incidentally of the Lawton-Gordon-Evans brigade, Army northern Virginia (Jesup Georgia, 1898) 116-117
[6] Mathews, William C., The Central Georgian, July 9th, 1863
[7] Death Notice of Capt. Charles Hawkins, Augusta Chronicle, July 30th, 1862, p. 2
[8] Mathews, W. C., Wrights Legion, The Sunny South, January 10th, 1891
[9] Mathews, William C., The Central Georgian, July 9th, 1863
[10] Bradwell, Isaac, Confederate Veteran, Vol 30, (S.A. Cunningham 1922) page 331
[11] Mathews, W. C., The Central Georgian, July 9th, 1863
[12] Americus, Augusta Chronicle Newspaper, July 30th, 1863

Chapter 11 **The Road to Gettysburg**

[1] Hudgins, F. L., With the 38th Georgia Regiment,"Confederate Veteran, (Vol. XXVI, pp. 161-163, 1918
[2] Lester George L., War Record of the Tom Cobb Infantry, Company E, 38th Georgia Regiment, Oglethorpe Echo newspaper, April 10th, 1879
[3] Hudgins, F. L., With the 38th Georgia Regiment,"Confederate Veteran, (Vol. XXVI, pp. 161-163, 1918)
[4] Gordon,, John B.. Reminiscences of the Civil War, New York: Charles Scribner's Sons, 1904, p. 150
[5] Hudgins, F. L., With the 38th Georgia Regiment,"Confederate Veteran, (Vol. XXVI, pp. 161-163, 1918)
[6] Hudgins, F. L., With the 38th Georgia Regiment,"Confederate Veteran, (Vol. XXVI, pp. 161-163, 1918)
[7] Letter to Mrs. Angeline McCleskey, Georgia State Archives, July 18th, 1863
[8] Confederate Service Record of Pvt William J. Powell
[9] Official Records of the War of the Rebellion, No. 477 Report of Brig. General J. B
[10] Official Records of the War of the Rebellion, No. 477 Report of Brig. General J. B. Gordon
[11] Campbell, Jeptha, Jep Campbell Spins a Reminiscence, Elberton Star Newspaper, February 2nd, 1889.
[12] Campbell, Jeptha, Jep Campbell Spins a Reminiscence, Elberton Star Newspaper, February 2nd, 1889
[13] Report of Brig. Gen. J. B. Gordon, C. S. Army, commanding brigade. JUNE 3-AUGUST 1, 1863.--The Gettysburg Campaign. O.R.-- SERIES I--VOLUME XXVII/2 [S# 44]
[14] Mathews, W. C., Wrights Legion, The Sunny South, January 10th, 1891
[15] Hudgins, F. L., With the 38th Georgia Regiment,"Confederate Veteran, (Vol. XXVI, pp. 161-163, 1918)
[16] Mathews, W. C., Wrights Legion, The Sunny South, January 10th, 1891
[17] Imboden, John D., "The Confederate Retreat from Gettysburg," included in Robert U. Johnson and Clarence C. Buel, eds. Battles and Leaders of the Civil War, 4 vols. (New York: Century Publishing Co. 1884-1904), 3:423.

[18] Lester George L., War Record of the Tom Cobb Infantry, Company E, 38th Georgia Regiment, Oglethorpe Echo newspaper, April 10th, 1879
[19] Official Records of the War of the Rebellion, No. 477 Report of Brig. General J. B. Gordon
[20] Harrison, W. H. Norcross Herald, Col Ed Atkinson, Nov. 13th, 1908, page 3,
[21] Hudgins, F. L., With the 38th Georgia Regiment,"Confederate Veteran, (Vol. XXVI, pp. 161-163, 1918)
[22] Reminiscences of the Civil War, Gordon, John B., New York Charles Scribner's Sons Atlanta The Martin & Hoyt Co. 1904, pages 173-174
[23] Letter to Mrs. Angeline McCleskey, Georgia State Archives
[24] Hudgins, F. L., With the 38th Georgia Regiment, "Confederate Veteran, (Vol. XXVI, pp. 161-163, 1918)

Chapter 12 **The Battle of Wilderness, Day One – May 5th, 1864**

[1] Mathews, W. C., Wrights Legion, The Sunny South, January 10th, 1891, Page 5
[2] Gordon, John Brown. Reminiscences of the Civil War. New York: Charles Scribner's Sons, 1903, p. 229-230
[3] JCC, Augusta Chronicle Newspaper, Sept 5th, 1863, Letters From Virginia, page 2
[4] Post war letter referenced provided by Ms. Connie Welch, descendant of John Cottle Chew.
[5] Sergeant, Augusta Chronicle Newspaper, Sept 5th, 1863, Letters From Virginia, page 2
[6] Illegible
[7] Letter Courtesy of Mrs. Lois Grubbs of Atlanta, GA, descendant of Lieutenant Simpson A. Hagood
[8] Mathews, W. C., Wrights Legion, The Sunny South, January 10th, 1891, Page 5
[9] Lester, George L., War Record of the Tom Cobb Infantry, Company E, 38th Georgia Regiment, Oglethorpe Echo Newspaper, April 10th, 1879
[10] Hudgins, F. L., With the 38th Georgia Regiment, Confederate Veteran, (Vol. XXVI, pp. 161-163, 1918)
[11] Gordon, John Brown. Reminiscences of the Civil War. New York: Charles Scribner's Sons, 1903. p. 238-239.
[12] Ibid.
[13] George L. Lester, War Record of the Tom Cobb Infantry, Company E, 38th Georgia Regiment, Oglethorpe Echo Newspaper, April 10th, 1879
[14] Hudgins, F. L., With the 38th Georgia Regiment, "Confederate Veteran, (Vol. XXVI, pp. 161-163, 1918)
[15] Gordon, John Brown. Reminiscences of the Civil War. New York: Charles Scribner's Sons, 1903, p. 239
[16] Bradwell, Isaac, Confederate Veteran, December 1919,Vol 27, P. 458-459
[17] Nichols, G. W., A Soldier's Story of His Regiment (61st Georgia), and incidentally of the Lawton-Gordon-Evans Brigade, Army Northern Virginia, (Jesup Georgia, 1898) p.80
[18] Ibid.
[19] Gordon, John Brown. Reminiscences of the Civil War. New York: Charles Scribner's Sons, 1903 p. 239.
[20] Gordon, John Brown. Reminiscences of the Civil War. New York: Charles Scribner's Sons, 1903, p. 237-241. Note: These two drawings have been slightly modified to show correct position of each regiment as General Gordon described in his battle report found in the *Official Records of the War of the Rebellion.*
[21] Ibid.
[22] Nichols, G. W., A Soldier's Story of His Regiment (61st Georgia), and incidentally of the Lawton-Gordon-Evans Brigade, Army Northern Virginia, (Jesup Georgia, 1898) p. 143-144
[23] Lester, George L., War Record of the Tom Cobb Infantry, Company E, 38th Georgia Regiment, Oglethorpe Echo Newspaper, April 10th, 1879

[24] Hudgins, F. L., With the 38th Georgia Regiment,"Confederate Veteran, (Vol. XXVI, pp. 161-163, 1918)
[25] Gordon, John Brown. Reminiscences of the Civil War. New York: Charles Scribner's Sons, p. 242
[26] Bradwell, Isaac, Confederate Veteran, December 1919, Vol 27, P. 458-459
[27] Roster of the Confederate soldiers of Georgia, 1861-1865 compiled by Lillian Henderson

Chapter 13 **The Wilderness, Day Two - May 6th, 1864**

[1] Gordon, John B. Reminiscences of the Civil War, New York: Charles Scribner's Sons, 1904, p. 242-243.
[2] Ibid, p. 244-249.
[3] Hudgins, F. L., With the 38th Georgia Regiment, "Confederate Veteran, (Vol. XXVI, pp. 161-163, 1918)
[4] Bradwell, Isaac, Confederate Veteran, Second Days' Battle of the Wilderness, May 6th, 1864, January 1920, Vol 28, p 20.
[5] Hudgins, F. L., With the 38th Georgia Regiment, "Confederate Veteran, (Vol. XXVI, pp. 161-163, 1918)
[6] Bradwell, Isaac, Confederate Veteran, Second Days' Battle of the Wilderness, May 6th, 1864, January 1920, Vol 28, p 20.
[7] Hudgins, F. L., With the 38th Georgia Regiment, "Confederate Veteran, (Vol. XXVI, pp. 161-163, 1918)
[8] Gordon, John B., Reminiscences of the Civil War, New York: Charles Scribner's Sons, 1904, p. 249
[9] Hudgins, F. L., With the 38th Georgia Regiment, "Confederate Veteran, (Vol. XXVI, pp. 161-163, 1918)
[10] Gordon, John B., Reminiscences of the Civil War, New York: Charles Scribner's Sons, 1904, p. 249-250.
[11] Hudgins, F. L., With the 38th Georgia Regiment, "Confederate Veteran, (Vol. XXVI, pp. 161-163, 1918)
[12] Gordon, John B., Reminiscences of the Civil War, New York: Charles Scribner's Sons, 1904, p. 265
[13] The century illustrated monthly magazine, 1897, Volume 53, P. 230
[14] Hudgins, F. L., With the 38th Georgia Regiment, "Confederate Veteran, (Vol. XXVI, pp. 161-163, 1918)
[15] Ibid.
[16] Henderson's Roster of Georgia Confederate Soldiers, Complied by Lillian Henderson, Georgia State Archives & Confederate Service Records, Georgia State Archives

Chapter 14 **The Battles of Spotsylvania Court House - May 10th, 1864**

[1] Lester, George L., War Record of the Tom Cobb Infantry, Company E, 38th Georgia Regiment, Oglethorpe Echo newspaper, April 10th, 1879
[2] Gordon, John B., OR, Numbers 287. Page1077 Chap. XLVIII, Report of Brigadier General John B. Gordon, C. S. Army
[3] Bradwell, Gordon, Confederate Veteran Magazine, 1920, 28 56-57
[4] Lyman, Theodore, Col Theodore Lyman, Meade's headquarters, 1863-1865:letters of Colonel Theodore Lyman from the Wilderness to Appomattox, The Atlantic monthly press, 1922, page 100, Letter from May 18th, 1864
[5] Gordon, John B., OR, Numbers 287. Page1077 Chap. XLVIII, Report of Brigadier General John B. Gordon, C. S. Army
[6] Bradwell, Gordon, Confederate Veteran Magazine, 1920, 28 56-57

[7] Ibid.
[8] General John B. Gordon, Reminisces of the Civil War, page 287-288
[9] Ibid.
[10] Nichols, G. W., A Soldier's Story of His Regiment (61st Georgia), and incidentally of the Lawton-Gordon-Evans brigade, Army Northern

Chapter 15 **Spotsylvania Court House – May 12th 1864**

[1] Gordon, John B., OR, Numbers 287. Page1077 Chap. XLVIII, Report of Brigadier General John B. Gordon, C. S. Army
[2] Ibid.
[3] John B. Gordon, Reminiscences of the Civil War, New York: Charles Scribner's Sons, 1904, P. 274-275
[4] Gordon, John B., OR, Numbers 287. Page1077 Chap. XLVIII, Report of Brigadier General John B. Gordon, C. S. Army
[5] Campbell, Jep E., The History of Co. H, 38th Georgia Regt, The Elberton Star, March 30, 1889; Elberton, GA
[6] Evans, Clement A., General C.A. Evans, Sketch of Captain F.N. Graves, Confederate Veteran Magazine, January 1897, pages 5-6
[7] Campbell, Jep E., The History of Co. H, 38th Georgia Regt, The Elberton Star, March 30, 1889; Elberton, Georgia
[8] Caldwell, J. F. J., Caldwell, J. F. J., The history of a brigade of South Carolinians, known first as "Gregg's" and subsequently as "McGowan's brigade." King & Baird Printers, Philadelphia, 1866, pages 143-145.
[9] Gordon, John B., Reminiscences of the Civil War, New York: Charles Scribner's Sons, 1904, P. 274-275
[10] Gordon, John B., OR, Numbers 287. Page1077, Chap. XLVIII, Report of Brigadier General John B. Gordon, C.S. Army
[11] Reynolds, William B., Report of Major William B. Reynolds, 17th Vermont, The War of the Rebellion: A Compilation of the Official Records of the Union and Confederate Armies, U.S. Government Printing Office, 1891 - Confederate States of America, Series I, Vol XXXVI, Part 1, Reports, pages 935-936
[12] Gould, Joseph, The Story of the Forty-eighth, Published by the Authority of the Regimental Association, (Philadelphia, 1908) Pages 179-180
[13] Nichols, G. W., A Soldier's Story of His Regiment (61st Georgia), and incidentally of the Lawton-Gordon-Evans brigade, Army northern Virginia (Jesup Georgia, 1898) P. 151-154
[14] Mathews, W. C., Wrights Legion, The Sunny South, January 10th, 1891
[15] Ibid, p. 157-159
[16] Benson, Berry, Berry Benson's Civil War Book: Memoirs of a Confederate Scout and Sharpshooter. Edited by Susan Williams Benson. Athens: University of Georgia Press, 1961, page 77
[17] Nichols, G. W., A Soldier's Story of His Regiment (61st Georgia), and incidentally of the Lawton-Gordon-Evans brigade, Army northern Virginia (Jesup Georgia, 1898) Page 160
[18] Bomar, Thomas H, Thomas H. Bomar, Letter from Confederate Service Record
[19] Gordon, John B., Reminiscences of the Civil War, New York: Charles Scribner's Sons, 1904, P. pages 290-292

Chapter 16 **From Spotsylvania To Cold Harbor, May - June 1864**

[1] Lester, George H., Lieutenant George H. Lester, The War Record of the Tom Cobb Infantry, 38th Ga., Oglethorpe Echo Newspaper (April 10th, 1879)

[2] Nichols, G. W., A Soldier's Story of His Regiment (61st Georgia), and incidentally of the Lawton-Gordon-Evans brigade, Army northern Virginia (Jesup Georgia, 1898) Page 168
[3] Roster of the Confederate soldiers of Georgia, 1861-1865 compiled by Lillian Henderson
[4] Bradwell, Isaac G., Confederate Veteran, S. A. Cunningham, Nashville, TN, Volume 28, 1920, page 138
[5] Rankin, John G, Letter to George Riley Wells, dated June 6th, 1864, courtesy of Mr. Lee Joyner
[6] Mathews, W. C., Wrights Legion, The Sunny South, January 10th, 1891
[7] Georgia. State Division of Confederate Pensions and Records, (1959 - 1964). The National Archives, Mirofilm roll: "M-266 Compiled Service Records of Confederate Soldiers Who Served in Organizations from Georgia - Thirty Eighth Infantry, Rolls 433 through 441. Confederate Pension Applications, Georgia Confederate Pension Office, RG 58-1-1, Georgia Archives.

Chapter 17 **From Cold Harbor to Washington, D. C., June - July 1864**

[1] Bradwell, Isaac G., Confederate Veteran, S. A. Cunningham, Nashville, TN, Volume 28, 1920, page 138
[2] Ibid
[3] Lester, George H., Lieutenant George H. Lester, The War Record of the Tom Cobb Infantry, 38th Ga., Oglethorpe Echo Newspaper (April 10th, 1879)
[4] Bradwell, Isaac G., Confederate Veteran, S. A. Cunningham, Nashville, TN, Volume 28, 1920, page 138
[5] Ibid
[6] Lester, George H., Lieutenant George H. Lester, The War Record of the Tom Cobb Infantry, 38th Ga., Oglethorpe Echo Newspaper (April 10th, 1879)
[7] Dickerson, F. L., Sketch of the Campaign of 1864 to Surrender, Henderson's Roster of Georgia Confederate Soldiers,
[8] Lester, George H., Lieutenant George H. Lester, The War Record of the Tom Cobb Infantry, 38th Ga., Oglethorpe Echo Newspaper (April 10th, 1879)
[9] Nichols, G. W. Nichols, A Soldier's Story of His Regiment (61st Georgia), and incidentally of the Lawton-Gordon-Evans brigade, Army Northern Virginia (Jesup Georgia, 1898) P. 170-175
[10] Ibid
[11] Gordon, John B., General John B. Gordon, Official Records of the War of the Rebellion, Page 350 Chap. XLIX, Operations In Shenandoah Valley, Etc
[12] Bradwell, I. G., Bradwell, Confederate Veteran, "Early's March to Washington in 1864," (28:176-177, 1920)
[13] Gordon, John B., General John B. Gordon, Official Records of the War of the Rebellion, Page 350 Chap. XLIX, Operations In Shenandoah Valley, Etc
[14] Bradwell, Isaac G., Confederate Veteran, S. A. Cunningham, Nashville, TN, Volume 28, 1920,.page 138
[15] Nichols, G. W. Nichols, A Soldier's Story of His Regiment (61st Georgia), and incidentally of the Lawton-Gordon-Evans brigade, Army Northern Virginia (Jesup Georgia, 1898) P. 170-175
[16] Gordon, John B., General John B. Gordon, Official Records of the War of the Rebellion, Page 350 Chap. XLIX, Operations In Shenandoah Valley, Etc
[17] Langford, Josiah C., Confederate Pension Application, Georgia Confederate Pension Office, RG 58-1-1, Georgia State Archives
[18] Nichols, G. W. Nichols, A Soldier's Story of His Regiment (61st Georgia), and incidentally of the Lawton-Gordon-Evans brigade, Army Northern Virginia (Jesup Georgia, 1898) P. 170-175
[19] The National Archives, Mirofilm roll: "M-266 Compiled Service Records of Confederate Soldiers Who Served in Organizations from Georgia - Thirty Eighth Infantry, Rolls 433 through 441
[20] Nichols, G. W. Nichols, A Soldier's Story of His Regiment (61st Georgia), and incidentally of the Lawton-Gordon-Evans brigade, Army Northern Virginia (Jesup Georgia, 1898) P. 170-175

[21] Gordon, John B., General John B. Gordon, Official Records of the War of the Rebellion, Page 350 Chap. XLIX., Operations In Shenandoah Valley, Etc
[22] Nichols, G. W. Nichols, A Soldier's Story of His Regiment (61st Georgia), and incidentally of the Lawton-Gordon-Evans brigade, Army Northern Virginia (Jesup Georgia, 1898) P. 170-175
[23] Gordon, John B., General John B. Gordon, Official Records of the War of the Rebellion, Page 350 Chap. XLIX, Operations In Shenandoah Valley, Etc
[24] Bradwell, I. G., Confederate Veteran, "Early's March to Washington in 1864," (28:176-177, 1920)
[25] Gordon, John B., General John B. Gordon, Official Records of the War of the Rebellion, Page 350 Chap. XLIX, Operations In Shenandoah Valley, Etc
[26] Bradwell, I. G., Confederate Veteran, "Early's March to Washington in 1864," (28:176-177, 1920)
[27] Gordon, John B, General John B. Gordon, Official Records of the War of the Rebellion, Page 350 Chap. XLIX, Operations In Shenandoah Valley, Etc
[28] Nichols, G. W. Nichols, A Soldier's Story of His Regiment (61st Georgia), and incidentally of the Lawton-Gordon-Evans brigade, Army Northern Virginia (Jesup Georgia, 1898) P. 170-175
[29] Mathews, William C., Captain William C. Mathews, W. C., *Wrights Legion*, The Sunny South, January 10th, 1891
[30] Nichols, G. W. Nichols, A Soldier's Story of His Regiment (61st Georgia), and incidentally of the Lawton-Gordon-Evans brigade, Army Northern Virginia (Jesup Georgia, 1898) P. 170-175
[31] Confederate Service Record of G.W. Graves, The National Archives, Mirofilm roll: "M-266 Compiled Service Records of Confederate Soldiers Who Served in Organizations from Georgia - Thirty Eighth Infantry, Rolls 433 through 441.
[32] Editorial, Sharpsburg and Gettysburg, Richmond Examiner, August 4th, 1864, Page 3
[33] Nichols, G. W. Nichols, A Soldier's Story of His Regiment (61st Georgia), and incidentally of the Lawton-Gordon-Evans brigade, Army Northern Virginia (Jesup Georgia, 1898) P. 170-175
[34] Mathews, William C, Captain William C. Mathews, W. C., *Wrights Legion*, The Sunny South, January 10th, 1891
[35] Gen. Jubal Early, Lieutenant General Jubal Anderson Early C.S.A.: Autobiographical Sketch and Narrative of the War Between the States, Pages 389-294, Philadelphia; London: J. B. Lippincott Company, 1912
[36] Captain Pearce, The Maryland Expedition, The Charleston Mercury, July 24th, 1864, page 1
[37] Grant, U.S., Gen. U.S. Grant, Personal Memoirs of U. S. Grant, New York, Charles L. Webster & Co., 1885, page 265

Chapter 18 **Return to the Shenandoah Valley, Aug - Sept 1864**

[1] Lester, George L., War Record of the Tom Cobb Infantry, Company E, 38th Georgia Regiment, Oglethorpe Echo newspaper, April 10th, 1879
[2] Dickerson, Francis A., 38th Ga. Sketch of Campaign of 1864 to Surrender, Henderson's Roster of Confederate Soldiers of Georgia, 1861-1861, Compiled by Lillian Henderson, US National Archives
[3] Ibid
[4] Sunny South
[5] Dickerson, Francis A., 38th Ga. Sketch of Campaign of 1864 to Surrender, Henderson's Roster of Confederate Soldiers of Georgia, 1861-1861, Compiled by Lillian Henderson, US National Archives
[6] Miller, Francis Trevelyan, Robert Sampson Lanier, Photographic History of the Civil War, Vol III, The Schweinler Press, New York, 1911, pages 149-150
[7] Dickerson, Francis A., 38th Ga. Sketch of Campaign of 1864 to Surrender, Henderson's Roster of Confederate Soldiers of Georgia, 1861-1861, Compiled by Lillian Henderson, US National Archives
[8] Nichols, G. W., A Soldier's Story of His Regiment (61st Georgia), and incidentally of the Lawton-Gordon-Evans brigade, Army northern Virginia (Jesup Georgia, 1898) P. 178

[9] Dickerson, Francis A., 38th Ga. Sketch of Campaign of 1864 to Surrender, Henderson's Roster of Confederate Soldiers of Georgia, 1861-1861, Compiled by Lillian Henderson, US National Archives
[10] Letter courtesy of Mr. John Davis, descendant of Lt Hagood.
[11] Dickerson, Francis A., 38th Ga. Sketch of Campaign of 1864 to Surrender, Henderson's Roster of Confederate Soldiers of Georgia, 1861-1861, Compiled by Lillian Henderson, US National Archives
[12] Grant, U.S., Gen. U.S. Grant, Personal Memoirs of U. S. Grant, New York, Charles L. Webster & Co., 1885, page 395
[13] Gordon, John B., Reminiscences of the Civil War, New York: Charles Scribner's Sons, 1904, pages 320

Chapter 19 **The Battle of 3rd Winchester (or Opequon Creek) September 19th, 1864**

[1] Nichols, G. W. Nichols, A Soldier's Story of His Regiment (61st Georgia), and incidentally of the Lawton-Gordon-Evans brigade, Army Northern Virginia (Jesup Georgia, 1898), P. 187
[2] Bradwell, Isaac, G., Pvt. I. G. Bradwell, Early's Valley Campaign 1864, Confederate Veteran, 28:218-219, 1920
[3] Nichols, G. W. Nichols, A Soldier's Story of His Regiment (61st Georgia), and incidentally of the Lawton-Gordon-Evans brigade, Army Northern Virginia (Jesup Georgia, 1898) P. 191-195
[4] Bradwell, Isaac, G., Pvt. I. G. Bradwell, Early's Valley Campaign 1864, Confederate Veteran, 28:218-219, 1920
[5] Lester, George H., Lieutenant George H. Lester, The War Record of the Tom Cobb Infantry, 38th Ga., Oglethorpe Echo Newspaper (April 10th, 1879)
[6] Mathews, William C., Captain William C. Mathews, W. C., *Wrights Legion*, The Sunny South, January 10th, 1891
[7] Pension Application of Widow, Martha Gardner, Confederate Pension Applications, Georgia Confederate Pension Office, RG 58-1-1, Georgia Archives.
[8] Bradwell, Isaac, G., Pvt. I. G. Bradwell, Early's Valley Campaign 1864, Confederate Veteran, 28:218-219, 1920
[9] Confederate Service Record of Garvin Farrer; Augusta Chronicle Newspaper dated 11/11/1864
[10] Hardin Jordan, Georgia Confederate Pension Application, Confederate Pension Applications, Georgia Confederate Pension Office, RG 58-1-1, Georgia Archives
[11] Lester, George H., Lieutenant George H. Lester, The War Record of the Tom Cobb Infantry, 38th Ga., Oglethorpe Echo Newspaper (April 10th, 1879)
[12] The National Archives, Mirofilm roll: "M-266 Compiled Service Records of Confederate Soldiers Who Served in Organizations from Georgia - Thirty Eighth Infantry, Rolls 433 through 441.
[13] Christian, Benjamin B., Confederate Service Record, The National Archives, Mirofilm roll: "M-266 Compiled Service Records of Confederate Soldiers Who Served in Organizations from Georgia - Thirty Eighth Infantry, Rolls 433 through 441.
[14] Lester, George H., Lieutenant George H. Lester, The War Record of the Tom Cobb Infantry, 38th Ga., Oglethorpe Echo Newspaper (April 10th, 1879)
[15] Mathews, William C., Wrights Legion, The Sunny South, January 10th, 1891
[16] Gordon, John Brown. Reminiscences of the Civil War, New York: Charles Scribner's Sons, 1903, P. 323 - 326
[17] The National Archives, Mirofilm roll: "M-266 Compiled Service Records of Confederate Soldiers Who Served in Organizations from Georgia - Thirty Eighth Infantry, Rolls 433 through 441

Chapter 20 **The Battle of Fisher's Hill – September 21-22, 1864**

[1] Bradwell, I. G., Confederate Veteran, The Battle of Fisher's Hill, Vol 28, 339, 1920

[2] Lester, George H., Lieutenant George H. Lester, The War Record of the Tom Cobb Infantry, 38th Ga., Oglethorpe Echo Newspaper (April 10th, 1879)
[3] Buck, Samuel, Confederate Veteran Vol 2, 1894, Battle of Fisher's Hill, S. A. Cunningham, page 338
[4] Early, Jubal, Lieutenant General Jubal A. Early, Autobiographical Sketch and Narrative of the Civil War, J. B. Lippincott & Co., Philadelphia, 1912, pages 429-431
[5] Ibid.
[6] Ibid.
[7] Gen. John B. Gordon, P. 326
[8] Lester, George H., Lieutenant George H. Lester, The War Record of the Tom Cobb Infantry, 38th Ga., Oglethorpe Echo Newspaper (April 10th, 1879)
[9] Dickerson, F. L., Sketch of the Campaign of 1864 to Surrender, Henderson's Roster of Georgia Confederate Soldiers,
[10] Private I. G. Bradwell, Confederate Veteran, The Battle of Fisher's Hill, Vol 28, 339, 1920
[11] Nichols, G. W. Nichols, A Soldier's Story of His Regiment (61st Georgia), and incidentally of the Lawton-Gordon-Evans brigade, Army Northern Virginia (Jesup Georgia, 1898) P. 198-200
[12] Evans, Clement A., Intrepid warrior: Clement Anselm Evans, Confederate General from Georgia; life, letters, and diaries of the war years, edited by Robert Grier Stephens, Jr., Dayton, Ohio: Letter to Allie, September 26th, 1864, Dayton, Morningside Press, 1992, page 455
[13] Nichols
[14] Lester, George H., Lieutenant George H. Lester, The War Record of the Tom Cobb Infantry, 38th Ga., Oglethorpe Echo Newspaper (April 10th, 1879)

Chapter 21 **The Valley in Flames, September - October, 1864**

[1] Evans, Clement A., Intrepid warrior: Clement Anselm Evans, Confederate general from Georgia; life, letters, and diaries of the war years, edited by Robert Grier Stephens, Jr., Dayton, Ohio: Letter to Allie, September 28th, 1864, Dayton, Morningside Press, 1992, pages 457-458
[2] Nichols, G. W. Nichols, A Soldier's Story of His Regiment (61st Georgia), and incidentally of the Lawton-Gordon-Evans brigade, Army Northern Virginia (Jesup Georgia, 1898) P. 200-201
[3] Payton, John Lewis, History of Augusta County, Virginia, Samuel M. Yost & Son, 1882, page 239
[4] Lester, George H., Lieutenant George H. Lester, The War Record of the Tom Cobb Infantry, 38th Ga., Oglethorpe Echo Newspaper (April 10th, 1879)
[5] Nichols, G. W. Nichols, A Soldier's Story of His Regiment (61st Georgia), and incidentally of the Lawton-Gordon-Evans brigade, Army Northern Virginia (Jesup Georgia, 1898) P. 202
[6] Gordon, John B., Reminiscences of the Civil War, New York: Charles Scribner's Sons, 1904, P. 332
[7] Mathews, William C., Captain William C. Mathews, W. C., *Wrights Legion*, The Sunny South, January 10th, 1891

Chapter 22 **Battle of Cedar Creek – October 19th, 1864**

[1] Early, Jubal A, Jubal Early's Memoirs: Autobiographical Sketch and Narrative of the War Between the States, J.B. Lippincott Company, 1912, P. 452
[2] Ibid, P. 452
[3] Gordon, John B., Reminiscences of the Civil War, New York: Charles Scribner's Sons, 1904, Pages 333-336
[4] Nichols, G. W. Nichols, A Soldier's Story of His Regiment (61st Georgia), and incidentally of the Lawton-Gordon-Evans brigade, Army Northern Virginia (Jesup Georgia, 1898) Pages 202-203
[5] Davant, P. E., Lt Col P. E. Davant, Casualties in the 38th Georgia Regiment, Evans Brigade, Macon Telegraph, November 1st, 1864

[6] Lester, George H., Lieutenant George H. Lester, The War Record of the Tom Cobb Infantry, 38th Ga., Oglethorpe Echo Newspaper (April 10th, 1879)
[7] Nichols, G. W. Nichols, A Soldier's Story of His Regiment (61st Georgia), and incidentally of the Lawton-Gordon-Evans brigade, Army Northern Virginia (Jesup Georgia, 1898) Pages 194-195
[8] Bradwell, Isaac G, The Battle of Cedar Creek, VA, Confederate Veteran, S.A. Cunningham, 1914, Vol 22, Page 315
[9] Nichols, G. W. Nichols, A Soldier's Story of His Regiment (61st Georgia), and incidentally of the Lawton-Gordon-Evans brigade, Army Northern Virginia (Jesup Georgia, 1898) Pages 194-195
[10] Lester, George H., Lieutenant George H. Lester, The War Record of the Tom Cobb Infantry, 38th Ga., Oglethorpe Echo Newspaper (April 10th, 1879)
[11] Nichols, G. W. Nichols, A Soldier's Story of His Regiment (61st Georgia), and incidentally of the Lawton-Gordon-Evans brigade, Army Northern Virginia (Jesup Georgia, 1898) Pages 204
[12] Gordon, John B., Reminiscences of the Civil War, New York: Charles Scribner's Sons, 1904, Pages 341-342
[13] Lester, George H., Lieutenant George H. Lester, The War Record of the Tom Cobb Infantry, 38th Ga., Oglethorpe Echo Newspaper (April 10th, 1879)
[14] Nichols, G. W. Nichols, A Soldier's Story of His Regiment (61st Georgia), and incidentally of the Lawton-Gordon-Evans brigade, Army Northern Virginia (Jesup Georgia, 1898) Page 205
[15] Ibid, P. 205
[16] Ibid, 196-197
[17] Childers, William J., Confederate Pension Applications, Georgia Confederate Pension Office, RG 58-1-1, Georgia Archives.
[18] House, Thomas J, Confederate Pension Applications, Georgia Confederate Pension Office, RG 58-1-1, Georgia Archives.
[19] Lester, George H., Lieutenant George H. Lester, The War Record of the Tom Cobb Infantry, 38th Ga., Oglethorpe Echo Newspaper (April 10th, 1879)
[20] DeForest, John W, A Volunteer's Adventure, by Captain John W. De Forest James H. Croushore, Ed., (New Haven: Yale University Press, 1949), 190 & 231
[21] Early, Jubal A, Jubal Early's Memoirs: Autobiographical Sketch and Narrative of the War Between the States, J.B. Lippincott Company,1912, P. 451

Chapter 23 **In the Trenches at Petersburg, Dec 1864 -April 1865/The Battle of Hatcher's Run – February 5 - 6, 1865**

[1] Lester, George H., Lieutenant George H. Lester, The War Record of the Tom Cobb Infantry, 38th Ga., Oglethorpe Echo Newspaper (April 10th, 1879)
[2] Nichols, G. W., A Soldier's Story of His Regiment (61st Georgia), and incidentally of the Lawton-Gordon-Evans brigade, Army Northern Virginia (Jesup Georgia, 1898), P. 211
[3] Gordon,, John B.. Reminiscences of the Civil War, New York: Charles Scribner's Sons, 1904
[4] Nichols, G. W., A Soldier's Story of His Regiment (61st Georgia), and incidentally of the Lawton-Gordon-Evans brigade, Army Northern Virginia (Jesup Georgia, 1898)
[5] Gordon,, John B.. Reminiscences of the Civil War, New York: Charles Scribner's Sons, 1904
[6] Lester, George H., Lieutenant George H. Lester, The War Record of the Tom Cobb Infantry, 38th Ga., Oglethorpe Echo Newspaper (April 10th, 1879)
[7] Mathews, W. C., Wrights Legion, The Sunny South, January 10th, 1891
[8] Lester, George H., Lieutenant George H. Lester, The War Record of the Tom Cobb Infantry, 38th Ga., Oglethorpe Echo Newspaper (April 10th, 1879)
[9] Nichols, G. W., A Soldier's Story of His Regiment (61st Georgia), and incidentally of the Lawton-Gordon-Evans brigade, Army Northern Virginia (Jesup Georgia, 1898) P. 218
[10] Ibid, P. 217
[11] Mathews, W. C., Wrights Legion, The Sunny South, January 10th, 1891

[12] Lester, George H., Lieutenant George H. Lester, The War Record of the Tom Cobb Infantry, 38th Ga., Oglethorpe Echo Newspaper (April 10th, 1879)
[13] Nichols, G. W., A Soldier's Story of His Regiment (61st Georgia), and incidentally of the Lawton-Gordon-Evans brigade, Army Northern Virginia (Jesup Georgia, 1898) P. 218
[14] Confederate Service Record of James C. Daniel, NARA M266, Records group 109, Compiled service records of Confederate soldiers from Georgia
[15] Lester George L., War Record of the Tom Cobb Infantry, Company E, 38th Georgia Regiment, Oglethorpe Echo newspaper, April 10th, 1879
[16] Confederate Service Records, NARA M266, Records group 109, Compiled service records of Confederate Soldiers from Georgia
[17] Jones, John W., Personal Reminiscences, Anecdotes, and Letters of Gen. Robert E. Lee, New York: D. Appleton and company, 1875, p. 352
[18] Gordon, John B., Reminiscences of the Civil War, New York: Charles Scribner's Sons, 1904, P. 382-388
[19] Davant, Phillip E, letter contained in Confederate Service Record of James C. Daniel, NARA M266, Records group 109, Compiled service records of Confederate soldiers from Georgia
[20] Letter Courtesy of Mr. John Davis, descendant of Lt. Simpson A. Hagood
[21] Lester George L., War Record of the Tom Cobb Infantry, Company E, 38th Georgia Regiment, Oglethorpe Echo newspaper, April 10th, 1879
[22] Bradwell, I. G., Fort Steadman and Subsequent Events, Confederate Veteran, S. A. Cunningham, January 1915, Volume 23, pages 20-23
[23] Gordon, John B., Reminiscences of the Civil War, New York: Charles Scribner's Sons, 1904 p. 403
[24] Lester George L., War Record of the Tom Cobb Infantry, Company E, 38th Georgia Regiment, Oglethorpe Echo newspaper, April 10th, 1879
[25] Nichols, G. W., A Soldier's Story of His Regiment (61st Georgia), and incidentally of the Lawton-Gordon-Evans brigade, Army Northern Virginia (Jesup Georgia, 1898) P. 220
[26] Mathews, W. C., Wrights Legion, The Sunny South, January 10th, 1891
[27] Lester George L., War Record of the Tom Cobb Infantry, Company E, 38th Georgia Regiment, Oglethorpe Echo newspaper, April 10th, 1879
[28] Confederate pension application
[29] Mathews, W. C., Wrights Legion, The Sunny South, January 10th, 1891
[30] Henderson's Roster
[31] Evans, Clement A., edited by Robert Grier Stephens, Jr., Intrepid warrior: Clement Anselm Evans, Confederate General from Georgia; life, letters, and diaries of the war years. Dayton, Ohio: Morningside Press, 1992, P 534-536
[32] Gordon, John B., Reminiscences of the Civil War, New York: Charles Scribner's Sons, 1904
[33] Confederate Service Records, NARA M266, Records group 109, Compiled service records of Confederate Soldiers from Georgia
[34] Confederate Service Record of E. P. Stewart, NARA M266, Records group 109, Compiled service records of Confederate soldiers from Georgia
[35] Jones, John W., Personal Reminiscences, Anecdotes, and Letters of Gen. Robert E. Lee, New York: D. Appleton and company, 1875, P. 41
[36] Mathews, W. C., Wrights Legion, The Sunny

Chapter 24 **The Road to Appomattox Court House, April 2nd - 12th 1865**

[1] Dickerson, William F. A., Roster of Georgia Confederate Soldiers 1861-1865, Lillian Henderson, Sketch of the Campaign of 1864 until Surrender
[2] Lester, George L., War Record of the Tom Cobb Infantry, Company E, 38th Georgia Regiment, Oglethorpe Echo newspaper, April 10th, 1879

[3] Dickerson, William F. A., Roster of Georgia Confederate Soldiers 1861-1865, Lillian Henderson, Sketch of the Campaign of 1864 until Surrender
[4] Ibid.
[5] Gordon, John B., Reminiscences of the Civil War, New York: Charles Scribner's Sons, 1904, P. 429-430
[6] The National Archives, Mirofilm roll: "M-266 Compiled Service Records of Confederate Soldiers Who Served in Organizations from Georgia - Thirty Eighth Infantry, Rolls 433 through 441.
[7] Dickerson, William F. A., Roster of Georgia Confederate Soldiers 1861-1865, Lillian Henderson, Sketch of the Campaign of 1864 until Surrender
[8] Hudgins, F. L., The Last Charge at Appomattox Court House, April 9th, 1865, Watson's Jeffersonian Magazine - Sept. 1911, Volume 13, Pages 425-426
[9] Gordon, John B., Reminiscences of the Civil War, New York: Charles Scribner's Sons, 1904, P. 436 - 437
[10] Ibid, P. 437
[11] Ibid. P. 437
[12] Hudgins, F. L., The Last Charge at Appomattox Court House, April 9th, 1865, Watson's Jeffersonian Magazine - Sept. 1911, Volume 13, Pages 425-426
[13] Gordon, John B., Reminiscences of the Civil War, New York: Charles Scribner's Sons, 1904, P. 438
[14] Hudgins, F. L., The Last Charge at Appomattox Court House, April 9th, 1865, Watson's Jeffersonian Magazine - Sept. 1911, Volume 13, Pages 425-426,
[15] Bradwell, I. G., Confederate Veteran, The Last Days of the Confederacy, Confederate Veteran, S.A. Cunningham, 1921, Page 56-57
[16] Hudgins, F. L., The Last Charge at Appomattox Court House, April 9th, 1865, Watson's Jeffersonian Magazine - Sept. 1911, Volume 13, Pages 425-426
[17] ***War Ends and The Soldiers Return to Sandy Springs to Start Again***
http://www.reporternewspapers.net/2011/04/07/war-ends-soldiers-return-sandy-springs-start/Retrieved Jan 9th, 2017
[18] Lester, George L., War Record of the Tom Cobb Infantry, Company E, 38th Georgia Regiment, Oglethorpe Echo newspaper, April 10th, 1879
[19] Lee, Robert E., Athens Southern Banner Newspaper, Capitulation of Lee's Army, April 25th, 1865, page 2

Chapter 25 **The Surrender and Prisoners of War**

[1] Mathews, William C., *Wrights Legion*, The Sunny South, January 10th, 1891
[2] Hudgins, F. L., The Last Charge at Appomattox Court House, April 9th, 1865, Watson's Jeffersonian Magazine - Sept. 1911, Volume 13, Pages 425-426
[3] Bradwell, I. G., Fort Steadman and Subsequent Events, Confederate Veteran, S. A. Cunningham, January 1915, Volume 23, pages 20-23
[4] Hudgins, F. L., The Last Charge at Appomattox Court House, April 9th, 1865, Watson's Jeffersonian Magazine - Sept. 1911, Volume 13, Pages 425-426
[5] Bradwell, I. G., Fort Steadman and Subsequent Events, Confederate Veteran, S. A. Cunningham, January 1915, Volume 23, pages 20-23
[6] Ibid
[7] Roster of Georgia Confederate Soldiers 1861-1865, Lillian Henderson, Sketch of the Campaign of 1864 until Surrender, William F. A. Dickerson
[8] Mathews, William C., *Wrights Legion*, The Sunny South, January 10th, 1891
[9] The National Archives, Mirofilm roll: M-266 Compiled Service Records of Confederate Soldiers Who Served in Organizations from Georgia - Thirty Eighth Infantry, Rolls 433 through 441
[10] ***War Ends and the Soldiers Return to Sandy Springs to Start Again***

http://www.reporternewspapers.net/2011/04/07/war-ends-soldiers-return-sandy-springs-start/Retrieved Jan 9th, 2017

[11] The National Archives, Mirofilm roll: "M-266 Compiled Service Records of Confederate Soldiers Who Served in Organizations from Georgia - Thirty Eighth Infantry, Rolls 433 through 441

[12] Mathews, William C., *Wrights Legion*, The Sunny South, January 10th, 1891

[13] Mason, Emily Virginia, The Southern Poems of the War, John Murphy & Co., Baltimore, 1889, p. 426.

Chapter 26 **Deaths, Deserters & "Galvanized Yankees"**

[1] Co. I, Ala. Company: These statistics only include the period while assigned to the 38th Ga. Regt., from 5/2/1862 through 3/1/1863.

[2] Company L, the Chestatee Artillery: Assigned to the 38th Ga. Regt. First organized as artillery on 10/9/1861 and assigned to Wright's Legion, this company was detached 6/10/1862 from the regiment and remained at Savannah serving as artillery. On 5/5/1864 the company was serving in S.C. and was ordered to rejoin the 38th Ga. Regt, as infantry. They rejoined the regiment just after the battle of Spotsylvania Court House 5/12/1864 and served until the surrender at Appomattox Court House 4/9/1865.

[3] Co. M: This company as detached 6/10/1862 and never rejoined the regiment. Statistics included only show the period of service as part of the regiment, from approximately 9/26/1861 to 6/10/1862

[4] Official Records, Series 2, Vol. 6, Part 1, (Prisoners of War) P. 823

Chapter 27 **Reunions And Life After The War**

[1] Horrid Murder, Macon Telegraph, May 21st, 1869, P. 2

[2] Georgia News, The Atlanta Constitution, October 21st, 1884, P.1

[3] To The Officers and Soldiers of Evans' Georgia Brigade, Macon Telegraph & Messenger, Oct. 25th, 1871, P. 2

[4] Ex-Confederates, Proceedings of the Meeting of the Surviving Confederate Soldiers of Gordon's Brigade, Savannah Morning News, Oct 29th, 1874, P 1

[5] Confederate Soldiers Reunion, The Atlanta Georgian, August 31st, 1906, P. 7

Index

12th Ga., Battalion, 153, 164
13th Ga. Regiment, 9, 33, 42, 64, 84, 105, 143, 146, 213
25th Ga. Regiment, 9
26th Ga. Regtiment, 9, 36, 46, 47, 60, 64, 102, 107, 109, 122, 124, 143, 146, 162, 185, 206
31st Ga. Regiment, 9, 10, 16, 17, 19, 42, 43, 45, 50, 51, 58, 64, 66, 68, 69, 75, 85, 88, 93, 98, 119, 122, 126, 129, 135, 141, 143, 157, 173, 175, 182, 184, 190, 194, 219
60th Ga. Regt., 9, 64, 81, 122, 136, 141, 143, 161
61st Ga. Regt., 9, 12, 31, 46, 48, 51, 53, 60, 68, 93, 94, 122, 124, 125, 143, 145, 146, 165, 173, 174, 186, 188, 206
Ables, Andrew J., 7
Adams, Hiram P., 73
Adams, John A., 150
Adams, Salathiel, 201
Adams, Tinsley Rucker, 212
Akers, John H., 40
Akins, Charles Thomas or Aiken, 53
Ala., 6th Regt., 121
Ala., 61st Regt., 12
Ala., Henry County, 12
Alexander, Phillip W., 104, 105
Almand, A. W., or Almond, 199, 201
Almond, Isaac B., 231
Anderson, James M. V., 25
Andoe, Theodore M., 155
Andrews, A. J., 127
Andrews, William T., 3
Applewhite, George W., 40
Arrington, Hilliary W., 177, 181
Arrington, Joshua P., 151
Arrington, William J., 3
Arthur, James J., or James S., 182, 231
Atkinson, Alexander, 58
Atkinson, Edmond. N, CSA Colonel, 60, 64, 66, 68, 107, 185
Atkinson, William F., 72, 187
Augustus Shaw, 11, 20, 21
Austin, Alexander C., 23
Austin, John T., 224
Austin, John Thomas, 222
Austin, William F., 52
Autry, Isaac Washington, 72
Autry, William Allen, 182
Baggett, Seaborn F., 223
Bagwell, Nathan B., 49
Bailey, John A., 155
Ball, John W., 71, 221, 224
Ball, Peter M., 224
Bartow Avengers, 2, 4, 12, 25, 40, 53, 66, 72, 74, 85, 113, 120, 126, 151, 155, 169, 182, 187, 201, 223
Battey Guards, 2, 3, 12, 17, 25, 27, 39, 41, 43, 45, 52, 53, 56, 58, 70, 72, 74, 78, 82, 87, 90, 95, 100, 101, 106, 113, 133, 141, 147, 151, 155, 159, 165, 166, 177, 178, 179, 181, 187, 191, 205, 212, 220, 223, 224
Battey, William H., 2, 3, 19, 27, 30, 53
Baxter, Francis Marion, 53
Baxter, George, 207
Baxter, John, 8, 11, 50, 80, 82, 83, 104, 226, 235
Bayne, Hendley V., 226
Beasley, Benjamin, 223
Beasley, Benjamin "Ben", 225
Bell, Augustus C., 112
Ben Hill Guards, 2, 3, 11, 21, 24, 39, 41, 49, 53, 58, 72, 73, 78, 101, 113, 116, 149, 150, 152, 159, 179, 181, 186, 201, 216, 222, 234, 235
Benjamin H. Hickman, 71, 72, 74
Bennett, Elijah B., 223
Bennett, William J., 162
Bentley, John D., 223
Bentley, Joseph "Joe", 216
Bentley, McAlpin Anslem, 74
Berryman, Charles Tabor, 212
Biggs, Craig, 58
Black, John W., 39
Blackburn, W. M., 12
Boggs, James S., 133
Bolton, James S., 223

Bomar, Thomas H., 3, 4, 11, 12, 119, 149, 154, 159, 191, 199, 202, 224
Bond, George W., 25
Booker, Obediah P., 223
Booth, Gabriel H., 25
Booth, John C., 182
Borin, Francis M., 162
Bostick, Green B., 152
Bowers, Asa Marion, 182
Bowers, John H., 142, 223
Boyd, Abraham, 39
Boyd, James, 201
Boyd, John, 24
Boyd, John B., 187
Bradwell, Gordon, 44, 46, 52, 69, 85, 86, 94, 99, 122, 126, 130, 134, 135, 136, 152, 157, 160, 161, 164, 174, 176, 177, 183, 185, 194
Braswell, George, 154
Braswell, George A., 155
Braswell, Samuel House, 36
Brawner's Farm, Va., 35
Breckinridge, John, CSA General, 160, 173, 180, 183
Brice, John Ratliff, 182
Brinson, James Daniel, 150
Brinson, John W., 3, 27, 66. 70, 78
Brisendine, William H., 152
Brittain, Jabez M., 27, 225
Broadwell, William "Billy" Thomas, 236
Brooks, Aaron A., 40
Brooks, William, 223, 226
Brown, Allen, 223, 226
Brown, Andrew J., 41
Brown, Asa Chandler, 127, 133
Brown, Benager T., 4, 27
Brown, David S., 223
Brown, John D., 223
Brown, Lewis, 223, 226
Brown, William A., 225
Brown, William Arnold, 181
Bruce, John A., 181
Bruce, Mack V., 52
Bryant, Robert M., 231, 232
Burdett, John Sylvester, 201
Burgess, Lewis, 52
Burt, William T., 222
Butler, George W., 25
Butler, Thomas Newton, 201
Caldwell, John, 142
Callahan, James S., 24
Callaway, Jonathan I., 7
Camp Bartow, Ga., 5, 6
Camp Kirkpatrick, Ga., 1, 2, 5
Campbell, James C., 187
Campbell, Jeptha E., 104, 155, 224
Campbell, Reuben M., 12
Campbell, William G., 223
Candler, Milton A., 1
Carrouth, James E., 223
Carrouth, Joseph A., 223
Carruth, Gideon A., 223
Carruth, Thomas Sanford, 25
Carter, Martin, 73
Cash, Lewellin J., 222
Cash, Seaborn J., 6
Cedar Mountain, Va., 29, 30
Chafin, John C., 222
Chamblee, William Larkin, 222, 224
Chandler, James E., 53, 226
Chesnut, David A., 71, 222, 224
Chestatee Artillery, 3, 4, 8, 11, 12, 120, 149, 152, 155, 159, 162, 178, 179, 182, 199, 214, 223, 224
Cheves, Edward, 17, 18
Chew, John Cottle, 116, 235
Childers, William Jackson, 199, 202
Childers. John, 214
Childress, William Anderson, 201
Christian, Benjamin B., 179, 182
Christian, John W., 50
Cochran, Samuel "Sam" Wylie, 82, 222
Coggin, James F., 222
Coile, George Nicholas, 150
Coile, Joseph Warren, 39
Cold Harbor, or Coal Harbor, Va., 30, 77, 78
Coleman, Isaac Washington, 72

Coleman, Mathew M., 53
Cook, William J., 186
Cooksey, Thomas B., 201
Corbin, Silas S., 24
Cowan, Zachariah J., 223, 226
Crane, Solomon, 154, 156
Crane, William H., 202
Craven, Elijah J., 3
Crook, George, Union General, 169, 184
Cross, Kinchen R., 112
Curl, Mathew Jr., 201
Curl, Peter, 222
Curry, John G., 27
Custer, George A., Union General, 198, 216, 218, 232
Dacus, Paschal Hawkins, 182
Daniel, Amaziah C., 222
Daniel, James Cicero, 206, 221, 225
Daniel, John J., 4, 27
Davant, Phillip E., 119, 136, 137, 147, 154, 191, 194, 205, 206, 209, 221, 223
Dawes, Rufus, 44, 48, 51
Dawson Farmers, 9, 12, 26, 30, 40, 53, 72, 119, 141, 151, 156, 159, 182, 187, 191, 214, 223
Day, John Henry, 24
De Forest, John W., 200
Deadwyler, Henry R., 27, 80, 141, 154
Dickerson, William F. A., 159, 169, 215, 222
Dinsmore, William Washington, 39
Dix, James Frederick, 212
Dix, John H., 231
Dorough, Robert T., 27, 31, 70, 221
Dorris, John M, or Dowis, 25
Douglass, Marcellus, 43, 46, 52, 78
Dudley, James W., 39
Dufrees, Elias, 40
Duncan, James Washington, 181
Dunn, John, 35
Durden, Norris N., 216
Duryea's Brigade, 46, 48
Dutton, John William, 181
Early, Jubal, CSA General, 157, 159
Eavenson, Thomas M., 39
Eberhart, 3, 30, 225
Eberheart, Robert P., 3, 30, 80, 81, 225
Edwards, William A., 232
Eidson, John, 201
Eidson. Robert, 181
Elliott, Joseph E., 221
Emerson, Henry Columbus, 222, 235
Evans, Clement, CSA General, 19, 64, 65, 67, 70, 75, 79, 81, 131, 133, 134, 152, 160, 161, 164, 186, 188, 192, 193, 198, 200, 211, 212, 218, 236
Evans, William Nelson, 187
Ewell, Richard, CSA General, 15, 16, 17, 29, 36, 38, 43, 51, 61, 77, 94, 95, 100, 102, 120, 121, 122, 126, 128, 129, 134, 135, 139, 146, 148, 207, 216
Faircloth, Benjamin, 222
Farmer, Henry G., 222
Farmer, Levin W., 151, 154
Farmer, Sidney T., 151, 154
Farmer, William J., 72, 74
Farrer, Gavin Hamilton, 178, 181
Fingal, blockade runner, 5
Fitz-Porter, John, 57
Fleeman, John Wesley, 24
Fleeman, William Johnson, 222
Fleman, James R, or Fleeman, 24
Fleming, John, 40, 151
Fletcher, Richard H., 191, 234
Flowers, John Yancy, 1
Forbes, William M., 26
Fortson, William Thomas, 74
Freeman, John, 182
Gaddy, George W., 28, 29, 50, 52
Gardner, Andy J., 177, 181
Gardner, Charles, 155
Gardner, James, 154, 155
Garrett, Andrew Jackson, 162
Garrett, George W., 156
Gary, Martin W., 19
Gasaway, Francis Marion or Gazaway, 40

Gasaway, William M., or Gazaway, 231, 232
Gazaway, James H., or Gasaway, 226

Georgia, Atlanta, 1, 2, 3, 5, 199, 221, 225, 236
Georgia, Augusta, 2, 6, 77, 100, 114, 116, 188, 201
Georgia, DeKalb Co., 1, 23, 28, 36, 39, 52, 71, 73, 113, 150, 155, 181, 186, 200, 201, 216, 218, 222, 235
Georgia, DeKalb County, 1, 2, 4, 8, 11, 12, 23, 28, 36, 39, 52, 71, 73, 113, 129, 150, 152, 155, 181, 186, 200, 201, 214, 216, 218, 221, 222, 225, 226, 234, 235, 236
Georgia, Doraville, 1
Georgia, Elbert County, 3, 6, 12, 22, 25, 40, 47, 53, 71, 72, 74, 104, 105, 113, 127, 142, 151, 155, 179, 182, 187, 199, 201, 209, 223, 224, 232, 234, 235
Georgia, Forsyth County, 3, 4, 182, 202, 223
Georgia, Fulton County, 3, 24, 25, 39, 40, 45, 52, 53, 71, 72, 73, 74, 113, 150, 151, 155, 181, 182, 186, 187, 201, 222, 236
Georgia, Hart County, 4, 11
Georgia, Jefferson Co., 2, 3, 12, 25, 39, 52, 53, 72, 74, 113, 133, 151, 181, 187, 216, 223
Georgia, Milton County, 3, 11, 24, 39, 52, 71, 73, 113, 150, 152, 181, 186, 201, 222, 224, 236
Georgia, Oglethorpe Co., 2, 4, 11, 24, 39, 53, 72, 73, 113, 133, 150, 170, 187, 201, 222
Georgia, Rome, 1, 2, 3
Georgia, Savannah, 4, 5, 6, 7, 8, 9, 10, 11, 21, 27, 56, 58, 59, 149, 195
Georgia, Skidaway Island, 5, 6, 7
Georgia., Decatur, 1
Gibbs, Harvey Jasper, 222, 225
Gibbs, Robert D., 24
Gibson, John B., 156
Ginn, James Alfred, 39, 40
Ginn, Singleton S., 223
Glenn, James M., 24
Gober, Julius J., 4, 27
Goodwin, Charles Harris, 25
Goodwin, Gustin E., 4, 40
Gordon, John B., 51, 83, 84, 85, 88, 89, 90, 94, 98, 99, 103, 104, 105, 109, 110, 112, 114, 115, 120, 121, 122, 125, 126, 128, 130, 131, 132, 133, 134, 135, 137, 139, 140, 142, 143, 149, 152, 153, 158, 160, 161, 163, 164, 165, 173, 176, 177, 178, 184, 188, 190, 192, 194,퀜196, 197, 203, 204, 206, 207, 209, 211, 212, 213, 214, 216, 217, 218, 220, 235, 236
Goshen Blues, 3, 6, 12, 25, 40, 47, 50, 53, 66, 72, 74, 104, 105, 113, 127, 141, 142, 151, 155, 159, 179, 182, 187, 199, 201, 209, 223, 224, 230
Goss, John W., 231, 232
Goss, William Thomas, 182
Goswick, John W., 12, 72, 141, 191
Grant, William J., 39
Grantham, John D., 27
Green, John W., 212
Gregg, Maxcy, 63
Gregory, John Jacob, 40
Gresham, Josiah Wheeler, 53
Griffin, William F., 24
Grogan, Gideon, 45, 53
Grogan, Joseph, 45
Grogan, Joseph D., 53
Grogan, Josiah T., 45, 222
Grogan, Richard W., 231
Grogan, William D., 222
Gunn, James, 25
Gunn, Robert L., 39
Gunn, William L., 25
Hagood, Simpson A., 112, 117, 171, 209
Hall, Beverly W., 127
Hall, James C., 3, 27, 40

Hall, Joseph H. "Joe", 150
Hall, Martin, 50
Hall, William H., 182
Hambrick, Robert C., 24
Hammond, Joshua "Josh" T., 53
Hampton, Robert C., 73
Hancock, Andrew J., 39
Hancock, Winfield Scott, Union General, 103, 137, 139, 140, 141, 143, 144, 146
Hanleiter, Cornelius R., 2, 3, 5, 6, 10, 11
Hansard, William James, 155
Harbin, Jasper, 66, 105, 155, 225
Harbin, Marion, 225
Hardeman, Charles T., 222
Hardman, Elizabeth J., 83
Harmon, Francis M., 232
Harman, James P., 225
Harmon, Josephus B., 222
Harris, James C., 181, 187
Harris, Starling G., 154
Harris, Woodson B., 24
Harrison, Tip, 75, 77, 109
Hawkins, Charles A., 4, 36, 66, 70, 79, 96, 97, 100, 101
Hawkins, John A., 53
Hawkins, John M., 97, 101, 201
Hawkins, Thomas, 72
Hayes, John, 201
Haygood, Simpson A., 27
Henderson, Ezekiel L., 156
Hendricks, W. J., 155
Hendrix, William, 4, 154, 156, 221
Henry Light Infantry, 9, 12, 41
Henry, Jesse L., 25
Henry, William R., 27, 66
Herring, Wiley B., 71, 231
Hewell, Josiah Tuck, 47
Hickman, Benjamin H., 72
Hicks, Crawford or Hix, 25
Hidgon, John B., 155
Higdon, Ebenezer G., 222
Higdon, John B., 181
Higginbotham, Elijah "Bensie", 142, 154
Hill, Starling Emanuel, 187
Hix, Abner Harrison, 222
Holcombe, Azariah Milton, 221, 224
Holmes, James M., 181
Hopper, Booker W., 222
Hopper, Daniel H., 39
Hopper, Whit, 212
Hornbuckle, Horton H., 25
Hotchkiss, Jedediah, 192, 203
House, John T., 39
House, Thomas J., 199, 201
Howard, William Glenn, 24
Howland, Thomas Jefferson, 222
Hudgins, Francis L., 85, 92, 102, 103, 106, 110, 120, 126, 129, 130, 216, 217, 223, 225
Hudson, Hampton J., 225
Hudson, Hezekiah H, 231
Huff, J. H., 187
Huff, John Peter, 24
Huff, S. H., 185
Hunt, Benjamin Thornton, 24
Hunt, Judge Middleton Barrett, 223
Hunter, David, Union General, 158
Hutcherson, James M., 24
Imboden, John D., CSA General, 108
Iron Brigade, Union, 36, 37, 48, 122
Irwin Invincibles, 8, 12, 25, 66, 74, 82
Irwin, Andrew B., 12, 27, 66
Isler, John B., 55
Jackson, John Baughn, 39, 40, 206, 207
Jackson, John W., 24
Jackson, William Sanders, 222
Jackson, Willis B., 31, 126, 185
Jackson, Willis Benjamin, 187, 222
Jameson, William Thomas, 39
Jay, John Wesley, 156
Jefares, Bennett Rainy, or Jeffers, 216, 234
Jeffers, Eugene, 125
Jenkins, Elias Pearce, 187
Jenkins, William H., 103, 221

Jernigan, William H., 3
Jett, James S., 27
Jo Thompson Artillery, 2, 3, 6, 7, 8, 10, 11, 12, 27
John J. Carpenter, 216, 218, 224, 225
Johnson, William W., 150
Johnston, William J., 221
Jones, Henry L., 9, 12, 25
Jones, James E., 12, 74
Jones, Joseph L., 182
Jones, Robert Daniel Francis, 223
Jones, Robert F., 25
Jordan, Harden A., 178, 182
Jordan, Thomas J., 45, 53
Kellis, John W., 30
Kersey, John A., 24
Kershaw, Jospeh, CSA General, 190, 192, 195, 197, 216
King, James "Jim", 216, 225
King, John, 39, 105
Kirby, William, 105
Lafoy, John A., 154
Lamar, John H., 54
Landrum, Allen O., 222
Landrum, Francis Marion, 200, 201
Lang, William B., 231
Lanier, Pressley "Press", 72, 150, 154
Lawton, Alexander R., 7, 10, 11, 15, 16, 17, 18, 43, 48, 50, 68, 236
Lawton, Edward P., 16, 19, 20, 68
Lee, George W., 7, 8, 27
Lee, William, 81
Leemore, J. G., 223
Lester, George H., 2, 4, 6, 8, 18, 20, 27, 30, 31, 32, 33, 34, 36, 38, 40, 41, 59, 62, 64, 67, 82, 102, 121, 126, 134, 152, 157, 158, 159, 169, 177, 178, 179, 183, 185, 186, 189, 194, 195, 197, 203, 204, 205, 206, 211, 212, 215, 216, 219
Lewis, Joshua K., 41
Lincoln, Abraham, President, 61, 167
Little. William J., 214
Logan, John W., 216
Loggins, Ervin Alexander, 156
Loggins, Timothy S., 26
Lord, Russell D., 222, 225
Lowe, John, 50, 51, 220
Luckey, John R., 231, 232
Lumpkin, Joseph J., 18
Maddox, Henry G.-, 231
Maddox, Joseph J., 3
Manassas, Va., 27, 31, 32, 34, 35, 38, 39, 41, 42, 43, 54, 77, 78, 101, 114, 115, 122, 147, 162
Mann, Edwin Jackson "Jack", 223
Marabel, John J., 3
Marlow, William T., 71
Martin, Absalom B., 182
Martin, Jesse W., 152
Martin, John M, 73, 151
Martin, Robert A., 223, 225
Martin, Thomas, 25, 150
Martin, Van Buren, 72
Mashburn, John H., 120
Mashburn, John Wesley, 72
Mason, Camilus A., 154, 222
Mason, Ezekiel, 1
Mathews, James D., 10, 18, 27, 80, 191, 236
Mathews, William Collins., 27, 31, 35, 42, 43, 44, 52, 56, 58, 71, 87, 91, 96, 104, 106, 114, 119, 147, 155, 165, 166, 169, 177, 179, 191, 205, 212, 214, 220, 221, 224, 226
Mathis, George L. or Mathews, 133
Mattox, Joseph, 225
Maxwell, Benjamin "Martin", 39
Maxwell, Jackson O., 4, 155, 191
McAfee, William H., 191, 235
McCanless, Jesse A., 154
McCleskey, Francis C., 13, 22, 84, 104, 111, 113
McCleskey, George W., 3, 22, 24, 84, 104, 111
McCullough Rifles, 2, 8, 11, 47, 50, 53, 72, 82, 159, 169, 171, 177, 179, 185, 186, 215, 216, 222, 230
McCurdy, John W., 8, 11, 79
McCurdy, John Wilson, 47, 191

McDaniel, Egbert Baldwin, 156
McDaniel, Joseph W., 222
McDaniel, Phillip S., 162
McDaniel, William P., 4
McGahee, Samuel M. or MGhee, 151
McKinney, William L., 182, 232
McLean, John, 39
McLean, P. Hugh, 24
McLeod Artillery, 3
McLeod, William, 3, 19, 38, 61, 62, 65, 68, 78, 79, 80, 81, 85, 90, 93, 104, 106, 113

McMakin, Andrew J., 3, 27
McMakin, John Carr, 222
McMakin, William Jefferson, 222
Maryland, Fox Gap, 159
Maryland, Frederick, 22, 41, 53, 110, 113, 159, 160, 162, 166, 181
Maryland, Maryland Heights, 159
Maryland, Monocacy, 42, 159, 160, 162, 163, 165, 166, 167, 168, 172, 188
Maryland, Sharpsburg, battle of, also known as Antietam, 29, 41, 42, 44, 52, 53, 59, 77, 78, 83, 101, 102, 115, 147, 226
Maryland, South Mountain, 45, 122, 159, 167
Meeks, Elisha, 150
Mehaffey, Francis M., 154
Michigan, Fourth Regiment, 56
Miller, Andrew J., 71
Miller, David Young, 52
Miller, Newton R., 222
Miller, Robert, 154
Miller, Robert W, 150
Miller, William A. C., 3, 103
Millsaps, Isaiah Peter, 151
Milton Guards, 3, 11, 13, 22, 24, 39, 50, 52, 71, 72, 73, 84, 101, 104, 113, 117, 150, 152, 159, 171, 181, 186, 201, 209, 216, 222, 224, 235
Mitchell, James R., 74
Mitchell, William Harris, 216, 218, 224, 225
Moon, William P., 40
Moore, John M., 72, 73
Moore, Thomas A., 223
Morgan, Simeon, 156
Morris, Benjamin "Ben" T., 225
Morris, Benjamin T., 120
Morris, Enoch H., 27
Morris, Enoch H. C., 40
Moxley, Henry M., 222, 225
Mulling, Francis M., 179, 181
Murphey, Charles, 1
Murphy Guards, 1, 3, 8, 11, 39, 50, 52, 71, 73, 113, 150, 155, 159, 162, 177, 181, 186, 201, 214, 221
Murphy, Augustus A., 225
Murray, Benjamin, 222
Nash, George W., 24
Nash, Isaac Newton, 106, 108, 110, 226
Nash, Miles H., 154
Nash, William T., 181
Neal, James Richard, 222
Nelms, James C., 74
Nelms, William, Drummer, 223
New York, 57th Regt., 22
New York, 76th Regt., 37
Newsom, Moses H., 72
Newton, Andrew J., 152
Nichols, George Washington, 13, 32, 69, 91, 94, 122, 124, 137, 145, 146, 147, 148, 152, 159, 161, 162, 163, 165, 166, 174, 185, 188, 190, 194, 195, 198, 199, 205, 207
Nix, John W., 52
Odom, Elijah A., 234
Odom, Elijah Anderson, 201, 222
Oglesby, J ohn, 3, 80, 104, 113
Orange, Va., 27, 32, 59, 112, 114, 117, 119
Orr, John, 222
O'Shields, David, 223
Owen, Francis Marion, 182
Owen, Jesse P., 152
Owen, William Jasper, 156
Pa., 7th Reserves Regt., 125

Pa., 118th Regt., 57
Pa., Chambersburg, 102, 113, 170, 172
Pa., Gettysburg, 82, 102, 103, 104, 105, 107, 108, 109, 111, 113, 114, 115, 147, 162, 177, 197, 226, 235
Parham, John W., 40
Parr, Lewis J., 2, 3, 18, 27, 81
Partin, Reuben, 162
Patrick, Marsena, Union General, 71
Patterson, Enoch, 156
Patterson, Hiram, 182
Patterson, Samuel R., 25
Payne, Raymond B., 7
Payne, Zebulon C., 30, 141
Pearson, George W., 231
Peebles, Howell L. or Peeples, 133
Peeples, Thomas J., 216
Pendleton, William N., 54
Penrow, John William, 223
Perdue, John F., 25
Perlinski, Julius, 181
Perryman, D., 223
Phillips, Andrew J., 3
Phillips, James M., 202
Phillips, John W., 25
Phillips, Thomas L., 73
Phillips, William J., 181
Pittard, John T., 222
Pool, Alfred J. H., 3
Pool, Michael W., 25
Poss, James, 73
Powell, James Lewis, 22, 25, 234
Powell, Wiley, 234, 235
Powell, William Joseph, 104, 113, 234, 235
Powell, William R., 212
Power, Asbury Hull, 186
Prater, Benjamin G., 233
Pratt, Elijah, 24
Price, Skirving, 38th Ga. Surgeon, 170
Pruitt, Calvin, 224
Pughesly, Jacob “Jake”, 225
Pughsley, George W., 212
Pughsley, Jacob "Jake" Perry, 72
Ramseur, Stephen, CSA General, 144, 167, 173, 183, 184, 192, 193, 198
Rankin, John, 2, 8, 11, 93, 107, 154, 155, 159, 177, 179, 181, 191, 225
Reese, Jonas, 74
Reese, Reuben P., 222
Reeves, James A. J., 216, 218, 224
Register, Irwin, 25
Rice, William Jasper, 187
Richardson, Josephus S., 25
Ricketts, James B., Union General, 161
Roberts, Wiley T., 74
Robertson, Walter S., 36, 200, 201
Robinson, Alexander Webb, 109, 120
Robinson, George F., 155
Robinson, Robert M., 221, 224
Rodes, Robert, CSA General, 134, 135, 139, 141, 144, 166, 167, 173, 174, 176, 183
Rodgers, Alexander A., 151
Rodgers, George A., 25
Rodgers, James W., 25
Rosenthal, Adolphus, 133, 154
Rousey, James, 25
Sandford, Truman H., 4
Scisson, Larkin R., 39
Scisson, Wm. D., 206
Scoggins, Malcolm, 49
Seals, Benjamin, 223, 226
Second Manassas, Va, battle of, 36, 38
Sedgwick, John, Union General, 84, 85, 86, 87, 88, 91, 92, 128, 129, 131, 132
Sellers, John A., 52
Seymour, Truman, Union General, 131
Shaler, Alexander, Union General, 131
Shaw, Augustus A., 3, 11, 20, 21, 22, 27
Shearer, William S., 53
Sheridan, Phillip, Union General, 157, 158, 172, 173, 174, 176, 179, 180, 183, 184, 185, 186, 188, 189, 190,

192, 193, 195, 196, 197, 199, 200, 203, 204
Sherrod, John H., 3
Shiflet, James Monroe, 212
Shirley, Aaron, 201
Simmemon, David W., 214
Simpson, Robert M., 154
Sims, James A., 222
Singleton, James M., 154
Singleton, William L., 154
Smith, Armory Solon, 223, 225
Smith, Doctor Simon, 24
Smith, George W., 18, 24
Smith, Isaac H., 212, 226
Smith, James M., 25
Smith, John A., 74
Smith, John W., 74, 151
Smith, Jordan, 39
Smith, Luther C., 182
Smith, Patrick H. "Pat", 225
Smith, Sion Thomas, 24
Smith, Tilnon P. "Sally", 133, 154
Smith, Weems F., 39
Smith, William James "Willie", 53
Sorrow, Mathew E., 201
Spivey, Irvin, 122
Stamps, James J., 231, 232
Stansell, James, 155
Starling, Isaac, 71, 72, 74
Steele, John H., 222
Stephenson, Thomas C., 39
Stewart, Ebenezer P., 212, 214
Stewart, James B., 223
Stewart, Robert, 39, 41
Stewart, Thomas H., 152
Stewart, Thomas J., 72, 74
Stewart, William D., 150
Stiles, William, 65, 76
Story, Josiah, 181
Stubbs, George W., 4, 103, 155, 169
Swiney, James M., 223
Tatum, Horatio R., 223
Taylor, Moses Mulchy, 223
Taylor, Samuel E., 4
Taylor, William Henry, 223
Teasley, John H. H., 4, 27
Terry, Joseph Chandler, 216
Thomas, Wyett L., 154
Thompson, Andrew "Drew", 225
Thompson, Berry A., 231
Thompson, Joseph M., 24
Thompson, William A., 151
Thornton, James C., 223
Thornton, James William, 181
Thornton, Jeptha Mercer, 181
Thornton, John C., 4, 80
Thornton, Samuel A., 221
Thornton, Thomas D., 27, 221
Thornton, William M., 222
Thornton, William T., 66, 74
Thornton's Line Volunteers, 4, 11, 24, 39, 53, 66, 72, 74, 113, 127, 132, 133, 151, 155, 159, 181, 187
Tiller, John G., 212
Tillotson, John J., 222
Tom Cobb Infantry, 2, 4, 5, 6, 10, 11, 18, 21, 24, 32, 39, 40, 48, 51, 53, 59, 64, 70, 72, 73, 75, 82, 96, 101, 108, 109, 113, 121, 126, 133, 134, 150, 152, 157, 159, 169, 177, 178, 179, 183, 187, 194, 201, 206, 207, 216, 222, 230
Truitt, David, 152
Turner, James M., 222
Turner, Samuel, 24
Tweedell, Marion J., 182
Virginia, Appomattox Court House, 45, 154, 179, 214, 215, 216, 218, 219, 220, 221, 224, 234
Virginia, Cedar Creek, 183, 190, 192, 194, 197, 199, 200, 201, 202, 203, 204, 224
Virginia, Chantilly, 40, 42
Virginia, Charlottesville, 13, 25, 155, 204
Virginia, Clark's Mountain, 114, 120
Virginia, Deep Bottom, 85, 170
Virginia, Fairfax Court-house, 40
Virginia, Fisher's Hill, 170, 183, 184, 185, 186, 187, 188, 190, 192, 194, 195

Virginia, Fort Stedman, 209, 211, 212, 213, 214, 225, 226
Virginia, Fredericksburg, 59, 60, 61, 62, 64, 67, 68, 70, 71, 72, 73, 74, 75, 78, 79, 82, 84, 85, 86, 87, 88, 90, 94, 101, 115, 116, 137, 139, 147
Virginia, Gaines Mill, 15, 21, 23, 24, 25, 28, 30, 43, 54, 68, 111
Virginia, Gordonsville, 27, 28, 30, 39, 127, 158
Virginia, Hamilton's Crossing, 75
Virginia, Harper’s Ferry, 41, 42
Virginia, Hatcher's Run, 204, 206, 207
Virginia, Hupp's Hill, 183
Virginia, Lexington, 3, 158, 170
Virginia, Little North Mountain, 183, 184, 185
Virginia, Lynchburg, 13, 151, 157, 158, 216, 220, 235
Virginia, Massanuttan Mountain, 183
Virginia, Massanutten Mountain, 192, 193
Virginia, Middletown, 42, 169
Virginia, Mine Run, 120
Virginia, Mt. Jackson, 186
Virginia, Petersburg, 13, 78, 145, 157, 161, 168, 172, 203, 204, 205, 206, 208, 210, 211, 214, 215, 220
Virginia, Port Republic, 13, 59, 188
Virginia, Port Royal, 59, 61, 75
Virginia, Rapidan River, 31, 114, 116, 120, 132
Virginia, Richmond, 5, 11, 13, 15, 22, 24, 25, 27, 28, 30, 45, 53, 58, 59, 61, 66, 73, 74, 77, 84, 91, 92, 93, 110, 112, 113, 131, 133, 149, 150, 152, 154, 157, 166, 190, 205, 207, 208, 210, 214, 216, 220
Virginia, Shenandoah River, 59, 193, 194
Virginia, Shenandoah Valley, 13, 15, 169, 172, 183, 188, 200
Virginia, Spotsylvania Courthouse, 71, 132
Virginia, Strasburg, 169, 170, 183, 188, 189, 232
Virginia, Trevilians Station,, 157
Virginia, White Post,, 169
Virginia, Winchester, 94, 95, 96, 98, 100, 101, 102, 110, 115, 158, 159, 169, 171, 172, 173, 174, 176, 179, 180, 181, 182, 183, 185, 186, 188, 190, 195, 197, 201
Valkenburg, J. D. Van, 125
Van Valkinburg, John, CSA Lt Col, 162
Vaughan, Miles B., 66, 74
Vaughn, William “Bill” T., 223
Vaughn, William H., or Vaughan, 25
Vaughn. Isaac C., 154
Verdin, Ledford N., 222
Verdin, William, 222
Vinson, George W., 24
Wade, Enoch D., 223
Wade, George W., 223
Waits, Newton, 201
Wallace, George W., 233
Wallace, Lewis, Union General, 159, 161, 163
Wallis, David James, 40
Warnock, John C., 207
Washington, D. C., 72, 74, 157, 168
Washington, D.C.,, 159, 160, 166
Watson, George W., 223
Watson, Richard, 223
Weeks, Jesse W., 182
Wells, George Riley, 8, 11, 54, 154, 185, 187
Westbrook, Reuben T., 186
Westbrook, Samuel, 156
Westbrook, William R., 156
Wheeler, Amos, 151, 154
White, Garnett, 224
Whitmire, Elias Earl, 182
Wiggins, Ellis W., 53
Wiggins, Jackson C., 154, 155
Wiggins, James M., 223, 226
Wiggins, Mathew O., 201, 212, 226
Wiggins, Robert, 212, 226

Wilderness, 84, 85, 86, 114, 120, 121, 126, 127, 128, 133, 134, 150, 153, 176, 197, 203
Wilderness, Battle of, 128
Wilkes Francis, 201
Williams, Benjamin N., 40
Williams, T. John, 154
Willoughby, George T., 39
Wilson, Aaron Jordan, 25
Wilson, Benjamin L., 74
Winder, Charles S., 19
Wise, Chelsea S., 225
Wood, William Hasley, 182
Woodliff, Thomas Jefferson, 72, 84
Woodliff, Thomas Jefferson "Jeff", 71, 73
Woods, Archibald R., 222
Woods, John R., 150
Woods, S. J., 181, 185
Wright, Augustus R., 1, 2, 3, 7
Wright, James W., 19
Wright, Miller, 225
Wright, Miller A., 49
Wright, William, 2, 4, 27
Wright, William A., 216, 225
Wright, William Aron, 155
Wright's Legion, 1, 3, 5, 10
WV, Martinsburg, 41, 42, 54, 59, 100, 158, 159, 169, 170, 172, 174
WV, Opequon Creek, 174
WV, Shepherdstown, 42, 52, 54, 57, 58, 102, 159, 170, 172
York, Zebulon, CSA General, 161
Young, John Bennett, 151
